WASTEWATER CLEANUP EQUIPMENT

Wastewater Cleanup Equipment

Second Edition

NOYES DATA CORPORATION

Park Ridge, New Jersey London, England

1973

Library of Congress Catalog Card Number: 72-82618
ISBN: 0-8155-0487-X
Printed in the United States

Published in the United States of America by
Noyes Data Corporation
Noyes Building, Park Ridge, New Jersey 07656

FOREWORD

Water pollution is becoming more of a problem with every passing year. Plants engaged in all types of manufacture are being more and more carefully watched by federal, state and municipal governments to prevent them from pouring their untreated effluents into the nation's waterways as they used to do. The sewage treatment plants of many municipalities are becoming too small for the burgeoning population, and many communities once served by individual septic tanks are having to build sewers and treatment plants.

Water pollution will be solved primarily by application of techniques, processes and devices already known or in existence today, supplemented by modifications of these known methods based on advanced technology. This book gives you basic technical information and specifications pertaining to commercial equipment currently available from equipment manufacturers.

This second edition of "Wastewater Cleanup Equipment" supplies technical data, diagrams, pictures, specifications and other information on commercial equipment useful in water pollution control and sewage treatment. The data appearing in this book were selected by the publisher from each manufacturer's literature at no cost to, nor influence from, the manufacturers of the equipment.

It is expected that vast sums will be spent in the United States during the decade of the 1970's for water pollution control. Much of the expenditure will be for the type of equipment described in this book.

The descriptions and illustrations given by the original equipment manufacturer include one or more of the following:

(1) Diagrams of commercial equipment with descriptions of components.
(2) A technical description of the apparatus and the processes involved in its use.
(3) Specifications of the apparatus, including dimensions, capacities, etc.
(4) Examples of practical applications.
(5) Graphs relating to the various parameters involved.

Arrangement is alphabetically by manufacturer. A detailed index by type of equipment is included, as well as a company name cross reference index.

CONTENTS

Contents

Contents

Company Name Cross-Reference

Listed below are corporations, divisions and subsidiaries which have been alphabetically listed under alternate company names in the body of the book.

Aer-O-Flow-Yeoman Products — see Clow Corporation
Amercoat Corporation — see Ameron Corrosion Control Division
The Bauer Brothers Company — see C-E Bauer
Chicago Bridge & Iron Company — see Walker Process Equipment
Combustion Engineering — see C-E Bauer, C-E Natco and C-E Raymond
Denver Equipment — see Joy Manufacturing Company
Dicalite — see Grefco Inc.
Ecodyne Corporation — see Graver Water Conditioning Company
Envirotech Corporation — see BSP Division
Fairbanks-Morse — see Colt Industries
General Signal Corporation — see Aurora Pump
Harsco Corporation — see Can-Tex Industries
Lamson Division — see Diebold Inc.
The Mec-O-Matic Company — see Ecodyne Corporation
The National Tank Company — see C-E Natco
Shartle Division — see The Black-Clawson Company
Smith & Loveless — see Ecodyne Corporation
Sybron Corporation — see The Permutit Company
United States Filter Corporation — see Calfilco
Wallace & Tiernan — see Pennwalt Corporation

Ajax International Corporation

REVERSE OSMOSIS SYSTEM FOR WATER PURIFICATION

If pure water is placed in one compartment of a vessel and separated from a solution of salts in a second compartment by a barrier known as a semipermeable membrane, a natural pressure is created which causes the pure water to permeate through the membrane and dilute the salt solution. This behavior is known as osmosis.

A simple osmotic cell is illustrated in Figure 1. Water, driven by osmotic pressure, flows through the membrane, dilutes the concentrated solution on the other side and causes the concentrated surface to rise. When the cell reaches equilibrium, the pressure head of the concentrated solution above the fresh water level on the other side of the membrane is equal to the osmotic pressure.

Osmosis is a reversible process. If an external pressure greater than the osmotic pressure is exerted on the concentrated solution, the flow of water through the semipermeable membrane is reversed. This causes an increase in the volume of the fresh water. Because the dissolved salts in the concentrated solution do not pass through the membrane, this solution becomes more concentrated. Reverse osmosis occurring under these conditions is illustrated in Figure 2.

Consider now if more of the original concentrated solution is added continuously to the cell under the external pressure greater than the opposing natural osmotic pressure. Also consider that provision is made to bleed off the more concentrated solution in this compartment resulting from the reverse osmosis process. Then there is a continual flow of pure water through the membrane to the fresh water compartment.

Figure 1 — OSMOSIS

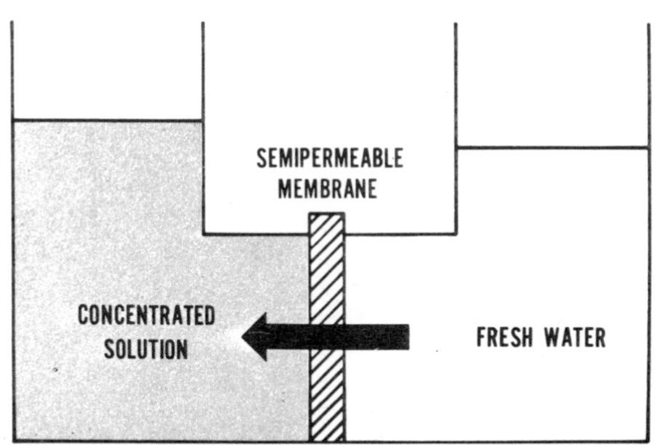

Figure 2 — REVERSE OSMOSIS

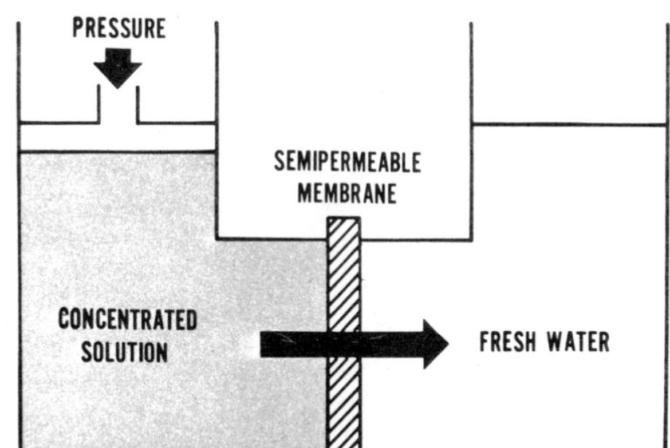

PARTIAL WATER ANALYSES FROM OPERATING INSTALLATIONS

	SAN DIEGO, CALIF. WELL WATER		BESSIE, OKLA. MUNICIPAL WATER		MIDLAND, TEXAS MUNICIPAL WATER	
	Feed	*Product*	*Feed*	*Product*	*Feed*	*Product*
OPERATING CONDITIONS						
Feed Pressure, psi.	600		600		500	
% Product Recovery		75		50		75
Hardness	1,880	20	1,950	8	560	8
Dissolved Solids	4,580	310	2,927	68	1,208	82
Alkalinity	280	10	108	10	60	11.6
Calcium	360	7.6	580	2.4	107	2.4
Magnesium	240	0.5	122	0.5	71	0.5
Sodium	900	110	96	7.5	172	21
Potassium	26	3.8	3.2	0.3	16	2
Sulfate	630	0	1,727	0	337	0
Bicarbonate	340	12	132	12	73	14
Chloride	2,020	170	92	10	400	34
Silica	34	7.6	35	5.3	91	18

ROGA® SPIRAL WOUND MEMBRANE MODULES*

Practical application of reverse osmosis to low cost water treatment has been made possible by the development of the ROGA® Spiral Wound Membrane Module.* The membrane used in these modules is a specially processed sheet of cellulose acetate. It has a higher flux rate (throughput of water per unit area of membrane) than any other type of membrane material that has been made at a reasonable cost. Even with this high flux rate, the application of this membrane would require very large equipment housings if it were used in flat sheets or in tubular form.

By incorporating the recently developed spiral wound concept, illustrated in Figure 3, a large membrane area (50 square feet) is contained in a very small volume (about 0.25 cubic feet).

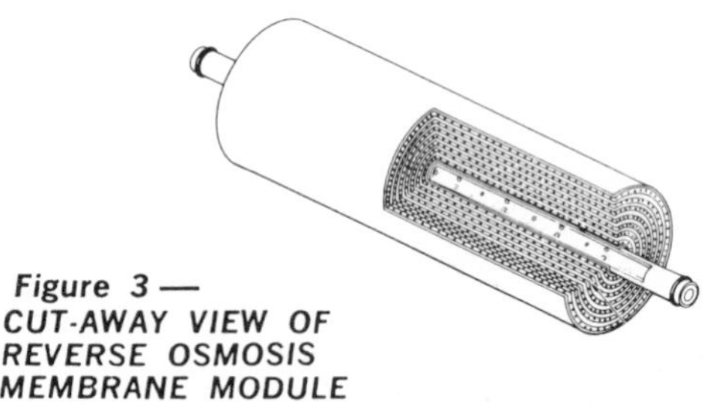

Figure 3 — CUT-AWAY VIEW OF REVERSE OSMOSIS MEMBRANE MODULE

The membrane in this uniquely compact cartridge is arranged to maximize the permeation of treated water. A mesh spacer is wrapped with a pouch, made from flat membrane around a central perforated tube. The mesh spacer extends towards the ends of the wrapped tube and allows the salty feedwater to pass right through the length of the roll. As it flows through the spacer gap of the roll, the feedwater, which is pressurized, passes over the outside of the membrane pouch. Some of the water, due to the reverse osmosis phenomenon, permeates the membrane and becomes trapped inside the pouch. The remaining water and salts pass on and out through the far end of the roll. Within the pouch is a porous, pressure-resistant backing layer. This porous layer is sandwiched between the flat sides of the pouch. The trapped permeated water flows freely within the porous layer to the perforated plastic tube which is sealed into the mouth of the pouch. The treated water, trapped within the pouch, passes through the perforations in the plastic tube. The ends of tube extend out of the pouch, allowing the water to escape. The entire roll or cartridge is known as a spiral wound membrane module.

The modules are inserted into tubular pressure vessels with a seal on the upstream end in the annular space between the module and the inner wall of the pressure vessel. This seal forces the feedwater through the module. Single modules may be used in very small systems, but usually several interconnected modules are assembled in a single pressure vessel, as is illustrated in Figure 4.

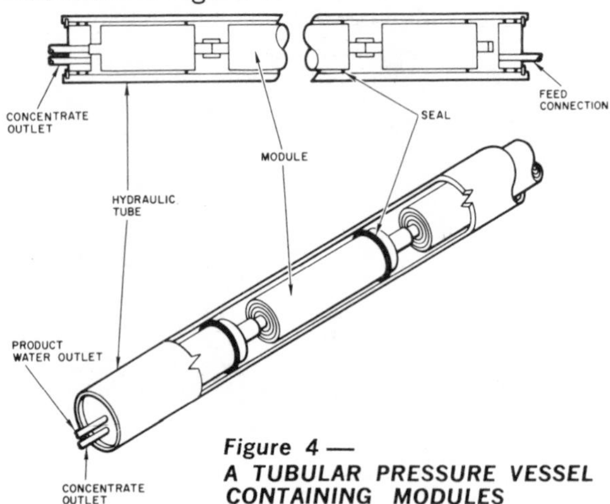

Figure 4 — A TUBULAR PRESSURE VESSEL CONTAINING MODULES

***MANUFACTURED UNDER PATENTS ISSUED OR APPLIED FOR IN MOST COUNTRIES OF THE WORLD**

40,000 (US) GALLONS PER DAY AJAX RO PLANT WITH CONDUCTIVITY MONITOR AND RECORDER

Ameron Corrosion Control Division

CORROSION-RESISTANT CONCRETE PIPE LINING

T-Lock Amer-Plate is a lining made specifically for concrete pipe,tunnels and structures, providing a continuous plastic lining which will resist hydrogen sulfide and other sewer gases, acids, alkalis and salts. It is made from high polymer vinyl chloride resins molded under high temperature and pressure to form an extremely dense and impervious liner. Projecting tee-shaped ribs securely anchor the lining to the concrete. Once T-Lock Amer-Plate is cast into place, it becomes a permanent and integral part of the structure, mechanically locked into the concrete surface.

T-Lock Amer-Plate prevents the disintegration of concrete structures. It will withstand severe chemical and physical abuse. It is resistant to a wide variety of chemicals, and is unaffected by fungus or bacterial action. Impervious to gas penetration and unaffected by continuous exposure to high humidity or water, T-Lock Amer-Plate shows no appreciable change under oxidizing conditions. Being thermoplastic, it can be welded to itself, thereby providing a continuous lining over the entire surface. It is flexible and extendable and, therefore, is able to bridge cracks which may develop in a concrete structure.

PROPERTIES

T-Lock Amer-Plate is composed of high molecular weight vinyl chloride resins combined with chemical resistant pigments and plasticizers. This completely inert mixture is molded at a temperature of 325°F. and pressure of 900 psi into a liner plate with a minimum thickness of .060 inches.

During the manufacture of T-Lock Amer-Plate, the resin mixture is actually flowed, and all portions of the composition are flowed within the die to eliminate any imperfections. To further assure that T-Lock Amer-Plate is uniform and free from blemishes, each sheet is tested with a high-voltage electronic holiday detector.

Because T-Lock Amer-Plate is formed in polished, stainless steel dies, the surface of the completed liner plate is smooth and dense. The color is glossy black.

T-Lock Amer-Plate is placed on the inner forms of concrete structures or pipe. When the concrete is poured, the tees are embedded and locked into the concrete, thus making the liner an integral part of the surface. Jointing sections of the liner is accomplished by fusing a welding strip of the same material over the joint using a hot air welding gun. At a temperature of approximately 400°F., perfect fusion of the two plastic surfaces is obtained, thus insuring continuity of the liner throughout the entire structure.

The dense, glossy surface of the T-Lock Amer-Plate will not absorb or retain precipitated or crystalline materials. Fungus growth and bacterial slimes do not adhere tightly. It is easily decontaminated and maintained in a sanitary condition.

Since T-Lock Amer-Plate is composed of thermoplastic vinyl chloride resins, it is permanently flexible. This property permits the liner plate to be shaped over intricate forms, corners, and edges, giving these difficult areas complete protection.

T-Lock Amer-Plate is entirely suitable for all temperatures encountered in sewage. It will also withstand continuous immersion in certain chemical solutions at temperatures as high as 180° F.

T-Lock Amer-Plate provides a molded, pore-free lining which forms an effective barrier to gaseous penetration. More than seventeen years of continuous exposure to hydrogen sulfide gas have shown no evidence of any sulfide penetration through the liner plate. Tests with hydrogen sulfide under pressure on one side of the sheet, and a lead acetate solution on the other, have shown no darkening in the lead acetate even after long exposure.

T-Lock Amer-Plate has been cast into concrete and then tested with water pressure to 40 psi on the back of the liner plate without any evidence of rupture in the Amer-Plate. 40 psi back-pressure would represent a ground water head of approximately 85 feet.

T-Lock Amer-Plate has a minimum elongation of approximately 200% and, since it is locked physically into the structure on 2½″ centers, any structural cracks which might occur are easily bridged.

Shore Durometer A is approximately 100 at ambient temperature.

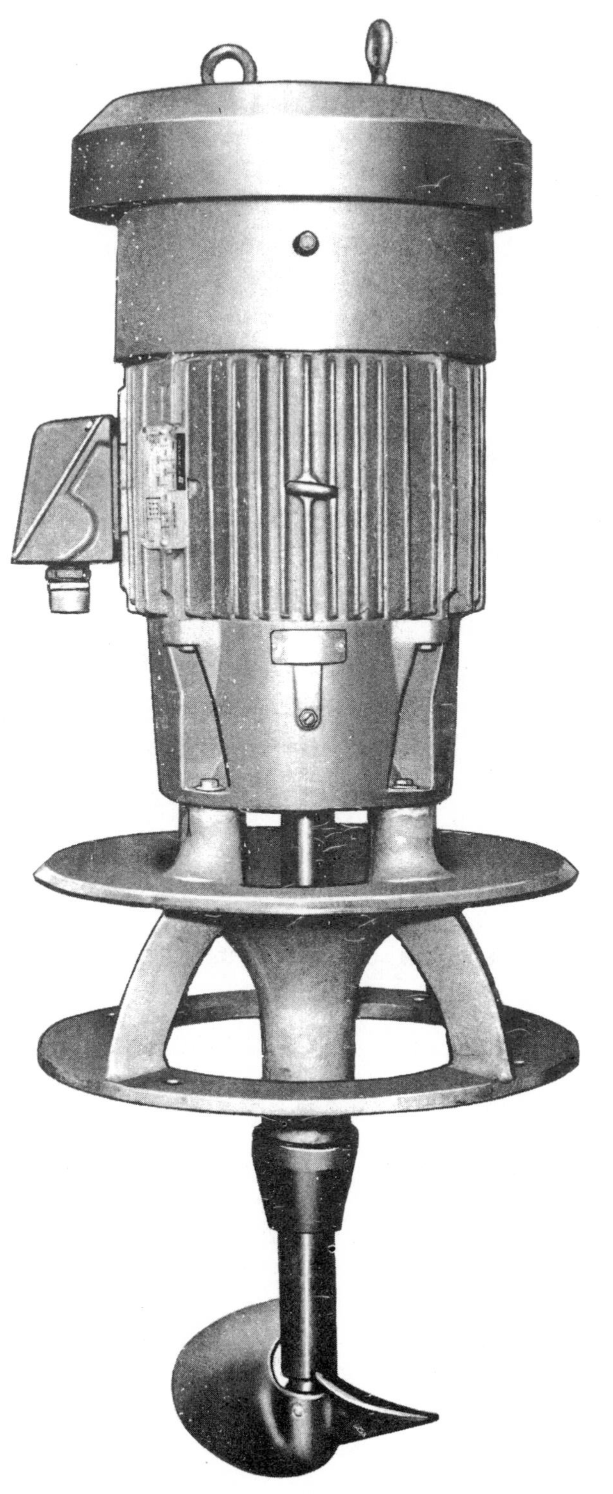

Motor with double engagement fluid deflector in place protecting motor from water ingestion. Also note 316 stainless steel non-fouling propeller.

The AQUA-AEROBIC SYSTEM'S AQUA-JET actually performs two very important functions simultaneously by combining high oxygen transfer rates with powerful circulation. Artificial aeration of wastes can occur at peak efficiency only when thorough mixing and circulation keep solids in suspension. The AQUA-AEROBIC SYSTEMS exclusive pumping system generates a high velocity circulation pattern causing a continual turnover from bottom to surface with complete intermixing. This mixing pattern assures complete dispersion of the dissolved oxygen throughout the tank or lagoon. High oxygen concentrations around the aerator, which tend to suppress oxygen transfer, are thus avoided and results in a marked improvement in oxygen transfer rates. The effluent passing through the aerator is discharged radially into the atmosphere in the form of a high velocity spray pattern.

This process results in a tremendous increase in interfacial exposure to atmospheric oxygen where a large percentage of transfer occurs. As the high velocity spray pattern impinges upon the surface of the reservoir, the transfer of kinetic energy increases total oxygen transfer and establishes the dynamic mixing pattern. This transfer of kinetic energy continues for a considerable distance gradually blending into the larger circulation pattern. Obviously, efficient aerobic digestion of sewage wastes can occur only when sufficient oxygen is available to support the desired bacteria population throughout the entire basin. For this reason, careful consideration should be given, not only to oxygen transfer rates but also to the mixing capacity of the aeration equipment. Mixing capacity is proportional to the kinetic energy produced by the aerator, and the resultant velocity and energy vectors.

ARCTIC-PAK LO-PROFILE DIFFUSER

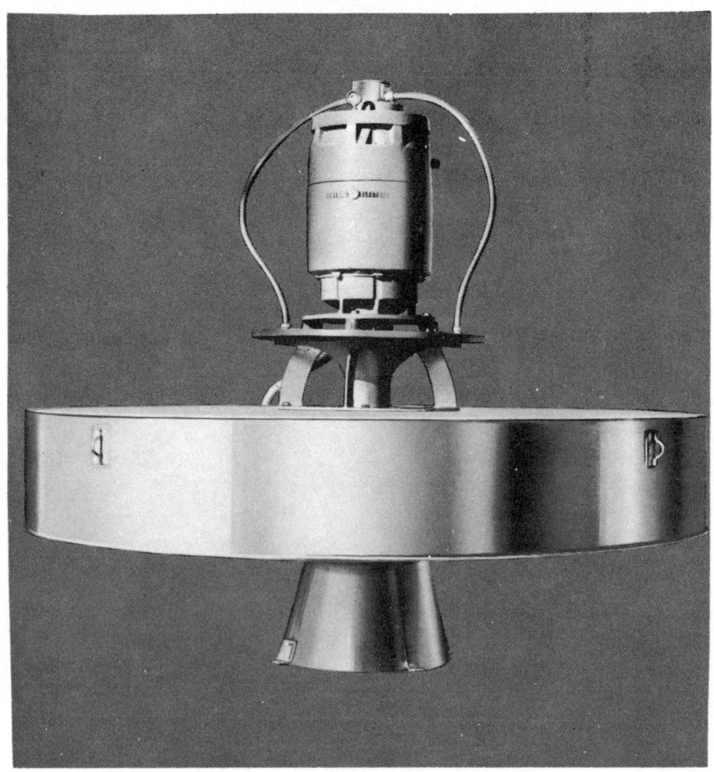

The normal Aqua-Jet Aerator produces a wide, diffused spray pattern to maximize oxygen transfer and mixing.

The Arctic-Pak lowers this pattern and makes it less susceptible to the effects of wind.

The Arctic-Pak does more than flatten the spray pattern. The circular ring contains thermal resistance heaters to keep the surface area of the heat ring and the cast 304 stainless steel diffuser-head surfaces (exposed surfaces) to a temperature of 40°+F. This heat denies the ice a base on which to form.

The Arctic-Pak comes complete with its own junction box (mounted on the aerator motor fan cover) and its own automatic controls and control panel. The Arctic-Pak is controlled by an ambient temperature thermostat. The Arctic-Pak is available for 110- or 230-volt single-phase, or 230-, 460-, and 575-volt three phase.

The Arctic-Pak anti-ice package consists of three basic parts: (1) A fiber glass ring which serves to flatten the water pattern, and which contains thermal resistance heaters to prevent ice from building up on the ring. (2) An electrical junction box, mounted on the aerator motor raincap, in which the two-conductor electrical service cable is attached to the leads of the thermal heaters. (3) A shore-mounted control panel to provide manual or automatic operation of the Arctic-Pak unit.

All components are precision crafted, and equal or exceed NEMA specifications.

AQUA-JET TFNI AERATORS

In many instances, the cost of unnecessary quality cannot be justified. When compared to materials like cast iron, carbon steel, and fiberglass, stainless steel can be considered a "quality" alloy. Its use in aerators is commonplace, but in many instances its additional cost is unnecessary.

Municipal wastes not contaminated with corrosive industrial elements and the effluent from many industrial processes themselves do not require the corrosion-resistant qualities of stainless steel. The TFNI Aerator was therefore developed for use in pollution control processes where the corrosive quality of the effluent is minimal.

The TFNI Aerator substitutes fiberglass and coated nodular iron for stainless steel in many of its components. It is designed to furnish the identical mechanical and hydraulic performance of the proven Aqua-Jet Aerator with significant savings in initial cost.

NO MECHANICAL OR STRUCTURAL DIFFERENCE

Structurally, the TFNI unit is on a par with the 304 stainless steel Aqua-Jet, which is structurally the best unit available. The use of heavy monolithic nodular iron castings for the diffuser head and the volute prevent the structural failures that are common to units featuring a nonmetallic diffuser head and/or volute. Fiberglass has proven to be a satisfactory float medium, if stresses are minimized. The exclusive design of the TFNI Aqua-Jet avoids any moment or shear loads on the fiberglass floats and limits the compressive load to approximately 5 psi.

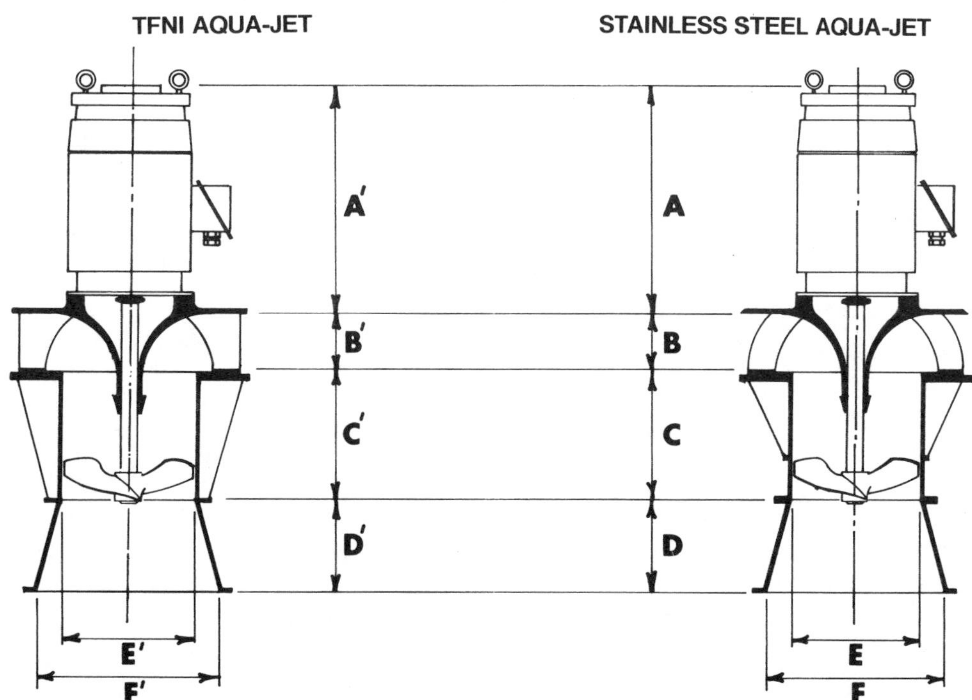

The TFNI Aqua-Jet Aerator is identical to the proven stainless steel Aqua-Jet Aerator in the areas shown above. The dimensions of the pumping chamber, both length and diameter, are identical, as are the dimensions of the diffuser head. The same propeller is used interchangeably in both units. THUS, THE TRIED AND PROVEN PERFORMANCE OF THE AQUA-JET AERATOR IS MAINTAINED WHILE SUBSTITUTING THE LESS EXPENSIVE NODULAR IRON FOR 304 STAINLESS STEEL. THE ABOVE ARRANGEMENT ALSO PRESERVES THE PROVEN RELIABILITY AS THE MOVING PARTS HAVE IN NO WAY BEEN CHANGED.

DESIGN MANUAL

Aqua-Aerobic Systems Inc. provides a unique catalogue which serves as a design manual. The manual is available to those engaged in the design and construction of pollution control systems along with two supporting slide rules. The manual discusses in detail process selection and process designs.

Process Selection: There are several widely used processes for secondary treatment of water-borne wastes. These processes all have one common factor in that all are dependent upon micro-organisms to convert offensive organic material to harmless and relatively inoffensive end products, plus more microbiology, which in turn is used to treat additional incoming raw waste. In some cases, excess sludge (microbial mass) is produced and necessitates a separate disposal scheme. Many factors affect the choice of treatment processes. In some cases, a single dominant factor, such as limited land space, will be the governing factor; however, the choice is generally made based upon economics or the type and character of the waste in question.

Economic factors include the cost of land, the initial construction cost and the operating costs. A good economic comparison will project all the costs over some number of years of operation, as a high initial cost plant may (or may not) prove to have lower operating costs which would more than offset the higher initial cost.

Oxidation Ponds: Oxidation or holding ponds have been used in many locales to treat domestic wastes and have been employed, but to a lesser extent, by some industries for treating industrial waste. The main disadvantages are related to the amount of land required for the lagoon itself and for the space relief required to the nearest dwelling. The degree of treatment, in the terms of BOD and suspended solids removal is not as reliable as conventional plants and ranges from 60% to 90% and varies with the season. Most states require that the pond be placed a minimum of 400 ft. from the nearest dwelling. The amount of solids removed can vary from 90% to a negative value: i.e., algal solids being produced creating downstream oxygen demands.

Aerated Lagoons: Aerated lagoons represent a compromise between activated sludge and oxidation ponds; consequently, aerated lagoons have some of the features of each. The aerated lagoon requires a great deal more space than an activated sludge plant, but much less space than an oxidation pond. Power consumption for aeration is roughly equivalent to the power consumption for aeration in an activated sludge system. Operation and supervision requirements are minimal, and no full-time operator is required.

A properly designed lagoon is capable of 90%+ BOD removals, and is much less prone to upset from shock loadings as compared to activated sludge. The outfall quality is usually considerably better than the outfall from an oxidation pond, and algae are not a problem, as the constant mixing continuously strips CO_2 and creates enough surface turbulence to greatly decrease photosynthetic activity. Lagoon retention times usually range from 3 to 15 days, and even longer in the northern latitudes. The longer retention time is necessary in cold climes if a high degree of BOD removal is to be maintained.

Aerated lagoons are generally followed by a polishing/settling pond to remove the solids from the outfall. Solids accumulations are minor, but some additional BOD can be removed by polishing and settling. Generally aerated lagoons are the choice when 1) land space is available; 2) real estate costs are not prohibitive; 3) a high degree of treatment is necessary, and 4) operating costs are to be minimized.

ACTIVATED SLUDGE AND ITS MODIFICATIONS

This widely known and widely used process has its primary advantages in the high degree of treatment that can be obtained and in the compactness of the plant size. Operating costs are relatively high and a full-time operator is generally required. Since an activated sludge plant produces excess sludge, sludge disposal facilities must be provided. Several schemes of sludge handling are available, including anaerobic digesters, aerobic digesters, vacuum filtration accompanied by sanitary land-fill operations or followed by incineration, sludge drying beds and various other methods. Initial cost and operating costs are increased and this system is one of the most expensive systems currently being utilized today.

Contact Stabilization or Biosorption: This is basically an activated sludge process and the same advantages and disadvantages that apply to the activated sludge process also apply to the contact stabilization process. Its main advantage lies in the two tanks (contact and reaeration) combined volume is usually less than the volume of an aeration tank for a similar activated sludge plant.

The contact stabilization process is basically a modification of the activated sludge process. Physically, a plant consists of a contact tank, a secondary settling tank, and a reaeration tank. This process allows a much greater BOD volumetric loading than does an activated sludge process, and herein has its advantage.

The influent to the plant is mixed with sludge from the reaeration tank in the contact tank. This results in a rapid adsorption of the particulate matter and assimilation of dissolved organic matter from the waste. When retention times are short (less than one hour) in the contact tank, the physical phenomenon of adsorption and the biochemical process of assimilation are the two major processes occuring. Oxygen requirements are generally limited to those of assimilation. The amount of HP required will generally be controlled by the HP necessary to completely mix the tank.

Longer retention times in the contact tank result in bacterial respiration also occuring in the contact tank, and the oxygen requirements become greater. Most state regulatory agencies govern the minimum time required for each tank, and generally these are based upon longer retention times. It is recommended that the total oxygen requirements be established for the system, on the basis of one pound of oxygen required per pound of BOD applied. In the event the contact tank does utilize a short retention time, approximately 80 HP per million gallons should be provided to insure good mixing.

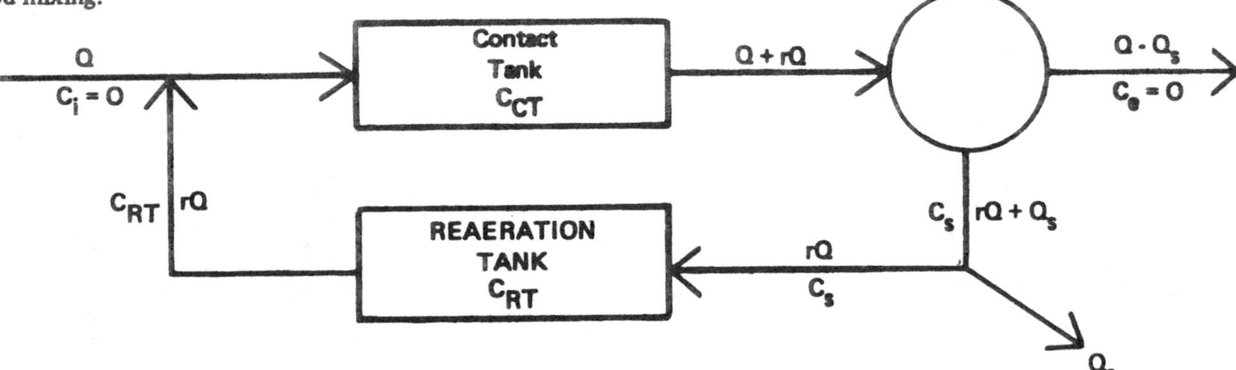

Extended Aeration: This process is designed to produce a minimum of excess sludge. The volume of the aeration basin for an extended aeration plant is greater than the volume of the aeration basin for an activated sludge process, which is reflected in increased costs. Since a minimum of sludge is wasted, sludge handling facilities are minimal and this reflects in both initial cost savings and savings in operating expenses.

Miscellaneous Modifications: There are other modifications of the activated sludge process such as step aeration, etc., but these modifications are generally refinements in the system in regard to the arrangement of the incoming flow and mixing capabilities, and can be considered as being subject to the same advantages and disadvantages as the activated sludge process.

Aurora Pump, Unit of General Signal Corp.

610 SERIES SINGLE STAGE SEWAGE PUMPS - "SPHER-FLO"

Spher-Flo Series 610 pumps are available in the following models:

MODEL 611

MODELS 611 are horizontally baseplate mounted with a driver flexibly coupled to the pump. This design is recommended where floor space is readily available and where flooding of the installation is not possible.

MODEL 612

MODELS 612 are vertically mounted and utilize flexible shafting between the driver and the pump. This model is frequently used on lift station applications where flooding of the installation is a possibility.

MODELS 613 are vertically mounted with an elevated driver coupled directly to the pump through a flexible coupling. Model 613 is very popular for installations where available floor space is limited and where flooding is marginal.

MODEL 613

STANDARD

3″ to 6″ sphere capacity
All iron fitted pump
Regreasable bearings
Double row outboard thrust bearing
Single row inboard radial brg.
Hardened stainless steel (450 min. Brinell) shaft sleeve
Removable split packing box
Interchangeable packing and mechanical seal inserts
External impeller clearance cartridge type adjustment
Taper shaft fit at impeller
Carbon steel shaft and impeller key
Front or back impeller pull out
Enclosed impeller with wiper blades on back shroud
Dynamically balanced impeller
Centerline discharge casing
Hydrostatic test all pumps
Discharge position No. 1
Interwoven graphited asbestos packing diagonally split
Lantern ring liquid seal for packed stuffing boxes
Gasket sealed pump shaft stuffing box extension

OPTIONAL

Stainless steel case wear ring
Stainless steel impeller wear ring
Suction nozzle with clean out (Model 611 only)
Double mechanical seal
Stainless steel shaft
Alloy shaft sleeve
Wear ring face flush line
External stuffing box piping with filter or valve
Automatic stuffing box grease seal lubricator
Stuffing box leakage collector (Model 612 and 613 only)
Spacer type coupling (Horizontal Model 611 only)
Coupling guard
Flexible shaft drive with or without guard (Model 612 only)
Water Seal Unit (See Bulletin 680 for details)
Constant liquid level system (Apco-Trol Variable Speed — See Bulletin 700)
Certified test report — witnessed or unwitnessed
Special alloy pump construction
Alternate discharge positions

PUMP SIZE	RPM (POWER SERIES)				
	1750	1150	875	700	580
4x5x7	4A	4A			
4x5x10	4A	4A	4A	4A	
4x5x12	5A	4A	4A	4A	
4x5x15	5B	5A	5A	5A	
6x6x12	5B	5A	5A	5A	
6x8x15	5B	5A	5A	5A	5A
6x8x18		5B	5A	5A	5A
8x8x15		5B	5A	5A	5A
8x10x18		6B	5C	5C	5C
8x10x22			6B	5C	5C
10x10x15			5A	5A	5A
10x12x22			6B	6A	5C
12x14x22			6B	6A	5C

SPHER-FLO pumps are designed for maximum interchangeability. Each model is available in 13 different sizes, offering a model and size precisely fitted to the installation requirements over a wide range of capacities. The 13 sizes are divided into 8 "power series." Within a given power series, all parts are interchangeable except for the liquid end and supports. The chart illustrates the degree of interchangeability achieved with the standard SPHER-FLO pumps.

BEARING LIFE

Tables 1, 2, and 3 will enable you to determine the minimum thrust and radial bearing life for any pump size. The minimum life shown is for the worst conditions of load or pump shutoff. Bearing life at any other point on the curve and at reduced impeller diameters will greatly exceed the minimum life shown. Average bearing life is equal to five times the minimum bearing life.

1. Determine the proper pump size and speed from the range charts illustrated on pages 8 and 9 of this bulletin.
2. Read K value by pump size and speed from Table 1. This value will be used for determining **both** outboard thrust and inboard radial bearing life in years.
3. Locate K on Table 2 and read across to the proper speed in RPM and then down for outboard thrust bearing life. If the K value from Table 1 is brown bold face, use the brown line. If the K value from Table 1 is light face, use the black line.
4. Locate K on Table 3 and read across to the proper speed in RPM and then down for inboard radial bearing life. If the K value from Table 1 is brown bold face, use brown line. If the K value from Table 1 is light face, use the black line.

EXAMPLE:

1. A 6 x 6 x 12 pump at 1750 RPM (**15.9** K value) would have
   - A. Continuous outboard bearing life of 4.48 years
   - B. Continuous inboard bearing life of 10.6 years
2. A 6 x 6 x 12 pump at 1150 RPM (19.1 K value) would have an
   - A. Continuous outboard bearing life of 13 years
   - B. Continuous inboard bearing life of 11.3 years

NOTE:

1. K values given are for worst condition of load or pump shutoff. (Max. impeller dia.)
2. One (1) year life is based on 8740 hours (continuous operation).
3. Additional bearing information can be found on page 7.
4. Detailed information on bearing life and shaft deflection can be obtained from the factory.

TABLE 1 — VALUES OF "K"

SIZE	4x5x7	4x5x10	4x5x12		4x5x15	6x6x12	6x8x15	6x8x18	8x8x15	8x10x18		8x10x22		10x12x22		12x14x22	
POWER SERIES	4	4	5	4	5	5	5	5	5	6	5	6	5	6	5	6	5
1750 RPM	26.2	15.2	16.6	–	**11.1**	**15.9**	**8.8**	–	–	–	–	–	–	–	–	–	–
1150 RPM	41	24.1	–	11.9	13.5	19.1	10.7	**9.8**	**10.9**	**9.2**	–	–	–	–	–	–	–
860 RPM	–	43.2	–	21.5	24.7	35.5	19.4	13.1	14.5	–	11.5	**9.8**	–	**8.3**	–	**8.3**	–
700 RPM	–	–	–	33	37.1	52.5	29.2	20	22.1	–	17.5	–	11	9.6	–	9.6	–
580 RPM	–	–	–	–	55.5	–	42.9	29.5	33	–	26	–	16.4	–	14.2	–	14.2
SPHERE SIZES	3	3	3	3	3	3	4	4	5	5	5	5	5	6	6	6	6

BOLD BROWN VALUES ARE FOR ROLLER BEARING CONSTRUCTION

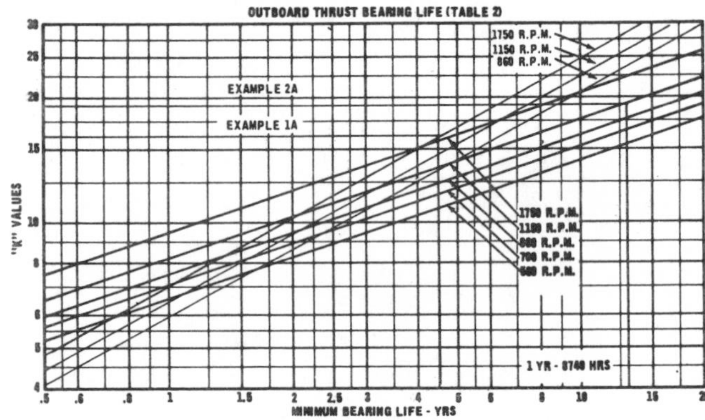

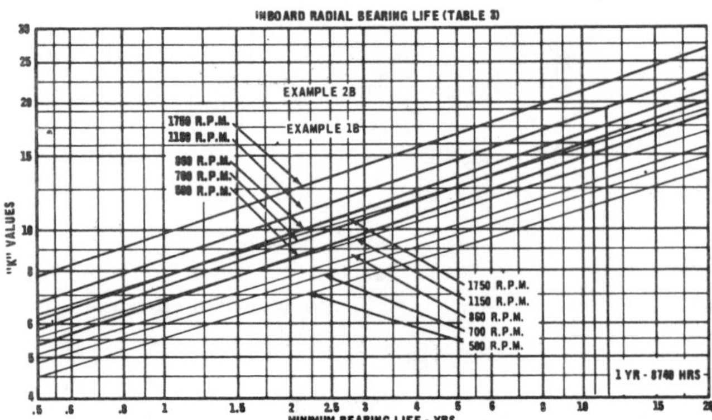

LIMITATIONS

Maximum hydrostatic test pressure.....150 PSI
Maximum recommended case working pressure.............................125 PSI
Maximum suction pressure.............125 PSI
Maximum Temperature °F .... Packing 250°F; Mech. Seal 225°F
Maximum operating speed...........1750 RPM

SUBMERGENCE

Air may be entrained in the pumped liquid if the pump suction is located too close to the free liquid surface in the suction source.

Pumping liquid with entrained air can cause a reduction of capacity, rough and noisy operation, vibration, loss of efficiency and wasted power. Excessive wear of close running parts, bearing stresses and shaft damage are also subsequent effects.

If the capacity in gallons per minute and the suction inlet size or area is known the minimum height of the liquid above the suction inlet (submergence) can be determined.

A properly designed suction inlet and sump can be accomplished with the help of the submergence chart illustrated.

Typical submergence requirements are shown on the chart.

EXAMPLES:

(1) When the pipe size is known, the minimum submergence required for 2000 GPM through an 8″ pipe is 9.6 feet.

(2) When the inlet area is known, the minimum submergence required for 2000 GPM through a 50 square inch outlet is 9.6 feet.

Minimum submergence requirements may exceed the available space requirements. When this occurs a larger pipe size or inlet will reduce the required submergence.

EXAMPLE:

(3) The minimum submergence required for 2000 GPM through a 10″ pipe is 4.6 feet (5.0 feet less than required for an 8″ pipe).

650 SERIES SINGLE STAGE SEWAGE PUMPS

MODEL 651

MODELS 651 are horizontally baseplate mounted with a driver flexibly coupled to the pump. This design is recommended where floor space is readily available and where flooding of the installation is not possible.

MODELS 652 are vertically mounted and utilize flexible shafting between the driver and the pump. This model is frequently used on lift station applications where flooding of the installation is possible.

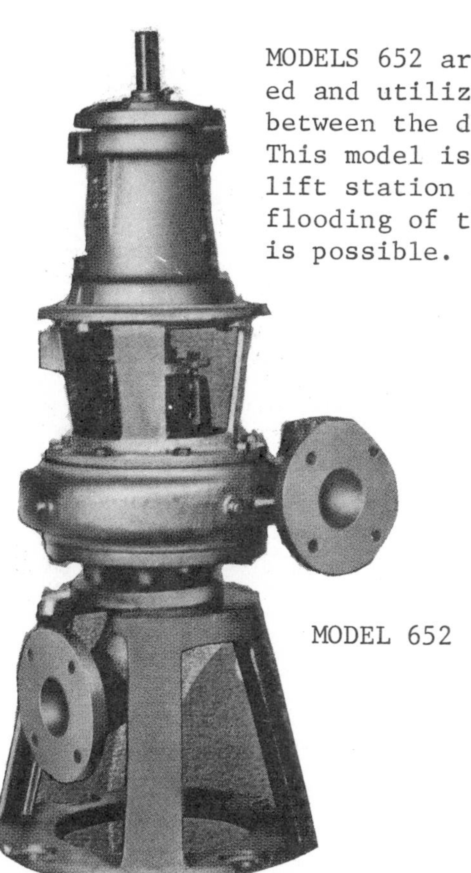

MODEL 652

MODEL 653

MODELS 653 are vertically mounted with an elevated driver coupled directly to the pump through a flexible coupling. Model 653 is very popular for installations where available floor space is limited and where flooding is marginal.

Aurora Pump, Unit of General Signal Corp.

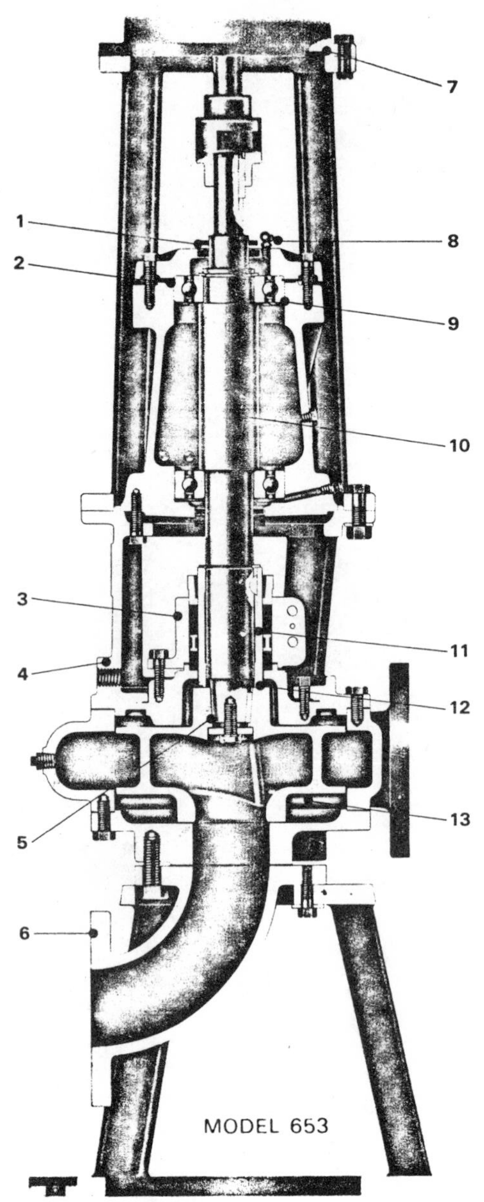

1 WATER SLINGERS and grease seals protect both bearings from moisture.

2 SHAFT ADJUSTMENT provides for renewing impeller clearance and maintaining pump efficiency.

3 SPLIT PACKING BOXES separate vertically through the packing insert to simplify packing replacement and shaft sleeve inspection.

4 BACK PULLOUT DESIGN for pump maintenance, does not disturb piping.

5 SHAFT EXTENSION with taper for easy impeller removal.

6 DISCHARGE and suction flanges can be located in 45° increments for 8 different positions.

7 NEMA-HI STANDARD "HP" face motors.

8 GREASE LUBRICATION purges old grease from bearings.

9 THRUST BEARINGS are rated for maximum loads. Average bearing life is 50,000 hours.

10 .002 MAXIMUM SHAFT DEFLECTION at stuffing box face extends packing and mechanical seal life.

11 HARDENED STAINLESS STEEL SLEEVES (450 minimum brinnel) are keyed to the shaft.

12 GASKETS protect shaft from pumped liquid.

13 OVAL CLEANOUT on 3" and 4" pumps are hand size and located to provide visibility and accessibility to the impeller and the casing cutwater.

Support of various pump components is important. Inadequate mounting designs impose unnecessary stress and strain on the entire pump and installation.

HORIZONTAL 651 UNITS are supported at both pump and coupling end for protection against pipe strain. The rear support foot greatly simplifies coupling alignment and is an important design feature.

On VERTICAL 653 UNITS the cast motor base has a registered fit at the motor end and is fastened to a separate pump adapter. This exclusive arrangement assures alignment and concentrates loads on the separate pump adapter thereby eliminating strain and misalignment of the bearing housing. On 652 and 653 units, the cast base is registered to the suction cover assuring adequate support for the complete unit.

BSP, Division of Envirotech

CARBON REGENERATION SYSTEM

Waste Water Treatment: Many processes have been evaluated for advanced treatment of waste water, but granular activated carbon adsorption has emerged as the most efficient in terms of organic removal and treatment cost. The key to low cost is the efficient regeneration of granular carbon. Spent carbon is dewatered, then fed to a multiple hearth furnace. Temperatures of 1600° to 1700°F. drive off organic matter. Using BSP systems, cost is approximately one-tenth that of new carbon. Regeneration of carbon is also important in industrial waste treatment plants that must absorb phenols, chlorophenols and organic resins.

Water Treatment: BSP carbon regeneration systems have also been used successfully in raw water purification. At the 15 MGD Nitro West Virginia facility, raw water from the Kanawaha River is treated to remove turbidity, taste, odor and dissolved organic contaminants. Replacing sand in the filtration beds with regenerated granular carbon was economical. Cost was reduced because carbon is reclaimed with a BSP carbon regeneration system.

Tertiary Treatment Lime Reclamation: BSP multiple hearth systems provide the lowest cost method of recalcining lime. Systems are now operating in raw water and waste water treatment plants and in industrial sugar refineries.

Basic Tahoe flowsheet

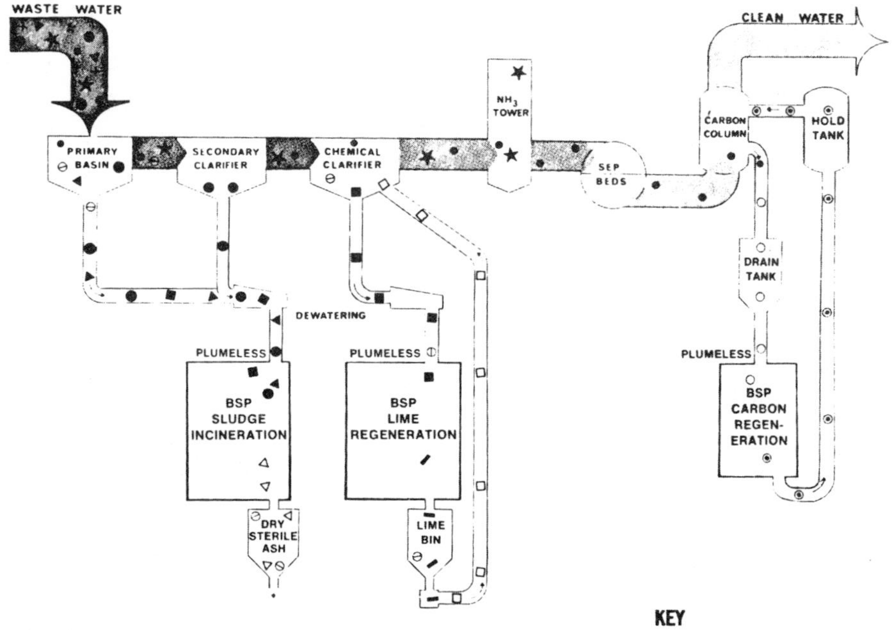

KEY

- ▬ Calcium Oxide CaO
- □ Calcium Hydroxide $Ca(OH)_2$
- ■ Calcium Carbonate $CaCO_3$
- ⊖ Calcium Phosphate $Ca(PO_4)$
- • "Light" Organics ABS etc.
- ● "Heavy" Organics
- ⦿ Activated Carbon
- ○ Spent Carbon
- ▶ Unclean Inorganics
- ◁ Sterile Ash
- + Other Phosphates
- ★ Ammonia NH_3

There are three plants producing potable drinking water from raw sewage. The largest is Lake Tahoe's 7.5 MGD tertiary sewage treatment plant, a plant which removes 98 percent of the organic material from secondary sewage effluent. BSP was responsible for solids handling and reclamation at Tahoe, including sludge incineration, lime recalcining, and carbon regeneration.

BSP, Division of Envirotech

PORTEUS PROCESS

The Porteous process is essentially a mechanical process which operates on the principal of syneresis, thereby breaking down the gelatinous structure of a sludge through heat and pressure. When this occurs, the bound water that is tied up in a sludge particle is released and the particle can readily settle and be dewatered on a vacuum filter, centrifuge, or as is commonly used in Europe, a filter press. The major advantage of this process is that sludge can be dewatered without any conditioning chemicals.

The heat mechanism is essentially achieved by raising the sludge temperature to over 300°F. This can vary depending upon the sludge. Surprisingly enough activated sludge is generally the sludge which will break down most readily. It has a small particle size and the bound water structure will release itself very well and may react at 300° to 600°F. Primary and digested sludges are more difficult, but the highest temperature used thus far has been 390°F. This is generally accomplished in a reactor vessel where the sludge is held for 30 minutes to an hour. The detention time required is simply a characteristic of a sludge. Some sludges react very quickly and others react much more slowly, but the end result produces a sludge that can be filtered without any chemicals.

Essentially sludge is taken from a holding tank, digestor or thickener, sent through a grinder or disintegrator, and then is pumped to 250 psig by a ram pump. The pump discharges at approximately 250 psig. The sludge is then pumped through a heat exchanger into the reaction vessel. Hot sludge which is being held anywhere from 30 minutes to an hour is then passed back through the heat exchanger and finally cooled. There is an interchange of heat, and this establishes the basic economics of the process. Assuming the sludge coming into the system is at 60°F., it is raised to approximately 350°F. as it leaves the heat exchanger and enters the reaction vessel. To reach the operating temperature (about 380°F.), steam is injected directly into the reaction vessel. This adds the necessary heat to attain the required process temperature. Steam also seems to have a beneficial effect upon the sludge itself. It was found in the laboratory that trying to treat a sludge indirectly does not produce filtering characteristics equal to that obtained by direct steam injection. Similar results were found in a large scale pilot test extending over several months.

From this heat exchanger the sludge then flows out through a discharge valve and into a decanter where it readily settles. The decant tank is very similar to a conventional picket type thickener. Assuming 5% solids going into this system, it will readily settle to from 10 to 15% solids within about a half an hour. These vessels are generally sized for approximately two hours to provide a buffer in front of the dewatering device. From the decant tank the sludge is pumped to a dewatering device. The material can then be incinerated or it can be disposed of on the land.

Since the sludge is heated to such a high temperature, it is sterile. It is free of pathogens, and seeds within the sludge from tomatoes will not grow as they will when found in digested sludge. The treated material appears similar to a peat, especially after it dries. Most dewatering equipment has been able to dewater this material to 50 to 55% solids without any conditioning chemicals. Gases released from the reactor are condensed, cooled, and liquified to avoid odor problems. The decanter tank is enclosed to prevent odors from that source. Odors can also be controlled by maintaining a temperature of roughly 75° to 100°F. at the dewatering device.

PORTEOUS FLOWSHEET

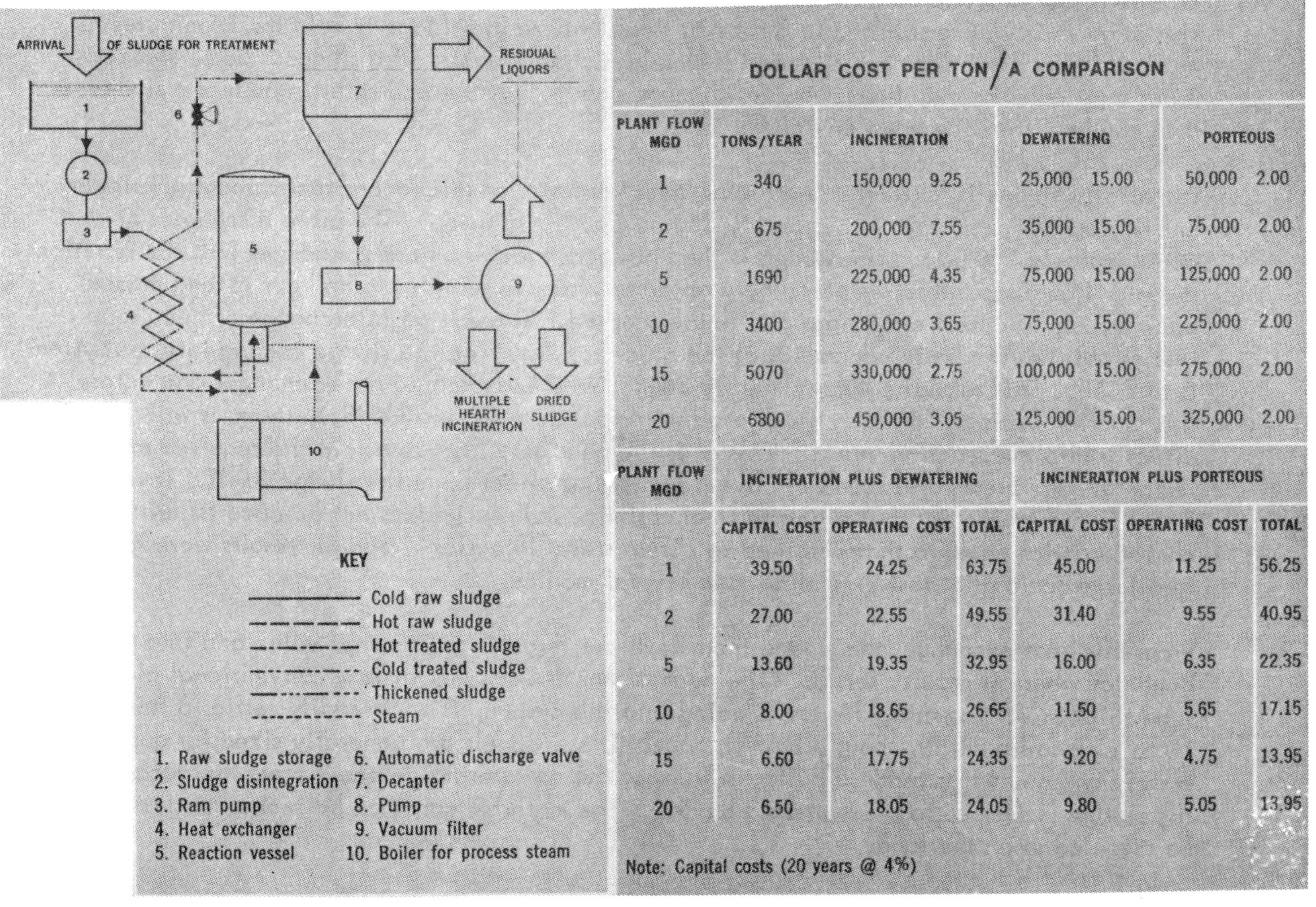

DOLLAR COST PER TON/A COMPARISON

PLANT FLOW MGD	TONS/YEAR	INCINERATION		DEWATERING		PORTEOUS	
1	340	150,000	9.25	25,000	15.00	50,000	2.00
2	675	200,000	7.55	35,000	15.00	75,000	2.00
5	1690	225,000	4.35	75,000	15.00	125,000	2.00
10	3400	280,000	3.65	75,000	15.00	225,000	2.00
15	5070	330,000	2.75	100,000	15.00	275,000	2.00
20	6800	450,000	3.05	125,000	15.00	325,000	2.00

PLANT FLOW MGD	INCINERATION PLUS DEWATERING			INCINERATION PLUS PORTEOUS		
	CAPITAL COST	OPERATING COST	TOTAL	CAPITAL COST	OPERATING COST	TOTAL
1	39.50	24.25	63.75	45.00	11.25	56.25
2	27.00	22.55	49.55	31.40	9.55	40.95
5	13.60	19.35	32.95	16.00	6.35	22.35
10	8.00	18.65	26.65	11.50	5.65	17.15
15	6.60	17.75	24.35	9.20	4.75	13.95
20	6.50	18.05	24.05	9.80	5.05	13.95

Note: Capital costs (20 years @ 4%)

The Baker Tube filter most nearly approaches the ideal filter. All filter designs are compromises between the ideal and the possible. The Baker spiral wound, continuously welded tube requires the minimum of compromise in the mechanical design.

CONTROLLED OPENING

The very closely controlled application of the spiral wire winding to the tube core results in the most uniform aperture between wires. The continuous welding of the wires in position ensures that the uniformity of the tube will exist throughout its long life. Very rapid clarity is established in the filtered liquid (with or without precoat) because the openings between wires can be selected for the solids to be collected. The aperture remains uniform throughout all tubes in the filter.

RAPID CLARITY

Clarity is established in the minimum time because the initial layer of solids is rapidly collected on the rigid wire winding and forms a most stable base for future filter cake build up.

Filtered liquid once inside the tube moves axially towards the discharge end displacing the contents of the tube in a short time. Flushing the first fill of liquid from the center of the tube cores contributes noticeably to the rapid establishment of clarity.

STRENGTH OF FILTER TUBES

It is seen from the details of the tube that the design can withstand high external pressure without deformation. Tubes have been tested to 400 psi without failure. The ability of the rigid tube to retain its exact shape while subject to increasing differential pressure ensures that there will be no cracking in the precoat and no disturbance of the collected solids. Deflection does not occur with pressure build up. The tube is even stronger in backwash, a feature unique with spiral wound tube elements. Most existing filter elements cannot use the effective cleaning available in backflush because this would destroy the elements themselves.

AREA INCREASES DURING FILTRATION

As solids are collected, the diameter of the tube is increased, thus presenting a larger surface area to the incoming flow resulting in lower applied flow rate at the media surface. The longest possible filter runs result from this advantage.

CAKE RETENTION ON THE TUBES

The tube design is an inherently strong design for the filter cake. During filtration the cake has the strength of a cylindrical column. In blowdown, the aperture between the wires helps meter the air or gas to apply a differential pressure to all parts of the collected cake. Again, the cylindrical form is of assistance.

If a final wash of collected solids is required, operation is improved by the resistance to cracking shown by the cylindrical cake. The rapid flushing of the core centers increases the effectiveness of the wash and reduces time required.

BACKWASH

The tube filter design can achieve the effective suspension of solids using a small volume of liquid. Using the "air bump" principle, liquid is used to compress air or gas in the domed top of the filter which drives the washing liquid back through the wire apertures at very high velocity when the dump valve is opened.

The continuous welding process ensures that, although subjected to considerable differential in backwash, the wires will not move or sag. Backwash technique avoids the problems encountered with element sprays, rotating leaves and unreachable zones for filter cleaning.

DRY CAKE DISCHARGE

For certain applications, backwash is not suitable and collected solids are required in the driest form possible. The tubular design using monofilament wire enables cake to be peeled or vibrated from the tubes effectively.

Filter tubes are 1 1/4" diameter and can be produced up to 72" long. 2 1/2" diameter tube is available for special applications.

Filter tubes are available in 12, 25, 50, 100 or 250 micron spacing between wires.

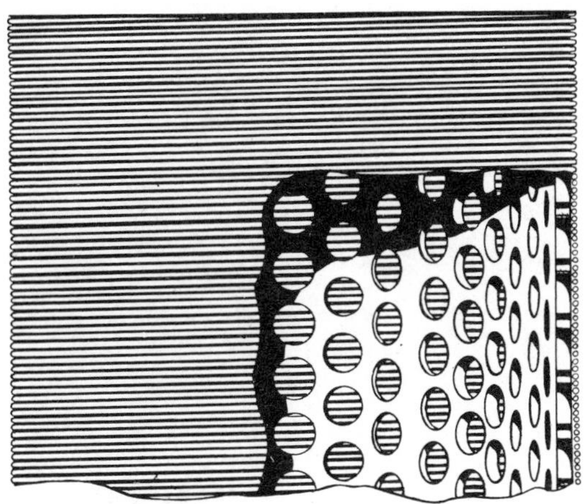

The continuously welded, spiral wire winding ensures a controlled, smooth opening at all points. The close spacing of perforations in the 316 stainless steel core ensures adequate open area.

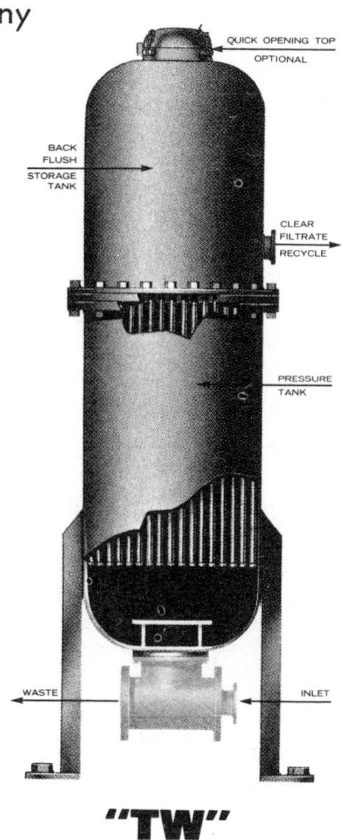

"TW"
SERIES FILTER

"**MODEL TW**" — the simplest model for straight forward filtration and backflush. Generally used for water filtration.

"**MODEL TF**" — This filter can displace the working fluid while retaining the cake in position and can rinse with a second liquid (or a third) before the cake is backflushed.

"**MODEL TS**" — Similar to the "TW," the tubes are spaced closer together for greater economy. This unit has less cake space.

"**MODEL TC**" — This filter can displace the working fluid and can rinse the collected cake if desired. The filter cannot backflush, but after drying, the cake is vibrated from the tubes resulting in a dry solids discharge.

"**MODEL TR**" — Rotating tube dry cake filter is similar to the "Model TC." It is used to discharge very sticky cakes from the filter surfaces.

CAPABILITIES

- Backflush cleaning in 5 seconds plus draining time.
- High velocity backflush of 25 foot per second through wire apertures.
- Filter tubes resist plugging because of the unique parallel wire design. Even sticky filter cakes are resuspended easily.
- Backflush discharge is concentrated into a small volume.
- Long tube life because filter tube wires are continuously welded to the tube core thus eliminating metal fatigue due to flexing.
- No moving parts gives filter great dependability.
- 10 micron slurries or coarser can often be filtered without filter aid by selecting the proper micron rated filter tubes.
- Short cycles—filter can be economically operated on very short cycles.
- Filter can easily be automated. Baker Filtration Company will provide customized full automatic controls as required.
- Hydraulic tube cleaning, if required, is possible because all tubes are accessible from the quick opening cover. A combined air/water ram cleaning jet can be inserted into each tube to hydraulically blast out any imbedded impurities caught between the filter wires.

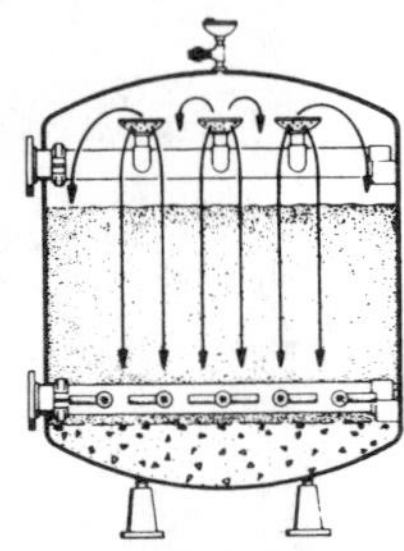

HI-RATE FILTER

Scientifically engineered over and underdrains allow high velocity parallel flow through the filter with a minimum of turbulence. Only two layers of media provide greater depth of filtering medium.

Fig. "A"

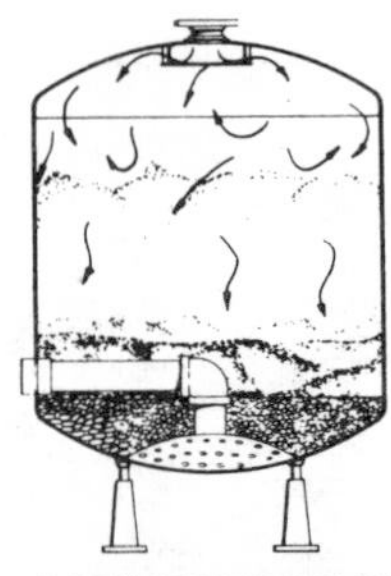

CONVENTIONAL SAND FILTER

Simplified over and underdrain in regular sand filter with its many layers of rock and sand causes turbulence and channeling at high flow rates

Fig. "B"

As the capacity charts indicate, the Hi-Rate permanent-media pressure-filter operates at rates seven to ten times higher than conventional, pressure sand filters.

Fig. "A" shows diagramatically the HI-RATE permanent media filter and Fig. "B" shows a conventional sand pressure filter. The units differ in two main respects:

1. Distribution and collection equipment is designed on scientific principles in the HI-RATE, whereas the sand filter uses a simplified form of baffling.

2. The sand filter has many grades of rock, gravel and sand, whereas the permanent media filter uses a greater depth of one filtering medium.

With the specific, balanced hydraulic flow design of the permanent media filter, water turbulence is reduced to very low limits, and flow paths at the media surface are almost wholly parallel and vertical. It is observed that flow rates in excess of 20 gpm/sq. ft. can be applied to the HI-RATE filter without displacing media or causing channeling which occurs in the conventional unit.

At high flow rates, collected solids are forced deep into the media, but selection of a small media particle size enables good filtration to be achieved. A conventional sand filter must retain the solids in the top 3/4" of the bed or difficulties are experienced in properly cleaning the media with normal backwash. The same volume of solids can be collected in a HI-RATE filter on a much reduced surface area, as the full depth of the media is utilized.

The permanent media filter can operate at differential pressures of 45 psi, but often the end of the filter cycle is controlled by penetration of media. Usually the filter is cleaned when the differential pressure reaches 25 psi. High differential pressure does not cause breakthrough in the HI-RATE filter.

The underdrain system of the HI-RATE filter is designed to cause strong agitation in the working media during backwash. Sand grains are rubbed together to release solids and circulation patterns are established to progressively present each particle of the media at the surface on approximately a 30 second cycle. The balanced-flow conditions induced by the collector system reduces water velocity to below that of the settling rate of the media grains. Bed expansion is observed to be about 6" at backwash rates of 15 gpm/sq. ft. It is further observed that collected solids are discharged during the first 90 seconds of backwash and that the quality of the discharged water approaches that of the backwashing water source within two minutes. The recommended backwash period is 2-1/2 minutes.

MODEL HF, HRA, HRB HI-RATE FILTERS

Two styles of Baker Vertical Hi-Rate filters are available. The Model HF comes in two small sizes with tanks having a low (22") side shell. Filter tanks for Models HRA & HRB are identical. The difference between the two is in the piping. HRA units have copper piping and gate valves as detailed in the specification. HRB units have C.I. fittings and butterfly valves with single lever operator.

MODEL HF

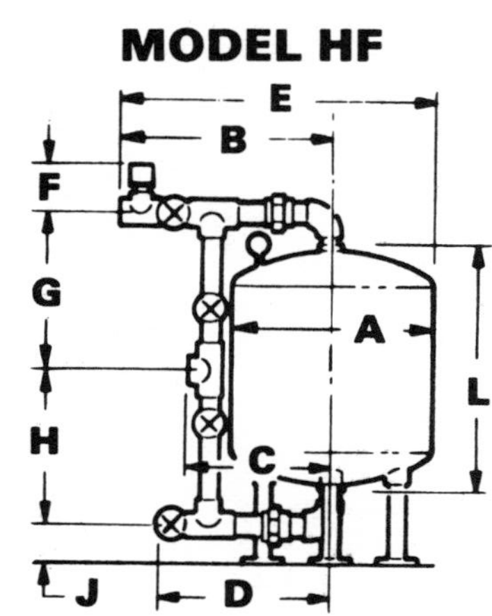

MODEL HRA

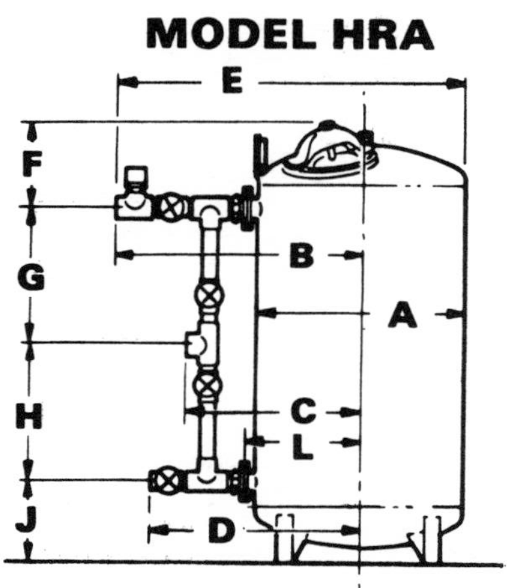

MODEL HRB

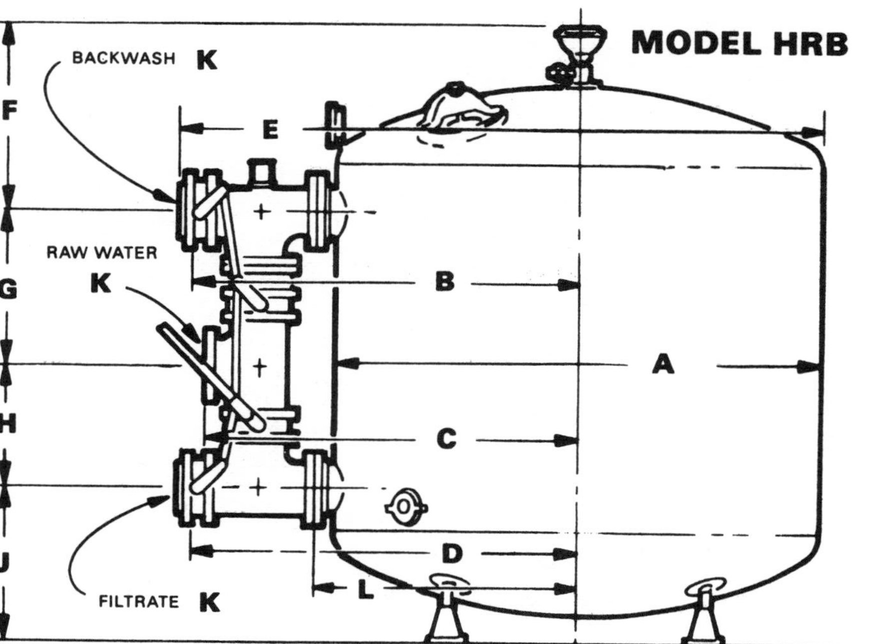

FLOW RATES AND PHYSICAL DIMENSIONS FOR MODELS HF, HRA AND HRB

Model No.	Max. Flow GPM	Area Sq.Ft.	A	B	C	D	E	F	G	H	J	K	L	*Concrete Cu. Ft.	Media Cu. Ft.
HF-20	44	2.2	20"	23-3/4"	15-1/4"	18-3/4"	33-3/4"	6"	17-15/16"	17-15/16"	4-9/16"	1-1/2"	31"	--	4
HF-24	63	3.1	24"	25-7/16"	17"	20-7/16"	37-7/16"	6-1/2"	18-1/2"	18-1/2"	4-1/4"	2"	33-1/2"	--	5
HRA-30	98	4.9	30"	36"	25-3/8"	30-5/16"	51"	12-3/4"	20-1/2"	20-1/2"	12"	2"	18-1/4"	--	13
HRA-36	142	7.1	36"	41-9/16"	29-9/16"	35-1/8"	59-9/16"	12-3/4"	20-1/2"	20-1/2"	12"	2-1/2"	21-3/16"	--	18
HRB-30	98	4.9	30"	28-7/8"	27-5/16"	28-7/8"	29-11/16"	13"	30-7/16"	10-9/16"	12"	2"	18-1/4"	1.5	13
HRB-36	142	7.1	36"	34"	32-1/4"	34"	35-3/16"	13"	28-1/4"	12-3/4"	12"	3"	21-3/16"	2.3	18
HRB-42	192	9.6	42"	39-5/16"	37-5/16"	39-5/16"	61-5/8"	20"	26"	15"	14"	4"	24-3/16"	3.5	26
HRB-48	252	12.6	48"	42-5/16"	40-5/16"	42-5/16"	67-5/8"	21"	26"	15"	14"	4"	27-3/16"	5	34
HRB-54	318	15.9	54"	45-5/16"	43-5/16"	45-5/16"	73-5/8"	22"	26"	15"	18"	4"	30-3/16"	6.8	43
HRB-60	392	19.6	60"	51-1/2"	49-3/8"	51-1/2"	83"	26"	22-15/16"	18-1/16"	23"	6"	33-3/16"	9	62
HRB-66	474	23.7	66"	54-1/2"	52-3/8"	54-1/2"	89"	27"	22-15/16"	18-1/16"	23"	6"	36-3/16"	10.6	76
HRB-72	566	28.3	72"	57-1/2"	55-3/8"	57-1/2"	95"	28"	22-15/16"	18-1/16"	23"	6"	39-3/16"	15	90
HRB-78	664	33.2	78"	60-1/2"	58-3/8"	60-1/2"	101"	29"	22-15/16"	18-1/16"	23"	6"	42-3/16"	18.8	106
HRB-84	770	38.5	84"	66"	63-7/16"	66"	109-11/16"	30"	20-1/2"	20-1/2"	28"	8"	45-5/16"	23.2	123
HRB-90	884	44.2	90"	69"	66-7/16"	69"	115-11/16"	31"	20-1/2"	20-1/2"	28"	8"	48-5/16"	39.6	141
HRB-96	1006	50.3	96"	72"	69-7/16"	72"	121-11/16"	32"	20-1/2"	20-1/2"	28"	8"	51-5/16"	47.6	161

**Concrete sub-fill is required only for applications where organic solids may decompose in the filter base. When concrete is omitted, increase the media requirement by the concrete volume.*

MODEL HRC HI-RATE

The Model HRC designates two or more filters, plumbed to provide their own backwash water. With two units, no filtered water is available for external use during the backwash period.

MODEL HRD HI-RATE

The Model HRD designates two or more filters, plumbed for backwash from an external supply. The water available for service is reduced by the capacity of one unit during backwash.

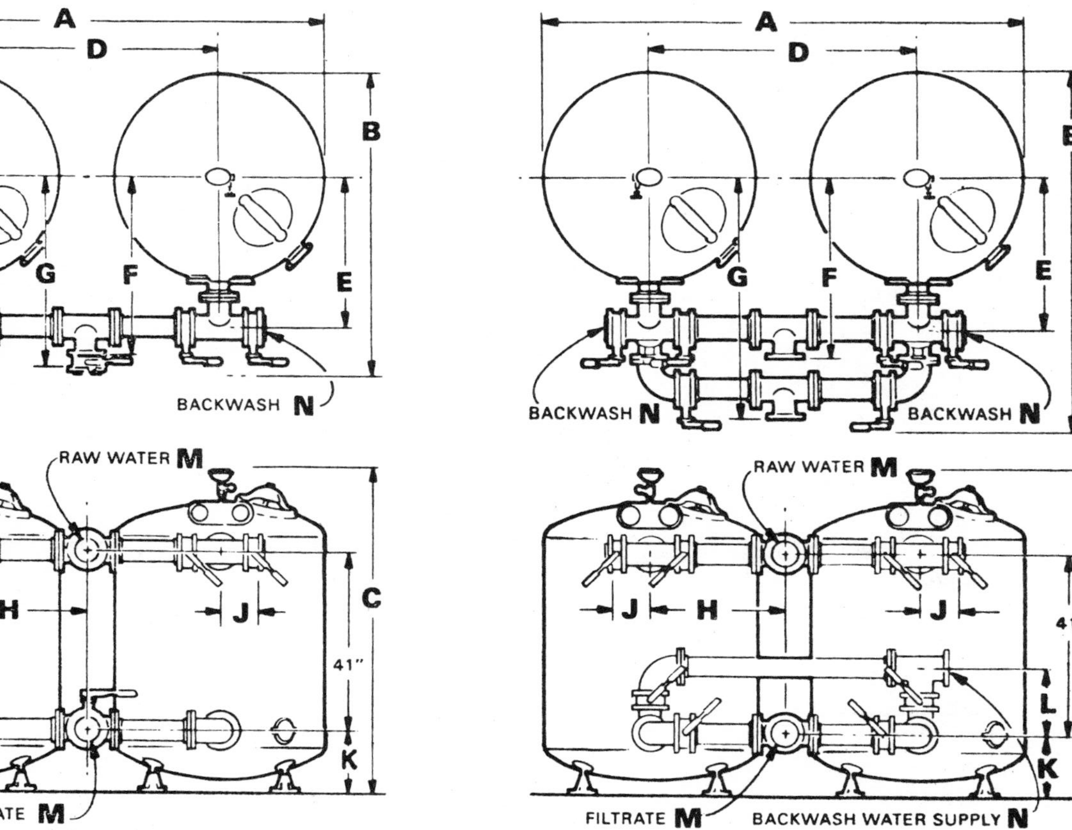

FLOW RATE AND PHYSICAL DIMENSIONS FOR MODELS HRC AND HRD

Model No.	Max. GPM	Area Sq. Ft.	Tank Dia.	A	B	C	D	E	F	G	H	J	K	L	M IPS	N IPS
HRC-230	196	9.8	30"	6'	45-5/16"	5'-6"	3'-6"	22-7/8"	28-3/8"	29-1/8"	1'-9"	6-1/16"	12"	—	3"	2"
HRC-236	284	14.2	36"	7'	54-13/16"	5'-6"	4'-0"	26-7/8"	33-3/8"	35-3/8"	2'-0"	7-1/4"	12"	—	4"	3"
HRC-242	384	19.2	42"	8'	61-13/16"	6'-3"	4'-6"	30-7/8"	37-3/8"	39-3/8"	2'-3"	8-1/2"	14"	—	4"	4"
HRC-248	500	25.0	48"	9'	69-7/16"	6'-4"	5'-0"	33-7/8"	41-7/8"	43-15/16"	2'-6"	8-1/2"	14"	—	6"	4"
HRC-254	636	31.8	54"	10'	75-7/16"	6'-9"	5'-6"	36-7/8"	44-7/8"	46-15/16"	2'-9"	8-1/2"	18"	—	6"	4"
HRC-260	784	39.2	60"	11'	83-7/16"	7'-6"	6'-0"	41-3/8"	49-3/8"	51-7/16"	3'-0"	10-1/16"	23"	—	6"	6"
HRC-266	952	47.6	66"	12'	90-9/16"	7'-7"	6'-6"	44-3/8"	53-3/8"	55-7/8"	3'-3"	10-1/16"	23"	—	8"	6"
HRC-272	1132	56.6	72"	13'	96-5/8"	7'-8"	7'-0"	47-3/8"	56-3/8"	58-7/8"	3'-6"	10-1/16"	23"	—	8"	6"
HRC-278	1328	66.4	78"	14'	102-5/8"	7'-9"	7'-6"	50-3/8"	59-3/8"	61-7/8"	3'-9"	10-1/16"	23"	—	8"	6"
HRC-284	1540	77.0	84"	15'	109-5/8"	8'-0"	8'-0"	54-7/16"	63-7/16"	65-15/16"	4'-0"	11-1/2"	28"	—	8"	8"
HRC-290	1768	88.4	90"	16'	117-7/8"	8'-3"	8'-6"	57-7/16"	68-7/16"	70-15/16"	4'-3"	11-1/2"	28"	—	10"	8"
HRC-296	2012	100.6	96"	17'	123-7/8"	8'-5"	9'-0"	60-7/16"	71-7/16"	73-15/16"	4'-6"	11-1/2"	28"	—	10"	8"
HRD-230	196	9.8	30"	6'	47"	5'-6"	3'-6"	22-7/8"	28-3/8"	39-1/2"	1'-9"	6-1/16"	12"	10-9/16"	3"	2"
HRD-236	284	14.2	36"	7'	66-1/2"	5'-6"	4'-0"	26-7/8"	33-3/8"	46-1/2"	2'-0"	7-1/4"	12"	12-3/4"	4"	3"
HRD-242	384	19.2	42"	8'	73-1/2"	6'-3"	4'-6"	30-7/8"	37-3/8"	50-1/2"	2'-3"	8-1/2"	14"	15"	4"	4"
HRD-248	500	25.0	48"	9'	84"	6'-4"	5'-0"	33-7/8"	41-7/8"	58"	2'-6"	8-1/2"	14"	15"	6"	4"
HRD-254	636	31.8	54"	10'	90"	6'-9"	5'-6"	36-7/8"	44-7/8"	61"	2'-9"	8-1/2"	18"	15"	6"	4"
HRD-260	784	39.2	60"	11'	97-1/2"	7'-6"	6'-0"	41-3/8"	49-3/8"	65-1/2"	3'-0"	10-1/16"	23"	18-1/16"	6"	6"
HRD-266	952	47.6	66"	12'	108"	7'-7"	6'-6"	44-3/8"	53-3/8"	71-1/2"	3'-3"	10-1/16"	23"	18-1/16"	8"	6"
HRD-272	1132	56.6	72"	13'	114-3/4"	7'-8"	7'-0"	47-3/8"	56-3/8"	74-1/2"	3'-6"	10-1/16"	23"	18-1/16"	8"	6"
HRD-278	1328	66.4	78"	14'	120-3/4"	7'-9"	7'-6"	50-3/8"	59-3/8"	77-1/2"	3'-9"	10-1/16"	23"	18-1/16"	8"	6"
HRD-284	1540	77.0	84"	15'	127-3/4"	8'-3"	8'-0"	54-7/16"	63-7/16"	81-9/16"	4'-0"	11-1/2"	28"	20-1/2"	8"	8"
HRD-290	1768	88.4	90"	16'	140"	8'-4"	8'-6"	57-7/16"	68-7/16"	90-9/16"	4'-3"	11-1/2"	28"	20 1/2"	10"	8"
HRD-296	2012	100.6	96"	17'	146"	8'-5"	9'-0"	60-7/16"	71-7/16"	93-9/16"	4'-6"	11-1/2"	28"	20 1/2"	10"	8"

MODEL "HH" HORIZONTAL HI-RATE FILTER

Larger size installations cannot be served satisfactorily with multiple vertical tank installations. The Horizontal Hi-Rate filter uses the same principles as for the vertical tank, plus the same synthetic internal distribution and collection and equipment. A large diameter tank is used to insure that the sand bed is sufficiently deep and that there is minimum change in area at the sides.

Horizontal Hi-Rate filters, up to 200 sq. ft., have proved themselves to be effective in field applications. No internal partitions are required.

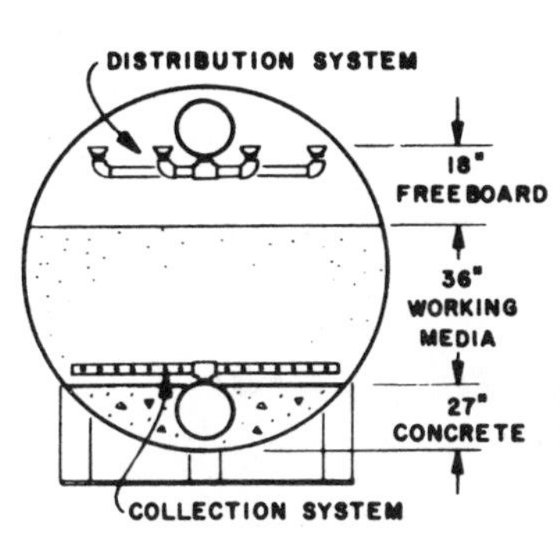

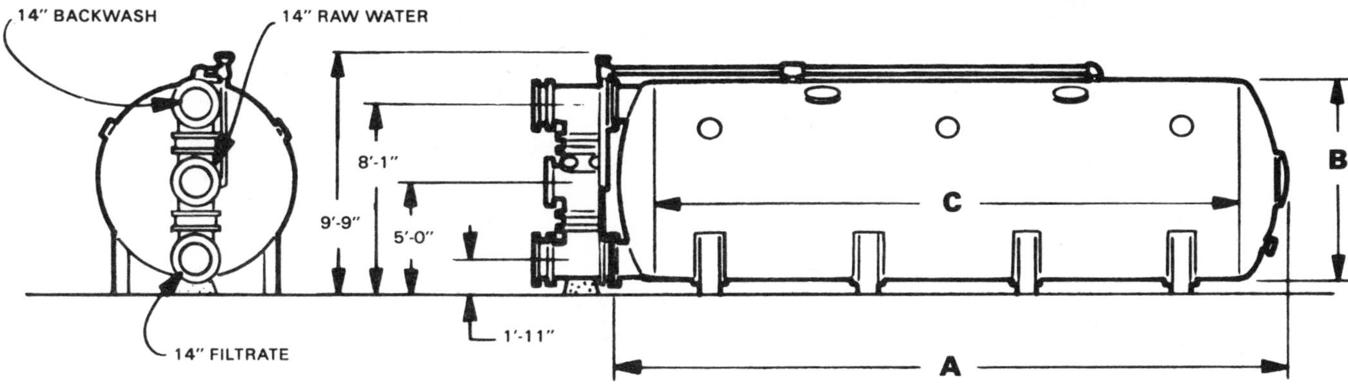

Model No.	Filter Area Sq.Ft.	Max. GPM	A	B	C	No. of Legs per Tank	*Concrete Cu. Ft.	Media Cu. Ft.
HH1500	75	1500	16'-6"	8'-0"	10'-0"	2	95	251
HH2000	100	2000	19'-6"	8'-0"	13'-0"	3	111	297
HH2500	125	2500	22'-6"	8'-0"	16'-0"	3	136	365
HH3000	150	3000	26'-6"	8'-0"	20'-0"	3	171	457
HH3500	175	3500	29'-6"	8'-0"	23'-0"	4	196	525
HH4000	200	4000	32'-6"	8'-0"	26'-0"	4	222	594
HH4002	400	8000	29'-0"	32'-0"	—	—	444	1188
HH4003	600	12,000	40'-0"	35'-0"	—	—	666	1782
HH4004	800	16,000	51'-0"	38'-0"			888	2376
HH4005	1000	20,000	62'-0"	42'-0"			1110	2970
HH4006	1200	24,000	73'-0"	45'-0"			1332	3564
HH4007	1400	28,000	84'-0"	48'-0"			1554	4158

**Concrete sub-fill is required only for applications where organic solids may decompose in the filter base. When concrete is omitted, increase the media requirement by the concrete volume.*

HI-RATE TWO STAGE SERIES FILTRATION

In situations where the collected solids vary widely in particle size, it is usually more efficient to remove the solids in two stages. The water is first passed through a Hi-Rate filter containing a fairly coarse media. Then the effluent from the first stage filter is passed to the second Hi-Rate filter containing a finer media. By using the two-stage method, the filters can carry heavier loads and increase the working cycle of both filters beyond the capacity of a larger single unit. Water quality is easily controlled by varying the size of the two media selected. Two-stage filtration is often necessary with heavy loadings from rivers, and is always required on sewage effluent recovery.

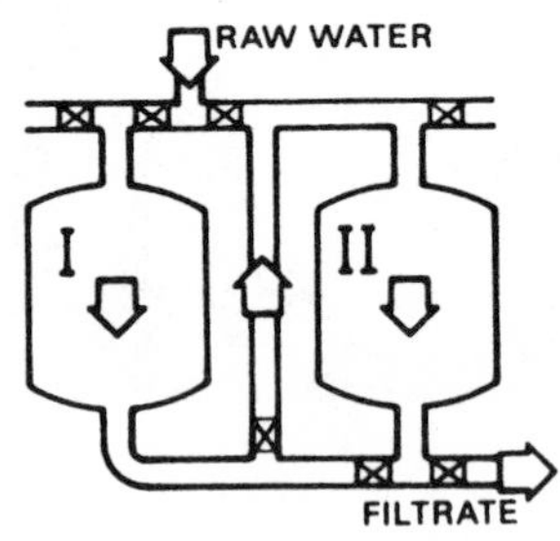

FILTER CYCLE

During filtration, water passes through the No. 1 unit with coarser media and then downwards through the No. 2 unit with finer media for two-stage filtration.

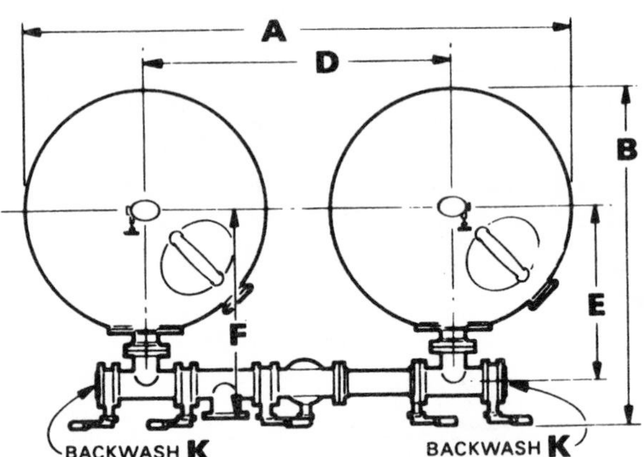

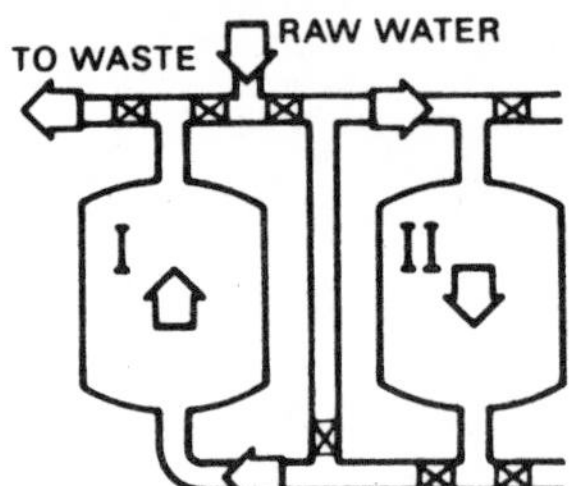

BACKWASH TANK 1

Unfiltered water is treated through the fine media of Tank 2 and is used to backwash Tank 1 for a period of two and one-half minutes.

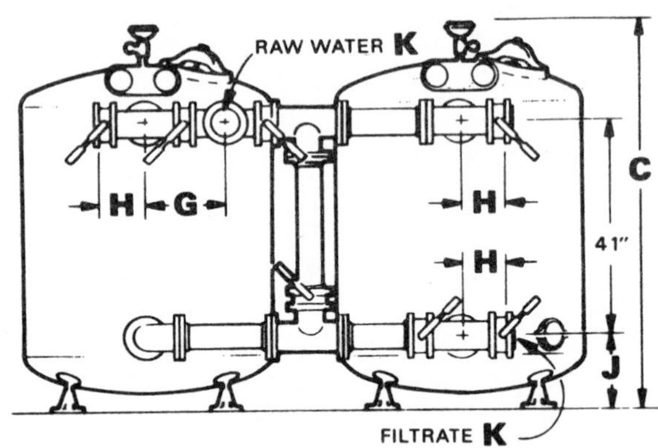

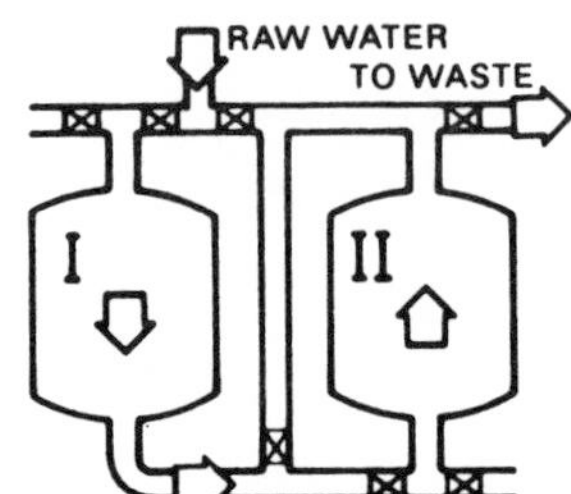

BACKWASH TANK 2

When Tank 1 has been cleaned, it is restored to filter service and water filtered by Tank 1 is used to backwash Tank 2 before ending the cleaning cycle.

Model No.	Max. Flow GPM	Area Sq. Ft.	Tank Dia.	A	B	C	D	E	F	G	H	J	K IPS
HRE-230	98	4.9	30"	6'	44-7/8"	5'-6"	3'-6"	22-7/8"	27-3/8"	10-9/16"	6-1/16"	12"	2"
HRE-236	142	7.1	36"	7'	52-7/8"	5'-6"	4'-0"	26-7/8"	32-3/8"	12-3/4"	7-1/4"	12"	3"
HRE-242	192	9.6	42"	8'	60-3/4"	6'-3"	4'-6"	30-7/8"	37-3/8"	15"	8-1/2"	14"	4"
HRE-248	250	12.5	48"	9'	66-3/4"	6'-4"	5'-0"	33-7/8"	40-3/8"	15"	8-1/2"	14"	4"
HRE-254	318	15.9	54"	10'	72-3/4"	6'-9"	5'-6"	36-7/8"	43-3/8"	15"	10-1/16"	18"	4"
HRE-260	392	19.6	60"	11'	81-3/4"	7'-6"	6'-0"	41-3/8"	49-3/8"	18-1/16"	10-1/16"	23"	6"
HRE-266	476	23.8	66"	12'	87-3/4"	7'-7"	6'-6"	44-3/8"	52-3/8"	18-1/16"	10-1/16"	23"	6"
HRE-272	566	28.3	72"	13'	93-3/4"	7'-8"	7'-0"	47-3/8"	55-3/8"	18-1/16"	10-1/16"	23"	6"
HRE-278	664	33.2	78"	14'	99-3/4"	7'-9"	7'-6"	50-3/8"	58-3/8"	18-1/16"	10-1/16"	23"	6"
HRE-284	770	38.5	84"	15'	108-7/16"	8'-3"	8'-0"	54-7/16"	63-7/16"	20-1/2"	11-1/2"	28"	8"
HRE-290	884	44.2	90"	16'	114-7/16"	8'-4"	8'-6"	57-7/16"	66-7/16"	20-1/2"	11-1/2"	28"	8"
HRE-296	1006	50.3	96"	17'	120-7/16"	8'-5"	9'-0"	60-7/16"	69-7/16"	20-1/2"	11-1/2"	28"	8"

ACTIVATED CHARCOAL

Barnebey-Cheney makes activated charcoal—in many forms and for most of its uses.

We manufacture a complete line of standard activated charcoals of all types in all sizes, develop charcoals for unique applications, and produce special grades to meet your particular specifications. We are the country's leading manufacturer of gas adsorbent charcoals, and the only manufacturer of a complete line of activated charcoals and related adsorption equipment for all air, gas, and liquid phase applications.

PARTICLE TYPES

GRAINS

Hard, dense grains provide the highest quality available for most air and gas phase applications. They provide the maximum internal surface area (micropores) and should be considered for applications requiring the maximum adsorption capacity per unit volume.

POWDER

Activated charcoal powders, manufactured from nut shells, coal, coke, and wood, are reduced to a variety of powders of uniform mesh size. All types have high external surface area and vary from low to high internal surface area depending on process requirements. They have low product retention, high filtration or settling rates, and are available with alkaline, neutral or acid pH of ash extraction. Barnebey-Cheney powders cover the full range of decolorizing and other liquid purification uses from low to high concentrations and small to large molecule sizes of contaminants.

COATING

Activated charcoal paint is made from high grade, finely powdered activated charcoal mixed with a binding agent which has a minimum of interference with the adsorptive properties of the charcoal. The charcoal has high capacity for adsorption of gases and odors. It can be applied to metal, concrete, stone, plastic, wood, fiber and paper surfaces and is water soluble so that it can be removed by water scrubbing. This feature is particularly useful for decontamination applications where surfaces must be restored to original finish following treatment. Activated charcoal paint is black.

Charcoal paint is furnished as a dry powder, which may be mixed with water to the desired consistency.

PELLETS

Charcoal pellets are manufactured from nut shells, coal, coke, peat, wood, and certain other special formulations. These are recommended for applications where high gross porosity (macropores) is desired. They are usually used where high concentrations of contaminants are involved and may be desirable for catalyst supports requiring large amounts of metallic additions. Ground pellets are similar to regular pellets except that they are usually in small random grain sizes. They are best suited for liquid phase applications where abrasion resistance and maximum hardness is not required.

FIBERS

Pure charcoal in an unusual new fiber form is available as bulk fiber, both activated and non-activated. The latter has found many applications and is supplied in various lengths and fiber diameters.

This is such a new product its full potential has only begun to be explored. Charcoal fiber is stable; it resists strong acids and alkalies. It can be burned for disposal or recovery of trapped materials. Properly protected it is an extremely good high temperature barrier and is finding use in rocket exhaust nozzles, missile nose cones and for insulating high temperature furnaces.

Charcoal fibers have been produced in a variety of useful material forms. We have developed charcoal (activated and unactivated) cloth, paper, and pads of various thicknesses in both flexible and rigid shapes for new and experimental applications.

FORMED

Molded forms offer another development in the use of activated charcoals. Self-supporting filter cartridges and elements can be molded to meet unusual requirements not readily served by other activated charcoal forms.

BELCO CLARATOR

APPLICATION:

The Belco Clarator offers the most efficient method for the initial treatment of large volumes of waste or raw water for the removal of suspended impurities, color and taste as well as certain dissolved impurities such as iron, manganese and calcium hardness.

FEATURES:

The Belco Clarator provides high rate clarification in minimum space. It has a capacity four times greater than the conventional coagulation and settling basins of equal size.

PRINCIPLES OF OPERATION:

The Belco Clarator operates on the principal of solids contact, utilizing dual recirculation of pre-formed precipitates for mixing with the chemically treated raw water to assist the formation of new floc particles.

Complete dispersion of chemicals throughout the raw water prior to coming in contact with previously formed precipitates takes advantage of the chemical reaction of the coagulant to the fullest extent for rapid formation of large tough floc particles which settle rapidly.

The principle of dual recirculation of precipitates assures complete formation of large floc particles thereby minimizing the possibility of fine floc being carried over into the treated water.

Completely formed heavy floc particles which settle at the bottom of the Clarator are partially recirculated up through the lower draft tube to act as nuclei in the formation of new precipitates.

Settled sludge is continuously moved to the center concentrator for removal by the automatic blow-off system. The blow-off system is adjusted for synchronization with the flow rate to maintain sufficient sludge as required for recirculation in the reaction to form new floc particles.

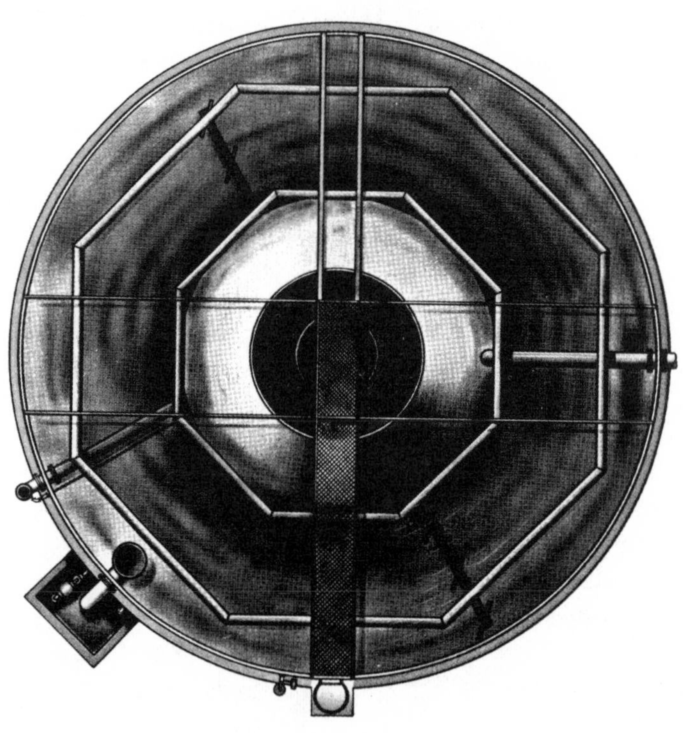

Belco Pollution Control Corporation

BELCO AIR-FLOW CLARIFIER

APPLICATION:

The Belco Air-Flow Clarifier is selected when the waste water to be treated contains floatable suspended solids such as greases, stringy fibers or solids as in paper mill white water waste, oils and materials of low specific gravity.

FEATURES:

The Air-Flow Clarifier operates at flow rates of 2 to 3 gpm/sq. ft. in the flotation zone, which is considerably higher than solids contact clarifiers. Adjustable skimmers rotating at a higher peripheral speed than the sludge scraper provide accelerated removal of floatable waste.

PRINCIPLES OF OPERATION:

Compressed air is introduced into the suction of the pressurizing pump. Using by-passed water from the pressure pump discharge, atmospheric air is injected into the pump suction line by an ejector.

As the waste water enters the dispersion chamber small air bubbles are released and attach themselves to the floatable material, which then rises in the flotation zone where it is removed by the continuous rotating skimmer.

The clarified water descends to the outer peripheral collector port, there rising in the clear water zone to uniformly spaced collecting orifices which discharge into a collector flume.

Particles of greater mass are hydraulically carried by the descending flow to the bottom of the Clarifier where the continuously rotating sludge scraper moves them to the concentrator for removal through the blow-off system. When the treatment of waste does not require the removal of settleable mass, the sludge scraper may be omitted.

NOTE: In many cases oils and solids can be removed by utilizing air alone in the flotation treatment. However, when treating oil emulsion wastes as encountered in metalworking plants, chemical treatment is necessary for reaction upon the emulsion to promote its flotation.

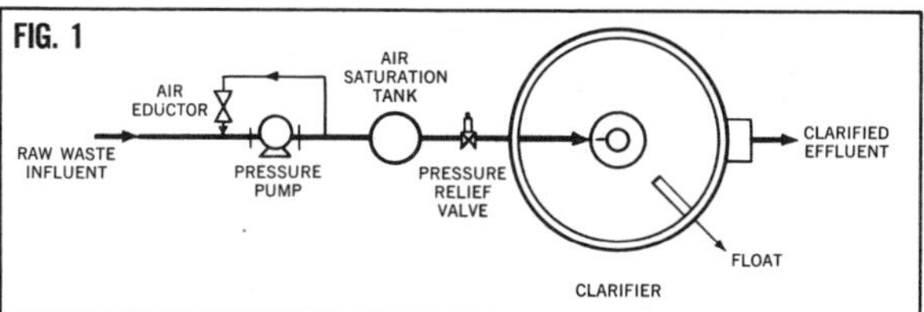

Fig. 1 Total Pressurized Flow

This system should be applied when stringy fibers or solids as in paper mill white water waste are being considered.

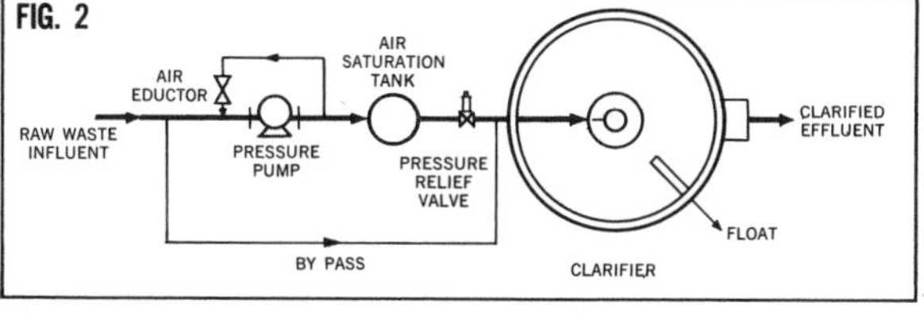

Fig. 2 Partial Pressurization of Flow

Where the waste water contains free oils and materials of low specific gravity, partial (50% or less) pressurization of the waste will produce satisfactory results and reduce pumping costs.

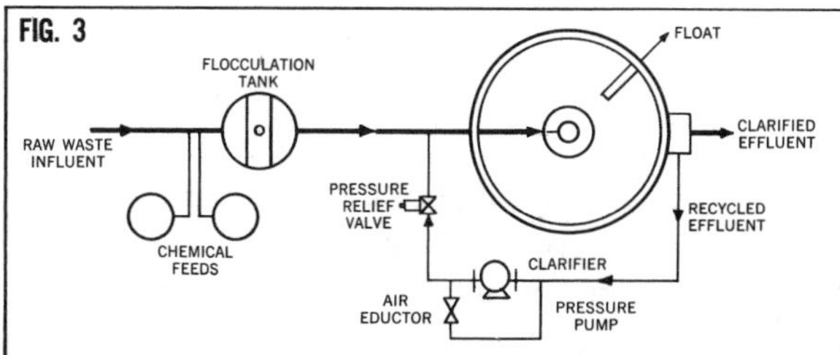

Fig. 3 Primary Chemical Treatment and Recycling

An arrangement particularly applicable to waste waters containing high concentrations of floatable suspended soilds, greases, etc., which might clog the pressurization valve. Flow from the floc-forming tank is by gravity and is blended with the pressurized effluent from the unit.

HYDRO-ARM THICKENER

The Belco Hydro-Arm Thickener is used to dewater the suspended matter by means of continuously rotating scraper arms designed to produce a sludge of high solids concentration.

The Thickener is designed to avoid the troublesome problems of rotating scraper stoppage often encountered in sludge thickening. By use of a pneumatic lifting device, either or both of the scraper arms can be lifted for ease of cleaning and maintenance. An adjustable skimmer rotating at higher peripheral speed than the scraper arms assures continuous, fast positive removal of coagulated and other light floating material from the water surface.

Principle of Operation: Sludge thickening is achieved by the propagation of the sludge which is promoted by the raking action of the continuously rotating scraper arms finalizing in the accumulation of a deep sludge bed at the bottom of the Hydro-Arm unit. The sludge is withdrawn at the point of its highest concentration.

AERATORS

Aerators are used to add oxygen to liquids for CO_2 removal and the oxidation of iron, manganese and hydrogen sulfide.

Belco aerators include an open type design with water cascading over trays of graded coke; the pressure air saturator type and the in-line air diffusors.

Principles of Operation:

(1) Coke Tray Aerator

Water is cascaded through a series of coke filled trays where additional aeration takes place during the free fall from tray to tray. Freshly precipitated iron or manganese on the coke produces a catalytic effect which increases iron and manganese removal from the water.

(2) Air Saturator

Air diffusion is accomplished by pumping air into a pressure tank and maintaining a blanket of air above a set water level thus controlling the amount of air introduced to prevent "white water."

(3) In-Line Diffusors

Air diffusion is accomplished by pumping air into a porous tube placed inside the raw water line.

CHLORINATION

Chlorination is used to kill bacteria, for the elimination of undesirable taste, odor and color, and for the oxidation and destruction of organic matter. Chlorination is used to oxidize and disinfect sewage and industrial waste; for slime control in cooling water and water distribution systems.

Chlorination equipment merely requires turning on the chlorine inductor water and setting the feed rate by the operator. Its operation is completely automatic.

"SLUDGE-ALL" SYSTEM

The Beloit-Passavant "SLUDGE ALL SYSTEM" is a one stage sludge dewatering process, capable of handling waste sludges from sanitary sewage treatment plants, water treatment plants and industrial treatment plants.

The process produces filter cakes from 45 to 55% solids concentration and a filtrate containing less than 20 ppm. The system will handle raw primary sludge, raw secondary sludge and straight waste activated sludge, or any combination of these wastes. Even partially digested and septic sludges may be dewatered successfully.

Beloit-Passavant "SLUDGE ALL SYSTEM" will handle with equal efficiency notoriously difficult materials, such as sedimentation basin sludge from water treatment plants, tannery wastes and many more of the complex industrial waste sludges.

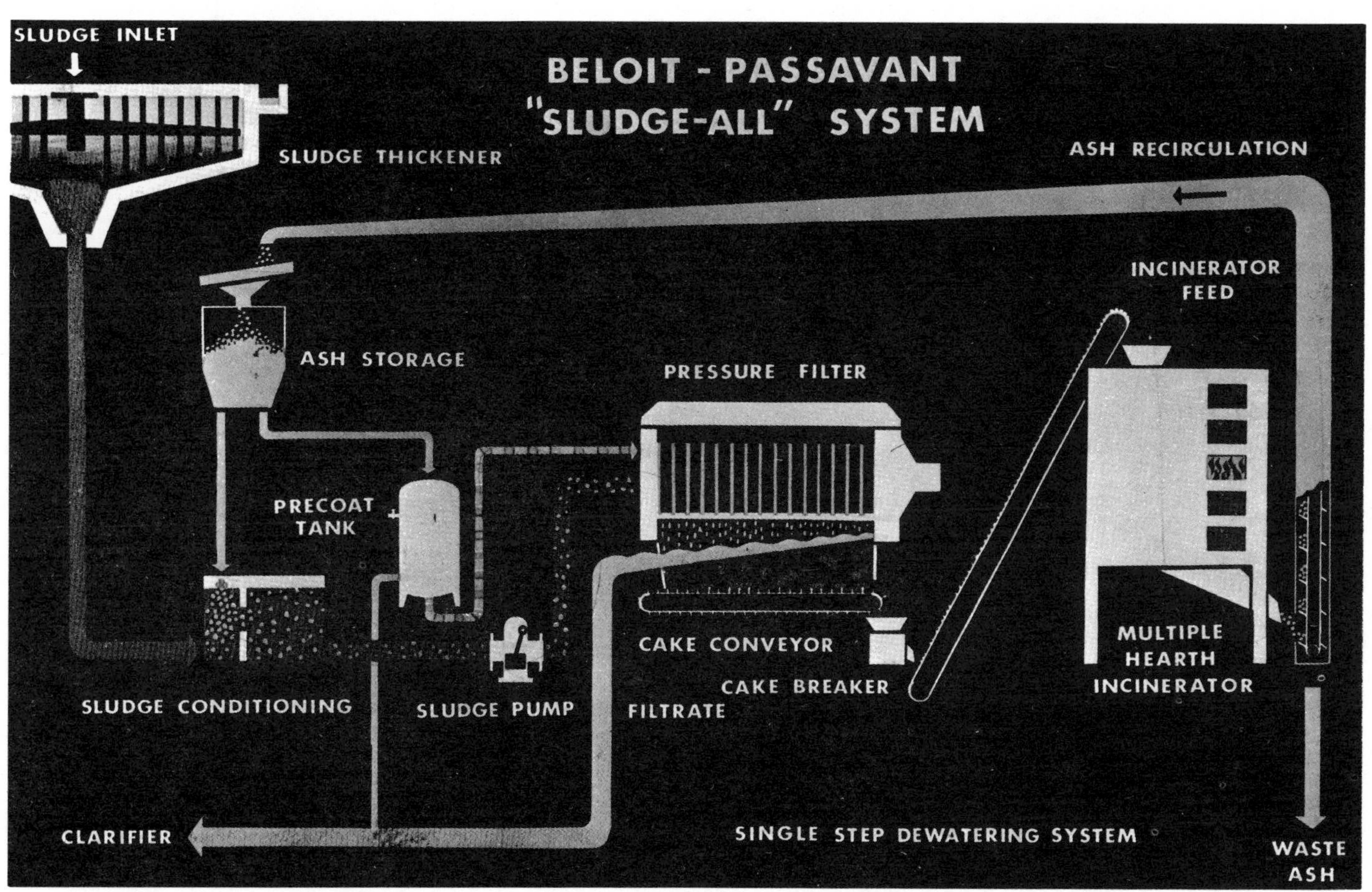

FILTERAIDS

The basic Beloit-Passavant "SLUDGE ALL SYSTEM" utilizes fly ash for precoating and as an admix to the sludge. While this material is inexpensive and widely used, it does not preclude the possibility of using other materials. Chemical flocculation has been used with great success. Diatomaceous earth is an excellent filter aid and may be used when the less expensive filter aid is not available or process reasons rule it out.

Paper fiber is also an excellent filter aid and precoat material. Woodflower is another excellent filter aid.

FILTERMEDIA

Since the filter medium is caulked into the filter plates as shown in the photo, it does not have to do double duty as a gasket material. There is a wide choice available of commercially synthetic filter media both woven and non-woven. Metallic filter media may also be furnished.

SLUDGE ALL PROCESS

With a predetermined amount of fly ash, sludge from a thickener is mixed in the contact tank and agitated to assure a uniform mixture. At the same time the precoat tank is filled with filtrate and a smaller proportion of fly ash. Before the actual start of filtration, the water-ash precoat mixture is pumped into the filter at a very high rate to insure turbulence and thus the entire filtration area is coated with a very thin layer of fly ash (or other materials as mentioned in preceding paragraphs).

The purpose of precoating is twofold. One, to protect the filter media against premature blinding and two, to provide a parting plane to assure a clean and complete filter cake discharge at the end of the filtration cycle. Precoating is accomplished in a matter of minutes and repeated every cycle.

As soon as a precoat is established, the sludge-ash mixture from the contact tank is admitted to the filter and the actual dewatering cycle begins. Average cycle time is less than two hours, completion of the filtration cycle is determined by the rate of filtrate flow. When a certain minimum flow is reached, the filter is ready for discharge.

An audible signal tells the operator to initiate the opening of the pressure filter. The hydraulic pressure in the cylinders is released, the oil flow reversed, and the filter pressure piece moved away from the stack of plates. The mechanized plate shifting device now moves one plate after the other through the opening created by the pressure piece.

As soon as a plate has moved away from the stack, the filter cake accumulated in the chamber drops out, without the aid of any mechanical or pneumatic devices, onto a conveyor belt beneath the filter.

The cake is usually broken up when it falls onto three or more taut steel wires and is conveyed in smaller pieces to a hammer mill. The mill reduces the pieces to a size small enough to be fed into a multiple hearth incinerator.

Fly ash and the bottom ash from the incinerator is recirculated to the ash hopper and re-introduced into the process. Only the excess ash is wasted and represents together with the clear filtrate the end product of the Beloit-Passavant process.

AUTOMATION

The entire process is completely automated; i.e., the precoat cycle, filter cycle and discharge. Ash admix to the slurry and ash addition for precoating is automatically controlled. All process valves are operated by this system and if need be the process may be manually controlled.

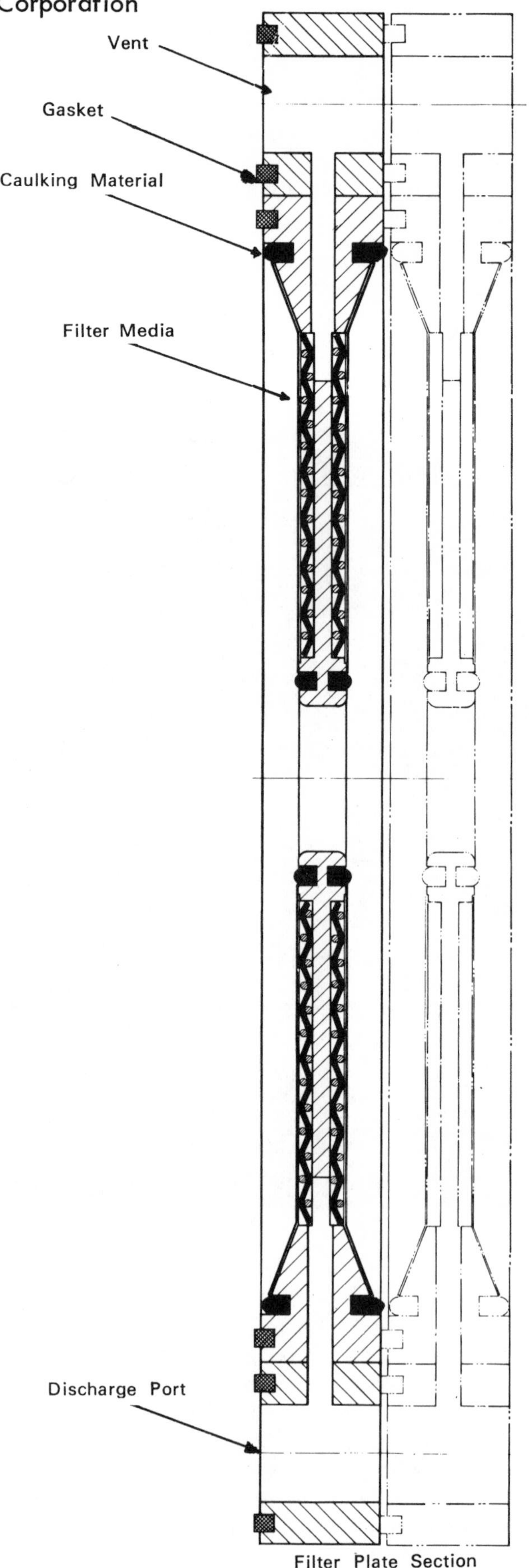

Filter Plate Section

OPERATION

1. Debris removed from screens is collected in a trough behind the screen.
2. The debris is conveyed to the Beloit-Passavant screenings press.
3. The debris is compressed to a compact, de-watered cube.
4. The compacted screenings are conveyed to ultimate disposal.

RESULT

1. Screenings debris is reduced to a compact, easy-to-handle solid.
2. Excess water held by the screenings is removed, providing a clean and sanitary operation.
3. Ultimate disposal of the dewatered compact debris can be by sanitary land fill, or by remote or on-site incineration.
4. Better plant effluent.

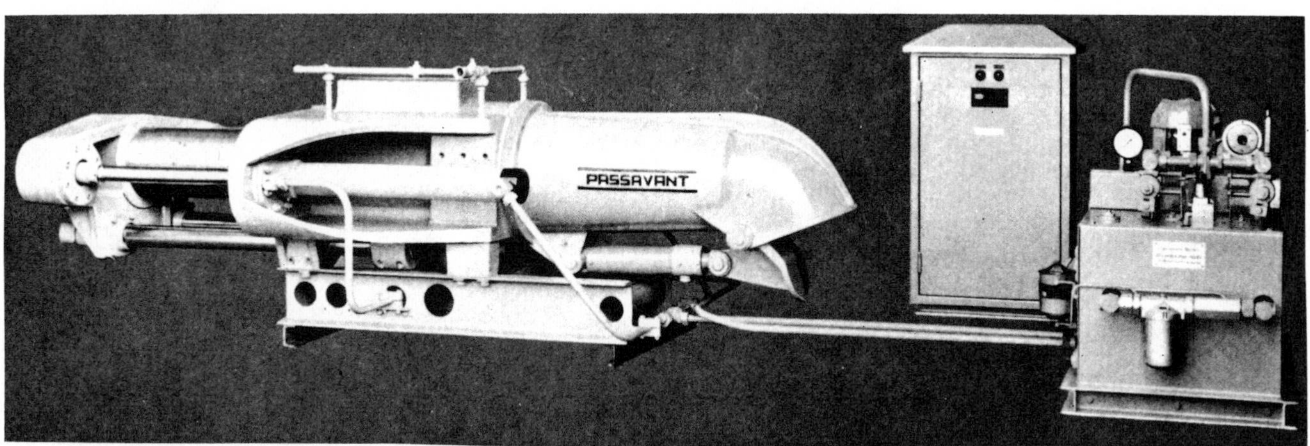

TECHNICAL DATA AND SPECIFICATIONS

MODEL NUMBER		SP - 10	SP - 11	SP - 20	SP - 21	SP - 30	SP - 31
Capacity cu ft/hr Screenings 90% Moisture		60 - 80	70 - 90	35 - 45	50 - 60	10 - 15	25 - 30
Percent Moisture of Screenings Cake		50 - 60	50 - 65	50 - 65	50 - 65	55 - 65	55 - 65
Inlet Opening Size		10″ x 16″					
Inlet Height from Operating Floor		3′ 6″					
Press Length		10′ 0″					
Press Width		4′ 6″					
Piston Diameter		10″					
Piston Stroke		2′ 6″					
Applied Pressure PSI		950	660	660	600	450	450
Hydraulic Pressure PSI		3000					
Hydraulic Unit Motor Hp		10	10	7.5	7.5	2	5
Hydraulic Pump Capacity GPM	Low PSI	2.6	2.6	2.1	2.1	1.0	2.2
	High PSI	22	22	8	15	1.0	2.2
Approximate Weight Screenings Press and Hydraulic Unit		7000#	7000#	6000#	6000#	5000#	5000#

Beloit-Passavant Corporation

WATER INTAKE SCREENING SYSTEMS

Screening systems for water intake facilities consist of two separate screening operations. The first screen is a roughing bar screen or trash rack to remove heavy debris from the intake water. It is designed with heavy metal bars assembled in a vertical position within a steel supporting frame. The bar screen section is cleaned of debris by an automatically operated cleaning rake suspended from the drive assembly by wire cable.

Beloit-Passavant Corporation manufactures two types of mechanically cleaned bar screens as illustrated below.

The choice of bar screen will depend upon the amount and type of debris and the water depth.

Series 1260 provides 22" opening between rake and bar screen, and screen widths to 10'0".

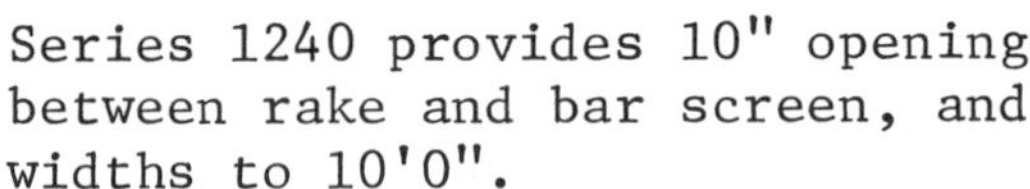
Series 1240 provides 10" opening between rake and bar screen, and widths to 10'0".

SERIES 1400 TRAVELING BAND SCREEN

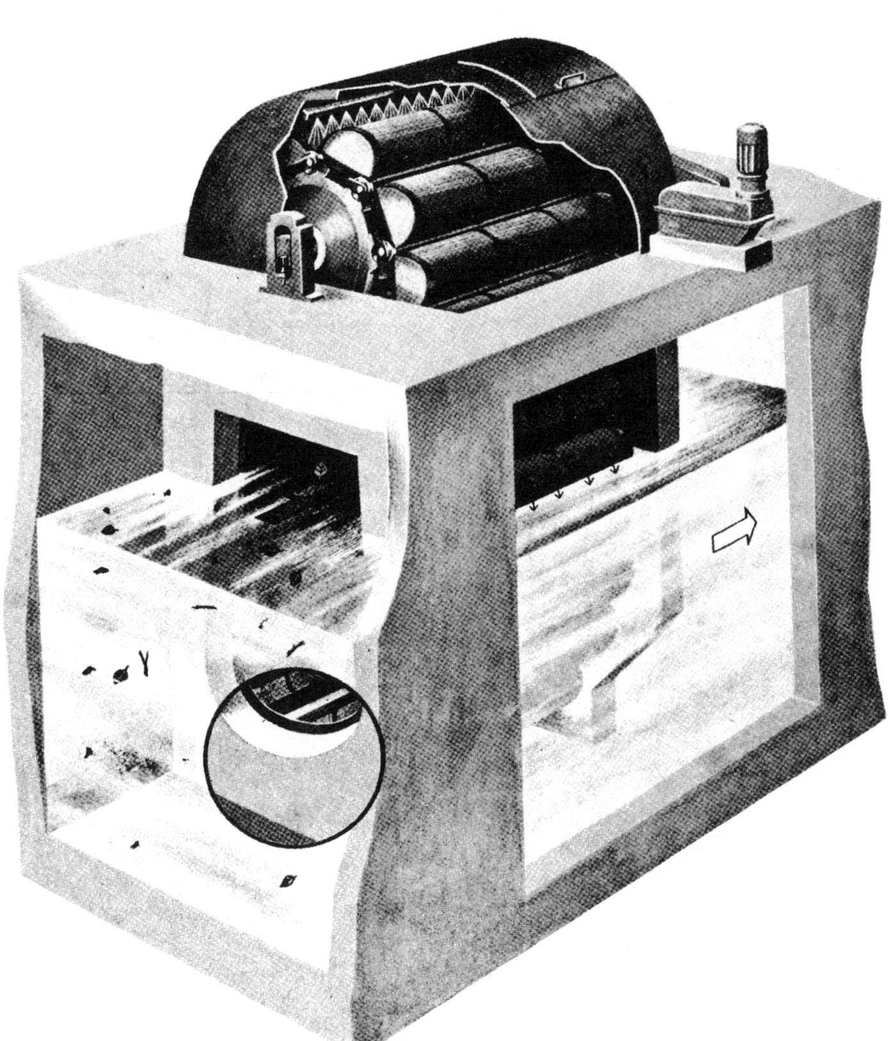

Figure 3

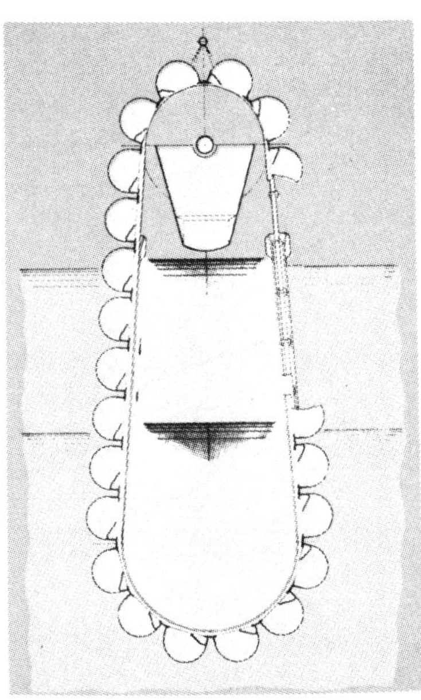

Cross Section of TBS

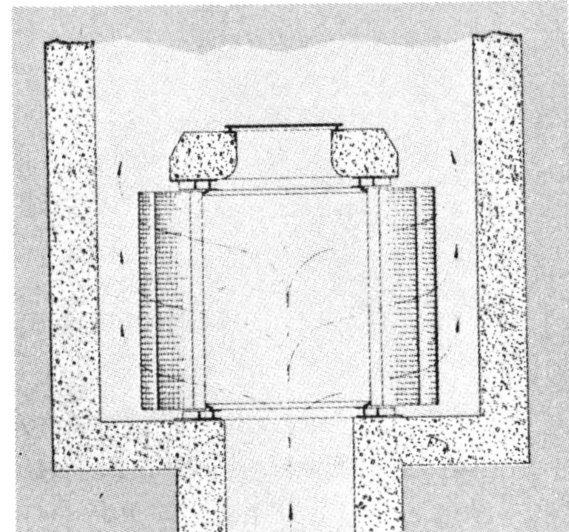

Figure 4

The traveling band screen (TBS) is used in the water intake channel following the rough and fine bar screens. The traveling band screen is used to remove debris remaining in the water following the pre-screening operations.

The traveling band screen, Figure 3, is a series of screening baskets connected to two strands of endless chain suspended from drive sprockets supported at the operating floor. The bottom portion of the screen is held in position by two steel guides which eliminates the need for underwater sprockets, shafting and bearings.

The traveling band screen is designed for channel depth exceeding 6′0″.

1. The TBS is installed in the intake channel with the screening basket parallel to the water flow.
2. Intake water enters the interior of the TBS and passes out through both sides. Figure 4
3. When the screening cloth becomes blind and causes a different water level between screen interior and exterior the screen will automatically rotate, clean the debris from the screen cloth and stop until blinding reoccurs.
4. The screen basket is cleaned by water spray nozzles located on the exterior of the screen which washes debris into a collecting trough located on the interior of the screen.

ROTODISINTEGRATOR

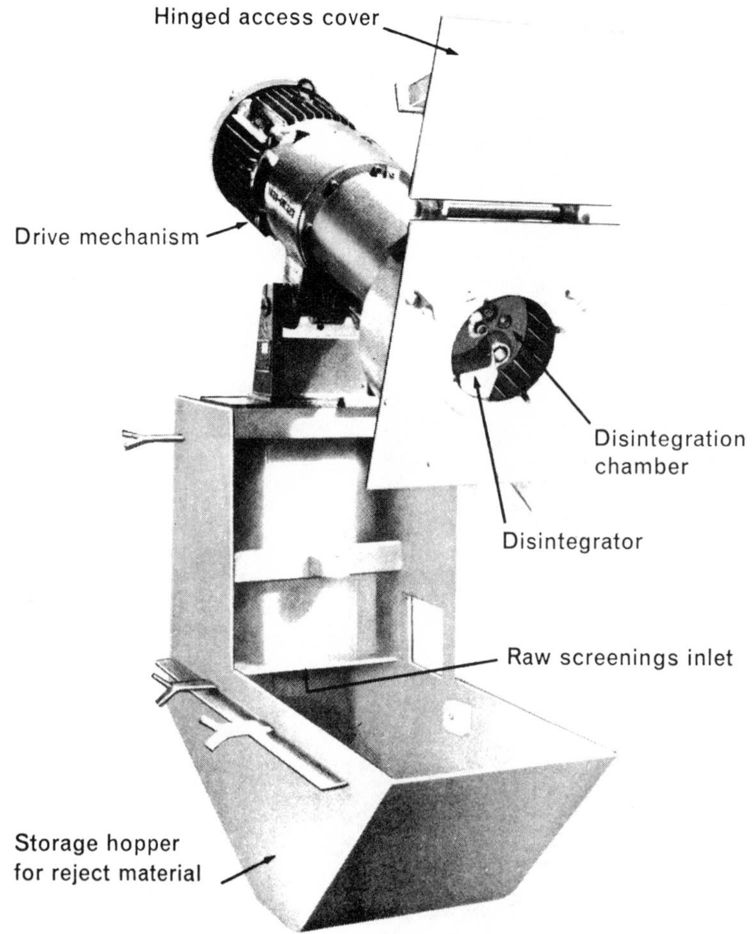

Debris is delivered to the primary chamber of the Rotodisintegrator and is carried to the secondary or disintegration chamber by hydraulic flow. In the secondary chamber, stones, metal, or hard, heavy materials of similar nature will separate and fall to the bottom where they are stored for periodic removal. Solid materials to be reduced in size are carried to the disintegration chamber by the hydraulic flow.

The disintegration chamber consists of a fixed grate-like slotted cage and a freely rotating disintegrator connected directly to the drive assembly. Debris carried to the disintegration chamber is collected on the walls of the slotted cage and is reduced in size by the centrifugal action of the disintegrator against the slotted cage.

The action of the disintegrator against the slotted cage constantly resharpens the aperture edges and thereby maintains the highest cutting efficiency.

The Rotodisintegrator is driven by a 10-hp, 900-rpm motor which enables the unit to handle any type of debris. The total available horsepower is used only for the most stubborn materials. During other operating periods the required horsepower is only a fraction of that provided.

MICRO-SIEVE

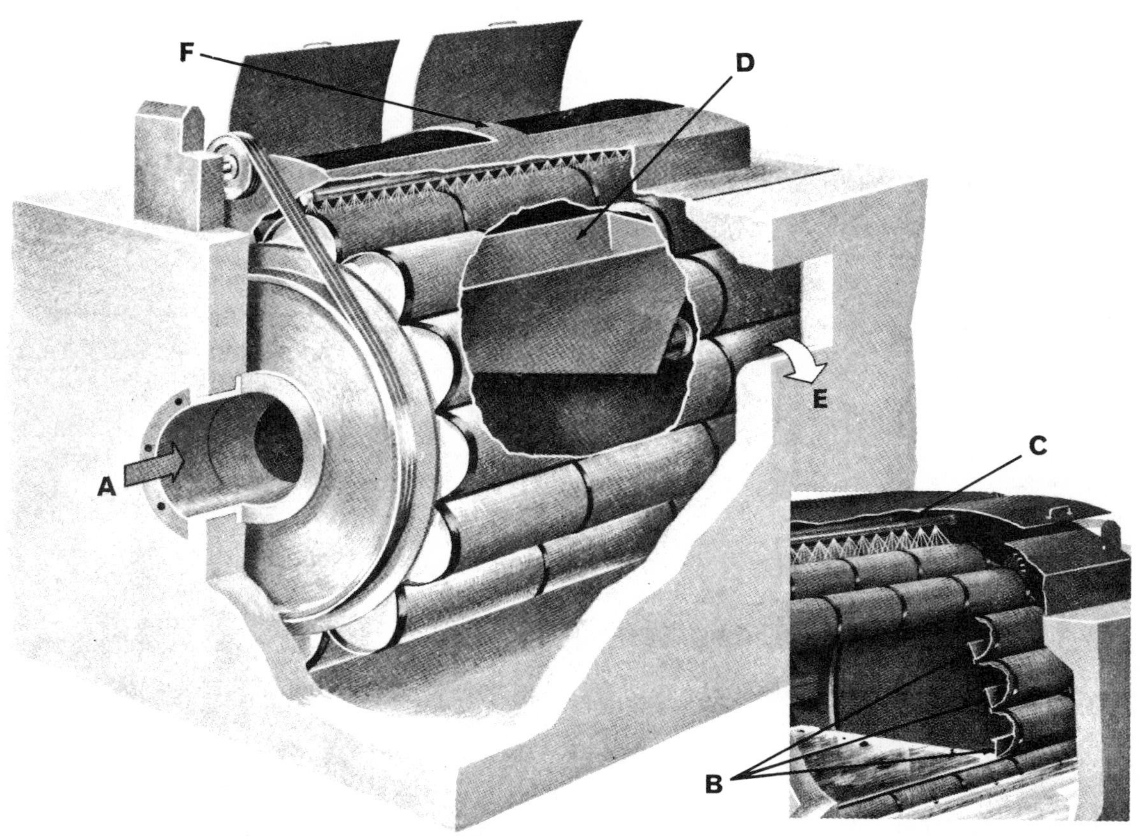

Beloit-Passavant Micro-Sieve installed in concrete structure.

Principles of Operation: Raw water enters the Micro-Sieve through the inlet port A on the upstream side of the unit. Flow of raw water through the Micro-Sieve is by gravity, and, as the water passes through the screen basket media, solids are intercepted and retained on the media. The screen drum automatically rotates, carrying the solids to the top of the unit where a water spray C washes the solids from the screen into a screenings trough D for ultimate removal from the system. Screened water passes from the Micro-Sieve through the outlet E. Above the operating floor, the Micro-Sieve is enclosed by a housing F equipped with hinged inspection doors.

Sewage Treatment Plant Effluent Polishing: The Micro-Sieve can be used as tertiary treatment for sewage treatment plant effluent. The Micro-Sieve will remove algae and organic floc carryover, thereby increasing the overall plant efficiency. Results are variable as to the screening efficiency and economy of operation, depending upon the characteristics of the effluent and degree of removal required. Such applications are investigated on an individual basis, and equipment requirements established accordingly. The practical applications of the Micro-Sieve are many and varied.

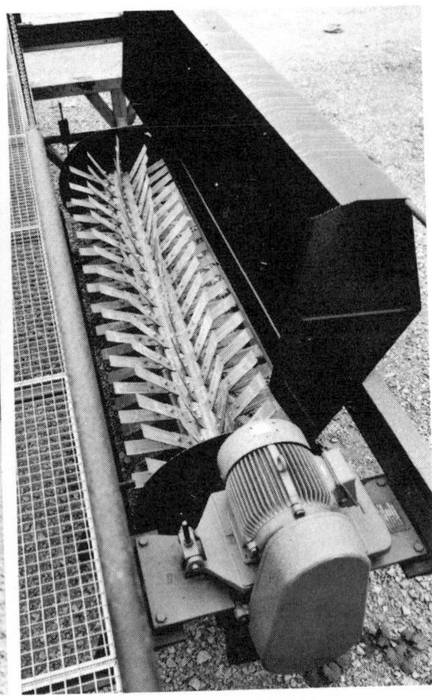

Mammoth Rotor Typical Design Calculations

Following is a typical design calculation:

Population: 2000 persons
Daily Flow: 100 gallons per person
Average Rate: 140 GPM
Design Rate: 350 GPM
Sewage Strength: 200 mg/L
Daily B.O.D.: 334 lb
Design Basis: Modified activated sludge
Aeration System:
Organic Loading: 12.5 lb B.O.D. per 1000 cu ft
Required Volume: 26,720 cu ft
Structure: 20′ bottom width x 32′ water level width x 155′ long center to center with semi-circular ends and 45 degree side wall slope. Free-board above the liquid level 2′ - 0″.
Aeration: Supply 2.3 lb 0_2 per lb. B.O.D. applied per day.
OC Required Per Hour: 32 lb per hour.
Equipment: Two Series 5200 Mammoth Rotors each having 9′ rotor length and 10 hp drive unit.
Circulation: 1500 cu ft per ft of rotor, (maximum 3000 cu ft per ft of rotor).
Required hp: From curve 4.5 hp. Use 10 hp drive motor to cover full range of rotor submergence.
Final Clarifier:
Surface settling rate: 400 gallons per sq ft
Required Surface Area: 500 sq ft
Required Average Detention: 4 hours
Required Volume: 4500 cu ft
Structure: 26′ diameter x 8′ side liquid depth
Sludge Return:
Pump Capacity: 100% average daily rate
Equipment: Duplex pumps each rated 140 GPM
Sludge Drying Beds (if required):
Area: 1 sq ft per capita
Area Required: 2000 sq ft
Structure: Two beds each 20′ x 50′

The following should be added to the design if required:

1. Raw sewage lift station
2. Flow recording
3. Grit removal
4. Sewage screen or grinder
5. Chlorinator
6. Laboratory and control building
7. Fence and landscaping
8. Effluent Micro-Sieve

Beloit-Passavant Corporation can furnish all mechanical and electrical equipment. This service provides the design engineer and owner one-source responsibility.

Beloit-Passavant Corporation

MAMMOTH ROTOR AERATOR SERIES 5200

Description

The Series 5200 Mammoth Rotor is a standard unit manufactured with rotor length of 6, 9, and 12 feet. The support bridge has a length of 18 feet, 6 inches. The support bridge and rotor are factory assembled and shipped as a unit. At the installation site, the bridge and rotor assembly are placed on its supports in the oxidation canal. The installation is complete after handrails and electrical are installed.

General

The Series 5200 Mammoth Rotor is used on oxidation canal treatment systems based upon the design concepts of modified activated sludge. This design concept has proven to be an efficient, easily operated system which will produce an effluent having 90 percent or better B.O.D. removals at low construction and operating cost.

Application

The following chart can be used as a guide for normal domestic sewage treatment plants. The canal dimensions are given as a guide and other tank configurations can be used which will provide the equivalent liquid volume.

Population	Gallons Per Day 100 GPCD	Average Flow GPM	B.O.D. Lb Per Day 200 mg/L	Canal Volume Cu Ft	Canal Liquid Depth	Canal Bottom Width	Canal Liquid Level Width	Canal Straight Side Length	Mammoth Rotor Length	Nonoverload Mammoth Rotor HP	Final Clarifier Diameter	Clarifier Side Wall Liquid Depth	Return Sludge Pump Capacity	Combined Sludge and Sewage Pump Capacity
500	50,000	35	83	6,640	6′	12	24	45′	6′	7.5	12′	8′	35	120
750	75,000	53	125	10,000	6′	12	24	80′	6′	7.5	16′	8′	53	185
1,000	100,000	70	167	12,360	6′	20	32	60′	9′	10	18′	8′	70	245
1,250	125,000	87	210	16,800	6′	20	32	90′	9′	10	20′	8′	87	300
1,500	150,000	105	250	20,000	6′	25	37	85′	12′	15	22′	8′	105	360
1,750	175,000	122	290	23,200	6′	25	37	105′	12′	15	24′	8′	122	420
2,000	200,000	140	334	26,720	6′	20	32	155′	2-9′	10	26′	8′	140	490
2,250	225,000	157	375	30,000	6′	20	32	180′	2-9′	10	26′	8′	157	560
2,500	250,000	175	417	33,350	6′	20	32	205′	2-9′	10	28′	8′	175	610
2,750	275,000	190	460	37,000	6′	20	32	225′	2-9′	10	30′	8′	190	670
3,000	300,000	210	500	40,000	6′	25	37	195′	2-12′	15	32′	8′	210	730
3,250	325,000	230	540	43,000	6′	25	37	215′	2-12′	15	33′	8′	230	800
3,500	350,000	245	583	46,640	6′	25	37	235′	2-12′	15	34′	8′	245	850
3,750	375,000	260	625	50,000	6′	25	37	255′	2-12′	15	35′	8′	260	900
4,000	400,000	280	667	53,360	6′	25	37	275′	2-12′	15	36′	8′	280	980
4,250	425,000	300	700	56,000	6′	25	37	285′	2-12′	15	36′	8′	300	1,050
4,500	450,000	315	750	60,000	6′	25	37	310′	3-12′	15	38′	8′	315	1,100
5,000	500,000	350	834	66,720	6′	25	37	340′	3-12′	15	40′	8′	350	1,230
5,500	550,000	385	917	73,360	6′	25	37	380′	3-12′	15	42′	8′	385	1,350
6,000	600,000	420	1,000	80,000	6′	25	37	420′	4-12′	15	44′	8′	420	1,500
6,500	650,000	450	1,084	86,640	6′	25	37	450′	4-12′	15	46′	8′	450	1,560
7,000	700,000	485	1,167	93,360	6′	25	37	490′	4-12′	15	48′	8′	485	1,700
7,500	750,000	520	1,250	100,080	6′	25	37	530′	4-12′	15	50′	8′	520	1,800
8,000	800,000	535	1,334	106,720	6′	25	37	565′	4-12′	15	52′	8′	555	1,950

Canal Design:
Detention: 24 hr
Loading: 12.5 lb B.O.D./1000 cu ft
Side Walls: 45 deg
Longer Length Canals can be separated into two equal size canals

Aeration Equipment:
OC: 2.3 lb OC/lb B.O.D. applied
Hp: Each rotor nonoverloading
Clarifier:
Surface Rise Rate: 400 G/sq ft/day
Detention: 4 hr

The oxygenation capacity and required horsepower at various depths of submergence are analyzed in Figure 1. As shown, an oxygenation capacity of 5 lb per foot of rotor length can be easily obtained at 1.24 hp per foot of rotor length and at 10.2″ rotor submergence. Similarly, other values of oxygenation capacity, horsepower, and submergence can be obtained from the curves.

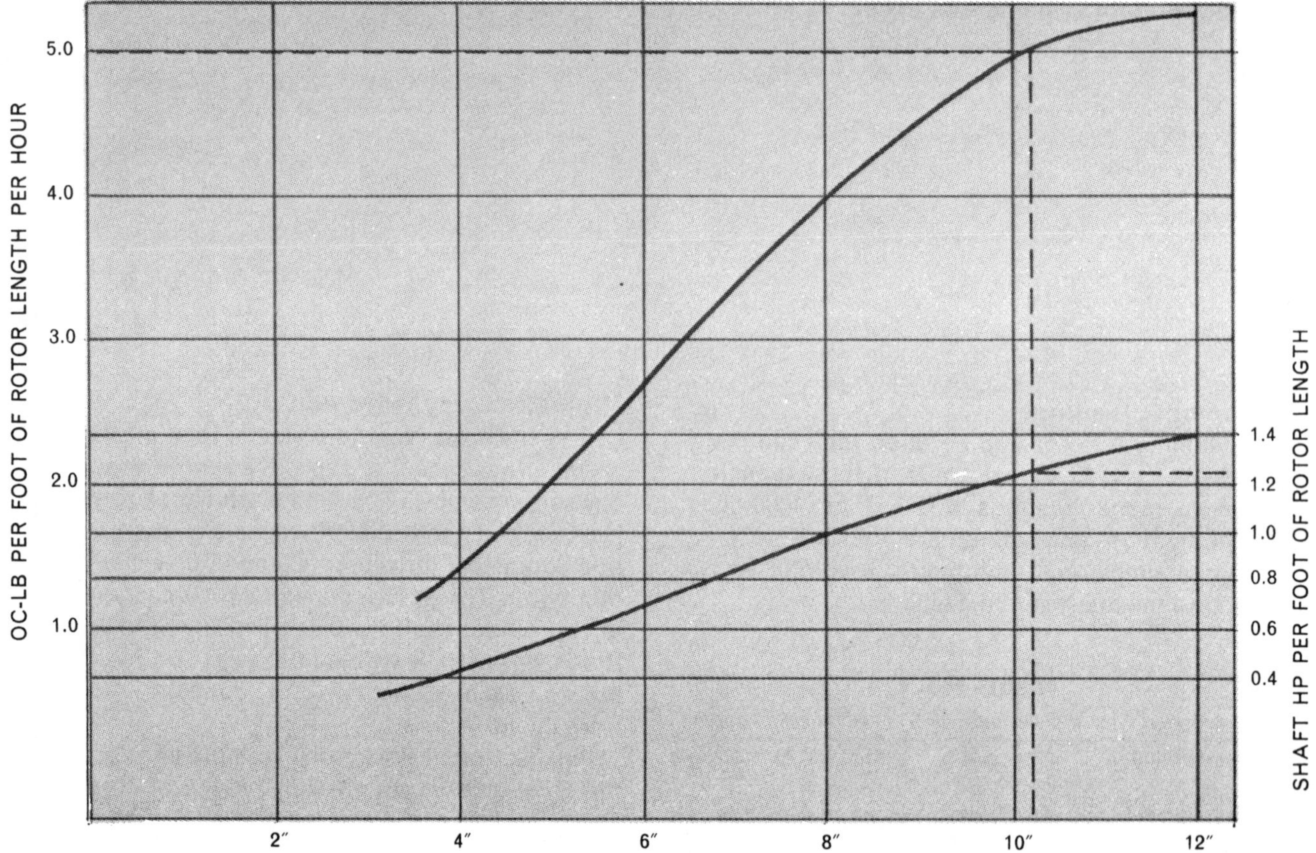

Figure 1

As this graph indicates, the Mammoth Rotor has an unusually high oxygenation capacity. The data represented by these curves indicate the rate of oxygen transfer in pound per hour, in tap water at 20°C, when oxygen concentration is zero. The sodium sulfite/cobalt ion catalyst test procedure, as generally accepted, was followed. All test data was confirmed at actual operating installations.

Shock loading

Because the aeration tank is a completely mixed plug flow system, B.O.D. in the influent is immediately absorbed by the mixed liquor suspended solids resulting in full B.O.D. reduction within a few feet of travel. The system can therefore absorb a shock loading with no change in effluent quality.

Conventional aeration systems produce a gradual reduction of B.O.D. between the influent and effluent of the aeration tank. Therefore, a shock load will reflect in a decrease of effluent quality.

Effect of shock loading on the Beloit-Passavant aeration system and conventional aeration system is illustrated by Figure 2. The same volumetric loading for both systems was used.

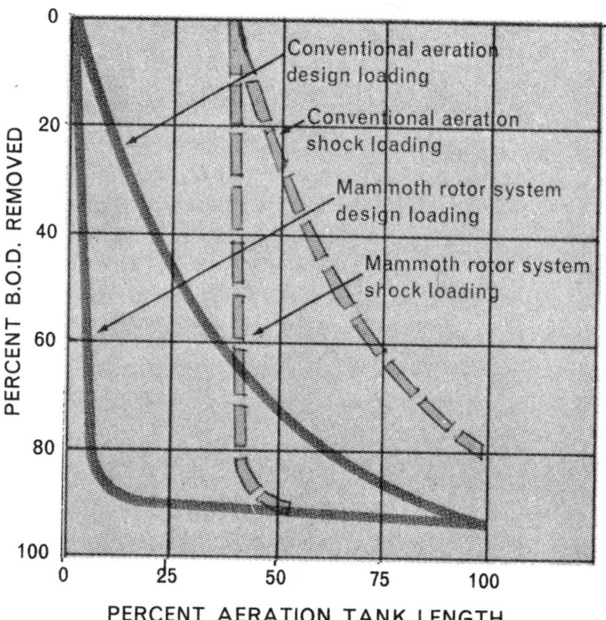

Figure 2

Volumetric loadings

The Mammoth Rotor and aeration tank can maintain higher solids concentrations in suspension; therefore, higher loadings of B.O.D. per 1000 cubic feet of aeration tank volume are possible.

The recommended volumetric loadings of the aeration tank are given in Table No. 1.

Table No. 1

Treatment	B.O.D. Reduction	Volumetric Loading
Activated Sludge	90% to 95%	112 # B.O.D./1000 cu ft
Complete Stabilization*	95% to 98%	60 # B.O.D./1000 cu ft

*Stable sludge (55% volatile solids) without separate digestion

The recommended oxygenation capacity of the Mammoth Rotor is given in Table No. 2.

Table No. 2

Treatment	Oxygenation Capacity
Activated Sludge	1.2 to 1.5 lb O_2 per lb B.O.D. applied*
Complete Stabilization	2.0 lb O_2 per lb B.O.D. applied

*Depending upon the waste

High-efficiency drive unit

One of the most important elements of mechanical aeration equipment is the drive unit and gear reduction assembly. The Mammoth Rotor drive unit is of rugged construction, especially designed for this equipment, driven by a standard vertical electric motor. The drive unit can be provided with one or two output shafts and used with one Mammoth Rotor, or centrally located between two Mammoth Rotors, each rotating in the opposite direction from its counterpart.

The Mammoth Rotor drive shaft is connected to the drive unit output shaft by means of a sphaer-elastic coupling, thereby preventing any vibrations at the rotor being transmitted to the drive unit bearings and gearing.

The drive unit is equipped with a sealed expansion and contraction bellows which prevents moisture accumulation in the gearbox from normal breathing due to ambient temperature change. The open-type breather cap normally used for venting gearboxes does not provide this protection.

LAGOONMASTER

The aerator section of the LAGOONMASTER consists of a series of blades connected to a horizontal drive shaft which, in turn, is supported at both ends by cylinder roller bearings. It is manufactured in convenient lengths up to a length of 25 feet per rotor. On large installations, two such sections can be connected with one centrally located drive unit to form a floating aerator spanning 50 feet of lagoon area.

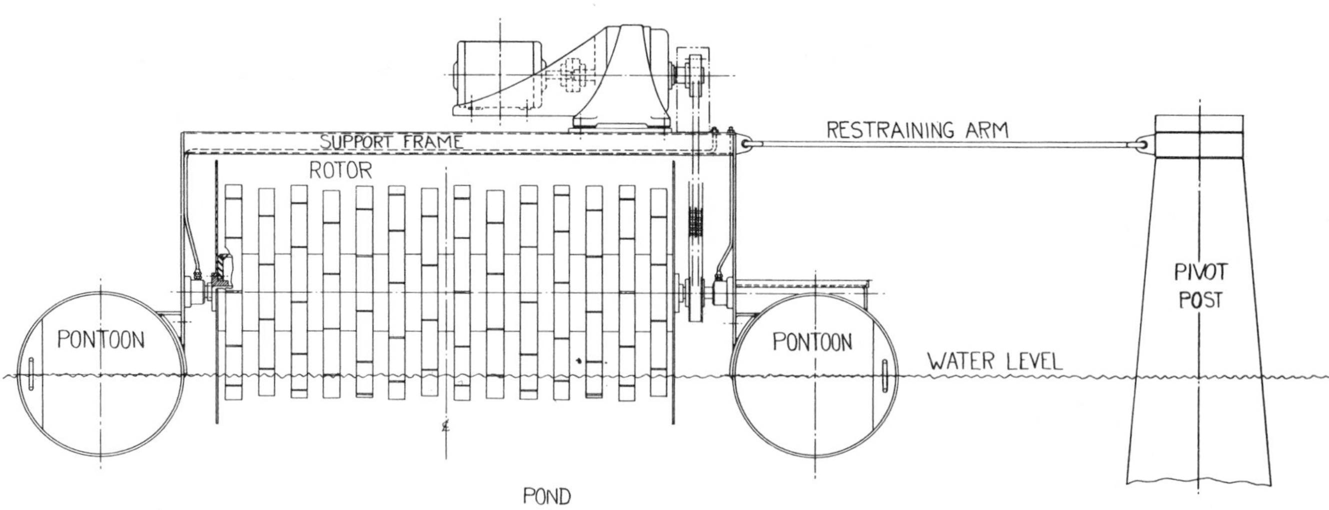

The LAGOONMASTER is mounted on specially designed pontoons. Positive flotation is assured at all times by means of foamed-in-place polyurethane placed inside each pontoon. Accidental punctures, or those resulting from acts of vandalism (amateur marksmen included) will not adversely affect the buoyancy of the unit.

The oxygenation capacity and required horsepower of the Mammoth Rotor are variable and are controlled by the rotor submergence. A simple, but effective ballast system on the LAGOONMASTER permits adjustment of the rotor submergence throughout the entire range —thus providing the most economical use of power consistent with the desired dissolved oxygen residual of the lagoon.

CLARIFIERS

BELOIT-PASSAVANT CORPORATION

CIRCULAR CLARIFIER

The traveling bridge is rotated by a traction drive unit conveniently located on the outer peripheral wall. Because of this location, speed changes governing the rate of rotation of the traveling bridge and sludge scraper blades can be easily made. In addition, this accessibility of the drive mechanism has made normal servicing and maintenance procedures safer. The traction drive is equipped with rubber wheels which provide the greatest friction with concrete.

Figure A

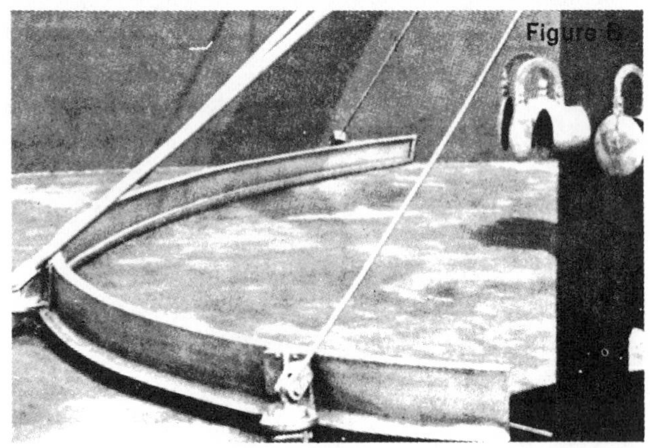
Figure B

Scraper blades are easily raised by a mechanical hoist for inspection and maintenance purposes (Figure A). This feature eliminates the necessity of dewatering the settling basin. The clarifier can be kept in service even during maintenance and repair periods.

The sludge scraper blades are connected to, and rotated by, the traveling bridge. There are no long movement arms – operating torque is kept to a minimum (Figure B). Positive bottom cleaning is insured by the hinged sludge scraper design. The scraper blade compensates for any uneveness in the basin floor. Should the scraper blades encounter an immovable object, they automatically ride up and over without increasing operating torque or causing equipment damage.

STENGEL INLET BAFFLES

These photographs of one type of inlet center well show the Stengel Inlet Baffles which distribute the inlet flow equally throughout the entire clarifier basin. Equal flow distribution provides better clarifier operation in the following ways:

- Controlled influent velocities – more efficient settling rates
- No hydraulic short circuiting because Stengel Inlet Baffles effectively dissipate inlet flow energy
- No channel flow patterns
- Completely nonclogging

RECTANGULAR CLARIFIERS

In multiple installations, full advantage may be taken of common wall construction techniques, through the use of single-span traveling bridge structures with double, triple, or quadruple sludge scraper assemblies.

Standard units are available in a full range of sizes up to 100 ft wide, for virtually any load requirement. These units may be modified to cover unusual conditions or special applications.

Fig. 1. View of installation showing Stengel Inlet Baffles, sludge scraper blade, traveling bridge, and traction drive.

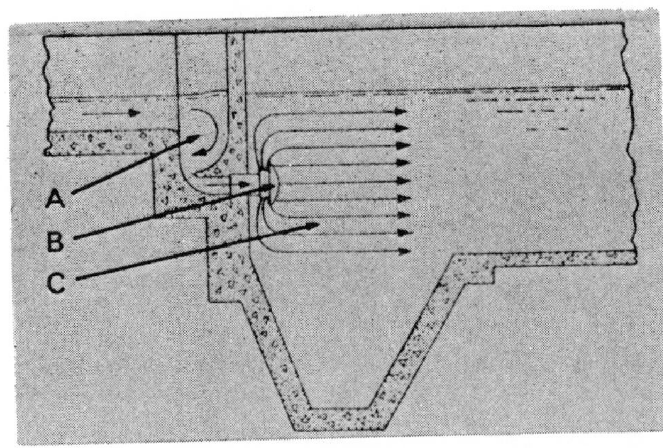

Fig. 2. Sectional view of a Beloit-Passavant Rectangular clarifier, showing (A) influent mixing chamber; (B) Stengel Inlet Baffle; (C) liquid flow pattern.

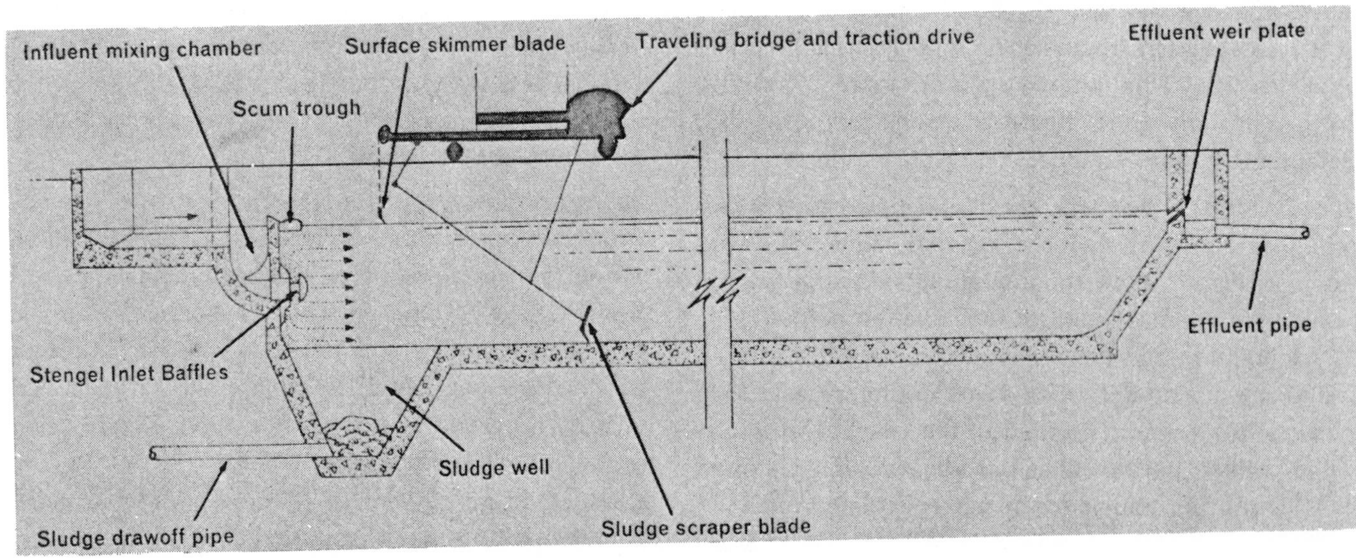

Fig. 3 Typical section of Beloit-Passavant Rectangular Clarifier.

KOAGULATOR

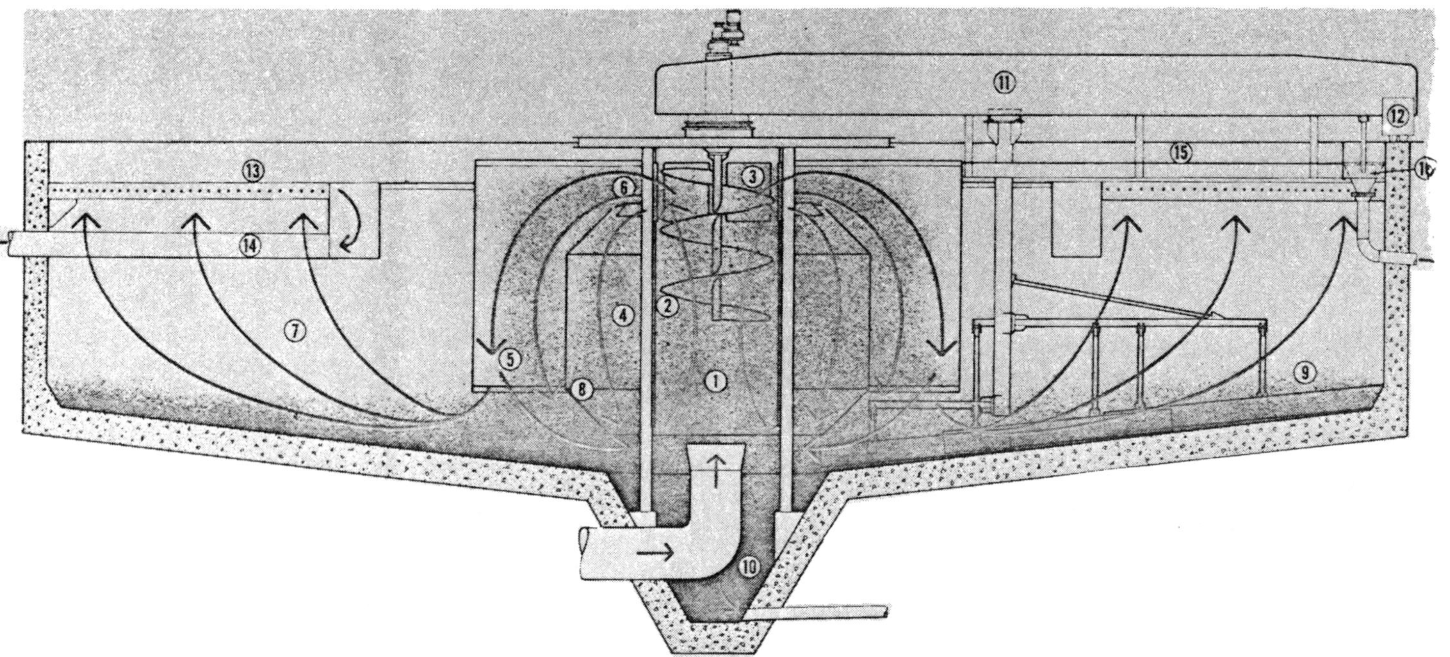

Raw water enters the Koagulator from below into the central mixing-activation zone (1). There, the incoming water containing the flocculating chemicals is gently but thoroughly mixed with large quantities of previously formed recirculated floc by the variable-speed circulation spiral (2). The influent is discharged from the mixing-activation zone through discharge ports (3) located at the liquid surface.

From the discharge ports, the liquid flows to the recirculation zone (4) and the reaction zone (5). The amount of flow to each zone is regulated by the height at which the upper adjustable recirculation baffle (6) is set. All liquid within the recirculation zone is recirculated to the mixing-activation zone and mixed with the incoming raw water. The fluid in the reaction zone is divided, with a portion of it passing into the clarifier zone (7), and the remainder being recirculated back to the mixing-activation zone. The amount of liquid flowing to the clarifier zone and mixing-activation zone is controlled by the lower adjustable recirculation baffle (8). Recirculation ratios as high as 6:1 can be achieved by baffle adjustment. This is **controlled recirculation.**

- **Controlled recirculation** provides flexibility to compensate for changes in flow rates and incoming water characteristics.
- **Controlled recirculation** provides complete utilization of flocculation chemicals.
- **Controlled recirculation** insures a positive, intimate contact between floc formation and raw water.
- **Controlled recirculation** provides the optimum in suspended solids removal.

Settled sludge or floc from the clarifier zone is removed from the tank bottom by the sludge scraper (9) and moved to the sludge well (10) for removal from the system. The sludge scraper is connected to, and rotated by, the traveling bridge (11) assembly. The speed at which the traveling bridge and sludge scraper rotate is easily adjusted at the traction drive (12) to provide maximum removal of settled solids.

Clarified liquid effluent is removed from the clarifier zone by radial effluent troughs (13), flowing to a common effluent pipe (14).

Convenient removal of floating material or scum is provided by a skimming arm (15) and scum effluent box, where required (16). The skimming arm is connected to, and rotated by, the traveling bridge.

Series	Diameter	Influent (gpm) gpm/sq ft 1.0	1.5	2.0	Circulation Spiral hp	Traction Drive hp	Number of Effluent Troughs	Maximum Influent Pipe Diameter
2206	6′-6″ to 10′-0″	27 63	41 94	54 125	3/4	1/3	6	3″ to 5″
2208	10′-0″ to 23′-0″	63 345	94 520	125 690	1	3/4	8	6″ to 10″
2210	23′-0″ to 36′-0″	345 841	520 1260	690 1680	1-1/2	1	10	12″ to 18″
2212	36′-0″ to 56′-0″	841 2065	1260 3100	1680 4125	5	1-1/2	12	18″ to 30″
2216	56′-0″ to 72′-0″	2065 3390	3100 5075	4125 6780	7-1/2	1-1/2	16	30″ to 36″
2220	72′-0″ to 85′-0″	3390 4715	5075 7050	6780 9400	7-1/2	1-1/2	20	36″ to 42″
2224	85′-0″ to 100′-0″	4715 6600	7050 9900	9400 13,200	10	1-1/2	24	42″ to 48″

TABLE 1—Design information pertinent to size, capacity, and horsepower requirements of various Koagulator models.

Detention Minutes	gpm/sq ft Clarifier Overflow Rate 0.75	1.00	1.25	1.50	1.75	2.0
60 min	4′-9″	6′-6″	8′-3″	10′-0″	11′-6″	13′-0″
70 min	5′-9″	7′-9″	9′-9″	11′-9″	13′-3″	15′-3″
80 min	6′-6″	8′-9″	11′-0″	13′-3″	15′-3″	17′-6″
90 min	7′-3″	9′-9″	12′-3″	15′-0″	17′-0″	19′-9″
100 min	8′-3″	11′-0″	13′-9″	16′-6″	19′-0″	22′-0″
110 min	9′-0″	12′-0″	15′-3″	18′-3″	21′-0″	24′-0″
120 min	9′-9″	13′-0″	16′-6″	19′-9″	23′-0″	26′-6″

TABLE 2—Koagulator sidewall liquid depth for various detention times and clarifier overflow rates.

BIRD-YOUNG ROTARY FILTER

Bird-Young Rotary Filter — Far greater capacity than conventional drum filters of equivalent size. Handles very dilute slurries at up to 400 gals./min./sq. ft. Unique single-cell design eliminates internal piping, hence there is nothing to obstruct filtrate flow or limit hydraulic capacity. Air blow-back cake removal permits discharge at high drum speeds. Cake removal is more effective and absence of scrapers lengthens filter cloth life. Can be supplied with multi-stage, counter-current wash that covers entire cake with a uniform flow of wash liquor. Wash separating pans in drum under each wash zone provide for sharp separations. SIZES: From 1 sq. ft. and up to meet every requirement.

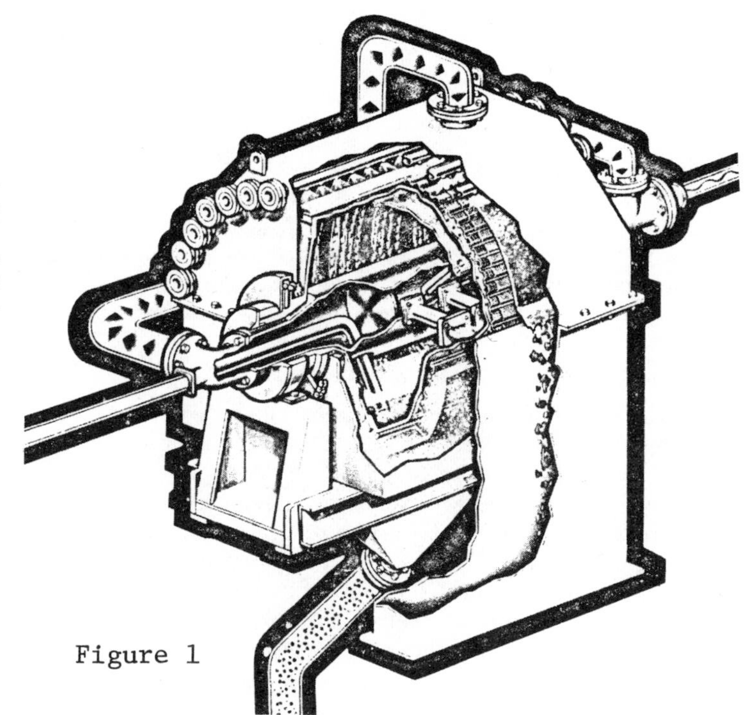

Figure 1

The Bird-Young Rotary Filter is a simple, rugged, compact unit. It is available with open tank, splash closure, a completely vapor-tight hood or pressure-tight enclosure. All air and vapor exhausted by the vacuum pump may be returned via the blow-back or into the hood to provide a constant pressure balance.

The unit may be constructed of any machinable metal with trip options available in FRP and other plastics.

BIRD-YOUNG FILTER TABLE OF SIZES*

Machine Size	Area Sq. Ft.	Installation Dimensions Width x Length
12" x 4"	1	22 1/2" x 39"
12" x 8"	2	26 1/2" x 43"
18" x 12"	5	56" x 36"
18" x 24"	10	56" x 48"
2' x 2'	13	56" x 60 "
2' x 3'	20	56" x 72"
3' x 2'	19	55" x 90"
3' x 3'	29	55" x 111"
4' x 1'	15	68" x 56"
4' x 2'	28	68" x 68"
4' x 3'	40	68" x 80"
4' x 4'	53	68" x 92"
5 1/2' x 4'	70	76" x 95"
5 1/2' x 6'	106	76" x 119"
5 1/2' x 8'	140	76" x 143"

*Dimensions do not include drive or piping beyond end of center girt.

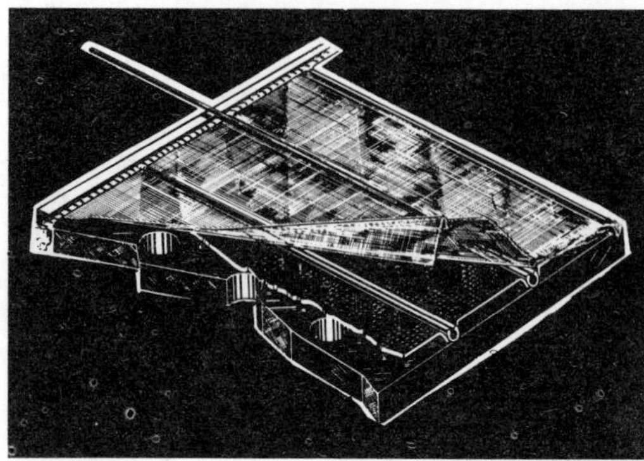

Figure 2
Close-up of the drum drainage deck illustrates the bridge construction of the drainage plates and the method of holding filter cloth firmly in place. This unique, patented design allows free flow of filtrate and maximum service from the filter medium.

The pictures tell the story. In the Bird-Young Rotary Filter, up to ninety-three per cent of the filter area is constantly active. There are no filtrate lines, valves or other hindrances to filtrate flow. All interior parts are stationary and rigidly attached to the sturdy center girt.

Pneumatic removal of the cake and cleaning of the filter medium is made completely effective without scraper, doctor or pick-up device of any kind. The blow-back shoe or valve is fitted with very close clearances to the accurately machined inside surface of the cylinder, thus sealing off vacuum at the point of cake discharge. The shoe has a narrow longitudinal slot in its sliding surface and across the entire cylinder length through which a large volume of low pressure air (usually obtained from the vacuum pump discharge) effects complete removal of the cake. Full capacity is constantly maintained. The vacuum is applied to the inside of the cylinder through a large trunnion and full vacuum is exerted directly under the filter cover without passing through any piping or valve.

The perforated cylinder is divided into multiple sections each 2 to 2½″ wide or about one-tenth the usual width of ordinary filter compartments. Fig. 3 shows the detail of this patented drainage panel construction and cloth covering application. Bridge construction of each section means full and free flow of filtrate. The filter medium is snugly and securely positioned and locked at each of the narrow panels. No wire winding is used. The life of the filter medium is exceptionally long even when very light fabrics are employed. Replacements are easily and rapidly made.

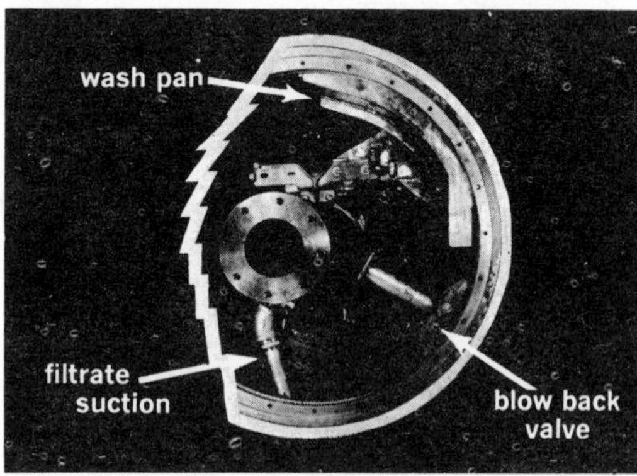

Figure 3
This interior view of the rotary drum illustrates the freedom from internal piping gained from the single cell design and points out wash collector arrangement.

The tank is specifically designed to prevent settling of the slurry without the use of mechanical agitation. The correct level of liquid for most efficient operation is automatically maintained. Wash collectors (Fig. 4) may be placed in the cylinder to secure the desired number of separations of wash liquor, and separations are exceedingly sharp.

The filtrate collecting in the bottom of the drum is continuously removed through a suction pipe connected to a self-priming pump.

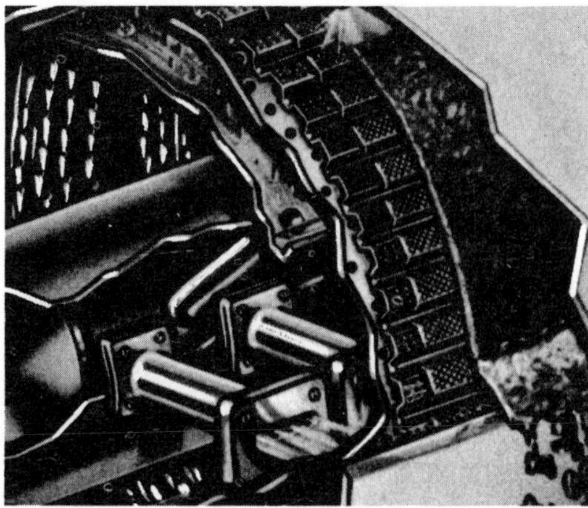

Figure 4
Close-up of cutaway section showing washed cake, bridges and filter cloth, wash collection pan, and air blow-back outlets.

BIRD-SIMPLEX HIGH INTENSITY AERATION PROCESS

The most effective method for purifying sanitary and industrial wastes is the activated sludge process. This versatile biological treatment process can be adapted to handle a wide variety of wastes and effluent requirements at high efficiency and low cost.

Its effectiveness depends, however, on two factors: the proper circulation and inoculation of the influent waste and activated sludge to achieve a thorough intermixing, plus absorption of sufficient atmospheric oxygen to support the biological activity involved. The more completely oxygen is brought into intimate contact with the tank contents, the more thorough the aeration and the more effective the process.

The Bird-Simplex Aerator — consisting of a heavy-duty drive unit, a high intensity aerating cone and a stationary draft tube — provides unusually efficient and economical aeration and BOD reduction. This is accomplished by (1) a circulation action so thorough that it actually creates a self-cleaning, scouring effect on the interior surfaces of the tank and (2) by cone action that maintains intense surface agitation for complete oxygenation. Optimum efficiency results from the unit's unique variable immersion depth feature which provides for a 300% variation in aeration capacity.

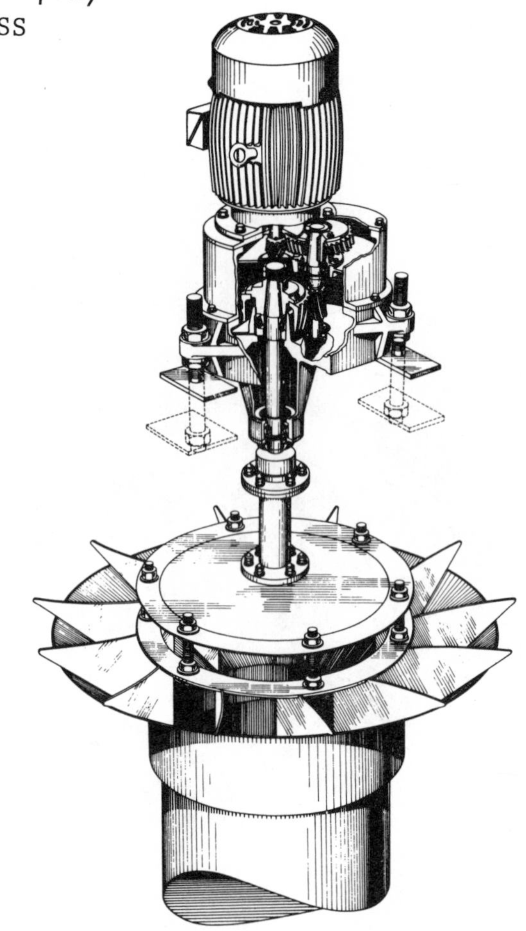

The Bird-Simplex Cone is available in sizes from two feet to more than ten feet in diameter, with power requirements determined by process conditions. The Cone is constructed with rugged, specially designed contoured blades and is, in effect, a high efficiency, low-lift pump normally operating at speeds of 30 to 60 r.p.m. Cone speed limits have been developed to maintain the activated sludge floc in the ideal condition to insure proper separation in the final clarifier. Cone designs can be tailored to meet special requirements such as icing conditions or super intensity aeration to reduce aeration detention time.

A unique and important feature of the Bird-Simplex Cone is its submergence ability. By adjusting the depth of immersion, a variation of up to 300% of the aeration capacity may be realized. The incorporation of a simple adjustable outlet weir allows the intensity of aeration to be infinitely varied between wide limits to accommodate rapidly changing process loading conditions. Optimum plant efficiency is thus maintained.

In operation, the "pump action" draws the liquid through the Draft Tube, discharging it in a torrent over the lip of the cone, striking the surface liquid with great force. This action causes intense agitation of the surface, resulting in a high uptake of oxygen. Oxygenation also occurs as the discharge from the Cone passes through the air.

To support the biological action, oxygen absorbed at the surface must be made available to the entire contents of the aeration tank. An additional feature of the Bird-Simplex Aerator with Draft Tube is its positive ability to draw liquor from the sides and bottom of the tank to be aerated at the surface. The intense circulation of the liquid causes a constant scouring action on the floors and walls making cleaning unnecessary. Due to the complete absence of "dead spots" and consequent septicity, the initial high standards of operation are sustained throughout the long life of the plant.

The Bird-Simplex Aerator is manufactured by Bird Machine Company under license from Ames Crosta Mills & Company Limited, Lancashire, England.

Bird Machine Company

CONCURRENT FLOW CENTRIFUGAL

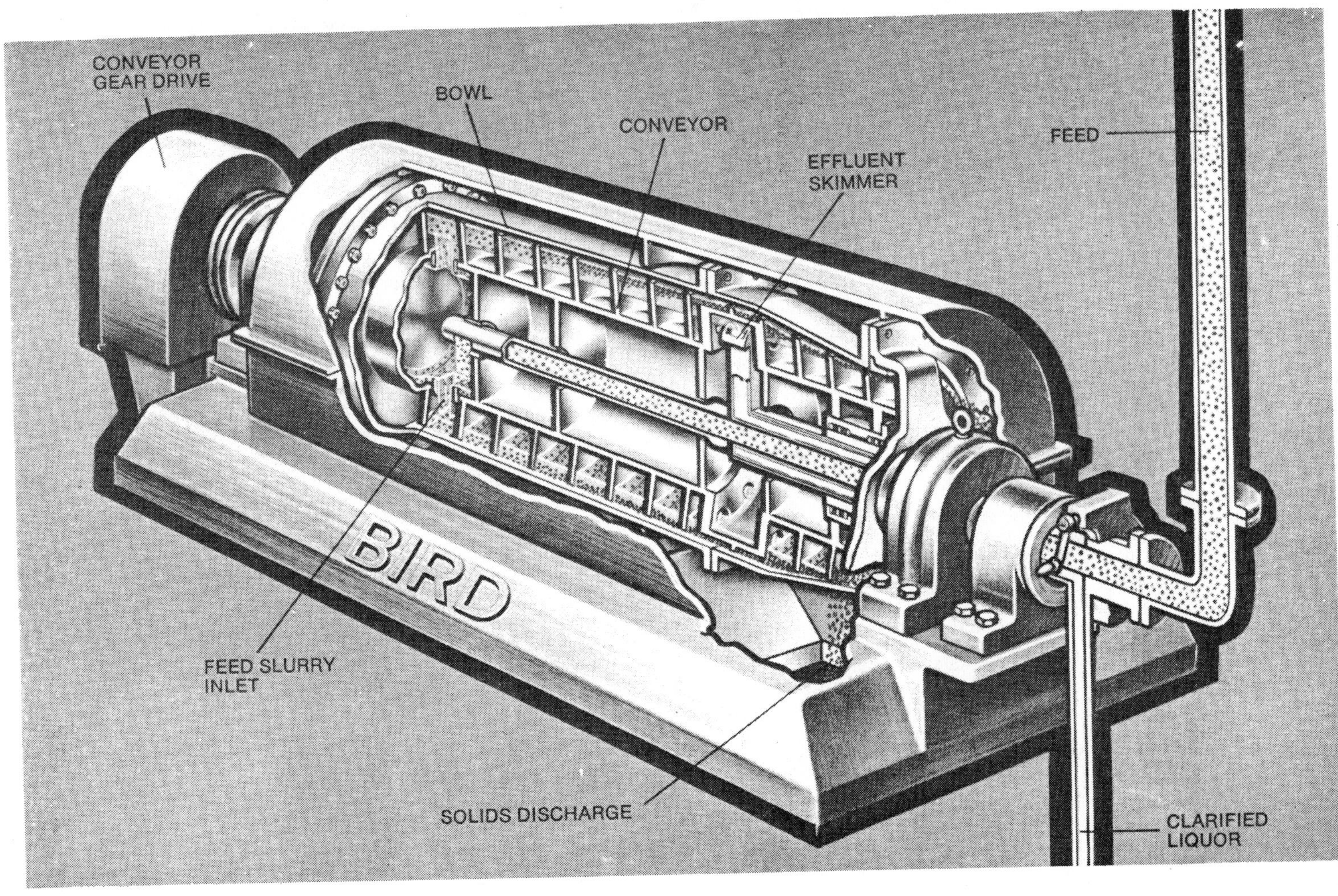

This cutaway drawing shows the unique, high-efficiency design of the new Bird Concurrent Flow Centrifugal. The machine is engineered throughout for improved dewatering of sludges in sanitary and industrial waste treatment.

As the cutaway indicates, incoming sludge is carried by the feed pipe to the far end of the bowl, away from the discharge end. As a result, settled solids are not disturbed by incoming feed. Solids and liquids pass through the bowl in a smooth parallel-flow pattern. Turbulence is eliminated entirely.

Note, too, that solids are conveyed over the entire length of the bowl before discharge to provide better compaction and drier cake.

The Concurrent Flow Centrifugal also incorporates a skimmer for discharge of effluent. The skimmer is designed for external adjustment while the machine is in operation, thus making it a simple matter to regulate pool depth for most efficient operation. The skimmer housing confines any splash or turbulence that might disturb settled solids. Effluent is discharged under pressure.

The ability of the Concurrent Flow Centrifugal to deliver better compacted, drier cake minimizes the need for costly chemical flocculants to obtain a clear effluent. The high efficiency of the unit also pays off in increased throughputs that, on activated waste sludge, range up to twice those of conventional centrifugals.

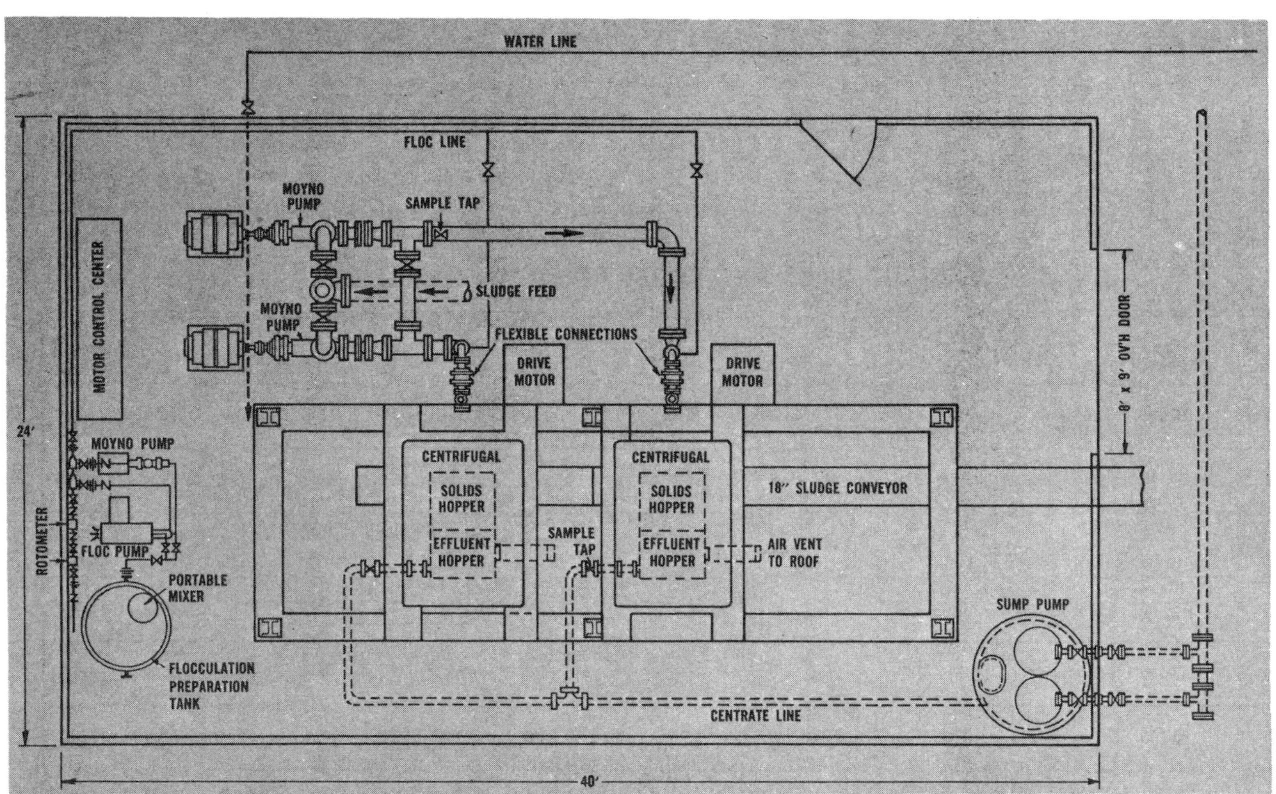

TYPICAL INSTALLATION OF TWO 24" x 38" CENTRIFUGALS FOR SEWAGE SLUDGE DEWATERING

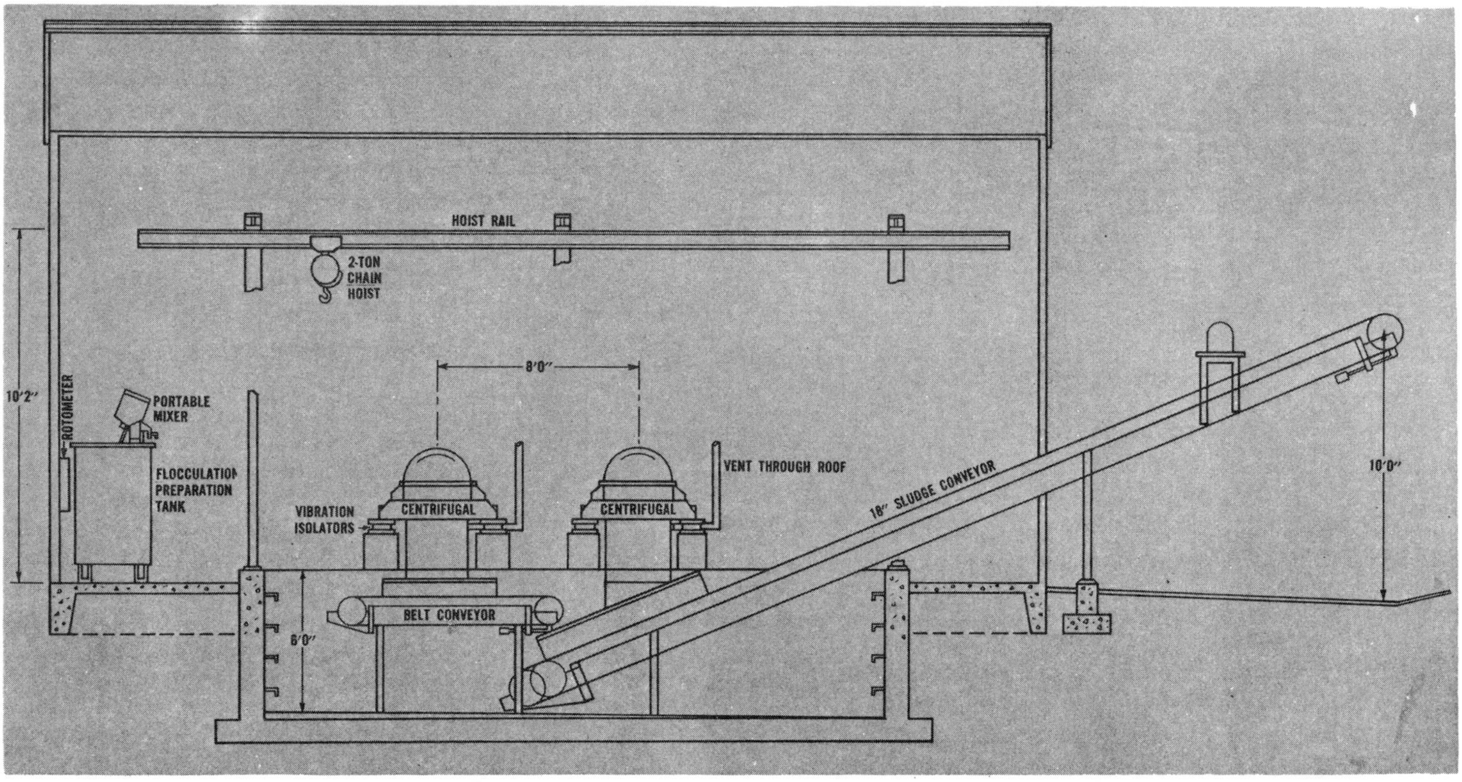

HOIST RAIL
2-TON CHAIN HOIST
8'0"
10'2"
ROTOMETER
PORTABLE MIXER
FLOCCULATION PREPARATION TANK
VENT THROUGH ROOF
VIBRATION ISOLATORS
CENTRIFUGAL
CENTRIFUGAL
18" SLUDGE CONVEYOR
10'0"
BELT CONVEYOR
6'0"

MACHINE SIZES AND DIMENSIONAL INFORMATION

SIZE (in.)	HP Required	TYPE	DRIVE (See diag.)	A	B	C	D	E	F	G	H	J	K	L	WEIGHT (lbs.)
12x 30	10-25	Counter-Current	1	75½	15½	4	47¾	5	9½	32¼	21¾	23⅝	60	41¾	1,800
18x 28	15-30	Counter-Current	1	92½	16 11/16	3¾	46½	5	9½	39¾	28⅛	33⅛	63	44¼	2,800
18x 42	20-40	Concurrent	2	117	23	5	68¾	5	9½	45½	32	37	86¾	55	4,500
18x 42	20-40	Counter-Current	1	120	23	5	68¾	5	9½	45½	32	37	86¾	55	4,275
24x 38	25-50	Counter-Current	1	118	23	5	64	5	11⅝	52½	35¾	40¾	92	64⅛	5,400
24x 60	50-125	Concurrent	2	129	24 3/16	5	81½	5	11⅝	52½	37 1/16	42 1/16	92	64⅛	6,500
24x 60	50-125	Counter-Current	1 or 2	129	24 3/16	5	81½	5	11⅝	52½	37 1/16	42 1/16	92	64⅛	6,500
24x104	125-200	Concurrent	1	195⅞	27½	5	132	5	11⅝	51½	40½	45½	112½	63⅛	8,000
24x104	125-200	Counter-Current	1	193⅜	27½	5	132	5	11⅝	51½	40½	45½	112½	63⅛	8,000
36x 72	75-150	Counter-Current	1	178⅛	32½	5½	103¾	7	16	73½	51¾	57¼	130¾	89½	13,500
36x 96	125-200	Counter-Current	1	207	32½	5½	127¾	7	16	73½	51¾	57¼	130¾	89½	18,000

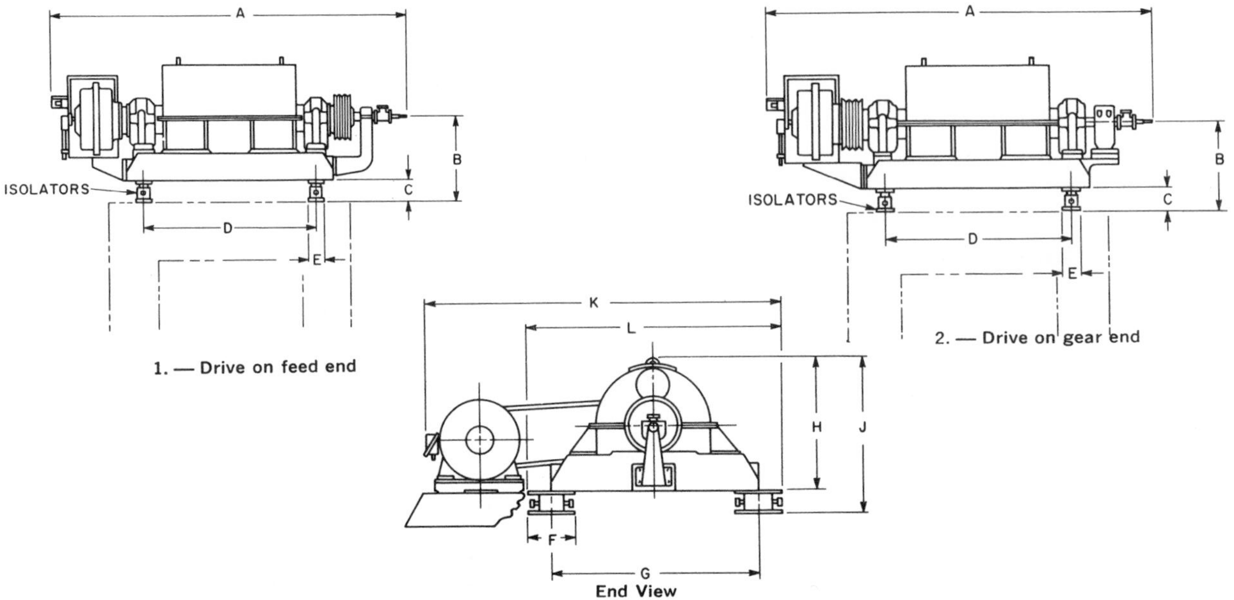

End View
(Dimensions same for machines with gear or feed end drive)

APPLICATIONS IN SANITARY AND INDUSTRIAL WASTE TREATMENT:
Dewatering of raw sludge in primary plants; Dewatering of digested sludge in primary and secondary plants; Thickening of digested sludge; Clarification of digester supernatant; Dewatering raw sludge to increase digester capacity; Thickening of waste-activated sludge; Dewatering sludge prior to trucking or incinerating; Dewatering of settled paper mill wastes, tanning wastes, foundry sludges, refinery wastes, cannery wastes, chemical process wastes, textile dyeing wastes, coal preparation refuse, neutralized waste pickle liquor from steel plants, dusts from basic oxygen furnaces in steel industry, etc.

Unifloat Flotation Save-All

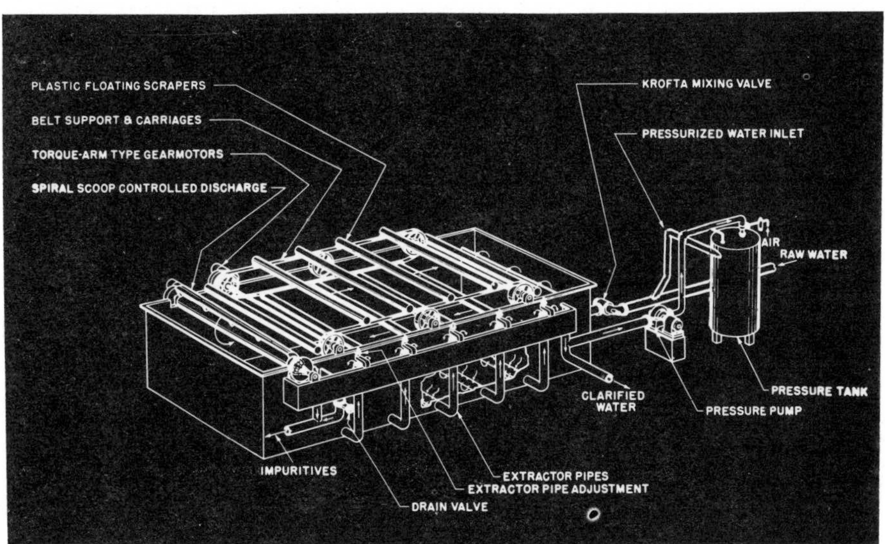

Vat type save-all for low headroom locations

Single-stage unit

Movable scrapers

Well suited for small installations

Capacities to 1800 GPM

DESCRIPTION

The Unifloat Flotation Save-All is a single stage unit with rectangular vat and movable scrapers. It is particularly useful for small installations where sufficient covered space is available and the recovered stock must be pumped to the point of re-use. Vat may be furnished in concrete, tile, steel or wood.

OPERATION

The influent after treatment with air and flocculants passes through a special friction valve where the pressure is released. It is then uniformly distributed over the width of the inlet of the Unifloat. The floated fibre and flocs are conveyed by floating scrapers to the spiral scoop which discharges into the reclaimed stock trough. Clarified water flows into the extractor pipes and is discharged at the adjustable overflow weir assemblies.

Type	Maximum Flow GPM	Length Ft.	Width Ft.	Height Ft.
2M3P	250	18'3"	9'2"	7'3"
2M4P	330	22'11"	9'2"	7'3"
2M5P	410	27'6"	9'2"	7'3"
2M6P	490	32'2"	9'2"	7'3"
2M7P	570	36'8"	9'2"	7'3"
2M8P	650	41'4"	9'2"	7'3"
3M3P	370	18'4"	12'6"	7'3"
3M4P	490	22'11"	12'6"	7'3"
3M5P	610	27'6"	12'6"	7'3"
3M6P	730	32'2"	12'6"	7'3"
3M7P	850	36'8"	12'6"	7'3"
3M8P	970	41'4"	12'6"	7'3"
4M3P	490	18'4"	15'9"	7'3"
4M4P	650	22'11"	15'9"	7'3"
4M5P	810	27'6"	15'9"	7'3"
4M6P	970	32'2"	15'9"	7'3"
4M7P	1130	36'8"	15'9"	7'3"
4M8P	1290	41'4"	15'9"	7'3"
4.5M12P	1600	58'0"	18'6"	7'3"
6M13P	2500	68'0"	23'6"	7'3"

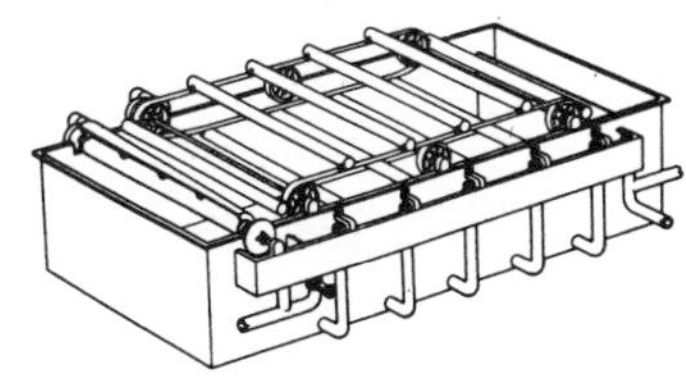

Airfloat
Flotation Save-All

OPERATION:

The Airfloat operates as a double stage flotation system with additional air-booster to enable operation without flocculants.

The influent, after treatment with air, is distributed uniformly by the upper pipe system fitted with nozzles and deflectors. Through the lower pipe system a small amount of clarified water is separately fed which has been treated with air under pressure. The fibres and flocs float and accumulate on the surface where they thicken to a high consistency. The stock is reclaimed at this consistency by a rotating spiral scoop. The clarified water flows towards the perforated bottom plates and as it passes the lower pipe system additional air bubbles reclaim the still remaining flocs. The clarified water passes the perforated plates and flows to the adjustable overflow which maintains the level in the Airfloat.

The Airfloat Flotation Save-All is used where operation without flocculants is requested because the water to be treated has such a high pH that the use of flocculants becomes uneconomical. Constructed of mild steel, the unit can be installed indoors or outdoors.

Type	Maximum Flow GPM	Outside Diameter Ft.	Height Ft.
A12	250	12′8″	8′10″
A15	330	15′	10′
A18	490	18′	10′6″
A20	630	20′	11′2″
A22	740	22′	11′6″
A24	860	24′	12′
A30	1,350	30′	13′
A36	2,000	36′	13′
A44	3,000	44′	13′
A49	3,650	49′	13′6″
A55	4,350	55′	13′6″

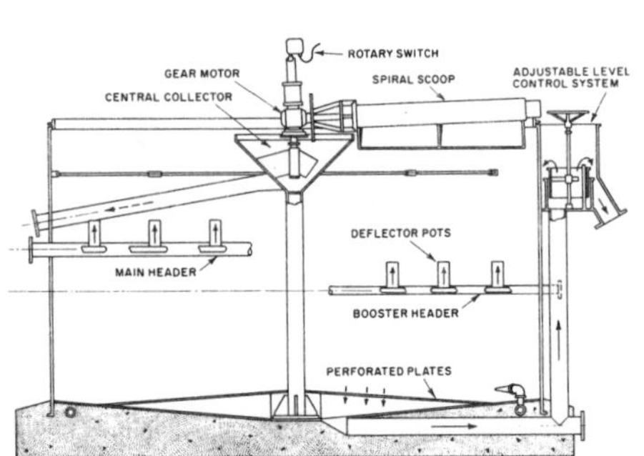

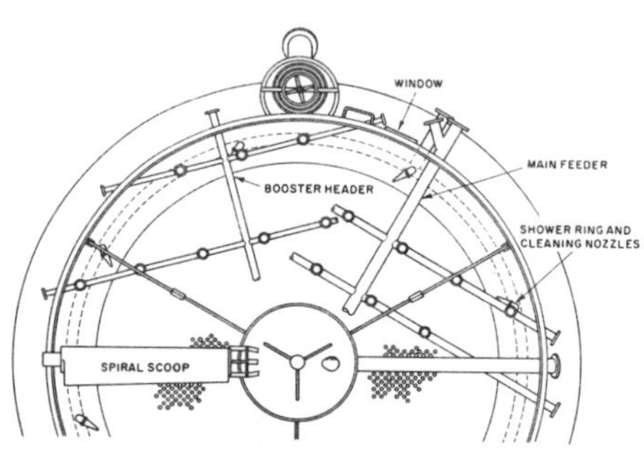

Superfloat Flotation Save-All

The influent, after treatment with air and flocculants, is uniformly distributed in the separate flotation of the Superfloat cells by means of the inlet pipes which cover the whole width of the cell. The flotation cells slope upward towards the reclaimed stock side. The fibres and flocs made more buoyant by the air bubbles slide along the ceiling of the flotation cell while the heavier clarified water reverses its flow and drops to the floor of the cell, and passes through a perforated vertical wall into the clarified water collector. The overflow of the clarified water side is adjustable and by this means the overflow of the reclaimed stock is also regulated and flows into the reclaimed stock outlet.

The primary condition for the pressure-flotation process is absorption of air by the effluent. Selection of the proper system depends on the solids content of the effluent, the size and capacity of the flotation installation and power costs.

Type	Maximum Flow GPM	Outside Dimensions			Number of Flotation Cells
		Width	Depth	Height	
7S6	370	7′ 3″	6′ 3″	8′ 4″	7
9S6	490	7′ 3″	6′ 3″	10′ 9″	9
7S9	560	7′ 3″	9′ 4″	8′ 4″	7
9S9	750	7′ 3″	9′ 4″	10′ 9″	9
7D6	750	13′ 4″	6′ 3″	8′ 4″	14
9D6	980	13′ 4″	6′ 3″	10′ 9″	18
7D9	1,120	13′ 4″	9′ 4″	8′ 4″	14
9D9	1,500	13′ 4″	9′ 4″	10′ 9″	18

Multiple flotation chambers built one over the other.

Compact construction—particularly suitable for use close to the paper machine with immediate return of the reclaimed stock.

Built fully pre-erected.

All parts in contact with water are stainless steel construction.

8 models—capacities to 1,500 GPM.

Sedifloat Primary Effluent Clarification

Highest clarification obtained by using sedimentation and flotation simultaneously.

Smallest space requirement because of selective settling and floating.

Highest sludge concentration produced by flotation and special spiral scoop removal.

Type (dia. in ft.)	Flow without Flocculants gpm	Flow with Flocculants gpm
8	78	95
10	133	145
12	200	250
15	270	332
18	342	492
20	500	610
22	590	735
24	685	840
30	1070	1310
36	1620	2000
44	2380	2910
49	2940	3620
55	3740	4600

The Sedifloat is designed for primary effluent clarification. A part of the clarified water in which air has been dissolved, is recycled and mixed with the influent. Tiny air bubbles carry the floating light material to the surface while the heavy particles, dirt and grit, settle on the bottom.

The clarified water is extracted at the outer rim halfway between the top and the bottom of the tank. Floating stock is skimmed off by a special revolving spiral-formed scoop and discharged at the center into the floated sludge tank. A revolving scraper moves the settled sludge along the bottom into the settled sludge well.

The primary condition for the pressure-flotation process is absorption of air by a recycled portion of the clarified effluent. Selection of the proper air system depends on the solids content of the effluent, the size and capacity of the installation and power costs.

49 ft. diameter Sedifloat processes 4.3 million gpd.

Flotator

DESCRIPTION

The Flotator will reduce the solids content in the clarified effluent down to between 0.5 lbs. and 0.1 lbs./1000 gallons and reclaim fibre at consistencies between 1% and 3½%. A circular tank with conical top forming a totally enclosed vessel, the Flotator is divided into two stages by an internal horizontal, conical-shaped division. Secondary stage is above the primary stage.

OPERATION

The influent, after treatment with air and flocculants, enters tangentially through the inlet-pipe into the inlet cone to insure uniform distribution. The airladen fibres and flocs then float up and slide along under the conical shaped division until they enter the upper section where they follow the wall of the cone to the stock overflow on the top. The heavier clarified water falls to the bottom in both flotation chambers and flows through perforated plates into the central collector pipes and overflows at the top.

The high consistency obtainable reduces the amount of water returned to the system, thus reducing the water in circulation. Because of the small volume of the Flotator and the relatively fast flow through the save-all, color changes can be quickly made.

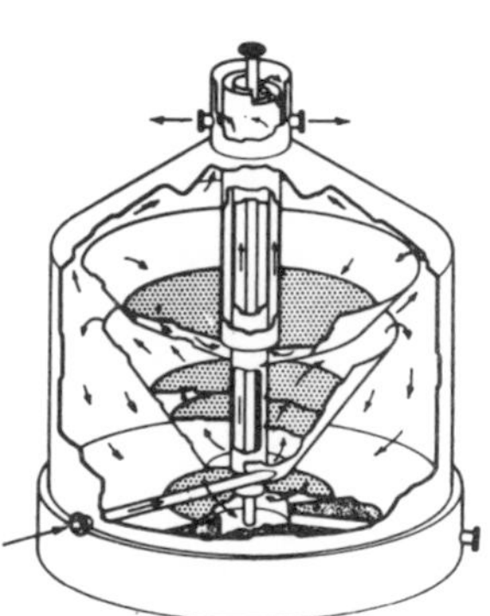

Where power economy is of prime consideration, a Flotator system is available in which only part of the white water is pressure treated, with the balance of the water being drawn into the system by an injector after the pump.

PERFORMANCE DATA

Paper Stock	% Filler T.O.	Clay	Feed lbs./1000 gals.	Clarified Water lbs./1000 gals.	Reclaimed Stock lbs./1000 gals.
Lightweight Bond	—	—	2.87	0.217	29.6
Regular Bond	1	5	10.17	0.374	60.2
Opaque Bond	7	—	7.67	0.469	54.2
Opaque Bond	6	—	9.95	0.441	62.0
Bond	1	—	3.95	0.495	42.8
Bond	1	—	2.97	0.309	35.1
Ledger	—	—	2.11	0.339	29.7
Ledger	—	—	2.05	0.492	23.6
Ledger Special	6	—	7.68	0.408	46.7

Test results from a Flotator installation at a rag paper mill; data represents averages of no less than 12 tests by filtration.

DIMENSIONAL DATA

Type		8	10	12	15	18	20	22	24	28	30
Capacity	US GPM	210	360	500	680	1,000	1,210	1,460	1,730	2,190	2,650
Diameter	Feet	8′	10′ 5″	12′ 8″	14′ 10″	18′	20′	21′ 10″	23′ 10″	27′ 8″	29′ 6″
Height	Feet	8′ 6″	11′	14′	14′ 10″	13′ 10″	15′ 9″	16′ 9″	17′ 8″	19′	19′ 8″

The Black Clawson Company

Biofloat

Description

The Biofloat provides biological activated sludge treatment and is based on the air-flotation principle. Flotation allows short separation time of clarified effluent from activated sludge. Reactivation of the sludge is not necessary.

Air-flotation thickens the sludge 2-4 times higher consistency than the conventional settling. This results in more economical sludge disposal. Clarified water is aerated under pressure before the final discharge guaranteeing a sufficient dissolved oxygen content at the point of discharge.

Type (dia. in ft.)	Aeration Power KW	BOD Reduction lbs/day	Flow gpm
8	2.3	214	49
10	4.0	364	87
12	6.0	531	128
15	7.9	708	168
18	11.9	1050	250
20	14.6	1280	304
22	17.5	1560	372
24	20.5	1810	435
30	32.0	2860	672
36	48.0	4280	1020
44	71.0	6300	1500
49	86.0	7700	1800
55	109.0	9800	2310

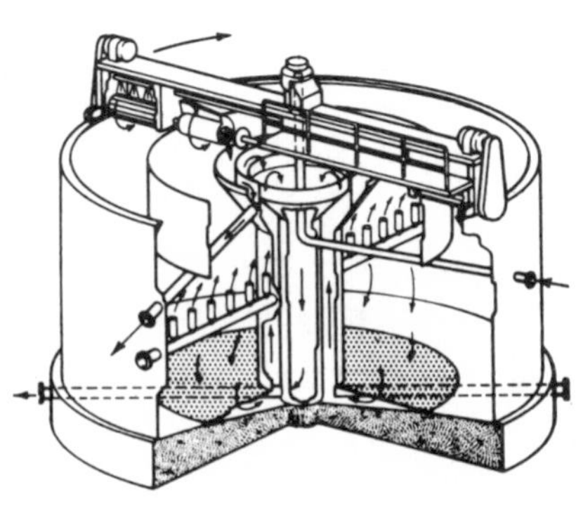

Operation: The effluent from the primary clarification installation enters the Biofloat at (1) for secondary biological treatment. The influent flows through the central pipe to the distribution header (2). It is discharged through spray pipes over the rotating aeration drums (3) for uniform distribution and immediate mixing with the aerated, activated sludge. The aeration drums are located on the bridge (4) which rotates slowly. As the floated, activated sludge rises to the surface, it is kept in motion by the rapidly rotating drums which are enclosed in the mixing chambers (5). At the same time new primary effluent is introduced into the mixing chambers. Perforated extraction plates are placed in the bottom of the unit to evenly lead the clarified water to the central overflow(7) which regulates the activated sludge level. The air is dissolved under pressure in the retention tank. This air-water solution is distributed uniformly throughout the Biofloat by a series of pressure release nozzles with deflector (8). Extraction of the clarified effluent is facilitated through the action of the microscopic bubbles, which carry the sludge particles to the surface.

A wall (9) encloses the central portion of the unit's surface where the surplus sludge is concentrated. The highly thickened sludge is extracted by the spiral scoop (10). The clarified overflow setting regulates the water level in order to give the best aeration intensity and most economical power consumption of the spray drums. An adjustable timer controls the rotation of the spiral scoop for the removal of the thickened sludge.

Disc Filter

"Pressure" filters paper machine water for fiber recovery and effluent clarification.

Strains and fine-filters fresh water or reused clarified water for further use.

Compact. Less floor space yet high filtration rate.

Stainless steel and non corrosive construction. Fully prefabricated. Easily installed. Easily moved.

Description

The Disc Filter is designed for cyclic **batch** rather than continuous operation in order to obtain highest capacity and clarification. Precoat stock is added **only at the start** of the filtering period until the basic filtering mat has been formed. From then on unclarified water, without additional sweetening stock, is filtered through the basic mat.

Filtration rate of the Disc Filter is definitely higher than that of a continuous filter of the same filtration surface. During the filtration period, the total filtration surface is used. Because of the purging-cleaning period, the net filtration time in the Disc Filter is at least 90% of the operational time.

Main tank of the unit is stainless steel and features a full length access door on one side. The hollow central shaft has a free discharge at the bottom and is suspended on a roller bearing. Single packing seals are located at the top and bottom of the housing. Filtration discs, each composed of 8 segments set in the openings of the hollow shaft and held outside by a ring, are horizontally mounted on the shaft. Each segment consists of a central supporting plate and two wire sides molded in plastic frames for easy assembly. A V-belt drive is located on top of the unit. The unclarified water inlet is located near the bottom of the tank; the reclaimed stock discharge is at the bottom of the tank. A self-cleaning shower is built into the door of the tank with nozzles spraying in between the discs on the wire surface.

Operation

The Disc Filter operates on the automatic purging principle which consists of the filtering and the purging-cleaning periods:

Filtering Period

The pump (13) feeds the unclarified water from the tank (16) into the inlet (5). Water is filtered through the discs (3) and discharged through the central shaft (2) at the outlet (6). The float operated butterfly valve (15) throttles the feed pump (13) proportionally to the unclarified water flow (17). During this filtering period the disc shaft is not moving and the shower pump (12) is not operating.

During this period the flow through the Disc Filter progressively decreases until it is smaller than the flow of the unclarified water. The level in the unclarified water tank starts to rise and then the automatic float control (18) starts the purging period.

Purging-Cleaning Period

Operational steps are automatically controlled. The feed pump (13) stops and the small water volume in the Disc Filter flows back into the tank (16). The spin-motor (1) starts and spins the stock off the filtration discs. The pneumatic valve at the reclaimed stock outlet (7) opens and discharges the filter mat into the stock chest. The shower pump starts, washing the rest of the stock off the wire. The stock outlet valve closes; the clarified water outlet valve closes. The spin motor and the shower pump stop and the feed pump starts. The precoat feedline valve (14) opens for a short period, building a mat, then the clarified outlet valve opens and a new filtration period begins.

The purging-cleaning period lasts 40-60 seconds depending on the timer setting. The filtering period lasts a minimum of 60 seconds and is automatically modified by the available flow of the unclarified water as well as by its consistency and filtration resistance.

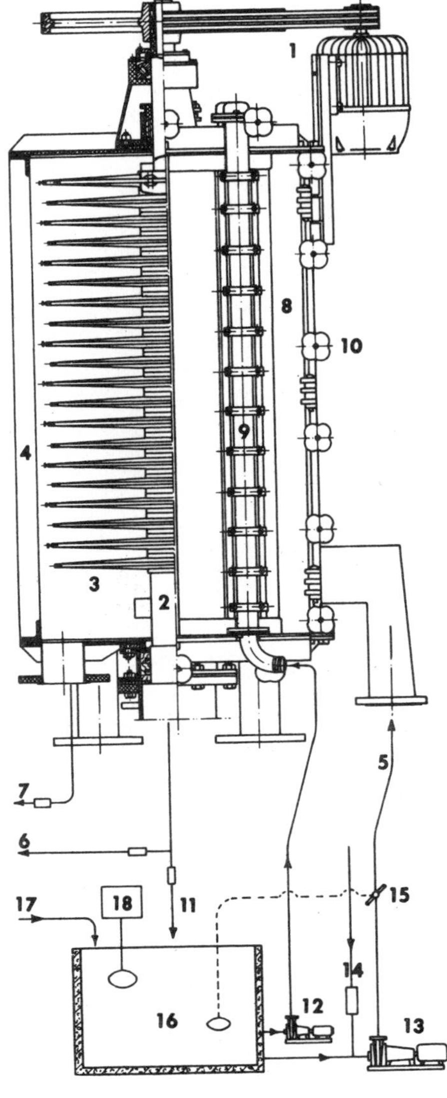

Type	Discs Dia.	Discs No.	Outside Dimensions Ht.	Outside Dimensions Dia.	Filtration Surface Sq. Ft.
10DF2	2′	10	4′	3′	61
20DF2	2′	20	5′7″	3′	122
30DF2	2′	30	7′2″	3′	183
20DF3	3′	20	8′3″	4′	280
30DF3	3′	30	11′	4′	420
40DF3	3′	40	12′6″	4′	560

DBC PLUS DRIED BACTERIA CULTURES

DBC Plus is a compound of dried bacteria cultures which have been combined with biological synergists and nutrients to produce a product which not only improves upon and speeds the action of the naturally occurring bacteria in sewage, but goes far beyond the capability of these organisms to digest greases, carbohydrates and proteins which are frequently difficult to break down otherwise. The bacteria which are used in DBC Plus are harmless saprophytes. They cannot damage pipes or equipment or harm people, animals, birds or plant life.

DBC Plus, Type A, has proven to be successful in several main areas of line and lift station maintenance in sewage collection systems. By introducing preactivated DBC Plus in a slurry form into sewer lines, it is possible to eliminate and/or prevent the accumulation of grease and other organic debris. The facilities are maintained at low cost and with minimal effort.

One form of automation available involves the use of an electrically-timed pump which measures a predetermined volume of water containing DBC Plus. Treatment is made by inserting a hose into the one-inch opening in a manhole cover. By the use of available formulas, a procedure can be easily set up to treat facilities quickly, effectively and at low cost.

The Bower Automatic Feeder holds enough DBC Plus to last for several hours or days. By programming the master control, it is possible to feed, on an automatic 24-hour cycle basis, the proper amount of DBC Plus for the facility being treated. The Automatic Feeder is ideally suited for use in out-of-the-way pumping stations, industrial situations and in the automatic feeding of DBC Plus at treatment plant locations.

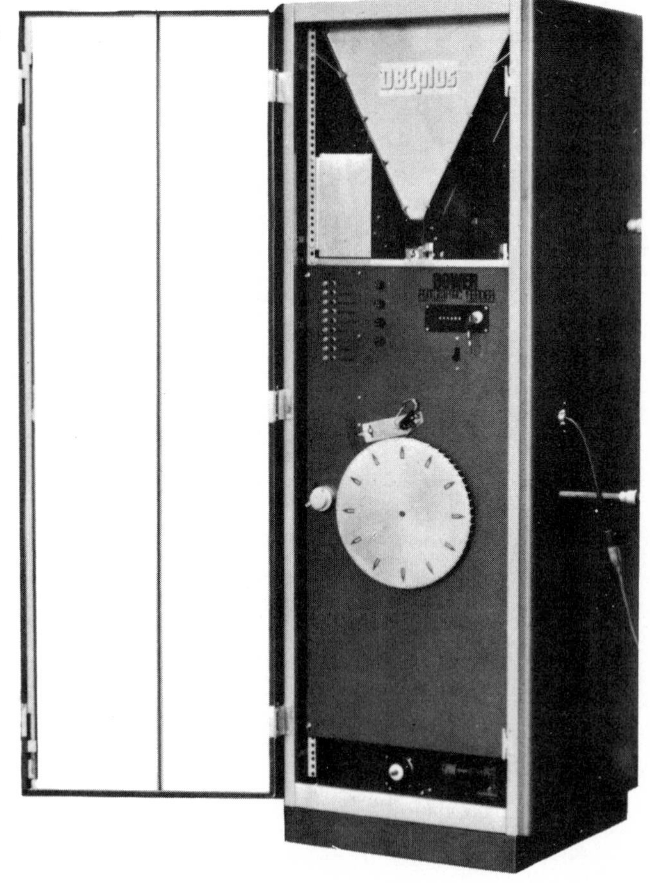

DBC Plus has been and is being used with great success in the treatment of sewerage lagoons, industrial wastewater, etc., as well as in line and lift stations.

The application of DBC Plus Dried Bacteria Cultures is made on the basis of the amount of material to be treated and the amount of grease in the system.

In maintenance of lines and lift stations, heavy initial treatment is called for, followed by gradual tapering off, and then small, periodically-applied amounts are used for trouble prevention.

BOSCO

High-Rate Trickling Filter Underdrain Block

DESIGNED TO ASTM C159 TYPE 1 H

APERTURE AREA: 30% of Block Face

CHANNEL AREA: 63% of End Section— 56 Square Inches per foot of filter width

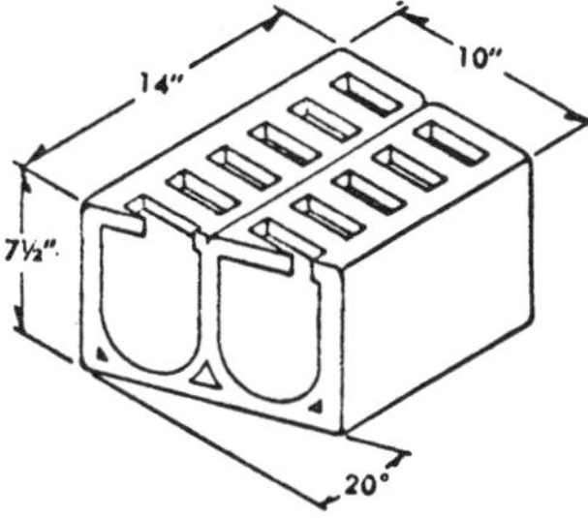

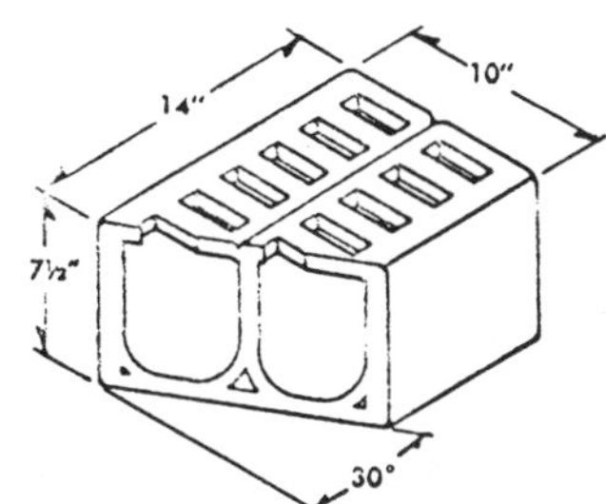

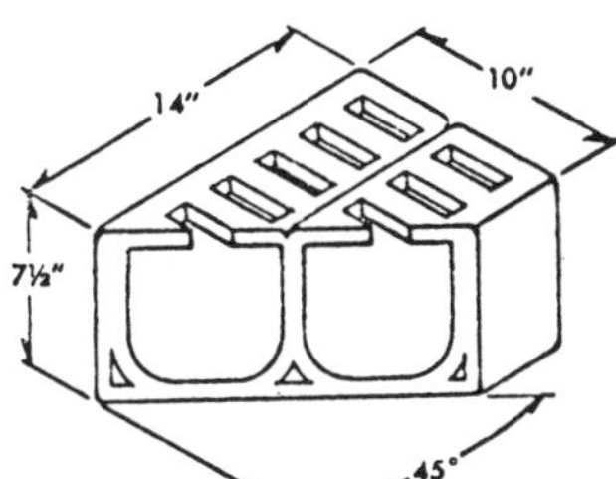

These special shapes allow block rows to conform to wall curvature—ventilation channels may be kept constant to allow free air flow to all block channels. Bosco provides installation diagrams for placement of these shapes in underdrain system.

Vent block shown provides support for ventilation pipe—allowing free flow of air into channel.

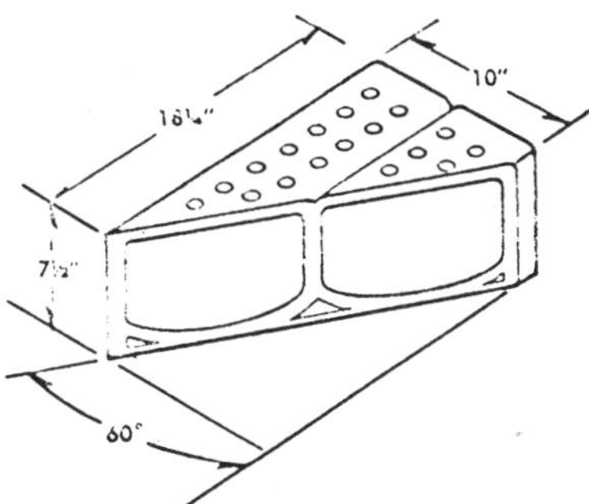

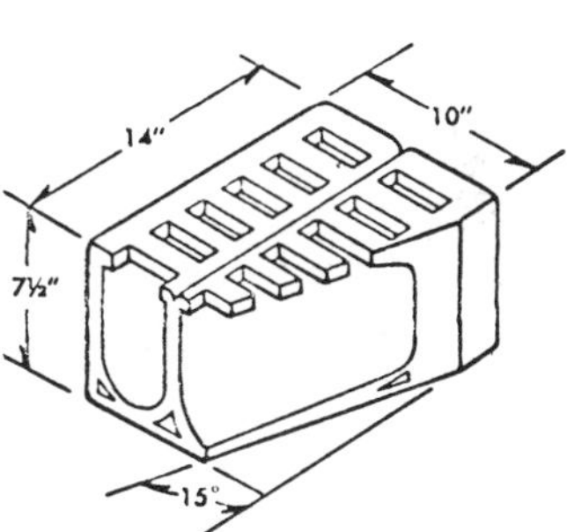

Channel alignment is assured with Bosco block. In installation, keep the groove in the face of the block in alignment with the groove in the preceding unit.

Care should be taken to see that all cover block are installed on even bearing. Quarter block or common brick may be used to provide support.

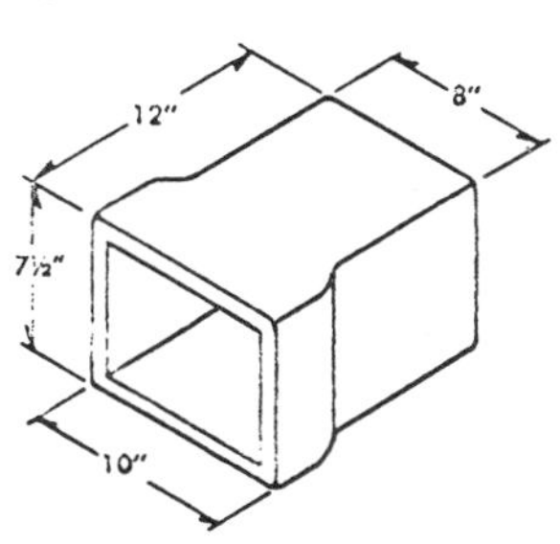

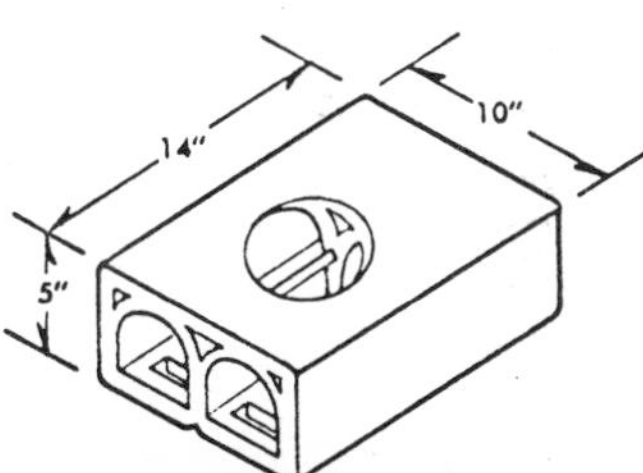

UNIT DIMENSIONS

Full Block **Half Block** **Quarter Block** **Half Block**	7½" x 10" x 14" 7½" x 10" x 7" 7½" x 10" x 3½" 7½" x 6" x 14"	USED IN FILTER BODY
Reducer Block **Reducer Extensions** **Angle "A" 20°** **Angle "B" 30°** **Angle "C" 45°** **Angle "60°"** **Angle "D" 75°**	7½" x 10" to 7½" x 8" 12" long 7½" x 8" x 6"-8"-12" lengths USED AT FILTER PERIPHERY	

- Cover Block in 14", 16", 18", 21", 24" and 30" Lengths used over center and ventilation channels
- Vent Block—14", 16", 18" Lengths to support vitrified clay vent pipes

NOTE: Block apertures are 3" x 1¼"

BOSCO

Standard Trickling Filter Underdrain Block

DESIGNED TO ASTM C159 TYPE 1 S
APERTURE AREA: 30% of Block Face
CHANNEL AREA: 50% of End Section— 33.5 Square Inches per foot of filter width

Channel alignment is assured with Bosco block. In installation, keep the groove in the face of the block in alignment with the groove in the preceding unit.

Care should be taken to see that all cover block are installed on even bearing. Quarter block or common brick may be used to provide support.

These special shapes allow block rows to conform to wall curvature—ventilation channels may be kept constant to allow free air flow to all block channels. Bosco provides installation diagrams for placement of these shapes in underdrain system.

Vent block shown provides support for ventilation pipe—allowing free flow of air into channel.

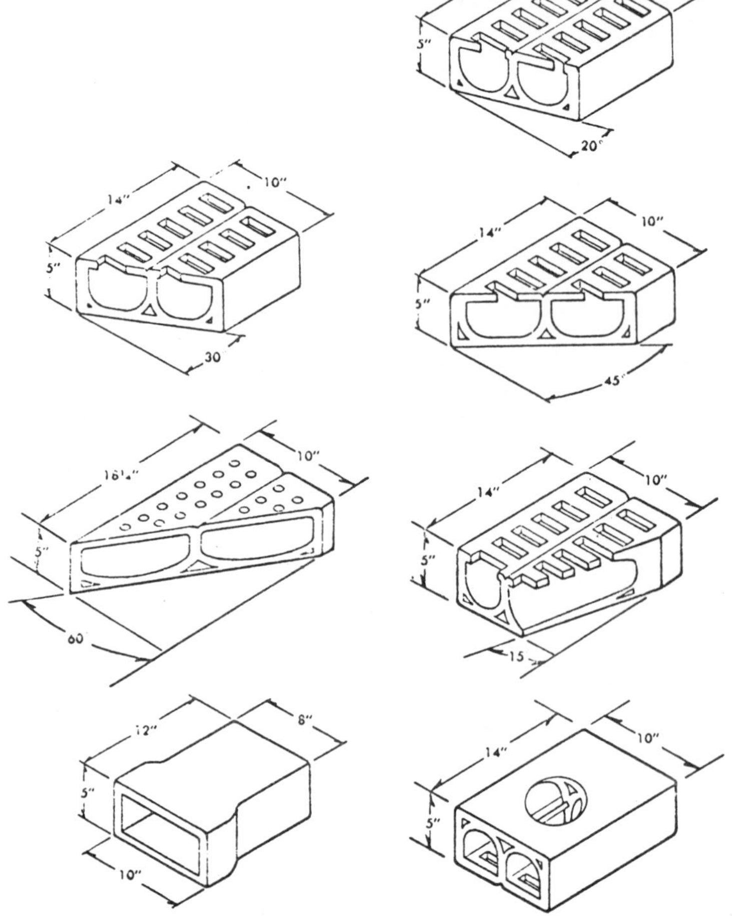

UNIT DIMENSIONS

Full Block **Half Block** **Quarter Block** **Half Block**	5" x 10" x 14" 5" x 10" x 7" 5" x 10" x 3½" 5" x 6" x 14"	USED IN FILTER BODY
Reducer Block **Reducer Extensions** **Angle "A" 20°** **Angle "B" 30°** **Angle "C" 45°** **Angle "60°"** **Angle "D" 75°**	5" x 10" to 5" x 8" 12" long 5" x 8" x 6"-8"-12" lengths USED AT FILTER PERIPHERY	

- Cover Block in 14", 16", 18", 21", 24" and 30" Lengths used over center and ventilation channels
- Vent Block—14", 16", 18" Lengths to support vitrified clay vent pipes

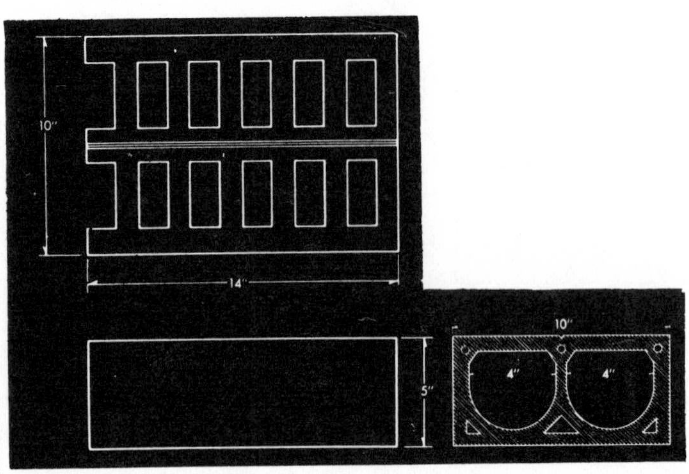

OZONEAIR

Economical, efficient system for water treatment and purification

Ozone is widely used in the treatment of drinking water.
As an oxidizer or purifier, Ozone is unequalled in effectiveness and economy.
It destroys disease-carrying germs and odors, usually in about one minute.
Ozone improves the taste of water by eliminating odor and color and by supersaturating it with oxygen. Its sterilizing action produces high quality water.
Ozone can be used in water treatment for:

- Disinfecting
- Deodorizing
- Decolorizing
- Taste removal
- Algaecide and slime control
- Iron and Manganese removal

TYPICAL APPLICATION METHODS

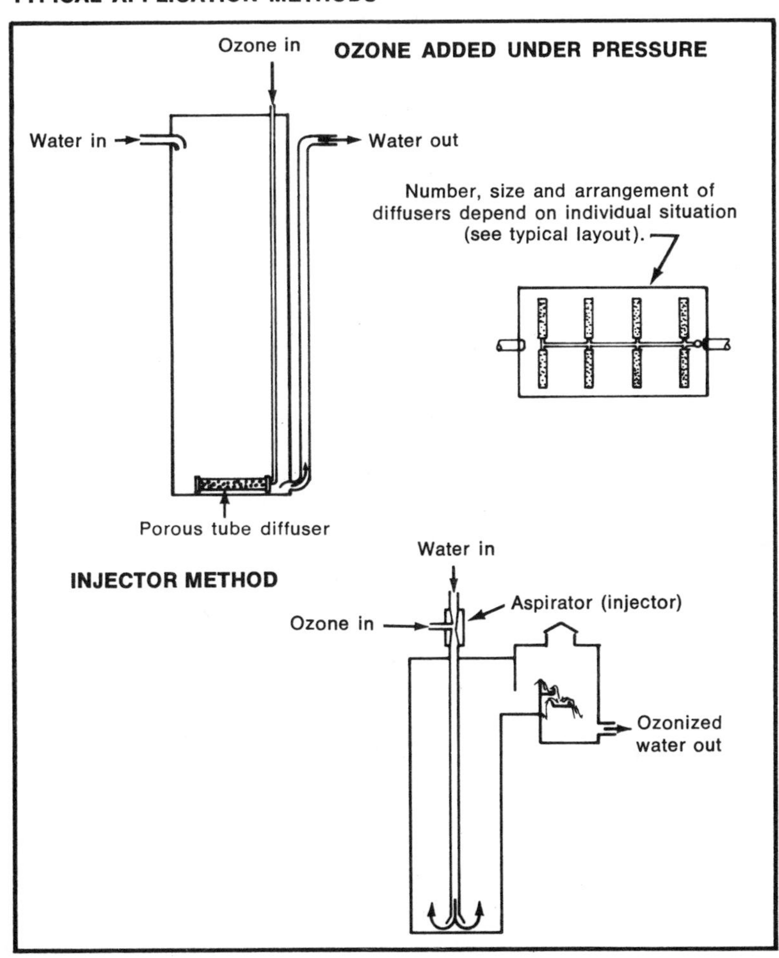

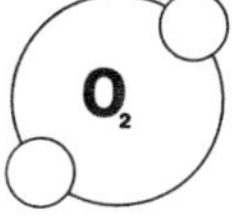

1. Oxygen molecule has 2 atoms (O_2).

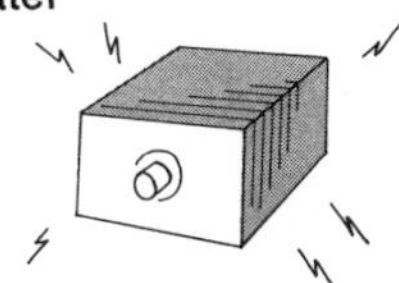

2. Silent electrical discharge is produced by applying controlled voltage to ozone generator.

3. Electrical discharge strikes oxygen molecule (O_2); splits it into 2 free atoms (O+O).

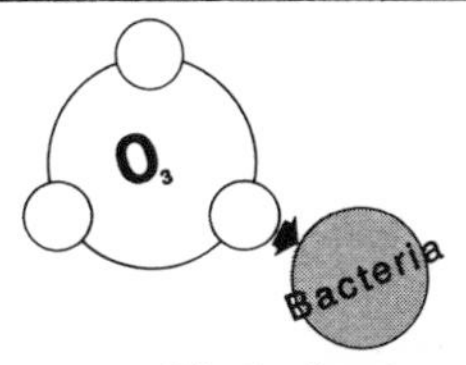

4. Free atom (O) attaches to an unsplit oxygen molecule (O_2) producing ozone (O_3) which attacks and oxidizes bacteria.

5. Bacteria is eliminated Remaining ozone (O_3) leaves a beneficial oxygen residual as a reaction product.

C–E Bauer

CENTRI-CLEANER LIQUID CYCLONES

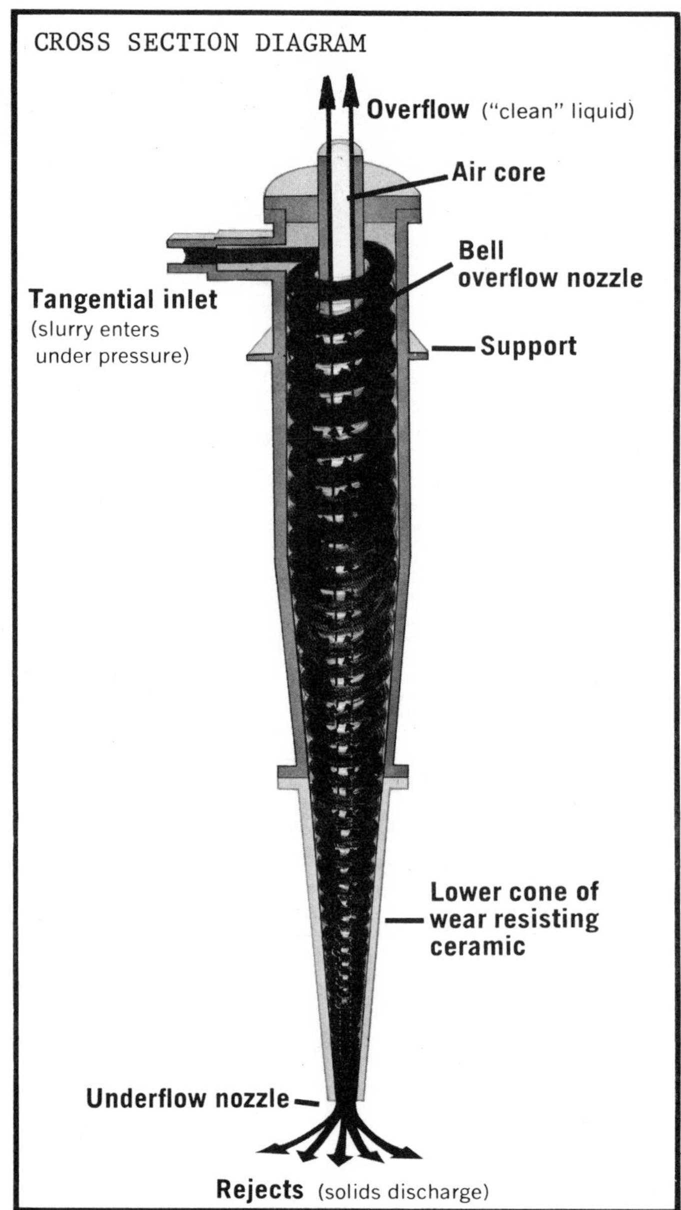

PRINCIPLE OF OPERATION

Bauer Centri-Cleaner Liquid Cyclones operate on the free vortex principle to take advantage of highest possible centrifugal forces. The phenomenon of the free vortex causes fluid velocity to accelerate rapidly and uniformly in the open conical section. Since velocity increases with relation to radius ($V_1R_1 = V_2R_2$), the force producing classification is multiplied tremendously.

Maximum centrifugal force is reached at the bottom orifice where heavier "rejected" material is discharged. In a 3" cleaner this force approaches 7500 times that of gravity. These smaller units effectively remove settleable matter as fine at 5 microns.

Bauer cleaners consist essentially of a hollow truncated cone joined at the top to an overflow nozzle. The upper cone has a tangential inlet where the slurry enters under pressure. During operation, incoming material spirals downward. As separation takes place, the lighter fractions are entrained in an inner vortex, spiral upward and exit through the centered overflow nozzle. Heavier fractions (solids) are forced down the wall in an accelerating outer vortex and are discharged through the underflow nozzle.

Located at the truncated apex of the lower cone, the underflow nozzle may discharge to atmosphere or be submerged.

C-E Bauer

HYDRASIEVE

The patented C-E Bauer Hydrasieve is a simple, highly efficient screening device for removing solids from low consistency slurries. Exclusive design features and the unit's ability to operate continuously for extended periods without attention make the rugged Hydrasieve a reliable profit builder in a broad range of fluid removal applications.

SEWAGE TREATMENT PLANT—Sewage influent in a suburban water pollution control plant is screened on three Hydrasieves to remove solids which would otherwise cause operational problems. The original objective was to remove at least 20% of the coarse, raw suspended solids before the influent reached the oxidation tower.

After parallel rotary screens failed to meet the plant's requirements, a Hydrasieve was tested. This was so successful that three No. 552-72" units are now in operation, handling 5 MGD. Average suspended solids removal is 33% and the reduction in B.O.D. is 31%. The need for a primary clarifier has been eliminated.

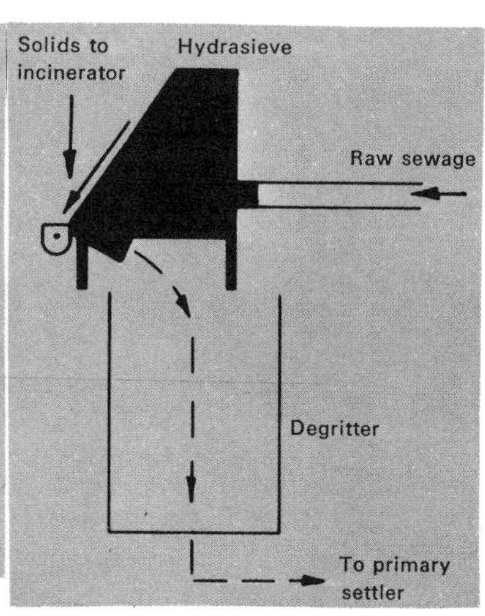

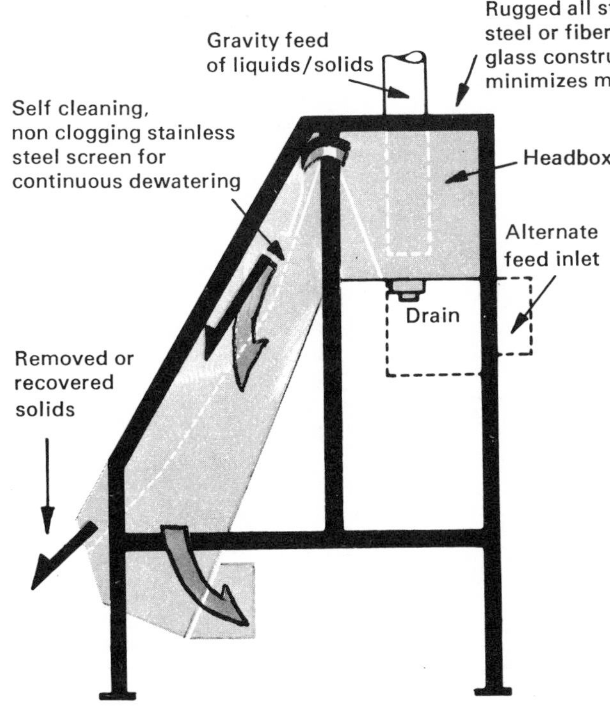

"Cleaned" effluent improves plant pollution control efficiency, lowers B.O.D., reduces sewage rates

Installations. Waste water treatment and pollution control systems. Municipal sewage plants. Chemical, plastic, and ceramic classification. Synthetic and natural fiber recovery. Salvaging rubber fines. Processing soup ingredients, fish, citrus fruits, etc. Recovering hog hair and other valuable solids for meat and hide processors, and similar operations.

CAPACITY CHART U.S. GPM
(Figures are approximate)

MODEL (and size)	552-18″	552-36″	552-48″	552-60″	552-72″
Storm Water (0.060″ screen)	150	350	600	800	1000
Kraft Pulp Mill Effluent (0.020″ screen)	100	200	300	450	600
Packing Waste (0.040″ screen)	70	150	280	420	550
Sanitary Sewage .05% cons. (0.060 screen)	150	300	500	650	800
Canning Waste (0.060″ screen)	120	200	350	500	650
White Water 0.1% cons. (0.020″ screen)	130	320	500	625	775
Kraft Rejects 0.5% consistency (0.040″ screen)	100	200	300	450	600
Kraft Liquor 0.5% consistency (0.030″ screen)	40	60	100	125	150
Groundwood Screen-ings 1.0% cons. (0.040″ screen)	75	175	300	400	500
Press Liquor 0.5% consistency (0.010″ screen)	70	170	260	350	425
Sulfite Pulp 1.5% cons. 350 CSF (0.010″ screen)	50	155	240	310	335
Secondary News 2% cons. 200 CSF (0.030″ screen)	45	115	185	240	300
Secondary Fibers Mixed .5% cons. (.020″ screen)	90	180	260	330	400
Average Nominal Capacity (Based on 0.040″ screen)	100	200	300	450	600

BRIEF SPECIFICATIONS

Model	552-18″	552-36″	552-48″	552-60″	552-72″
Width	22″	42″	54″	66″	78″
Height	57″	60.5″	84″	84″	84″
Depth	42″	44.5″	61″	61″	61″
Inlet	16″x18″	16″x36″	18″x48″	20″x60″	12″x72″
Outlet	8″ diam.	10″ diam.	10″ diam.	12″ diam.	14″ diam.
Weight	350 lb.	550 lb.	650 lb.	800 lb.	1000 lb.

Note: The largest Hydrasieve requires less than 50 square feet of floor space.

OPTIONAL CONFIGURATIONS (48″, 60″, 72″ models only)

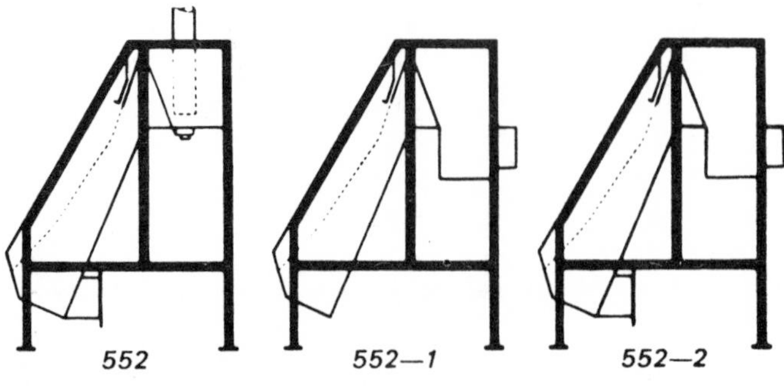

STANDARD 552 Standard headbox, flanged effluent discharge
552—1 Deep headbox, wide effluent discharge
552—2 Deep headbox, flanged effluent discharge

STRIPPING TOWER FOR OXYGEN REMOVAL

Many processes require oxygen-free water. National Tank Company has fully developed several deaerator vessels for this application.

Oxygen content in water can be reduced to 0.5 ppm by using natural gas or an inert gas stripping effect in a tray type tower, counterflowed with the water. This unit is simple to operate and has lowest initial cost.

Chemical scavenging with sodium sulfite or hydrazine will provide oxygen free water.

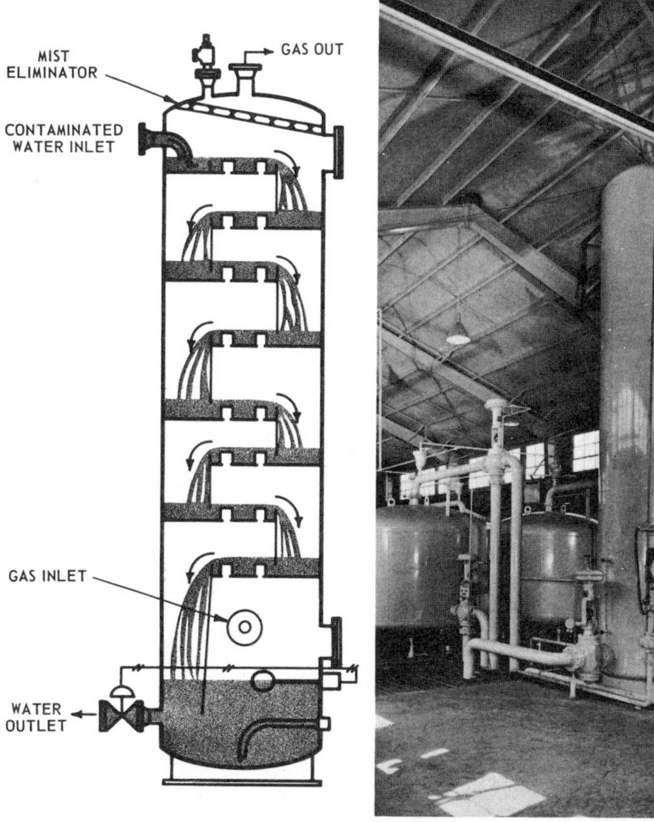

National Deaerator utilizing natural gas to strip oxygen from water.

STRIPPING TOWER FOR HYDROGEN SULFIDE REMOVAL

A unique application of the stripping gas (or desorption) phenomena was utilized to remove hydrogen sulfide gas from water solution for a West Texas oilfield requirement.

A "sweetened" methane gas stream was used to effect the water stripping effect in lieu of oxygen. It minimized corrosion and maintained a closed system. Hydrogen sulfide content above 300 ppm was reduced to less than 5 ppm in this process.

Alloy tray parts, plus full vessel internal corrosion-resistant coating of carbon steel water-wetted surface was furnished as part of the system.

Chemical scavenging with chlorine can replace the hydrogen sulfide content to zero ppm.

SPRAY TYPE DEAERATING HEATER

Also for noxious gas removal, National Tank Company offers a spray type deaerating heater. Water temperature is raised by intimate contact with steam. The initial heating reduces the gas content below 0.3 ml/liter. The final scrubbing reduces the gas content to 0.005 ml/liter or less. Carbon dioxide methane, and nitrogen are removed along with oxygen, and a closed system is maintained. This type of equipment is most economic if waste steam heat is available in sufficient quantities.

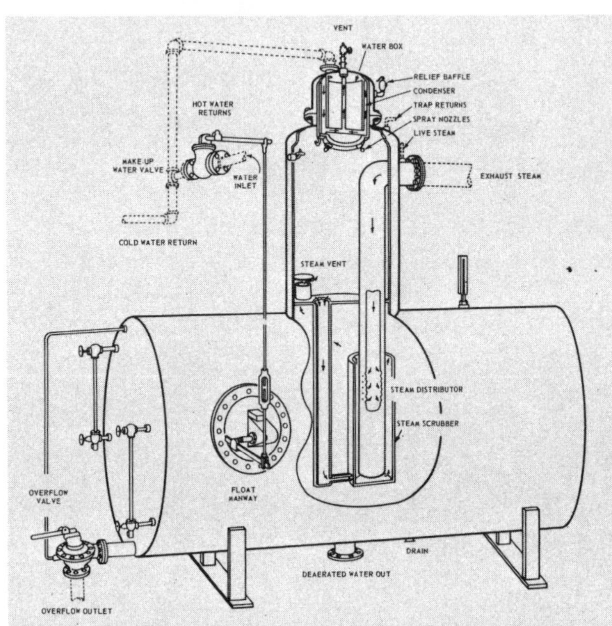

VACUUM TOWER SYSTEM

Processes requiring oxygen-free water where no stripping gas or waste steam heat is available can be supplied by utilizing an evacuated type deaerating vessel. Oxygen content is lowered to less than 0.5 ppm by reducing the internal "atmosphere" of the vessel, in which water is passed downward over an arrangement of packing to provide a continuous cycle of thin surfaces.

A vacuum pump, or steam jet ejector reduces the vessel's internal atmosphere to within a few-hundredths mm Hg of the vaporization point of the water bath, depending upon water temperature, to effect solution gas removal.

The evacuated type deaerating vessel requires elevation of the vessel a minimum of 35 feet to provide NPSH for pump withdrawal from the tower. This element plus the vacuum pump component makes the evacuated system slightly more costly and more complicated to operate than a stripping gas system.

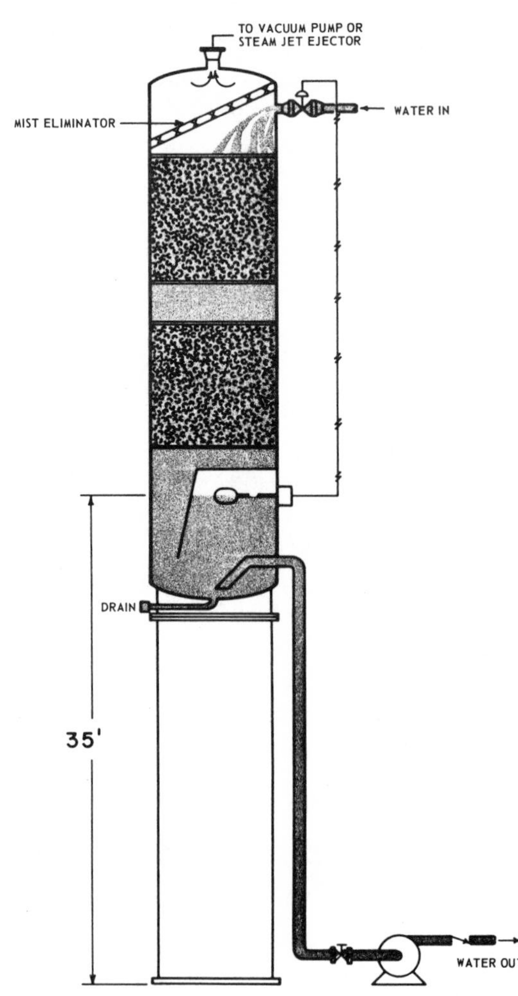

MECHANICAL SPX - SUSPENDED PARTICLE EXTRACTOR

C-E Natco's Mechanical SPX is a four stage flotation system in which flotation and removal of oils and suspended particles take place within each of the four sections. Air or other gases are used as the flotation media.

In each of the stages, the flotation gas is conducted to the impeller-diffuser section through an annular area between the impeller shaft and the surrounding casing. This gas is then recycled within the unit by forced and controlled means to ensure the proper quantity is introduced into the system. When the gas reaches the bottom of each flotation section, an impeller shears this gas into a range of small bubble sizes. As the impeller-diffuser system creates the gas bubbles it also agitates and circulates the water in each cell to provide exposure of the particles to a great number of gas bubbles. This system promotes the attachment of gas bubbles to the suspended particles. The particles with attached gas particles are buoyant and will rise to the surface. Water continues onto the next section where the process is repeated. Each section removes additional suspended particles.

By momentary interruption of the outlet water stream on an automatic and adjustable time cycle basis, the floated oil and other suspended particles are conducted into the oil compartment for discharge from the unit.

Because of the agitation within each individual section, the particles too heavy to float receive abrasion and scouring among themselves. This action can remove binding compounds on the particle that would help keep it oil-wetted. Without the binding agent present, the particle becomes water wetted, and can be suspended in the water system. Since the agitation within each section does not permit accumulation of the heavier particles, they progress from compartment to compartment with the water and receive the scouring action as they move to the outlet. Due to this action, the heavy solid particles leaving the unit with the water are freed of much of the oils that wetted these particles before they entered the unit.

The unit may require the addition of a flocculant or coagulant to produce the desired quality effluent water. Conditions, both inlet and desired outlet, dictate the use of chemical aids. The proper type, quantity and method of introduction are all important in producing the desired results.

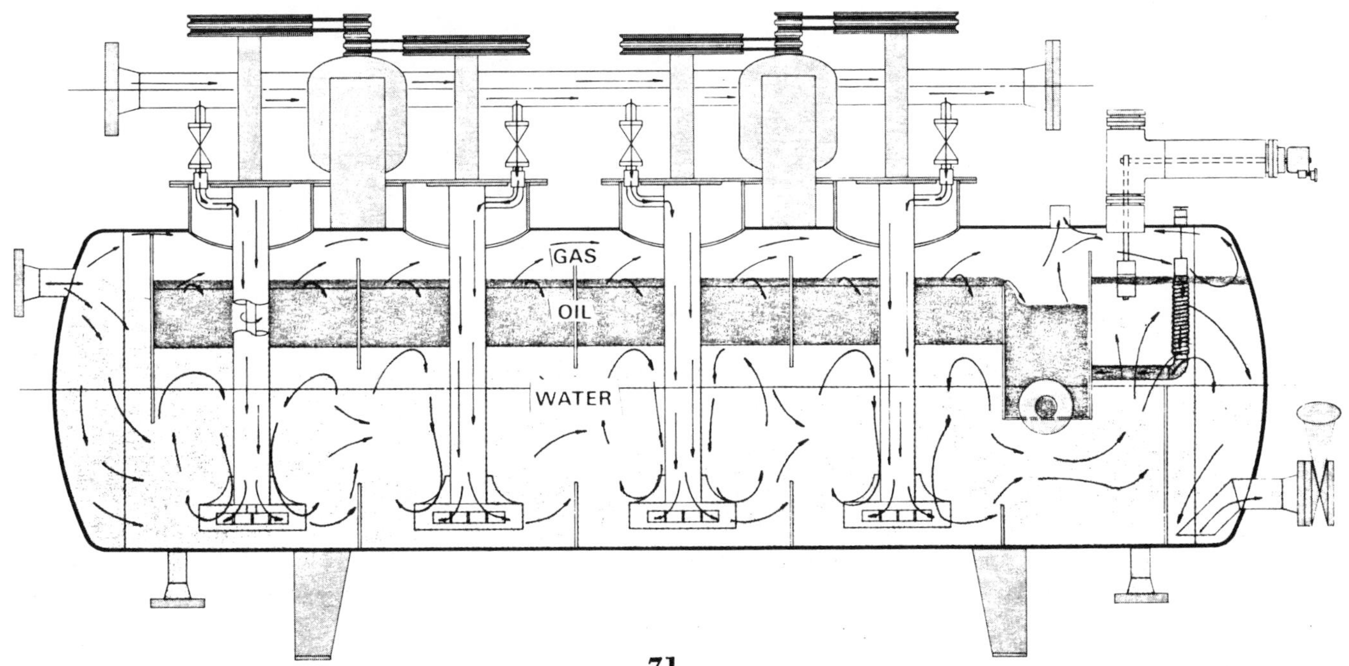

SUSPENDED PARTICLE EXTRACTOR (SPX)

OPERATION

In the National Tank Company SPX System each gallon of polluted water is saturated with millions of microscopic sized gaseous bubbles to assure a high probability of bubble attachment to each and every particle suspended in the water. Bubbles are generated by a unique process including a pressurized gas saturated recycle stream.

Polluted water enters a central flume of the vessel (Refer to flow diagram). A coagulant may be added prior to entry at the flume when required. Polluted water travels up the flume and becomes saturated by the gas bubbles. Particulates are carried to the top by the attached bubbles where a rotary scraper pushes them into the oil outlet trough. Proper water level within the tank is controlled by means of the liquid level control and throttling valve in the effluent water line. The conical bottom and sand drain provide a discharge route for sand and non-floatable debris.

To generate the bubbles for flotation, a side stream of clean water is taken and pressured through a pump into the gasification tower. Air, Nitrogen, or Natural Gas is supplied to the tower and is absorbed by the water in the packing. The gas laden water from the bottom of the tower is applied to the flotation unit by a precision turbine flowmeter set point controller.

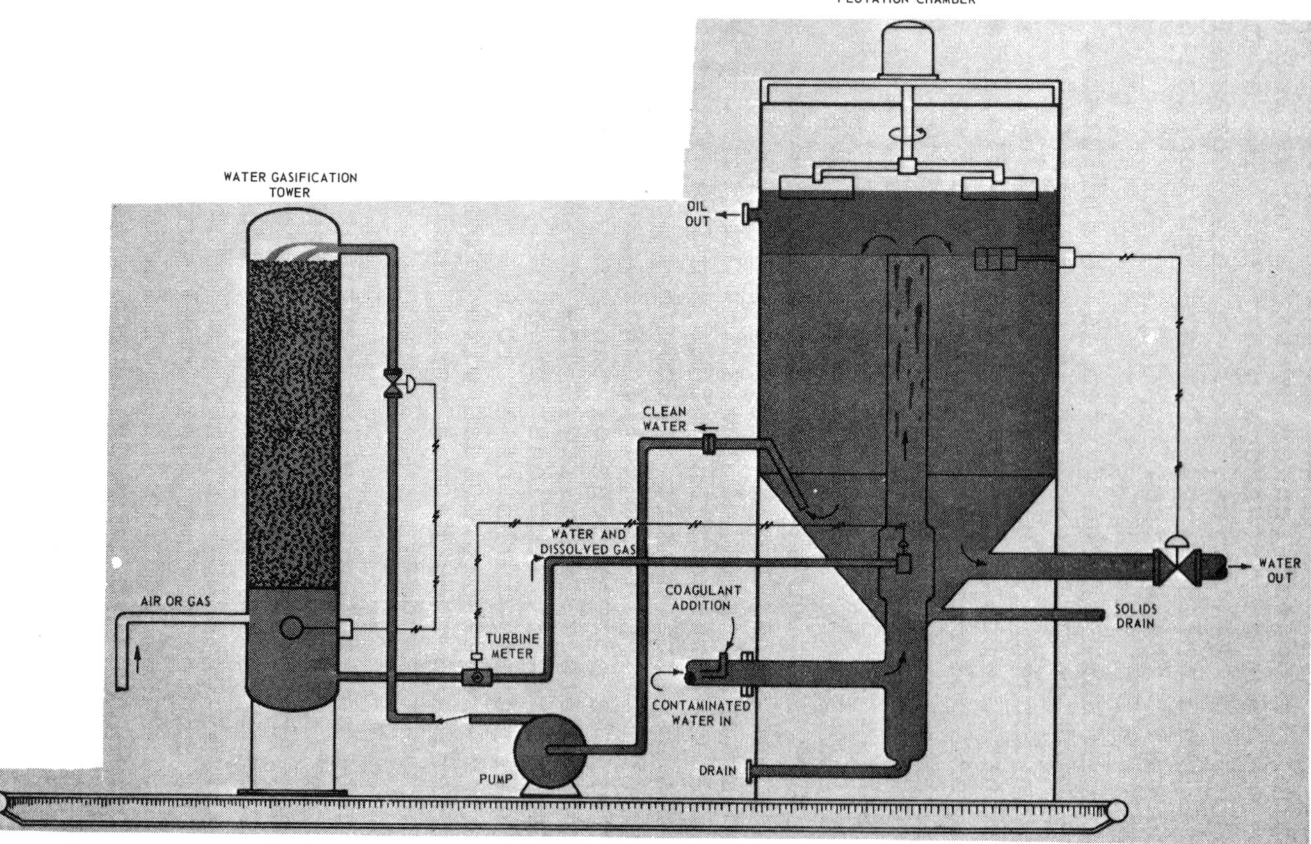

C-E Natco

DISSOLVED GAS SPX - SUSPENDED PARTICAL EXTRACTOR

This Dissolved Gas SPX system was developed by C-E Natco to overcome many of the difficulties and short-comings of other types of dissolved gas flotation systems. This system will ensure that the gas bubbles released are of a much smaller size and of greater quantity. It also ensures, through mechanical arrangement, that there is a greater contact time and opportunity to allow attachment of these minute gas bubbles to the suspended particles to be floated.

The gas required for this flotation process is introduced into the system by the recycle of a relatively small portion of the clean effluent which is pumped into a vertical packed absorber column. This absorber column is held at a suitable elevated pressure so that the required amount of gas can be absorbed for the flotation need. Column packing ensures that a high efficiency of absorption is attained, which minimizes pressure and quantity of recycle water.

The most effective gas bubbles are those that are the smallest and these are obtained by allowing the dissolved gas to come out of solution under the unique mechanical conditions of the system. It is important to allow these microscopic gas bubbles to attach to the particle to be removed as quickly as possible after the bubble forms and before they grow in size by coalescence. The SPX system is unique and exclusive in providing a period of isolation after expansion of the pressurized recycle stream during which bubbles of microscopic size are formed. After their formation, the small bubbles are placed in immediate contact with the particles in the inlet flume.

The high contact between the oils and suspended solids and the microscopic size gaseous bubbles take place within the inlet flume of the SPX. The vertical inlet flume also serves another important purpose in that the inlet water to the SPX is directed upward. The upward velocity imparted to the particle greatly aids in flotation by bringing the particle to the surface of the water. Without this flume, the particles must reach the surface through flotation alone.

In contrast to systems that pressurize the entire inlet stream, the benefits of the SPX system are that only the pump horsepower required to pressurize the recycle stream (normally less than 20% of the SPX design capacity) is needed. By using the clean effluent water from the process for recycle, the chance of additional dispersion of the suspended oils and solids that could take place within the recycle pump is minimized. If a chemical aid is required for flocculation, the floc in the inlet stream formed by the chemical aid is not degraded by severe nor violent handling.

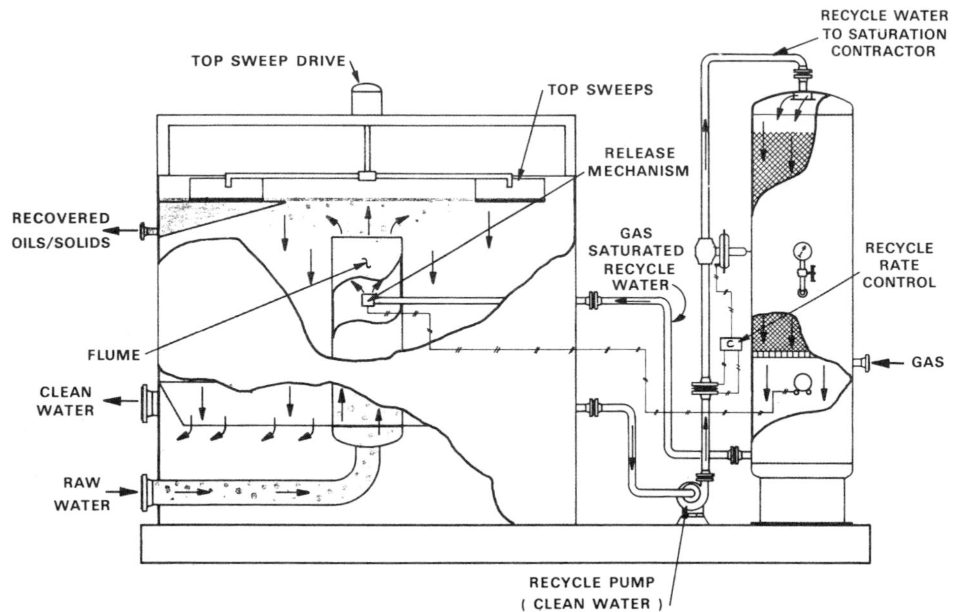

HIGH RATE FILTERS

C-E NATCO's High Rate Filter offers suspended solids filtration PLUS (1) high filtration rates, (2) full bed loading, (3) small backwash volume requirements, (4) downflow filtration, preventing possible bed expansion, and (5) excellent cleaning of the media upon backwashing.

The High Rate Filter is a downflow, single grade media bed filter that will remove suspended particles at throughput rates of approximately four times that of the conventional graded bed downflow filters.

The filter bed is divided into two sections with accelerator spheres surrounded by media in the upper section and media only in the lower section. A grate separates the two sections.

The accelerator spheres serve to increase the velocity of the particles before they contact the media interface. The resulting principle is that the particles gain sufficient kinetic energy to penetrate the media interface rather than building up a filter cake on the surface. Once within the media, the particles come to rest due to frictional forces. Particle deposition continues until the deep bed is literally saturated with the particles being filtered. The filter is ready for backwash just before this saturation point is reached.

For backwashing, air or other gas along with an upflow of backwash water is introduced below the bottom bed support. The gas flowing up through the bed causes severe agitation and scrubbing together of the media particles, releasing the solids removed during filtration. The backwash water merely carries the material, released by gas agitation, from the bed. A screening device prevents the media from leaving the filter, but allows the filtered solids to leave the filter. This method of severe agitation during backwash cleans the media and allows this type filter to be used for the filtration of sticky or tacky materials, such as paraffin particles. The backwash water may be filtered or non-filtered water; the rate is usually one-half that of the filtration rate. The backwash cycle is normally three to ten minutes in duration.

Depending upon the particulate matter to be removed from the water, one or more types of media may be recommended. Most types are high density, hard, granular material with angular, sharp edges. Pressure drop through the filter is minimal and normally reaches 10 to 15 psi differential before backwashing is required. The filter can be fully automated for automatic backwashing on a time, pressure drop, or turbidity measurement basis. The High Rate Filter is capable of removing solid particles greater than the 5 to 10 micron range in most cases without the use of chemical aids. The filter is capable of removing particles of substantially smaller size with the use of chemicals.

C-E Natco offers shop assembled and packaged water filtration systems with wide capacity ranges. Physical sizes can be tailored for most space requirements. Vessels, instruments, controls and piping can be manufactured to various codes, regulations and requirements.

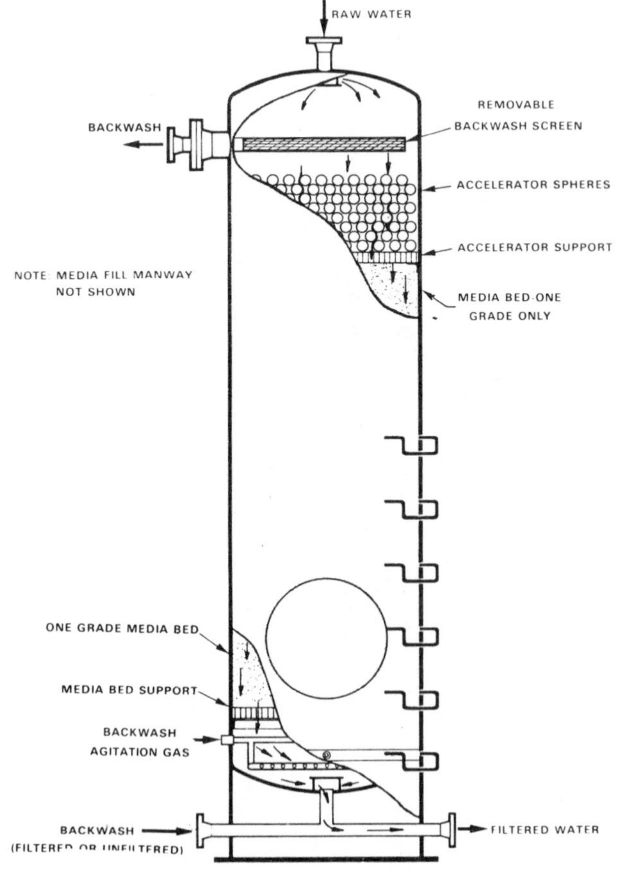

DOX - DISPERSED OIL EXTRACTOR

C-E Natco's DOX System is designed for the separation of oil-in-water dispersions which are normally considered too "tight and stable" to separate by conventional methods. The DOX process promotes the agglomeration of the fine oil droplets and then separates these large particles from the water.

Process water from various sources may contain highly dispersed oil droplets too small to rise to the surface of the water in a reasonable length of time. Some oil-water mixtures contain oil droplets of such small sizes that the particles fall into the classification of colloids. Although the difference in density of the two liquids should cause them to separate, the separation does not occur because of the minute particle size. The DOX can bring about the separation of the two liquids within minutes.

This DOX system consists of two vessels: the Column and the Separator. The DOX Column contains a fixed bed of treated granular media. Flow is downward through this material. The oil passes from the bottom of the column in globules many times the original droplet size.

The stream containing these oil drops then flows into the DOX Separator. Sufficient quiescence and fluid retention time within this vessel allows the lighter oil to rise to the top where it is collected and removed. The water is withdrawn from the lower part of the vessel.

Where larger oil droplets are needed to increase gravity separation efficiency, the DOX separation vessel may contain an optional secondary coalescing section. This section further increases the average size of the oil globules. The water containing these larger oil masses flows into the quiescent section of the vessel where gravity separation occurs.

The DOX media does not require regeneration or replacement unless fouled by foreign material.

The amount of oil remaining in the separator water outlet is a function of many variables, therefore each application must be evaluated individually. The system is capable of producing effluent water acceptable for most discriminating uses and for plant recycle uses.

Pressure drop through the system is quite nominal and may be designed for low pressure drop requirements. The system can be automated and may have pneumatic, electric or mechanical controls and valves.

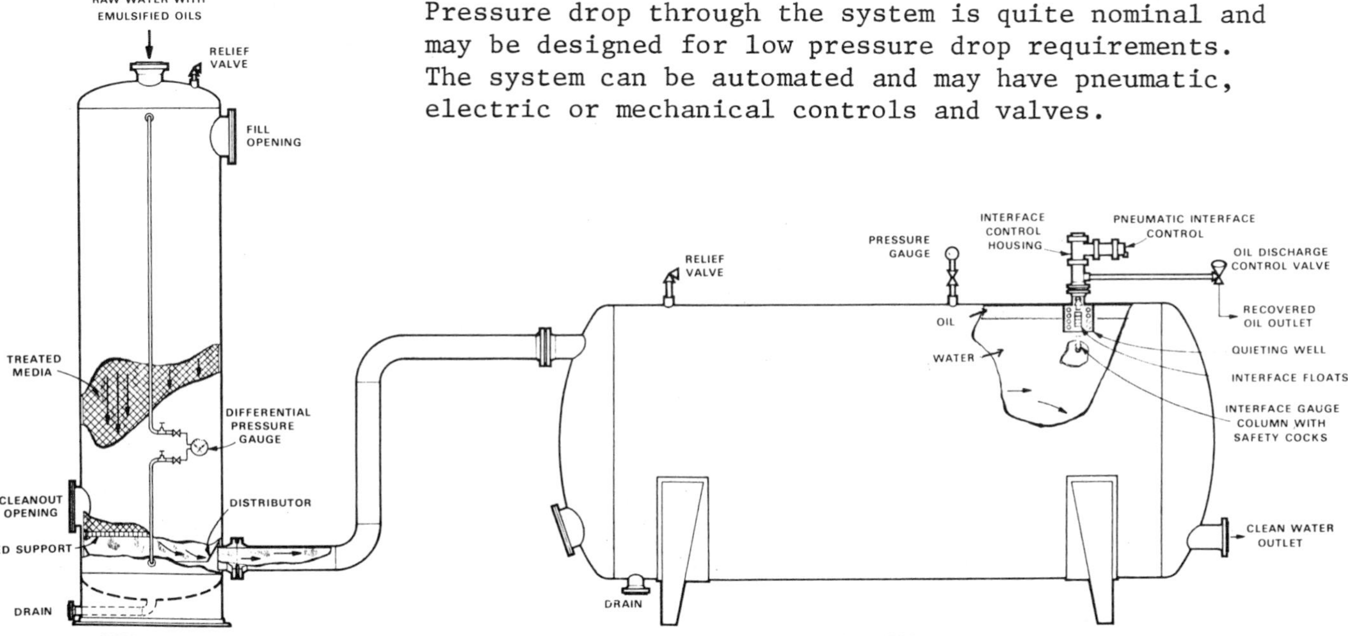

OIL REMOVAL

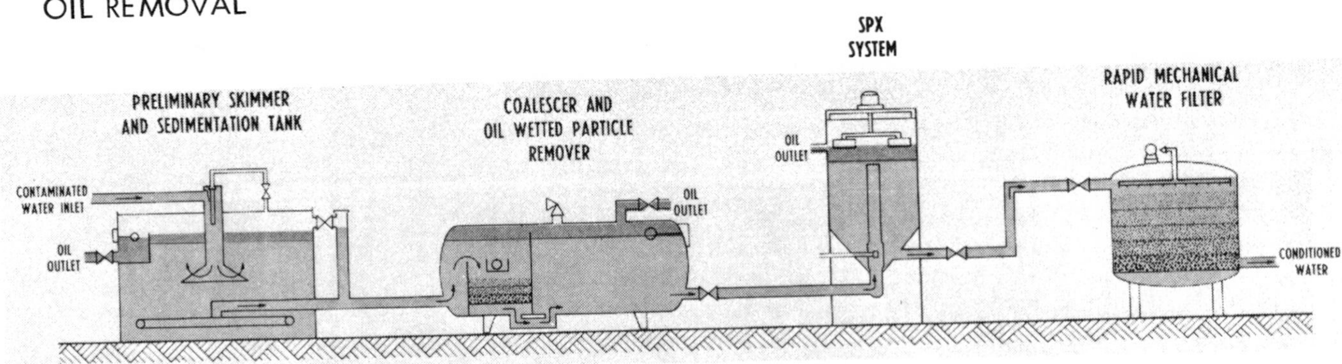

This equipment arrangement is suggested for removing suspended oil particulates from waters containing suspended (and oil wetted) solids such as iron sulfide or sand. The dual-compartment coalescer design illustrated is limited to throughout rates through 15,000 BPD.

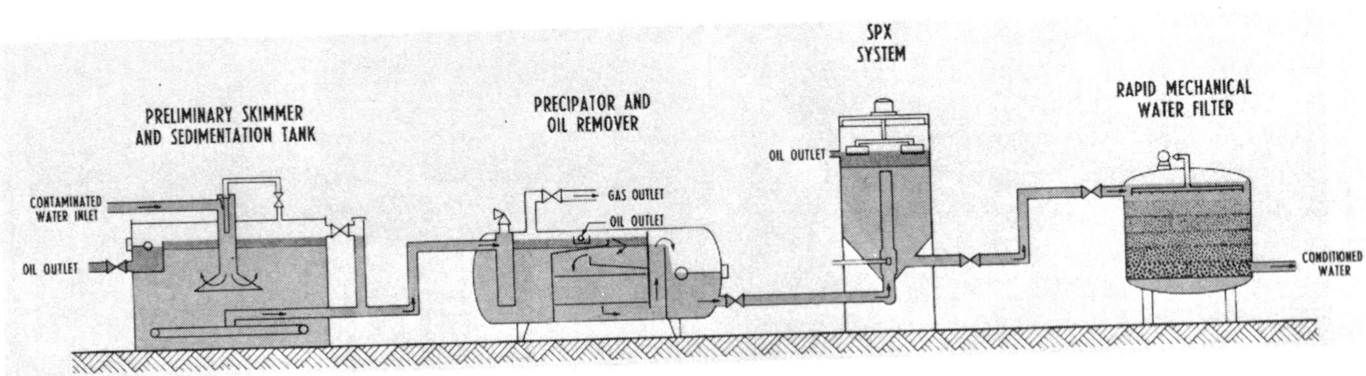

This equipment arrangement is suggested for removing suspended oil particulates from waters containing **no suspended solids or has any scale forming tendencies.** Each piece of equipment illustrated may be optional, depending upon the nature of suspended oil content in the raw water stream **and** the desired level of suspended oil content in the total system effluent.

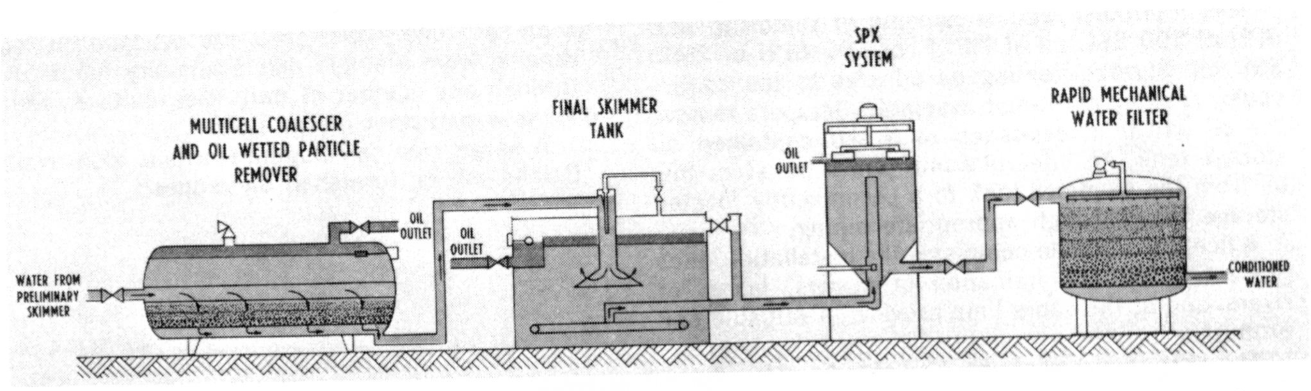

Here a repeat of the coalescer vessel equipment arrangement is illustrated for throughout rates above 15,000 BPD. These higher rates require the utilization of two separate vessels in coalescing and skimming normally provided in a single vessel at lower rates.

SKIMMER AND SEDIMENTATION TANK

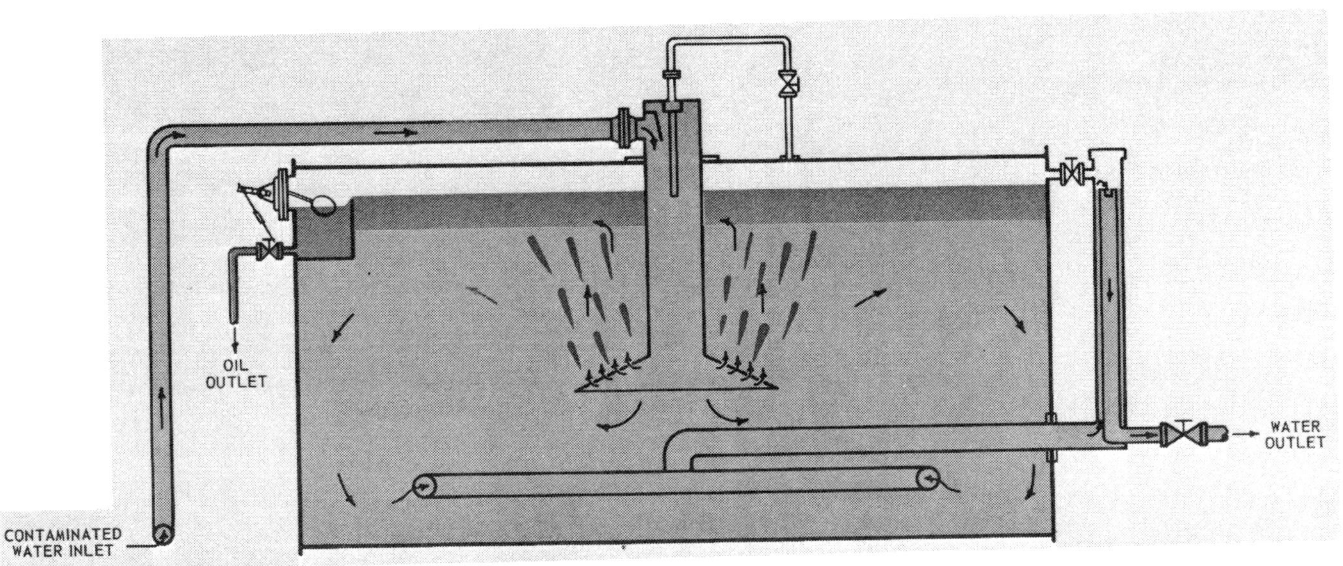

PRECIPITATOR AND OIL REMOVER

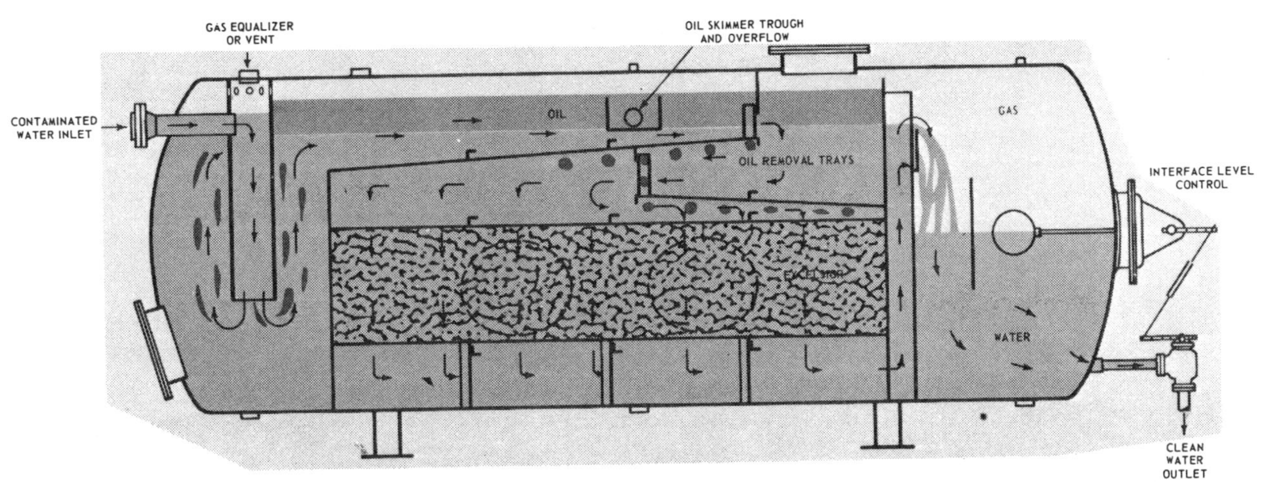

COALESCER AND OIL WETTED PARTICLE REMOVER

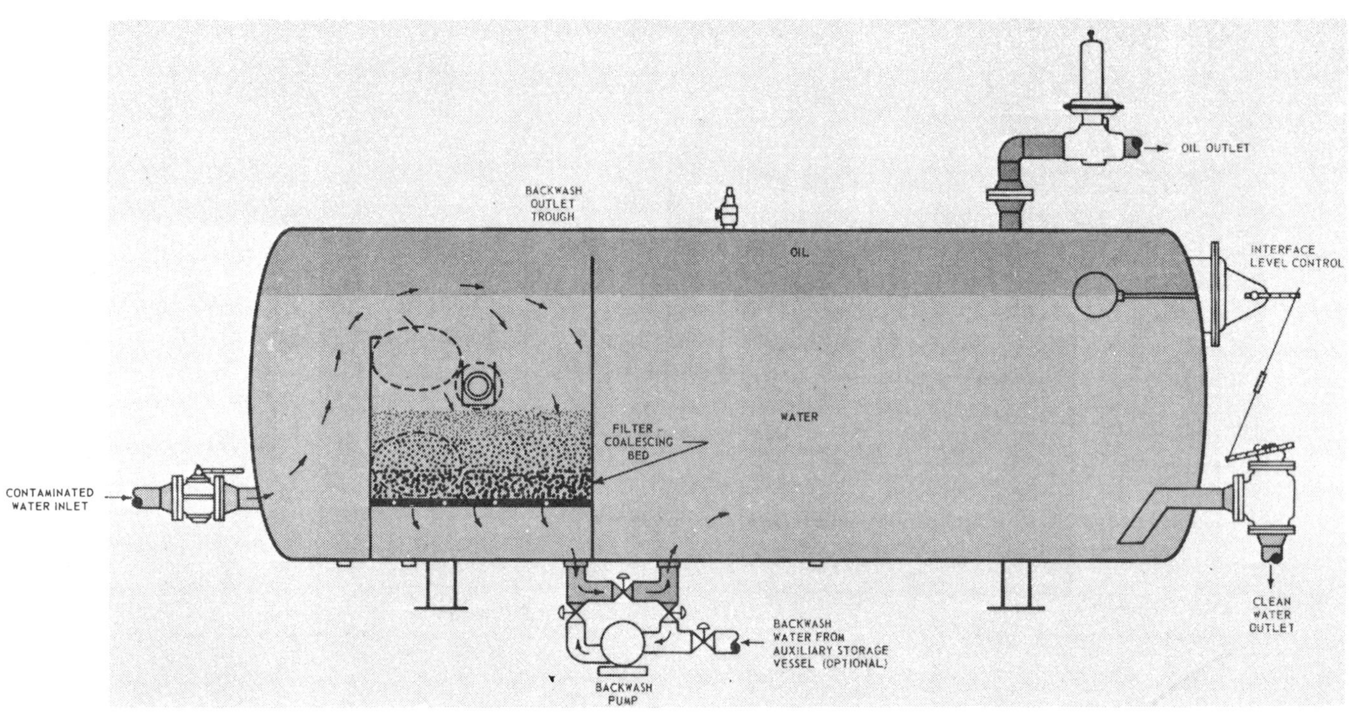

MULTICELL COALESCER AND OIL WETTED PARTICLE REMOVER

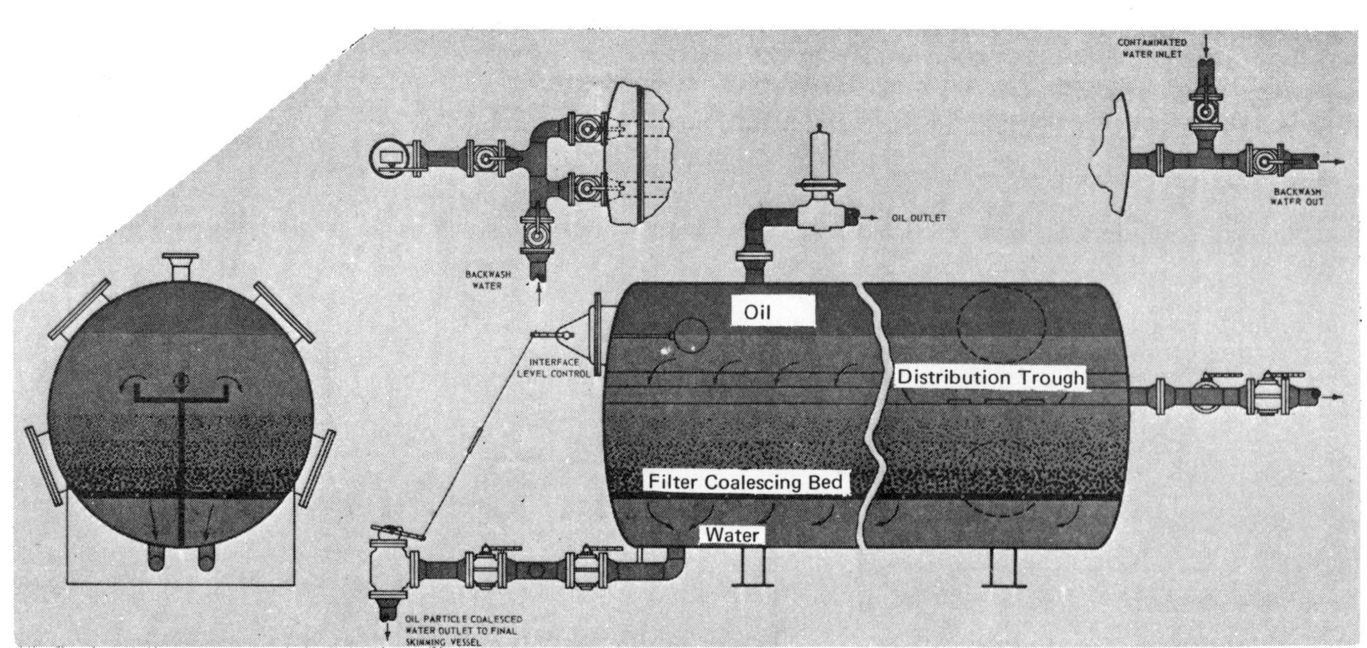

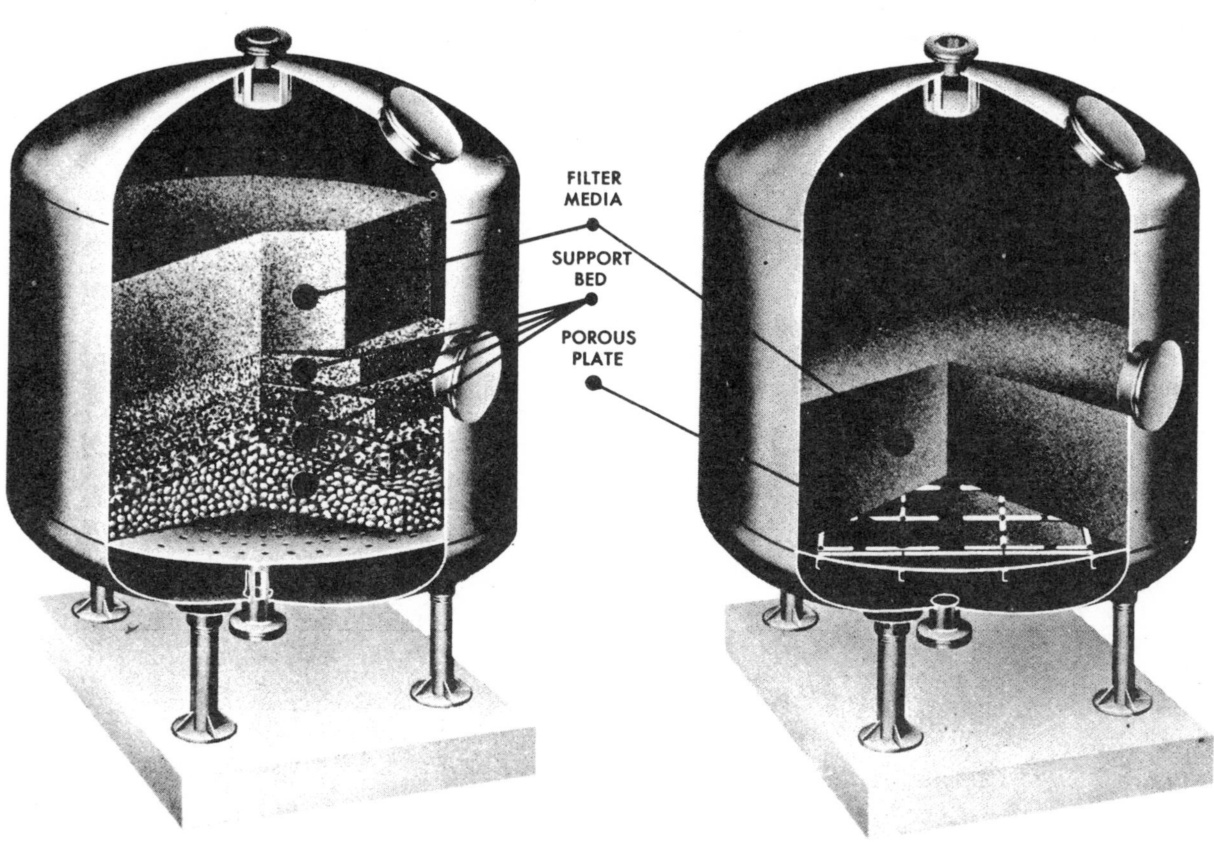

Cutaway views of the two common types of National Mechanical Pressure Filters show the graded support bed and the alternate fused porous aluminum oxide support. Alternate inlet and outlet connection locations are available to fit the customers' requirements.

Pressure Drop: Within the operational capacities of either type filter, the pressure drop through clean filters after backwash cycle is less than six inches of water. Pressure drop increases with fineness of top layer of filter bed media and filter cake deposition. This is essentially the same in either type.

Backwash Rates: Backwash rates should be ample to expand the fine media section by at least 50%. Supporting structures or media are not disturbed at such rates. This is the same for either type when the filter-media is the same material.

Quality of Filtered Water: Quality of water is associated with filtering ability. Filtering ability has to do with fineness of top layer filter bed media and porosity of filter cake. The finer the media, the better the quality of water. Supporting beds have no measurable effect on the quality of water. This is the same through either unit when using same filter-media.

If aeration and retention time are not sufficient to remove dissolved solids and stabilize the water under treatment, **it is necessary to add chemical.**

Chemical treatment is best accomplished in a tank. Rapid and thorough mixing of chemical with the raw water is essential.

To follow the flow: The raw water is introduced into a central flume or the **primary mixing zone** which houses an impeller, which serves as a pump and a mixer. The flume is contained inside a hood which serves to separate the zone of clear water from the zone of slurry.

The impeller is driven by a **variable speed drive** and circulates the slurry through the flume at a rate of several volumes to every volume of raw water.

After intimate contact of the raw water with the slurry in the flume, it travels into the **secondary zone** (the annulus between the flume and hood) in a downward path and at a reduced velocity, where precipitation and agglomeration of particles takes place.

The heavier or larger particles settle to the bottom and are drawn off automatically through the anti-channel drain boxes. **(A feature of the National Unit.)**

The slurry that was not used in the agglomeration or accretion process is circulated to be used again.

The addition of certain chemicals will cause undesirable solids to precipitate and be removed. The analysis of the water will determine what chemical to use. In the process a certain amount of chemical is used so additional chemical is added to the primary zone to compensate for that used.

The treated water travels under and around the hood into an area of reduced velocity permitting any particles in the stream to settle out. The result is a stabilized clear water being drawn off to the filters.

In a tank is the best way:
- Rapid mix
- Gentle motion floc formation
- Quiet zone for solid settling

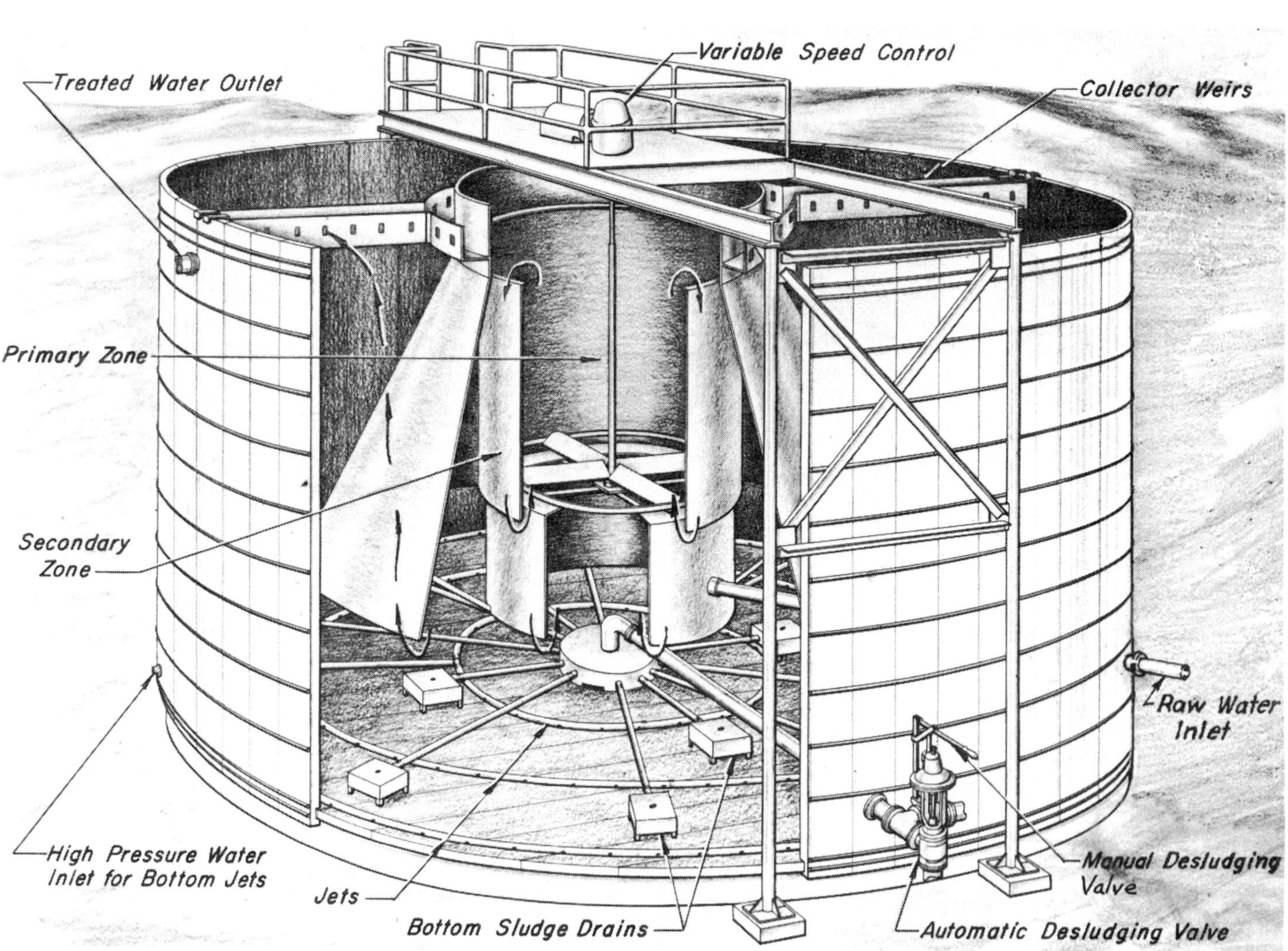

This cutaway view of a National Stabilization and Clarification system shows the various features of the unit, and the flow path of the water.

C-E Raymond

FLASH DRYING SYSTEM WITH SLUDGE INCINERATOR FOR PRODUCTION OF SOIL CONDITIONER

The C-E Raymond system of sewage-waste processing, with more than fifty-five plants installed to date, was developed by Combustion Engineering, Inc.. In this unique system, drying of the sludge is accomplished in a series of separate but closely coordinated steps which permit quick adjustment for handling different types of sludge, according to their moisture content and general characteristics.

The first step is preparation of the sludge. As the dewatered material delivered by the vacuum filters will not permit rapid, efficient drying, the sludge is conveyed to a mixer for proper preparation. Here the filter cake is mixed with previously dried sludge, with the percent moisture content being sufficiently reduced so that the resulting fluffy material can be rapidly dried.

Flash drying is the second step. This is accomplished in the turbulence of the cage mill where the new moist sludge is intimately mixed with drying gases at temperatures up to 1300°F. The thoroughly-dried sludge, together with the moisture laden gases, then goes to the cyclone collector, where it is separated from the spent gases. The entire drying action takes approximately six seconds, and the final dried sludge, delivered with no scorching or burning, is of the same chemical character as the dry solids of the original filter cake.

The third step is deodorization. During this drying period, some objectionable organic volatiles are vaporized and carried through the system with the spent gases. These odors, along with any organic particulate matter, are oxidized by raising the temperature of the gases to approximately 1300°F. in the special deodorization section of the furnace. Before emission to the atmosphere, these gases pass through heat exchangers where their heat is recovered by the system to obtain high system thermal efficiency.

Part of the dried sludge is returned to the mixer for blending with incoming wet filter cake. The rest of the dried sludge is then conveyed either to a holding bin to be used as soil conditioner or into the furnace to be used as auxiliary fuel.

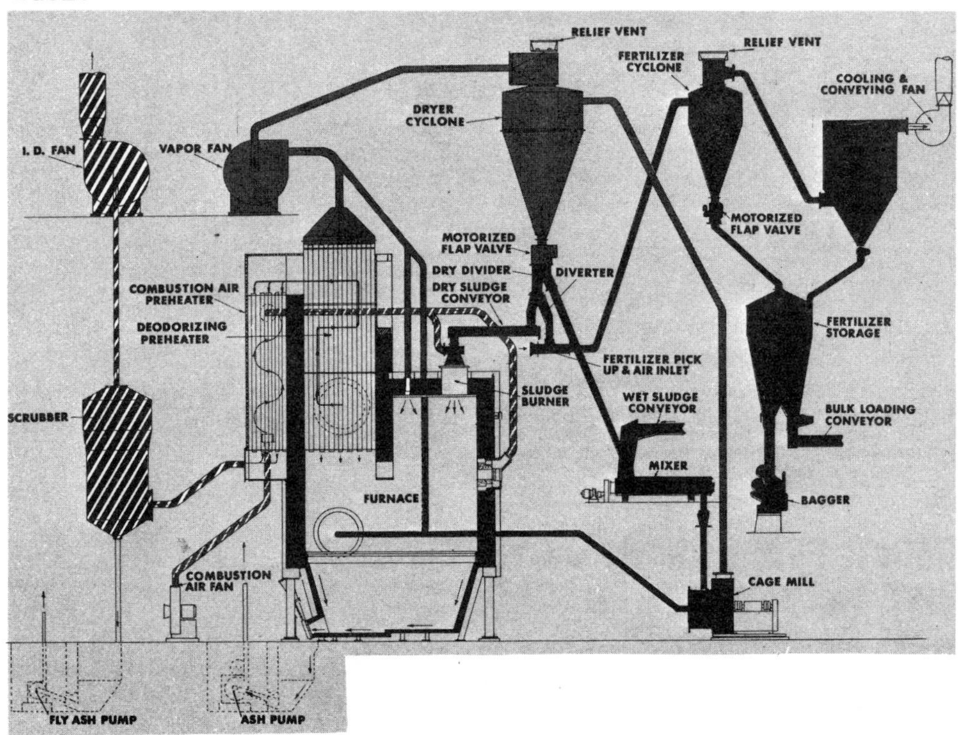

Calfilco

CALFILCO REACTIFIER --- TYPE R

COMPACT RECTANGULAR CLARIFIER

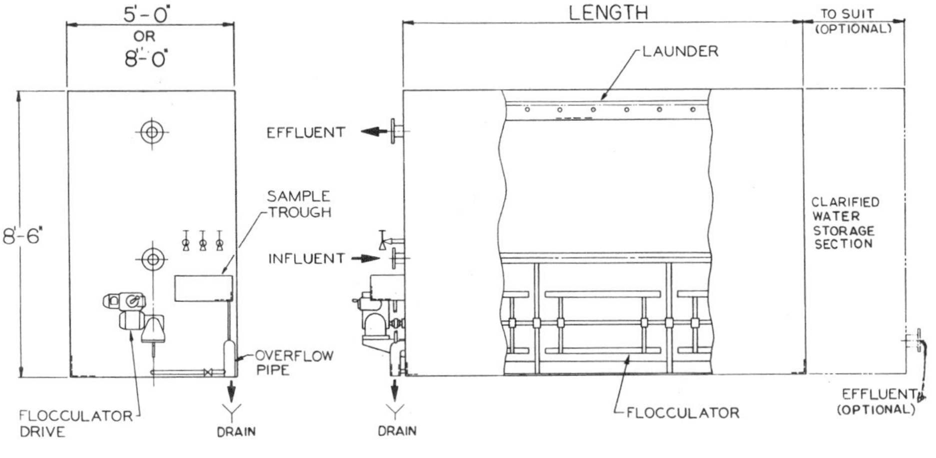

CALFILCO REACTIFIER — TYPE R

Size	G.P.M.	Width	Length	Influent Pipe Size	Effluent Pipe Size	Overflow Pipe Size
1	10	5'	5' 0"	1½"	2"	2"
2	20	5'	5' 0"	1½"	2"	2"
3	30	5'	7' 6"	2"	2½"	2½"
4	40	5'	10' 0"	2"	2½"	2½"
5	50	8'	8' 0"	2½"	3"	3"
6	75	8'	12' 0"	2½"	3"	3"
7	100	8'	16' 0"	3"	4"	4"
8	125	8'	20' 0"	3"	4"	4"
9	150	8'	24' 0"	3"	6"	6"

GENERAL DESCRIPTION:

The Calfilco "Reactifier, Type R" is a compact clarifier of the upflow, sludge blanket type. In a single rectangular vessel it provides efficient flash mixing, flocculation, upflow sludge blanket filtration and clarification together with sludge concentration and removal. This versatile unit is widely employed in the clarification of turbid surface waters for potable and process use, as well as for the removal of iron, manganese and color. It can also be used for lime softening and in the treatment of a variety of industrial wastes.

DESIGN FEATURES:

The rectangular configuration of the "Reactifier, Type R" results in a compact, easily arranged unit which is ideally suited for many installations. Standard units are available in sizes ranging from 10 to 150 gpm thruput. The dimensions of the "Reactifier" are such that it can be shipped completely shop assembled, thereby greatly minimizing field erection costs. Due to its completely steel construction, foundations can be of the simplest type. The design of the unit provides for stable operation over a wide range of operating conditions. Flocculation and the production of dense, readily settleable precipitates is promoted by means of fully adjustable, paddle type flocculators of proven design. Through the utilization of the principle of upflow sludge blanket filtration and a continuously increasing cross-section in the direction of the flow, stable operation and the production of a high clarity effluent is assured. Settled sludge is concentrated and removed from the unit on a semicontinuous basis.

OPERATION:

Required chemicals, fed in proportion to flow, are intimately mixed with the influent water as it enters the unit. A special distributor directs the water uniformly throughout the length of the flocculation chamber where paddle type flocculators promote intimate contact between the treating chemicals and the raw water. As the water leaves the flocculation chamber, it is directed upward into the clarification zone. Here it must pass through a blanket of sludge, held in suspension by the upward velocity of the water. As the water continues to rise, its effective velocity decreases due to the increasing cross-sectional area in the clarification zone and the greater distance between the individual sludge particles. A point is reached at which the clarified water disengages itself from the sludge and continues to rise to the effluent launder which removes it uniformly from the unit. Due to the intimate contact between the water containing suspended matter and the sludge, the individual particles gradually become heavier. This causes them to settle into the quiescent sludge concentrators from which they are blown down from the unit.

APPLICATION:

The rectangular "Reactifier" is normally employed as part of a complete treating system, consisting of chemical feeders, filters, pumps and required instruments and controls. When the water leaving the unit is to be pumped directly, a clarified water storage section is usually provided, integral with the unit. The size of this chamber varies depending on system requirements. Typical flow diagrams and layouts of plants employing the "Reactifier, Type R" are available on request. Detailed recommendations on specific applications can be readily obtained through your local Calfilco representative.

CALFILCO REACTIFIER --- TYPE C

HIGH RATE UPFLOW CLARIFIER

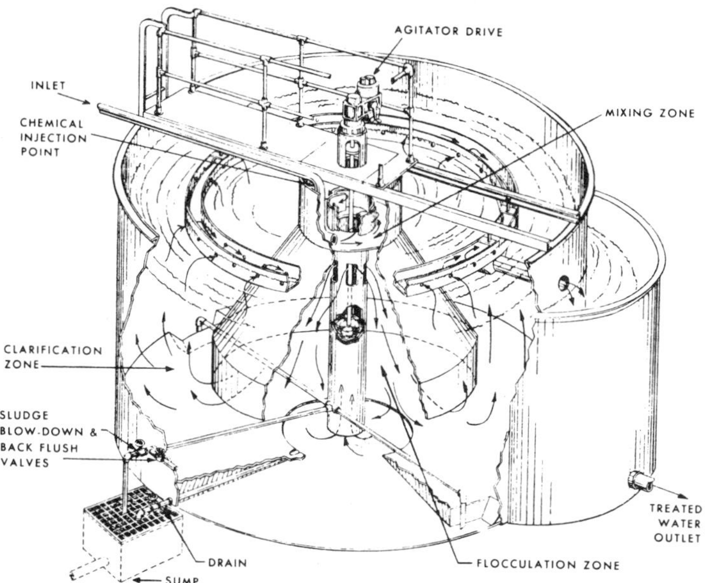

CALFILCO REACTIFIER — TYPE C

GENERAL DESCRIPTION:

The Calfilco "Reactifier, Type C" is a high flow rate, circular clarifier of the upflow, sludge blanket type. It combines in a single tank rapid mixing, flocculation, positive sludge recirculation, sludge blanket filtration and final clarification. Settled solids are also efficiently concentrated and removed from the unit as required. This highly efficient unit finds extensive application in the clarification of turbid surface waters for potable and process use. Because of its high degree of versatility, it is also commonly employed for iron, manganese and color removal, lime softening and in certain industrial waste treating applications.

DESIGN FEATURES:

Standard Calfilco "Reactifiers" of this type are available in a full range of sizes from 50 to 400 gpm thruput. Custom units can be furnished in higher capacities and to meet particular requirements. All steel units are most common, although tanks with concrete bottoms can also be utilized when desired. The careful hydraulic design of the "Reactifier" provides an extremely stable and compact means for achieving desired process results. Efficient flash mixing, coupled with continuously adjustable, positive sludge recirculation, assures optimum utilization of treatment chemicals and contributes greatly to the production of dense, rapidly settling sludges. These characteristics are further enhanced by the design of the clarification zone which provides optimum sludge blanket volume and a continuously increasing flow cross-section as the clarified water rises to the effluent launder. These features combine to insure a high clarity effluent over a wide range of operating conditions. Settled sludge is removed from the unit on a semicontinuous basis.

OPERATION:

Water enters the flash mixing section of the unit where it is brought into intimate contact with chemicals fed in proportion to flow. The water, together with the chemical reaction products, pass downward where they are immediately contacted by water containing previously formed precipitates, continuously recirculated at a high rate by the efficient propeller type circulator employed. The combined streams flow outward and downward under the flocculation hood, providing additional contact between the newly reacted chemicals and previously formed precipitates, thereby creating larger, more rapidly settling particles. As the recirculated stream approaches the bottom of the unit, a portion of the flow equivalent to the water entering the unit passes under the flocculation hood and upward into the sludge blanket zone. Here the water and associated precipitates contact similar particles suspended by the upflowing water. These particles gradually build up in size and settle into the sludge pockets from which they are blown down from the unit. As the water rises its velocity decreases due to the expanding cross-section of the clarification zone and the greater distance between particles in suspension. At this point the clarified water disengages from the sludge blanket and continues to rise to the launders, where it is drawn off uniformly and carried to the common effluent line.

Size	G.P.M.	Tank Dia. Standard	Tank Dia. Special	Tank Depth	Influent Pipe Size	Effluent Pipe Size	Ⓐ	Overflow Pipe Size	Ⓑ	Drain Pipe Size
1	50	8' 0"	8' 6"	10' 0"	2½"	3"	9"	3"	3"	2"
2	75	10' 0"	11' 0"	10' 0"	2½"	3"	9"	3"	3"	2"
3	100	11' 0"	12' 0"	10' 0"	3"	4"	9"	4"	3"	2"
4	125	12' 0"	14' 0"	10' 3"	3"	4"	12"	4"	4"	2"
5	150	13' 0"	15' 0"	10' 6"	3"	6"	12"	6"	4"	2"
6	175	15' 0"	16' 0"	10' 0"	4"	6"	12"	6"	4"	2½"
7	200	15' 0"	17' 0"	10' 9"	4"	6"	12"	6"	4"	2½"
8	250	17' 0"	19' 0"	10' 6"	4"	6"	15"	6"	4"	3"
9	300	19' 0"	21' 0"	10' 6"	4"	8"	15"	8"	6"	3"
10	350	20' 0"	22' 0"	11' 0"	6"	8"	15"	8"	6"	3"
11	400	22' 0"	24' 0"	11' 0"	6"	8"	15"	8"	6"	3"

SPECIFICATIONS

(Calfilco Reactifier, Type C)

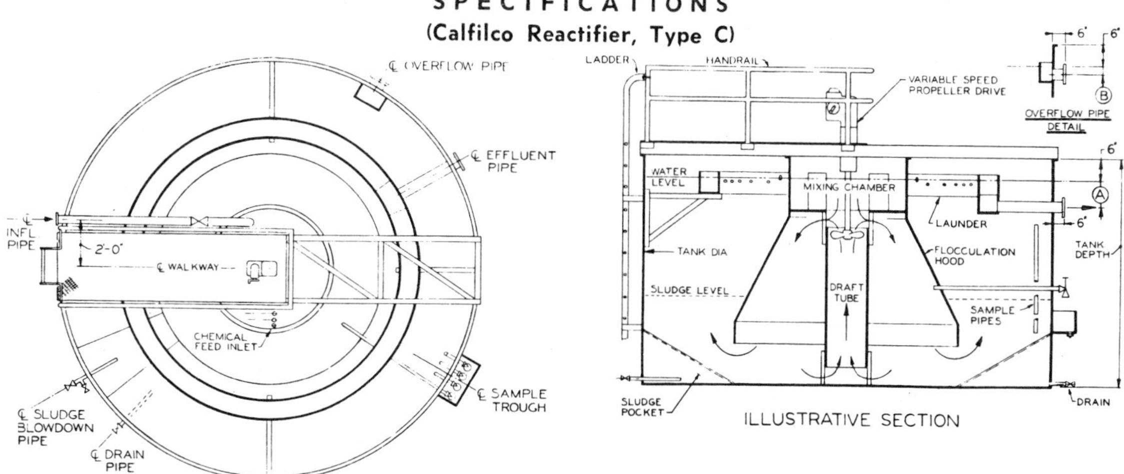

CALFILCO REACTIFIER—TYPE CA

High Rate Upflow Clarifier with Mechanical Sludge Rakes

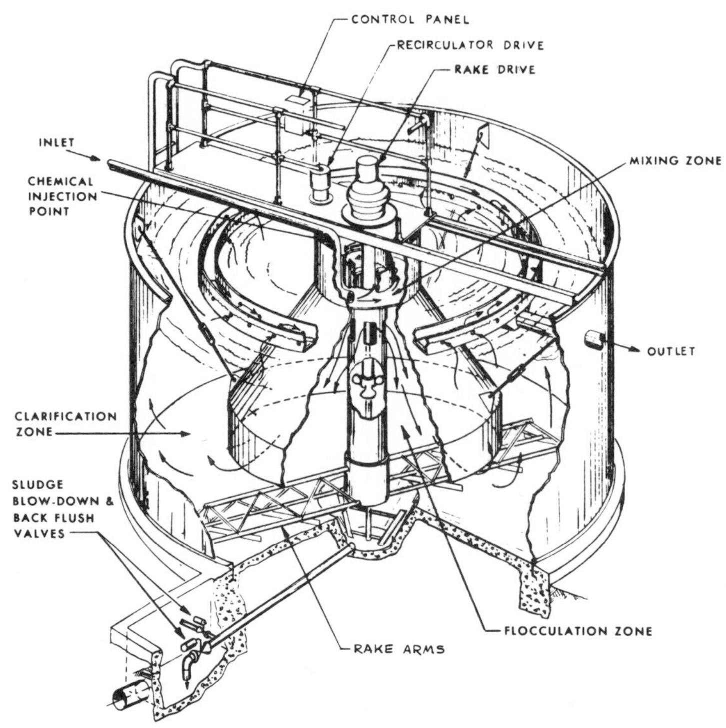

The Calfilco Reactifier, Type CA is a high flow rate, circular clarifier of the upflow, sludge blanket type, similar to Type C, except that it has mechanical sludge rakes. Settled sludge is continuously raked to the center of the unit by heavy duty, gear driven sludge rakes. From the sludge concentrator it is then removed from the unit on a semicontinuous basis.

SPECIFICATIONS: Calfilco Reactifier, Type CA with Mechanical Sludge Rakes

SIZE	GPM[1]	TANK DIAMETER	TANK DEPTH	INFLUENT PIPE SIZE	EFFLUENT PIPE SIZE	(A)	OVERFLOW PIPE SIZE	(B)	SLUDGE PIPE SIZE
1	50	10'-0"	10'-0"	2½"	3"	9"	3"	3"	2"
2	100	12'-0"	10'-0"	3"	4"	9"	4"	3"	2"
3	125	14'-0"	10'-0"	3"	4"	1'-0"	4"	4"	2"
4	150	16'-0"	10'-6"	4"	6"	1'-0"	6"	4"	2½"
5	200	18'-0"	10'-6"	4"	6"	1'-0"	6"	4"	2½"
6	250	20'-0"	11'-0"	6"	8"	1'-3"	8"	6"	3"
7	350	22'-0"	11'-0"	6"	8"	1'-3"	8"	6"	3"
8	400	24'-0"	11'-0"	6"	8"	1'-3"	8"	6"	3"
9	450	26'-0"	11'-0"	6"	10"	2'-0"	10"	9"	3"
10	500	28'-0"	11'-0"	6"	10"	2'-0"	10"	9"	3"
11	650	30'-0"	11'-6"	8"	10"	2'-0"	10"	9"	3"
12	750	32'-0"	11'-6"	8"	12"	2'-3"	12"	1'-0"	3"
13	850	34'-0"	11'-6"	8"	12"	2'-3"	12"	1'-0"	3"
14	900	35'-0"	11'-6"	8"	12"	2'-3"	12"	1'-0"	3"
15	950	36'-0"	11'-6"	10"	14"	2'-6"	14"	1'-3"	3"
16	1000	38'-0"	11'-6"	10"	14"	2'-6"	14"	1'-3"	3"
17	1100	40'-0"	12'-0"	10"	14"	2'-6"	14"	1'-3"	4"
18	1250	42'-0"	12'-0"	10"	16"	2'-9"	16"	1'-3"	4"
19	1350	44'-0"	12'-0"	10"	16"	2'-9"	16"	1'-3"	4"
20	1500	45'-0"	12'-0"	10"	16"	2'-9"	16"	1'-3"	4"

(1) Nominal flow rate at 1 gpm/ft² of rise area in clarification zone below launders. See engineering drawing CD-2-205

Calfilco

HORIZONTAL "MULTIPLEX" FILTERS

GRANULAR BED TYPE HMP

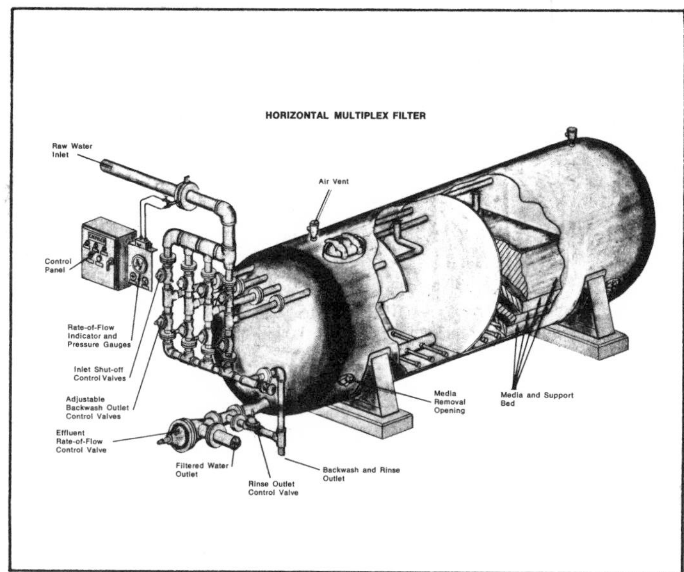

The Calfilco "Multiplex" filter is a medium capacity unit widely used for filtration of potable, industrial process and cooling tower water as well as tertiary sewage effluent. It consists of a horizontal vessel divided into several individual filter sections by vertical separating baffles.

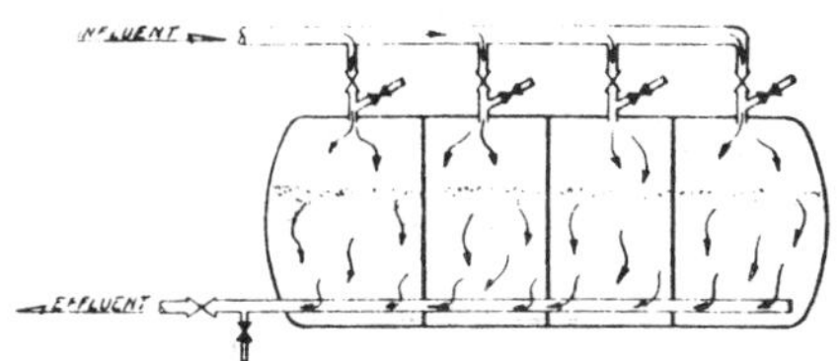

IN SERVICE

The influent water is distributed to each filter compartment, passes evenly through the media, is collected by the underdrain system and leaves the filter through the effluent valve.

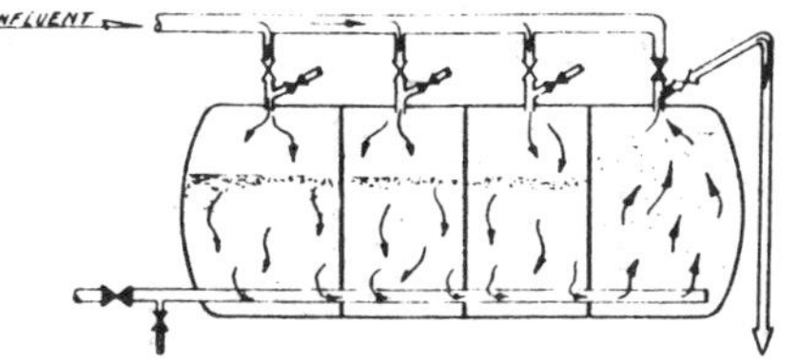

BACKWASHING

The effluent valve is closed. At the same time the influent valve to one compartment is automatically closed. The backwash outlet valve to that compartment is opened. The influent water continues to filter downward through the remaining three compartments and upward through the fourth. This is repeated sequentially until the entire filter has been washed.

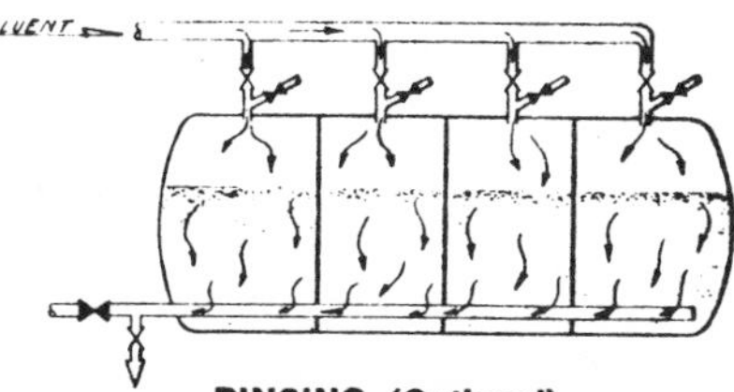

RINSING (Optional)

Before the filter is returned to service a rinse valve is opened allowing rinsing of the entire filter to waste. At the completion of this step the effluent valve is opened as the rinse valve closes, returning the unit to service.

PERFORMANCE AND SIZES

Filter Dia., Feet	Shell Length, Feet	Filter Bed Area, Sq. Ft.	Service Flow Rates 2 GPM/SQ. FT. GPM	2 GPM/SQ. FT. Gal/24 Hr	3 GPM/SQ. FT. ① GPM	3 GPM/SQ. FT. ① Gal/24 Hr	5 GPM/SQ. FT. GPM	5 GPM/SQ. FT. Gal/24 Hr	Normal② Backwash GPM Flow Rates Sand	Anthracite	Approximate Weights—Pounds Shipping Filter Tank③	Shipping Piping & Valves	Shipping Filter Bed Sand④	Total Operating	Normal Filter Piping Sizes, Inches Cell⑤ Inlet	Filter⑤ Outlet	Back⑥ Wash	Rinse⑥ to Waste
4′	8	32	64	92,000	96	138,000	160	230,000	80	48	2,300	800	8,250	18,650	1½	3	1½	1½
	10	40	80	115,000	120	175,000	200	288,000	100	60	2,700	950	10,700	20,250	2	3	2	2
	12	48	96	138,000	144	208,000	240	346,000	120	72	3,000	1,000	12,700	23,750	2	3	2	2
	14	56	112	162,000	168	242,000	280	402,000	140	84	3,300	1,050	14,700	27,250	2	3	2	2
5′	12	60	120	175,000	180	260,000	300	432,000	150	90	4,600	1,600	19,000	37,000	2½	4	2	2
	14	70	140	208,000	210	320,000	350	504,000	175	105	5,100	1,650	21,800	41,000	2½	4	2½	2
	16	80	160	230,000	240	346,000	400	576,000	200	120	5,600	1,725	24,700	46,200	2½	4	2½	2½
	18	90	180	260,000	270	388,000	450	648,000	225	135	6,200	2,000	27,700	51,800	3	4	2½	2½
6′	16	96	192	276,000	288	414,000	480	690,000	240	144	8,200	1,950	31,600	63,500	3	4	2½	2½
	18	108	216	312,000	324	466,000	540	776,000	270	162	9,100	2,100	35,400	71,050	3	6	3	2½
	20	120	240	346,000	360	518,000	600	864,000	300	180	9,800	2,125	39,100	77,750	3	6	3	3
	22	132	264	381,000	396	570,000	660	950,000	330	198	10,500	2,200	42,800	85,200	3	6	3	3
8′	18	144	288	414,000	432	620,000	720	1,070,000	360	216	16,200	2,550	67,600	132,300	3	6	3	3
	20	160	320	460,000	480	690,000	800	1,150,000	400	240	17,600	4,500	74,200	146,850	4	6	4	3
	22	176	352	506,000	528	760,000	880	1,265,000	440	264	18,800	4,600	80,700	159,500	4	6	4	3
8′	25	200	400	576,000	600	864,000	1,000	1,440,000	500	300	20,600	4,700	90,500	178,200	4	6	4	4
	27	216	432	620,000	646	930,000	1,080	1,556,000	540	323	21,800	4,775	97,100	190,800	4	6	4	4
	30	240	480	690,000	720	1,070,000	1,200	1,730,000	600	360	23,500	4,875	106,900	209.400	4	6	4	4

① Service flow rates for normal applications.
② Backwash flow rate based on 60°F water. Drains should be sized to carry 50% higher flow rates to allow for peak flows.
③ Based on 125 psig, ASME Code tanks.
④ For anthracite, divide weights shown by two.
⑤ Based on 4 to 6 ft/sec maximum velocity at 3 gpm/ft² flow rate.
⑥ Based on 100 psig available supply pressure.

Calfilco

VERTICAL PRESSURE FILTERS

Granular Bed Type VP, with Valve Nest Piping

General Description: One or more vertical filter vessels operated in parallel with a bed of granular media of select silica sand, anthracite coal, or a combination of these. Vertical filters will remove extremely fine particulate and colloidal matter when coagulated.

All filters are furnished complete with all internals, filter media, face piping, valves, standard instruments and control components. Available in two basic styles: MVN for manual operation and AVN for automatic operation.

Applications: Potable and Process Water Filtration: For removal of turbidity, softening process precipitates, oxidized iron and manganese, suspended and colloidal matter, algae and micro-organisms.

Recirculating Water Filtration: For the removal of suspended turbidity and organic matter from cooling tower circuits, once-through cooling systems, boiler plant condensate systems, swimming pools, reflection ponds, etc.

Chemical and Process Fluid Filtration: For the removal of particulate matter from such fluids as brines, caustics, acids and plating solutions.

Waste Water Filtration: For the removal of suspended particulate and colloidal matter and for the reduction of oxygen demand, so as to permit water reuse or to eliminate contamination of receiving waters by industrial wastes and sewage plant effluents.

PERFORMANCE AND SIZES

Filter Diameter, Inches	Filter Bed Area, Sq. Ft.	SERVICE FLOW RATES @ 2 GPM/FT²		SERVICE FLOW RATES @ 3 GPM/FT² (1)		SERVICE FLOW RATES @ 5 GPM/FT²		NORMAL (2) BACKWASH GPM FLOW RATES		APPROX. WEIGHTS—POUNDS SHIPPING			Total Operating	NORMAL FILTER PIPING SIZES, INCHES		
		GPM	Gal/24 Hr	GPM	Gal/24 Hr	GPM	Gal/24 Hr	Sand	Anthracite	Filter Tank (3)	Piping & Valves	Filter Bed (Sand) (4)		Inlet & Outlet (5)	Backwash (6)	Rinse to Waste (6)
14	1.0	2	2,900	3	4,300	5	7,200	10	6	200	60	450	900	¾	¾	¾
16	1.4	3	4,300	4	5,800	7	10,100	15	10	225	60	575	1,100	¾	¾	¾
20	2.2	4	5,800	7	10,100	11	15,800	22	15	250	60	750	1,450	¾	¾	¾
24	3.1	6	8,600	9	13,000	16	23,000	30	20	500	140	1,100	2,350	1½	1½	¾
30	4.9	10	14,400	15	21,600	25	36,000	50	30	700	150	1,950	3,900	1½	1½	¾
36	7.1	14	20,200	21	30,200	36	51,800	70	45	800	160	2,700	5,200	1½	1½	¾
42	9.6	19	27,400	29	41,800	48	69,100	95	60	1,150	200	3,850	7,500	2	2	¾
48	12.6	25	36,000	38	54,700	63	90,700	125	75	1,400	225	4,800	9,750	2	2	1
54	15.9	32	46,100	48	69,100	80	115,200	150	95	1,900	300	6,100	12,450	2½	2½	1
60	19.7	39	56,200	59	85,000	99	142,600	200	120	2,500	325	7,600	15,950	2½	2½	1
66	23.8	48	69,100	71	103,700	119	171,400	240	145	2,800	550	9,200	19,200	3	3	1½
72	28.3	57	82,100	85	122,400	142	204,500	285	170	3,500	600	10,800	22,900	3	3	1½
78	33.2	66	95,000	100	144,000	166	239,000	330	200	4,100	650	12,700	27,600	3	3	1½
84	38.5	77	110,900	116	167,000	193	277,900	385	230	5,200	1,200	14,600	33,000	4	4	1½
90	44.3	89	128,200	133	191,500	222	319,700	445	205	6,500	1,250	16,800	38,600	4	4	1½
96	50.3	101	145,400	151	217,400	252	362,900	505	300	8,000	1,300	19,100	44,800	4	4	2
102	56.6	113	163,000	170	244,800	283	407,500	565	340	9,000	1,350	21,600	51,600	4	4	2
108	63.5	127	182,900	191	275,000	318	457,900	635	380	9,700	1,400	24,300	57,800	4	4	2
114	70.9	142	204,500	213	306,700	355	511,200	710	425	10,600	1,550	27,000	65,000	4	6	2
120	78.5	157	226,000	236	340,000	393	565,900	785	470	11,500	1,650	29,800	72,150	4	6	2

1. Service flow rates for normal applications.
2. Backwash flow rates based on 60°F water. Drains should be sized to carry 50% higher flow rates to allow for peak flows.
3. Based on 125 psig, ASME Code tanks.
4. For anthracite, divide weights shown by two.
5. Based on 4 to 6 ft/sec maximum velocity at 3 gpm/ft² flow rate.
6. Based on 30 psig available supply pressure.

Calfilco

HORIZONTAL PRESSURE FILTER

Granular Bed Type HP, with Valve Nest Piping

GENERAL DESCRIPTION: The Calfilco horizontal pressure filter is a medium to high capacity unit widely used in filtration of potable, industrial process and cooling tower water. It is also successfully employed in tertiary sewage and industrial waste filtration.

A typical installation consists of one or more filter vessels operated in parallel with a bed of granular media of select silica sand, anthracite coal or a combination of these. When supplied with appropriate chemical feed equipment, these filters will remove extremely fine particulate and colloidal matter. They are also often used as roughing filters without chemical feed when high clarity effluents are not required.

All filters are furnished complete with internals, filter media, face piping, valves, standard instruments and control components.

APPLICATIONS: **Potable and Process Water Filtration:** For removal of turbidity, softening process precipitates, oxidized iron and manganese, suspended and colloidal matter, algae and microorganisms.

Recirculating Water Filtration: For the removal of suspended turbidity and organic matter from cooling tower circuits, once-through cooling systems, boiler plant condensate systems, swimming pools, reflection ponds, test pools and other recirculating systems.

Chemical and Process Fluid Filtration: For the removal of particulate matter from such fluids as brines, caustics, acids and plating solutions.

Waste Water Filtration: For the removal of suspended particulate and colloidal matter of recycled process or discharged waste water effluents.

PERFORMANCE AND SIZES

Filter Dia., Inches	Shell Length Feet	Filter Bed Area Sq. Ft.	SERVICE FLOW RATES: 2 GPM/Sq. Ft.		3 GPM/Sq. Ft.(1)		5 GPM/Sq. Ft.		NORMAL(2) BACKWASH GPM FLOW RATES		APPROX. WEIGHTS—POUNDS: SHIPPING				NORMAL FILTER PIPING SIZES, INCHES		
			GPM	Gal/24 Hr.	GPM	Gal/24 Hr.	GPM	Gal/24 Hr.	Sand	Anthracite	Filter Tank(3)	Piping & Valves	Filter Bed Sand(4)	Total Operating	Inlet & Outlet (5)	Back Wash(6)	Rinse to Waste(6)
48	8	32	64	92,000	96	138,000	160	230,000	320	192	2,000	800	8,250	15,800	3	3	1½
48	10	40	80	115,000	120	175,000	200	288,000	400	240	2,300	950	10,700	19,850	3	3	2
48	12	48	96	138,000	144	208,000	240	346,000	480	288	2,600	1,000	12,700	23,300	3	4	2
48	14	56	112	162,000	168	242,000	280	402,000	560	336	2,900	1,050	14,700	27,650	3	4	2
60	12	60	130	175,000	180	260,000	300	432,000	600	360	4,100	1,600	19,000	38,700	4	4	2
60	14	70	140	208,000	210	320,000	350	504,000	700	420	4,600	1,650	21,800	42,650	4	4	2
60	16	80	160	230,000	240	346,000	400	576,000	800	480	5,000	1,725	24,700	45,100	4	6	2½
60	18	90	180	260,000	270	388,000	450	648,000	900	540	5,900	2,000	27,700	50,700	6	6	2½
72	16	96	192	276,000	288	414,000	480	690,000	960	576	7,600	1,950	31,600	60,950	6	6	2½
72	18	108	216	312,000	324	466,000	540	776,000	1,080	648	8,400	2,100	35,400	67,950	6	6	2½
72	20	120	240	346,000	360	518,000	600	864,000	1,200	720	9,000	2,124	39,100	74,500	6	6	3
72	22	132	264	381,000	396	570,000	680	950,000	1,320	792	9,700	2,200	42,800	83,050	6	6	3
96	18	144	288	414,000	432	620,000	720	1,070,000	1,440	864	15,000	2,550	67,600	128,850	6	6	3
96	20	160	320	460,000	480	690,000	800	1,150,000	1,600	960	16,100	4,500	74,200	142,850	6	6	3
96	22	176	352	506,000	528	760,000	880	1,265,000	1,760	1,056	17,300	4,600	80,700	154,200	6	6	3
96	25	200	400	576,000	600	864,000	1,000	1,440,000	2,000	1,200	19,200	4,700	90,500	172,400	8	8	4
96	27	216	432	620,000	646	930,000	1,080	1,556,000	2,160	1,292	20,400	4,775	97,100	184,300	8	8	4
96	30	240	480	690,000	720	1,070,000	1,200	1,730,000	2,400	1,440	22,100	4,875	106,900	207,100	8	8	4
120	30	300	600	864,000	900	1,295,000	1,500	2,160,000	3,000	1,800	36,000	5,000	168,000	320,000	8	8	6
120	35	350	700	1,008,000	1,050	1,510,000	1,750	2,520,000	3,500	2,100	40,500	6,000	196,000	375,000	10	10	6
120	40	400	800	1,150,000	1,200	1,730,000	2,000	2,880,000	4,000	2,400	45,000	6,100	225,000	430,000	10	10	6
120	45	450	900	1,295,000	1,350	1,945,000	2,250	3,240,000	4,500	2,700	50,200	6,200	252,000	480,000	10	10	6
120	50	500	1,000	1,440,000	1,500	2,160,000	2,500	3,600,000	5,000	3,000	54,800	7,200	280,000	535,000	12	12	6

(1) Service flow rates for normal applications.
(2) Backwash flow rate based on 60° F water. Drains should be sized to carry 50% higher flow rates to allow for peak flows.
(3) Based on 125 psig, ASME Code tanks.
(4) For anthracite, divide weights shown by two.
(5) Based on 4 to 6 ft/sec maximum velocity at 3 gpm/ft^2 flow rate.
(6) Based on 100 psig available supply pressure and sand media.

Calfilco

VERTICAL ACTIVATED CARBON FILTERS

TYPE VCP

Installation utilizing Type VCP Carbon Filters in combination with pre-filter battery.

APPLICATION: Calfilco activated carbon filters are widely used for the removal of chlorine, odor and taste forming substances, dissolved organic matter and color bodies from water. Typical applications include preparation of chlorine-free, tasteless and odorless water for use in beverages and other food products. Additionally, they are employed to remove organic matter and chlorine from feedwater to ion exchange equipment. Final treatment of sewage and other industrial wastes to remove objectionable organics and odors is also becoming an increasingly important end use. Carbon filtration can also be tailored for specific removal of other substances from water, as well as other process fluids. This data sheet covers Calfilco's general purpose units.

ACTIVATED CARBON FILTERS
TABLE OF SIZES AND CAPACITIES — TYPE VCP

MODEL NUMBER	FILTER DIAMETER (1)	FILTER BED AREA SQ. FT.	FLOW RATE RANGE, GPM (2)		SERVICE PIPING SIZE (3)	SHIPPING WEIGHT (POUNDS)		
			LOW	HIGH		TANK (4)	PIPING	MEDIA
VCP-1	14	1.0	4	6	¾"	220	45	330
VCP-2	16	1.4	6	8	¾"	235	45	375
VCP-3	20	2.2	10	14	1 "	260	55	505
VCP-4	24	3.1	14	20	1½"	325	75	620
VCP-5	30	4.9	20	30	1½"	425	85	1000
VCP-6	36	7.1	30	45	2 "	710	105	1500
VCP-7	42	9.6	40	60	2 "	980	125	2020
VCP-8	48	12.6	50	75	2½"	1145	165	2480
VCP-9	54	15.9	65	95	3 "	1420	240	3220
VCP-10	60	19.7	80	120	3 "	1820	260	3970
VCP-11	66	23.8	90	140	3 "	2205	280	4865
VCP-12	72	28.3	115	170	4 "	2480	835	5780
VCP-13	78	33.2	130	200	4 "	3225	850	6820
VCP-14	84	38.5	155	230	4 "	3660	875	7700
VCP-15	90	44.3	175	265	4 "	4230	900	8700

(1) Standard filter side shell height is 60". Side shell can be reduced to 48" when low first cost is desired or when contamination is normally at low levels. 48" side shells are also utilized when strainer type underdrains are employed. Under these conditions, reduce stated overall height by 12".

(2) **Employ High Flow Rating** for removal of less than 5 ppm chlorine, low to moderate organic contamination, and where trace quantities of chlorine are tolerable in the effluent.
Employ Low Flow Rating for removal of chlorine above 5 ppm, high organic contamination and where ultimate removal is desired.

(3) Piping size based on four to six ft./sec. maximum velocity at peak service flow.

(4) Based on 125 psig ASME Code tanks.

CALFILCO GRAVITY FILTERS

Engineering Data Sheet

PERFORMANCE AND SIZES:

Filter Diameter	Filter Bed Area, (Sq. Ft.)	SERVICE FLOW RATES @ 3 GPM/FT^2 * GPM	SERVICE FLOW RATES GAL/24 HR	BACKWASH FLOW RATES SAND GPM	BACKWASH FLOW RATES ANTHRACITE GPM	FILTER PIPING SIZES, (INCHES) IN & OUT	FILTER PIPING SIZES, (INCHES) BACK-WASH	FILTER PIPING SIZES, (INCHES) WASTE
2'-6"	4.9	15	21,600	75	50	1-1/2"	1-1/2	3/4
4'-0"	12.6	38	54,700	190	125	2" 2"	3	4
4'-6"	15.9	48	69,100	225	150	2-1/2"	3	4
6'-0"	28.3	85	122,400	425	300	3" 2-1/2"	4	6
7'-0"	38.5	116	167,000	580	400	4" 3"	6	6
8'-0"	50.3	151	217,400	750	505	4" 3"	6	8
9'-0"	63.5	191	275,000	960	640	4" 4"	6	8
10'-0"	78.5	236	340,000	1180	790	6" 4"	8	10
11'-0"	95.0	285	410,000	1430	950	6" 6"	8	10
13'-0"	136.73	410	590,000	2000	1370	6" 6"	10	12
14'-0"	153.94	462	665,000	2320	1540	8" 6"	10	12
15'-0"	176.7	530	760,000	2550	1770	8" 6"	10	14
18'-0"	255	765	1,100,000	3820	2650	10" 8"	12	18

*Service flow rates for normal applications.

Calgon Corporation

CALGON-HAVENS SYSTEMS OSMOTIK PROCESS

The Calgon-Havens Systems OSMOTIK Reverse Osmosis Process has achieved success in the dairy industry for pollution control and by-product recovery from cheese whey. Calgon-Havens Systems offers ultrafiltration and reverse osmosis equipment for whey protein concentration. BOD removal and whole whey concentration. The processes are also effective for many other applications where water removal and solute separation are of paramount interest.

The Calgon-Havens Systems OSMOTIK Process offers significant advantages in pollution control and by-product recovery for various waste streams in the pulp and paper industry.

The OSMOTIK Process utilizes reverse osmosis and ultrafiltration techniques to separate solvents from solutes, and to separate molecules of significantly different molecular weights and sizes.

Normal osmosis occurs when a semi-permeable membrane is placed between solutions with two different concentrations. The difference in osmotic pressure in the two solutions will force water to flow through the semi-permeable membrane into the solution with higher concentration. The flow of water will continue until the concentrations of solutions, and hence the osmotic pressure, in the solutions are equalized. An ideal semi-permeable membrane will allow water to flow through the membrane with the exclusion of all solutes.

Reverse osmosis describes the flow of water through a semi-permeable membrane in a direction opposite that which is normal. When pressure in excess of the net osmotic pressure is applied to the solution with higher concentration, water molecules flow from it into the lower concentration fluid. The purified liquid can then be extracted from the system.

Ultrafiltration is similar to reverse osmosis, but utilizes membranes with larger pore sizes. This allows separation of larger molecules than reverse osmosis. Ultrafiltration requires lower pressures than reverse osmosis — generally less than 450 psi.

OSMOTIK MODULE CONSTRUCTION — The OSMOTIK Process utilizes a porous, filament-wound tube to support the selective membrane. Although this tube is only .030 inches thick, it has a burst strength exceeding 5000 psi. Eighteen of these tubes, with the membrane integrally cast to their inside walls, are combined to form an OSMOTIK Module.

This tubular support system, developed and patented by Calgon-Havens Systems, has proven to be ideal for the commercial processing of enzymes, organic salts, protein fractions and other substances where large concentrations of particulate matter are of concern.

The modular concept for system design allows great versatility for many applications. This tubular module system, developed and patented by Calgon-Havens Systems, has proven to be excellent for commercial processing of waste streams.

All materials of construction within the module are fiberglass, plastic or corrosion-resistant metal. The system piping and other hardware are generally stainless steel or other corrosion-resistant material.

Calgon Corporation
CATIONIC COAGULANT FOR WATER CLARIFICATION

Cat-Floc is a clear water-white to pale yellow viscous liquid, cationic polyelectrolyte. The product is completely soluble in water and functions as a coagulant in the clarification of water. It is accepted by the U. S. Public Health Service for use in the treatment of drinking water supplies.

Cat-Floc B is a clear, colorless to pale yellow, viscous liquid, cationic polyelectrolyte which is unaffected by chlorine or permanganate. The product is completely miscible with water and functions as a coagulant and a flocculent in the clarification of water supplies. It is accepted by the Environmental Protection Agency for use in the treatment of drinking water supplies.

Both Cat-Floc and Cat-Floc B are highly effective and economical for use as a primary coagulant in replacing inorganic coagulants such as ferric salts or alum in the clarification of municipal and industrial water supplies. Generally, superior results can be achieved with Cat-Floc and Cat-Floc B than with conventional materials, thus reducing bulk handling. Liquid formulation is easy to feed. Appreciable reduction in treatment costs is possible. When used in this manner, a specially selected clay may also be needed when mixing time is short. By eliminating the use of alum altogether, the carryover of soluble aluminum is prevented. Such carryover can foul zeolite softeners and result in troublesome deposits in distribution systems, water meters, heat exchangers and steam boilers. Laboratory tests have indicated that basin sludges are more easily dewatered when no inorganic coagulant is used.

When used in conjunction with another primary coagulant such as alum or iron salt, these Cat-Flocs will improve clarity, color removal, and the settling rate of the floc. In addition, the amount of inorganic coagulant required is reduced by 40 to 60%. Since the Cat-Flocs react more slowly than inorganic coagulants, it is usually added ahead of the alum or iron salt to get as much mixing time as possible.

Calgon Corporation can supply a series of corrosion-resistant diaphragm pumps which are capable of feeding concentrated liquid polymers directly out of the shipping container or as a dilute solution. Choice of a suitable pumping unit will depend on volume requirements and discharge pressure limitations.

Calgon Corporation

POLYMERS FOR LIQUID/SOLIDS SEPARATION OF MUNICIPAL AND INDUSTRIAL WASTES BY CENTRIFUGING

PREPARATION AND FEEDING OF POLYMER SOLUTION

Charge relationships are depicted for WT-2570 and WT-2580 with respect to each other and other Calgon polymers used for liquid/solids separation of wastewater. Calgon Corporation offers a complete line of synthetic polymers formulated for specific requirements. Their characteristics include a wide range of molecular weights and progressive degrees of cationic and anionic charge. Information on these and other products is available upon request.

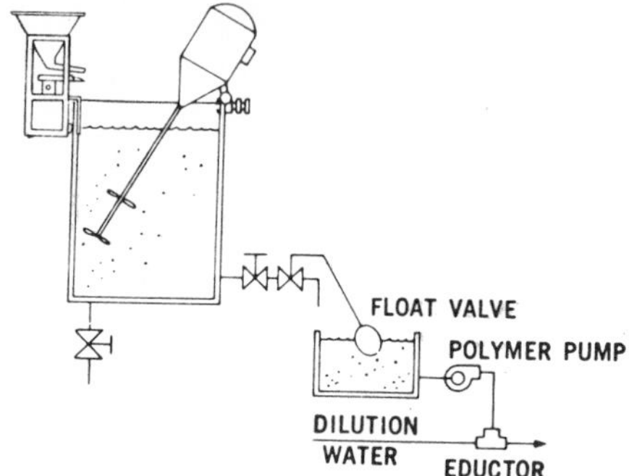

Diagram shows a typical arrangement of a tank with a powered mechanical mixing apparatus, dry feeder and associated equipment for the preparation and feeding of polymer solution.

Calgon® WT-2570 and WT-2580 are off-white, powdered, high molecular weight cationic polyelectrolytes which are completely soluble in water. They are similar except that WT-2580 possesses a stronger cationic charge.

WT-2570 and WT-2580 are formulated specifically for use in liquid/solids separation processes that utilize centrifuge equipment. These products are soluble, nonscaling and do not contribute bulk to the dewatered sludge. Of prime importance is their ability to condition sludge to withstand the high shear effect of centrifuging. This, in effect, removes a greater amount of solids and, also, results in a cleaner centrate. In the final analysis, these products help the plant to handle greater loads more efficiently because cleaner centrate is recycled to the head of the plant, thereby reducing the amount of recycled solids on the wastewater treatment facility. WT-2600, WT-2630 and WT-2640 should also be investigated as possible coagulants for centrifugation.

Dosage rates will depend upon conditions and type of sludge to be dewatered.

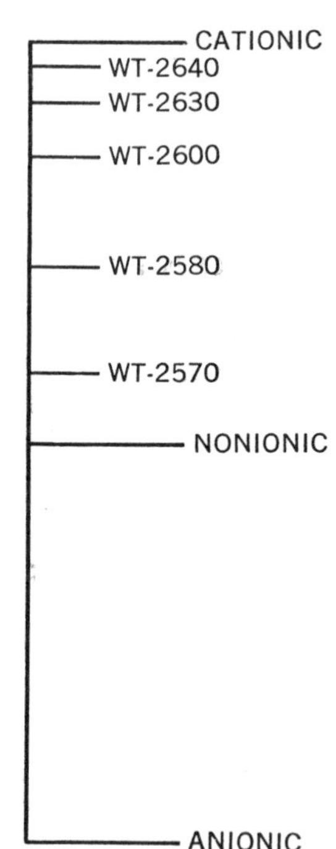

Calgon® WT-2600 and WT-2630 are off-white, flake-like, cationic polyelectrolytes which are completely soluble in water.

Calgon® WT-2640 is a clear, colorless, viscous, liquid, cationic polyelectrolyte that is completely soluble in water. It can be used as a coagulant or in combination with inorganic primary coagulants.

Calgon® WT-2660 is a dark amber viscous liquid, cationic polyelectrolyte which is completely soluble in water.

Calgon® WT-2690 is an off-white flake-like, nonionic polyelectrolyte which is completely soluble in water.

Calgon® WT-2700, WT-2900 and WT-3000 are off-white, flake-like anionic polyelectrolytes which are completely soluble in water.

They can be used as coagulants or in combination with inorganic primary coagulants.

WT-2870 is a clear-white to yellow viscous liquid, cationic polyelectrolyte. It is completely soluble in water.

Calgon Corporation's granular activated carbon products Filtrasorb 100 and 200 are made from select grades of coal that has been processed into products of superior hardness and longer life. Produced under rigidly controlled conditions by high temperature steam activation, they are very dense carbons with high surface areas. Their pore structures have been carefully controlled for the adsorption of both high and low molecular weight impurities from water. These coal-based carbons wet readily, do not float, and their high density facilitates backwashing under conventional flow rate conditions. The particle size of each differs to suit specific engineering design requirements.

Filtrasorb 100 and 200 granular activated carbon products are designed for efficient use in water filtration equipment for the removal of organic impurities found in potable and industrial water supplies. These impurities include tastes, odors, color, insecticides, detergents, phenols, and other contaminants from both natural or industrial sources. Furthermore, Filtrasorb 100 and 200 are effective filtration media for removal of turbidity and suspended matter. In addition to the adsorption and filtration functions, they are capable of concurrently removing excess oxidants such as chlorine and permanganate used in pretreatment application.

Granular activated carbon has been specifically manufactured for thermal reactivation and re-use. Numerous reactivation installations have demonstrated the feasibility and economy of granular carbon reactivation. The carbon is handled as a water slurry between the filters and the furnace area. No additional labor or labor skills are required to operate a reactivation facility which is automated to run with minimal surveillance.

physical properties

	FILTRASORB 100	FILTRASORB 200
Total Surface Area (N_2 BET method) m^2/g	850–900	850–900
Bed Density Backwash and Drained, lbs/ft^3*	30 approx.	30 approx.
Particle Density Wetted in Water g/cc	1.4–1.5	1.4–1.5
Effective Size mm	0.8–0.9	0.55–0.65
Uniformity Coefficient	1.9 or less	1.7 or less

*To be used to calculate volume requirements.

specifications

	FILTRASORB 100 (specifications)	FILTRASORB 200 (specifications)	FILTRASORB 100 (typical analysis)	FILTRASORB 200 (typical analysis)
Sieve Size U.S. Standard Series				
Larger than No. 8—Max. %	8	–	3	–
Larger than No. 14—Max. %	–	5	–	3
Smaller than No. 30—Max. %	5	–	1	–
Smaller than No. 40—Max. %	–	5	–	1
Mean Particle Diameter mm	1.5–1.7	0.8–1.0	1.6	0.9
Iodine Number—Minimum	850	850	875	875
Abrasion Number—Minimum	70	70	86	83
Ash, Maximum %	8	8	4	4
Moisture as packed, Maximum %	2	2	0.5	0.5

FILTRASORB 300 and 400

Filtrasorb 300 and 400 are two grades of granular activated carbon made by Calgon Corporation to remove impurities from municipal or industrial waste water. Both products are high density carbons with large surface areas. Pore structure and mesh size have been carefully controlled for the adsorption of both high and low molecular weight impurities from waste water. Filtrasorb wets readily and does not float. Its high density allows convenient backwashing under conventional flow rate conditions.

Granular carbon treatment has built-in flexibility since the waste water can be either upgraded partially to meet effluent treatment and receiving stream standards or upgraded completely for direct re-use in high quality water applications. In addition, valuable by-products sometimes can be adsorbed in granular carbon beds and then recovered from the carbon by a chemical regeneration process.

For unit processes dealing with ion exchange or inorganic re moval, granular carbon is used to protect the resins and/or membranes which would become fouled by the organic impurities in the waste.

Filtrasorb 300 and 400 are designed for the purification of waste water by removing organic contaminants whether measured as BOD, COD, TOC, color, odor, optical density, or by other analytical techniques. Its dual role of filtration/adsorption is a concurrent function that filters out suspended solids and adsorbes other dissolved organic compounds.

physical properties

	FILTRASORB 300	FILTRASORB 400
Total surface area (N_2, BET method) m^2/g	950-1050	1050-1200
Bulk density, $lbs/ft.^3$*	26	25
Particle density wetted in water g/cc	1.3-1.4	1.3-1.4
Pore volume cc/g	0.85	0.94
Effective size mm	0.8-0.9	0.55-0.65
Uniformity coefficient	1.9 or less	1.9 or less

*To be used to calculate volume requirements.

specifications

	FILTRASORB 300		FILTRASORB 400	
	Specification Value	*Typical Analysis*	*Specification Value*	*Typical Analysis*
U.S. Standard Series Sieve Size				
Larger than No. 8	Max. 8%	3	—	—
Smaller than No. 30	Max. 5%	1	—	—
Larger than No. 12	—	—	Max. 5%	3
Smaller than No. 40	—	—	Max. 5%	1
Mean Particle Dia. mm.	1.5-1.7	1.6	0.9-1.1	1.0
Iodine Number, min.	950	975	1050	1100
Abrasion Number, min.	70	80	75	80
Ash	Max. 8%	5%	Max. 8.5%	5%
Moisture	Max. 2.0%	0.5%	Max. 2.0%	0.5%

The following are major areas of water treatment where a process utilizing Filtrasorb can purify waste waters beyond levels normally obtainable by conventional biological secondary waste treatment systems:

*Calgon Polysorb Process**—This is a physical-chemical waste water treatment process that can produce secondary or tertiary quality water. The process clarifies raw sewage in the primary plant by utilizing water-soluble polymers to maximize removal of suspended solids. The clarified sewage is then percolated through Filtrasorb beds which remove the remaining suspended solids and adsorb the dissolved organics.

Advanced Waste Water Treatment—A process utilizing Filtrasorb can be incorporated as a final polishing step. It can be used to complement industrial and municipal waste water treatment systems whose sedimentation, clarification, and bacteriological techniques are inadequate for meeting higher standards of effluent quality.

Industrial Process and Waste Water Treatment—Filtrasorb can be incorporated into manufacturing process systems to remove particular types of pollution. It can be used in by-product recovery operations and in industrial waste treatment programs for the removal of phenols, colors, insecticides, TNT, detergents, polyols, dyes and other wastes.

* Polysorb Process is a Service Mark of Calgon Corporation.

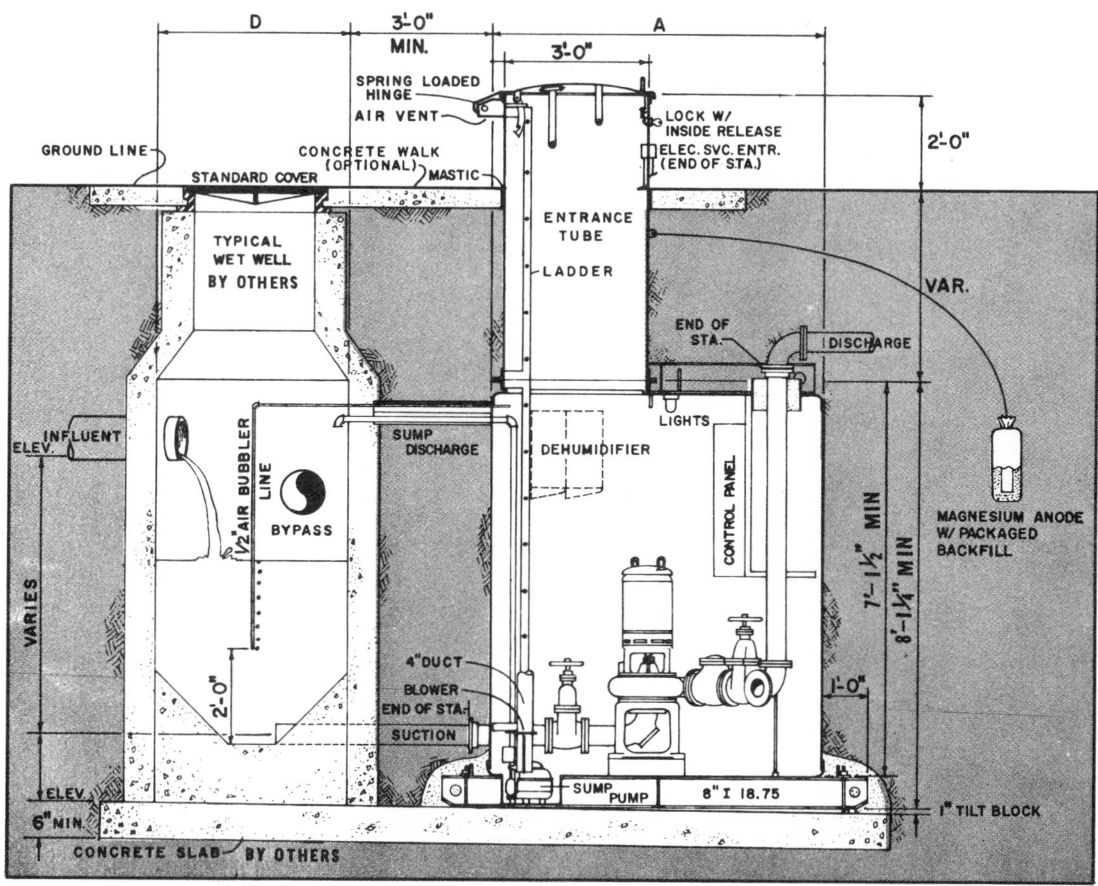

FACTORY-ASSEMBLED PACKAGE . . .

Offers advantages to the installing contractor. Requires only excavation and concrete slab with minimum electrical and piping connections.

RELIABILITY . . .

Assured by factory tests. Eliminates expensive field testing. One responsible source for service and spare parts reduces delays. Fully automatic controls require only routine operator attention.

FUNCTIONAL DESIGN . . .

Based upon years of experience. The application of time-proven components guarantees quality and performance, with efficient utilization of available space and environmental control features, which cannot be duplicated in field-fabricated stations. Interior chamber is attractively finished, dry, and well ventilated. Corrosion protection is provided through sacrificial anodes.

MAXIMUM COMFORT FOR OPERATORS . . .

Station interiors and operating equipment are attractively painted with reflective colors which brighten the working areas. A blower and dehumidifier keep the station free of odors and moisture-laden air.

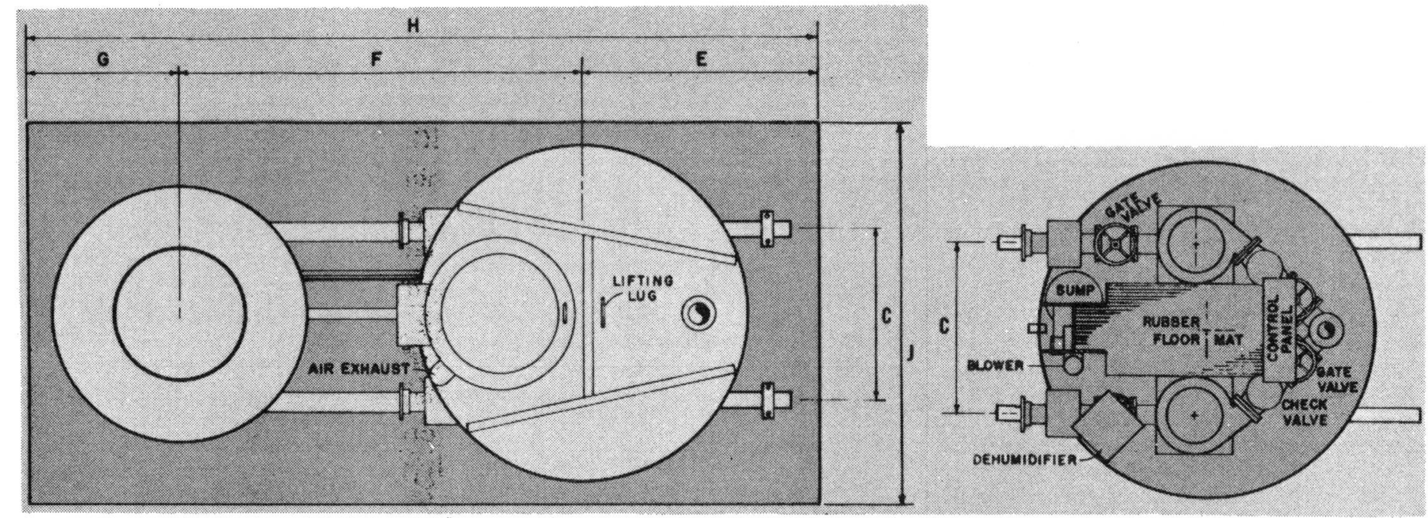

NOTE: Pipe sizes referred to in dimensional data chart include internal and discharge pipe sizes (Example: 4" = 4" x 4" x 4"). Combinations of pipe sizes (Example: 8" x 4" x 6") may change dimensional data! Refer to factory before final selection and design.

DIMENSIONAL DATA — 4" VERTICAL PUMPS

	4" Frame 2 Pumps		4" Frame 3 Pumps			6" Frame 3 Pumps
	4" Piping	6" Piping	4" Piping	6" Piping	8" Piping	10" Piping
A	7'-0" min.	7'-0" min.	7'-0" min.	7'-0" min.	8'-0" min.	9'-6" min.
B	1'-6"	1'-6"	1'-11⅞"	1'-11⅞"	1'-9⅞"	1'-8⅞"
C	3'-7⅝"	4'-0½"	3'-8⅞"	4'-1¾"	3'-5¼"	5'-3¼"
D MIN.	4'-0"	6'-0"	4'-0"	6'-0"	8'-0"	10'-0"
E MIN.	5'-0"	5'-0"	5'-0"	5'-0"	5'-6"	6'-3"
F MIN.	8'-6"	9'-6"	8'-6"	9'-6"	11'-0"	12'-9"
G MIN.	3'-2"	4'-2"	3'-2"	4'-2"	5'-2"	6'-2"
H MIN.	16'-8"	18'-8"	16'-8"	18'-8"	21'-8"	25'-2"
J MIN.	8'-0'	8'-0"	8'-0"	8'-0"	9'-0"	10'-6"

TEX-A-ROBIC CONTACT STABILIZATION

The Contact Stabilization Plant utilizes the ability of a healthy, well conditioned, activated sludge to adsorb and absorb the organic material in the incoming raw sewage.

Activated sludge, drawn from the reaeration zone, is intimately mixed with the incoming sewage for approximately 1½ hours based on a recirculation ratio of 1:1. Oxygen, both for stimulating bacteria growth and mixing, is introduced into the mass (mixed liquor) through specially designed diffuser heads.

The mixed liquor then flows through the clarifier inlet pipe to the clarifier loading well where the velocity is evenly dissipated and the liquor is diffused through the clarifier. The mixed liquor is retained in the clarifier approximately 4 hours providing ample time for the mixed sludge to settle and be collected in the clairfier, leaving a clear liquid in the upper portion of the clarifier.

The supernatant liquor passes over the adjustable effluent weir and is collected in the weir trough for discharge from the plant (chlorination is usually required before discharge into the receiving stream).

The sludge is then passed into the reaeration zone by an airlift pump for reactivation by aeration for a period of 7-10 hours. After this period of conditioning, the activated sludge is ready to be mixed with the raw incoming sewage. The mixing and reaeration is a continuous process.

Sludge digestion is not completed in the reaeration zone and at full design loads the solids concentration will increase until the plant efficiency is impaired. This condition is avoided by automatically wasting part of the sludge into the aerobic digester. Provision is also made for wasting scum directly from the clarifier to the digester.

The aerobic digester retains the sludge in the presence of oxygen until all putrescible matter is stabilized. In the digester stilling well the sludge and supernatant are separated. The supernatant liquor flows to the reaeration zone and the sludge is returned by air lift pump to the digester. The digested sludge is withdrawn periodically for disposal by tank truck or drying beds.

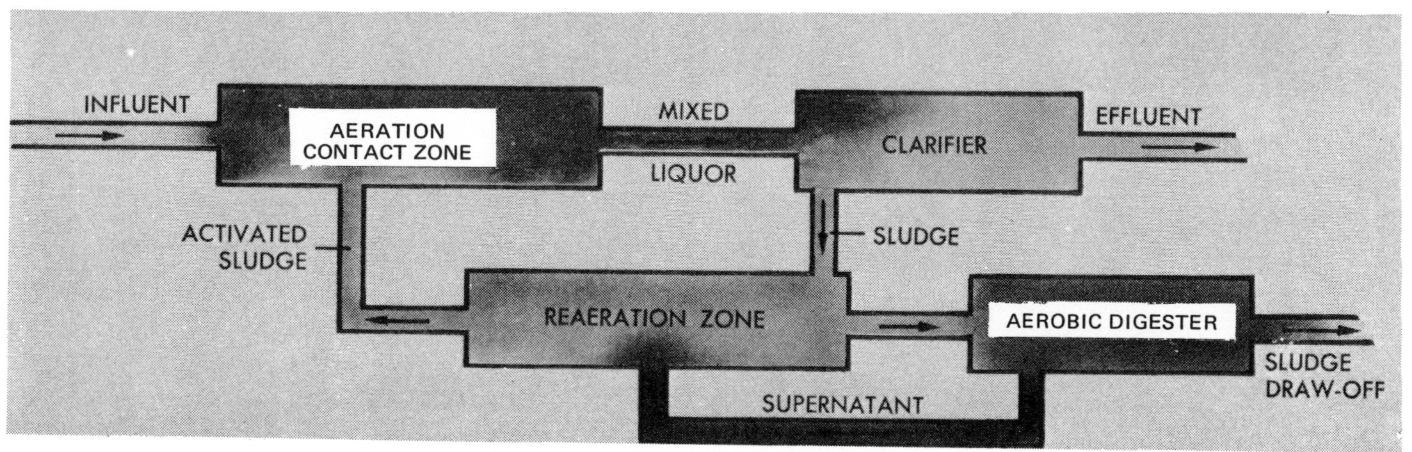

TEX-A-ROBIC CONTACT STABILIZATION

Population and Model Number	5-Day BOD lbs/day	FLOW		"A"	"B"	"C"	"D"	"E"	"F"	AIRLIFT PUMPS			PLANT PIPING			BLOWERS (6 PSI)	
		Max. GPM	Average GPD	Outside Dia.	Clarifier Dia.	Loading Well Dia.	SWD in Aeration Compartments	(Nominal) Clarifier SWD Ft. - In.	Overall Height Ft. - In.	Sludge Return	Sludge Waste & Digester Decant	Scum	Influent and Effluent	Air Main	Drain	Total Air Required (CFM)	No. of Motors & H. P.
500	85	87	50,000	22'-6"	11'-0"	3'-0"	15'-3"	13'-0"	16'-6"	3"	3"	3"	4"	4"	4"	109	2 @ 7½
600	102	104	60,000	24'-3"	11'-6"	3'-0"	15'-3"	13'-0"	16'-6"	3"	3"	3"	4"	4"	4"	126	2 @ 7½
700	119	122	70,000	26'-3"	12'-6"	3'-0"	15'-3"	13'-0"	16'-6"	3"	3"	3"	4"	4"	4"	144	2 @ 7½
800	136	139	80,000	28'-0"	13'-6"	3'-0"	15'-3"	13'-0"	16'-6"	3"	3"	3"	4"	6"	4"	162	2 @ 7½
900	153	156	90,000	30' 0"	14'-6"	3'-0"	15'-3"	13'-0"	16'-6"	4"	3"	3"	6"	6"	4"	182	2 @ 10
1,000	170	174	100,000	31'-6"	15'-0"	3'-0"	15'-3"	13'-0"	16'-6"	4"	3"	3"	6"	6"	4"	200	2 @ 10
1,250	212	217	125,000	35'-0"	16'-9"	4'-0"	15'-3"	13'-0"	16'-6"	4"	3"	3"	6"	6"	4"	244	2 @ 10
1,500	255	260	150,000	38'-3"	18'-3"	4'-0"	15'-3"	13'-0"	16'-6"	4"	3"	3"	6"	6"	4"	289	2 @ 15
1,750	297	304	175,000	41'-6"	19'-9"	4'-0"	15'-3"	13'-0"	16'-6"	6"	3"	3"	8"	6"	4"	348	2 @ 15
2,000	340	347	200,000	44'-3"	21'-0"	4'-0"	15'-3"	13'-0"	16'-6"	6"	3"	3"	8"	6"	4"	392	2 @ 15
2,500	425	434	250,000	49'-6"	23'-6"	4'-0"	15'-3"	13'-0"	16'-6"	6"	4"	3"	8"	6"	4"	481	2 @ 20
3,000	510	521	300,000	54'-3"	25'-9"	4'-0"	15'-3"	13'-0"	16'-6"	6"	6"	3"	8"	8"	4"	569	2 @ 20
3,500	595	608	350,000	58'-6"	27'-9"	5'-0"	15'-3"	13'-0"	16'-6"	6"	6"	3"	10"	8"	4"	658	2 @ 25
4,000	680	694	400,000	62'-6"	29'-9"	5'-0"	15'-3"	13'-0"	16'-6"	8"	6"	3"	10"	8"	4"	761	2 @ 30
5,000	850	868	500,000	69'-9"	33'-3"	5'-0"	15'-3"	13'-0"	16'-6"	8"	6"	3"	10"	8"	4"	938	3 @ 20
7,500	1,275	1,302	750,000	85'-6"	40'-6"	6'-0"	15'-3"	13'-0"	16'-6"	10"	6"	3"	12"	10"	6"	1,408	3 @ 30
10,000	1,700	1,736	1,000,000	98'-6"	46'-9"	6'-0"	15'-3"	13'-0"	16'-6"	10"	8"	3"	12"	10"	6"	1,850	3 @ 40
12,500	2,125	2,170	1,250,000	109'-9"	52'-3"	7' 0"	15'-3"	13'-0"	16'-6"	12"	8"	3"	12"	12"	6"	2,326	4 @ 30

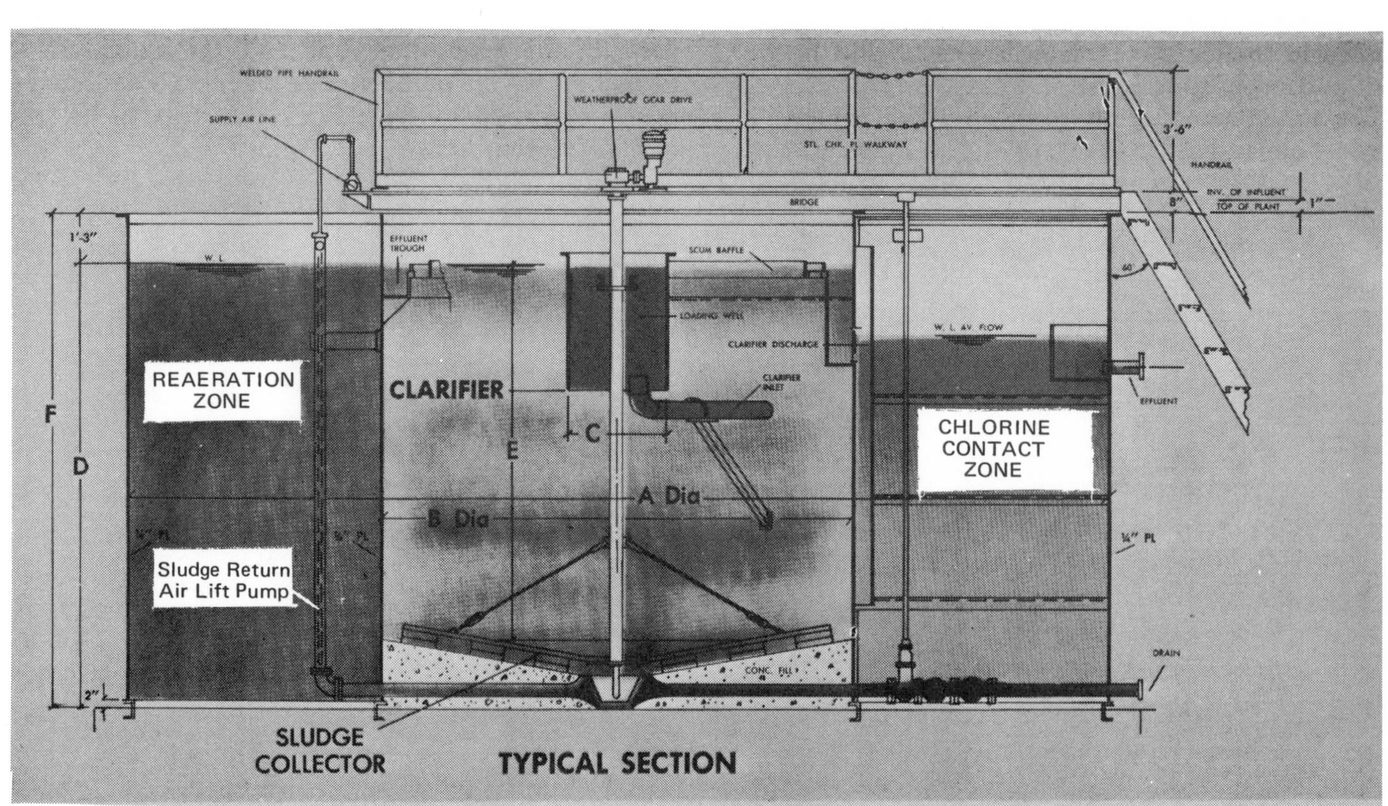

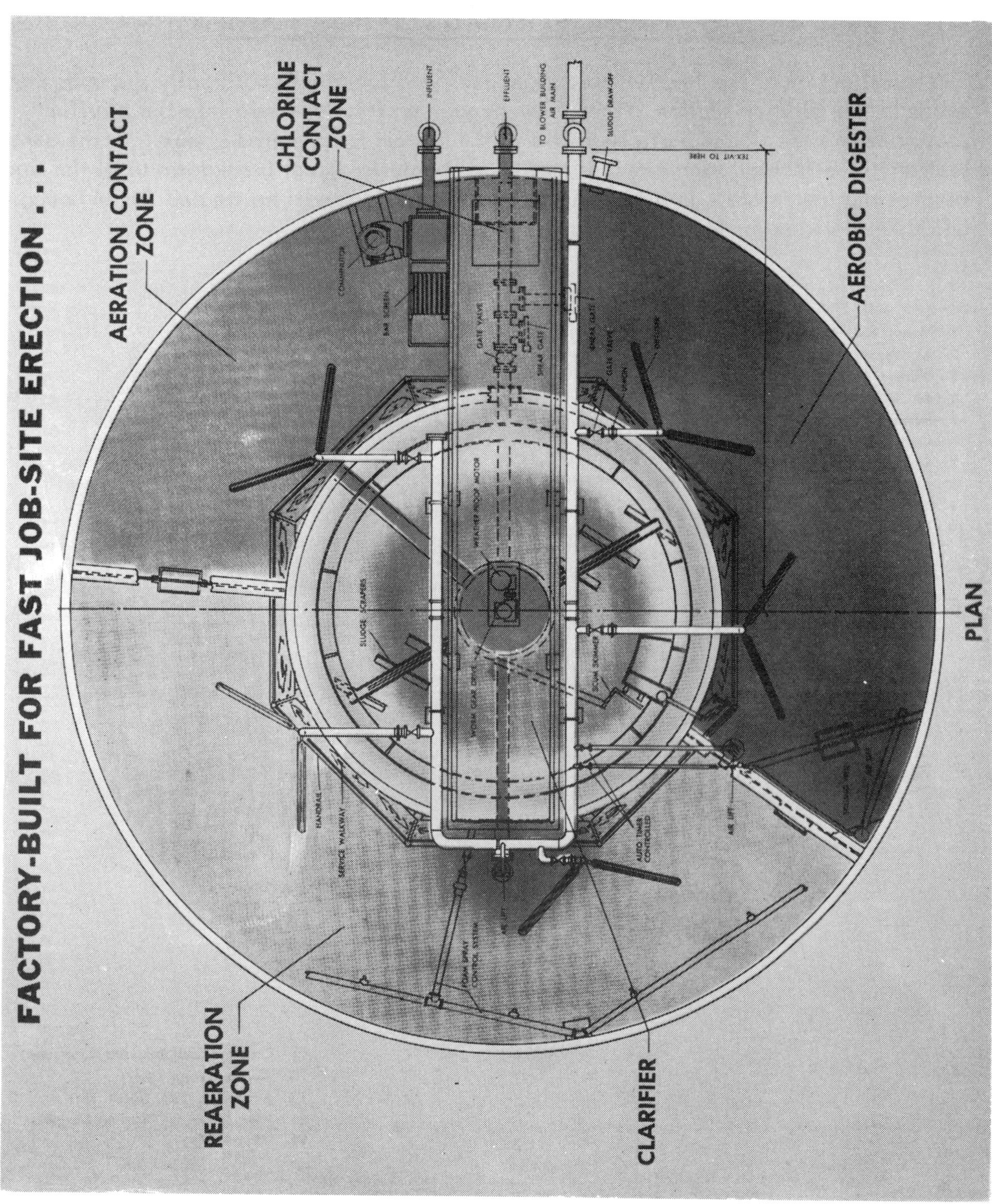
FACTORY-BUILT FOR FAST JOB-SITE ERECTION . . .
AERATION CONTACT ZONE
CHLORINE CONTACT ZONE
AEROBIC DIGESTER
PLAN
CLARIFIER
REAERATION ZONE

TEX-A-ROBIC EXTENDED AERATION PLANT

The spiral action of the Tex-A-Robic Extended Aeration Plant continually suspends the solids in the aeration section. They flow through an inlet aperture into the clarifier section where the sludge particles settle to the bottom to be pumped back into the aeration section. The recirculation provides additional bacteriological breakdown until the sludge reaches the inert state. In a typical installation, sewage will be treated for 24 hours, and then clarified for four hours in a single tank.

1 Motor, and Blower Housing
2 Air Supply Pipe
3 Diffuser Pipe
4 Air Lift Pump
5 Influent Pipe
6 Coated Steel Walls
7 Steel Braces
8 Bar Screen
9 Aeration Tank
10 Clarifier

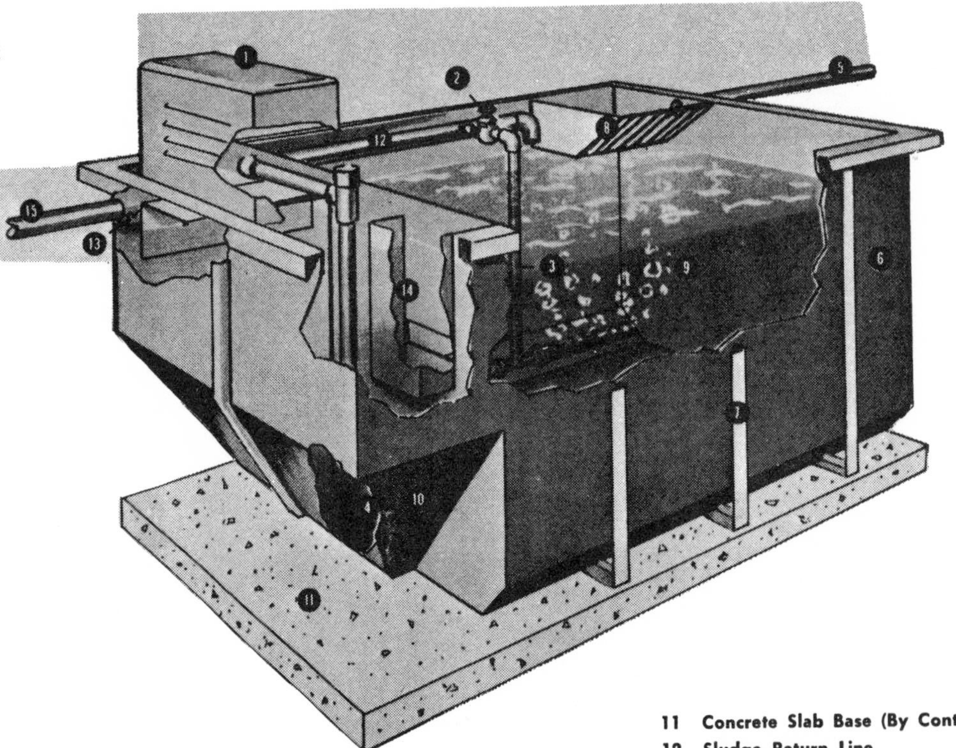

11 Concrete Slab Base (By Contractor)
12 Sludge Return Line
13 Adjustable "V" Notch Weir
14 Inlet Baffle to Clarifier Chamber
15 Effluent Pipe

STANDARD SELECTION DATA

Model	1	2	3	4	5	6	7
50M	5,000	833	240	190	1,666	1	8,700
75M	7,500	1,250	240	200	1,666	1.5	10,800
100M	10,000	1,666	240	297	1,250	1.5	13,600
125M	12,500	2,083	240	278	1,390	2	15,800
150M	15,000	2,500	240	278	1,365	2	17,666
175M	17,500	2,910	240	278	1,450	3	19,500
200M	20,000	3,330	240	274	1,666	5	18,100
250M	25,000	4,180	240	290	1,742	5	20,500

DIMENSIONAL DATA

A	B	D	E	G	H	K
13′- 5″	8′	9′- 3″	3′-6″	5′-6″	5′-6″	7′-4″
19′- 6″	8′	13′-10″	5′	5′-6″	5′-6″	7′-4″
23′- 8″	8′	18′- 6″	4′-6″	6′	5′	7′-4″
24′- 5″	9′- 8″	18′- 9″	5′	6′	5′	9′
29′- 2″	9′- 8″	22′- 6″	6′	6′	5′	9′
33′-11″	9′- 8″	26′- 3″	7′	6′	5′	9′
31′- 2″	11′-11″	24′	6′-6″	6′	5′	11′-3″
38′- 4″	11′-11″	30′	7′-8″	5′-6″	5′-6″	11′-3″

1. Aeration tank capacity in gallons
2. Clarifier capacity in gallons
3. Detention in clarifier in minutes
4. Rise rate in clarifier in gal./day/sq. ft.
5. Weir load gal./ft./day
6. Blower motor horsepower
7. Approximate weight

Design Criteria:

FLOW — 100 gal./day/capita
B.O.D. — 0.17 lbs of 5 day B.O.D./capita
AIR — 1,500 cubic ft. per lb. of 5 day B.O.D.
NOTE — Air requirements will vary. Check your particular state requirements.
DETENTION — Aeration tank — 24 hrs.
Clarifier — 4 hrs.

Notes:

1. In selecting models for other than domestic wastes contact factory, stating anticipated 24-hour flow and 5 day B.O.D.

Blower output in CFM is based on 4 PSI.

Concrete foundation by others.

Grate walkway on clarifier is standard with unit. Additional grating available on request.

Plants 100M, 125M, 150M, 200M, and 250M have two-hopper clarifiers with a sludge return pump for each hopper.

36 models are available: Consult factory for unlisted sizes and accessory equipment.

Communitor and chlorinator available on request.

Standby blower available on request.

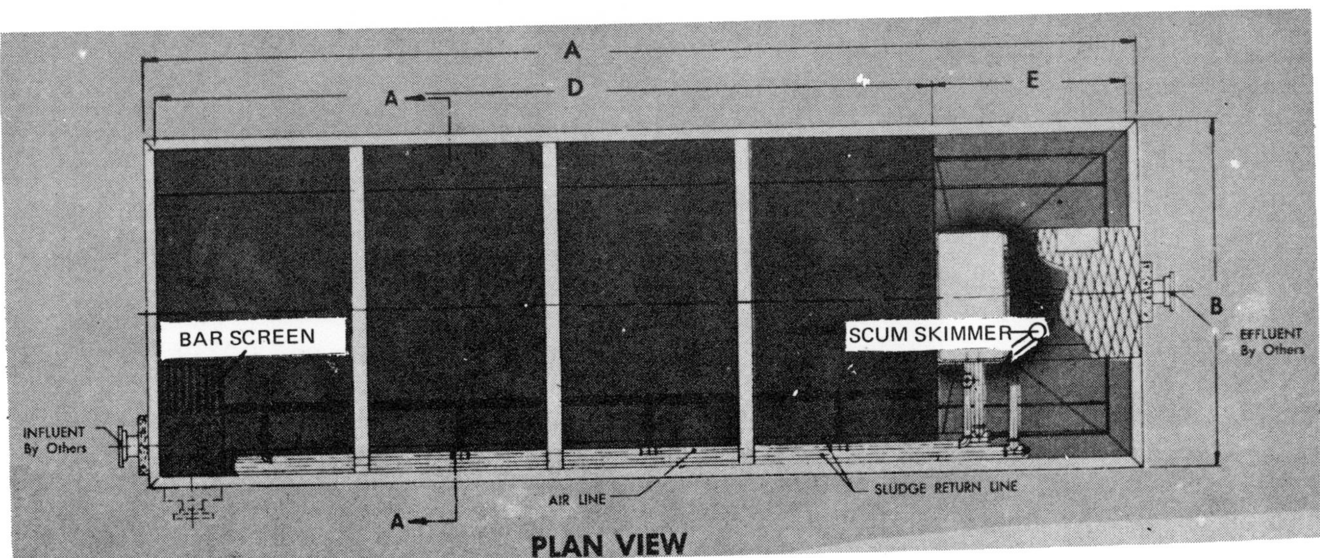

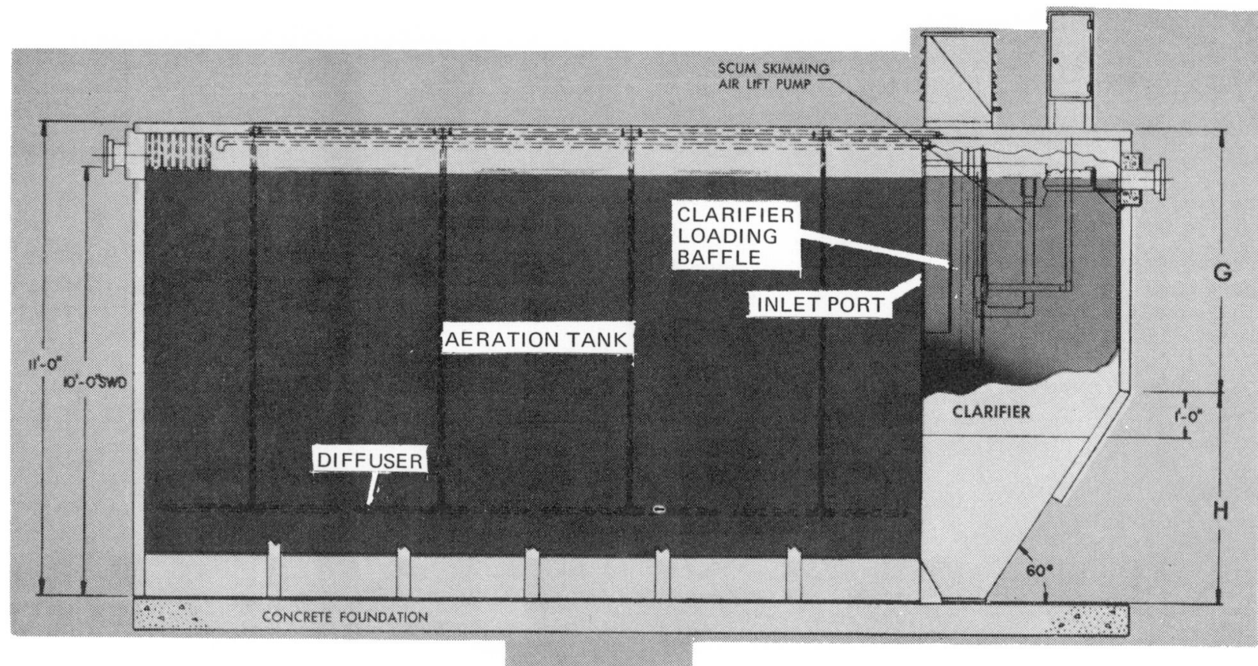
SCUM SKIMMING
AIR LIFT PUMP
CLARIFIER
LOADING
BAFFLE
INLET PORT
AERATION TANK
G
11'-0"
10'-0"SWD
CLARIFIER
1'-0"
DIFFUSER
H
60°
CONCRETE FOUNDATION
ELEVATION

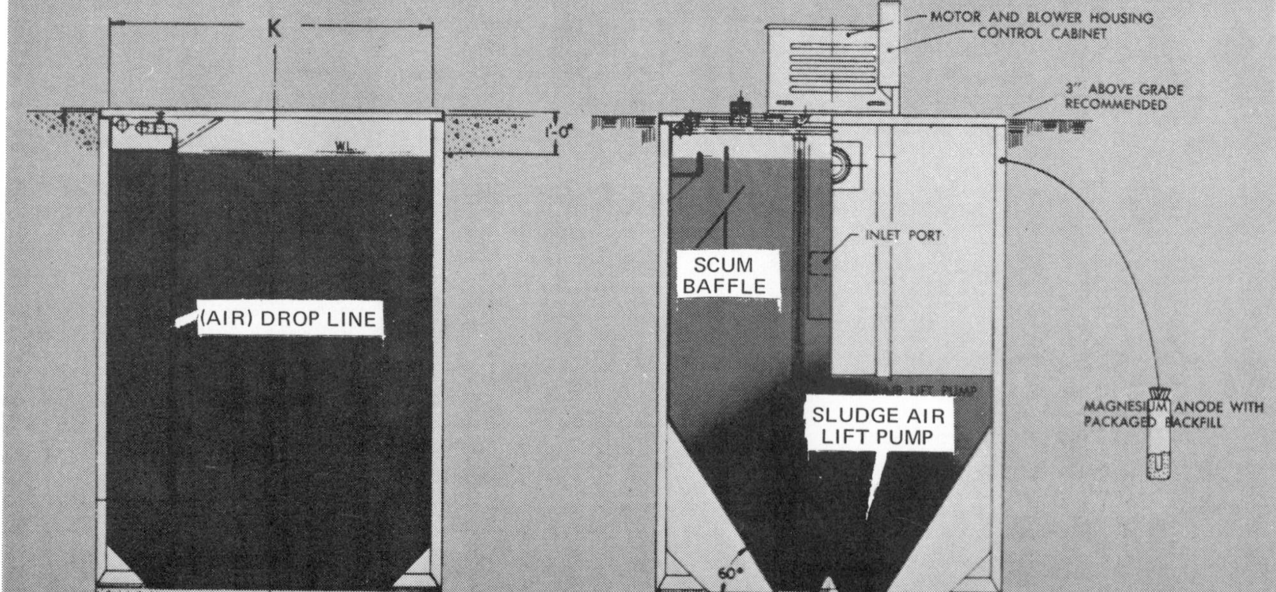
K
MOTOR AND BLOWER HOUSING
CONTROL CABINET
3" ABOVE GRADE
RECOMMENDED
1'-0"
W.L.
INLET PORT
SCUM
BAFFLE
(AIR) DROP LINE
SLUDGE AIR
LIFT PUMP
MAGNESIUM ANODE WITH
PACKAGED BACKFILL
60°
SECTION A-A
END VIEW

TF-2 TERTIARY FILTER

The Can-Tex TF-2 Tertiary Filter has been designed for the purpose of removing suspended solids from the effluent of an activated sludge wastewater treatment plant. The filter can be adapted to existing plants or plants of new construction of the aerobic process, or other conventional systems. The filter can be installed above or below grade depending on the need for the specific application. The TF-2 filter contains a dual compartmented deep-bed media filter and a singular filter effluent holding tank for backwash purposes. The controls are automatic, monitoring the operation of interconnecting valves to provide a completely self-contained automatically operated system.

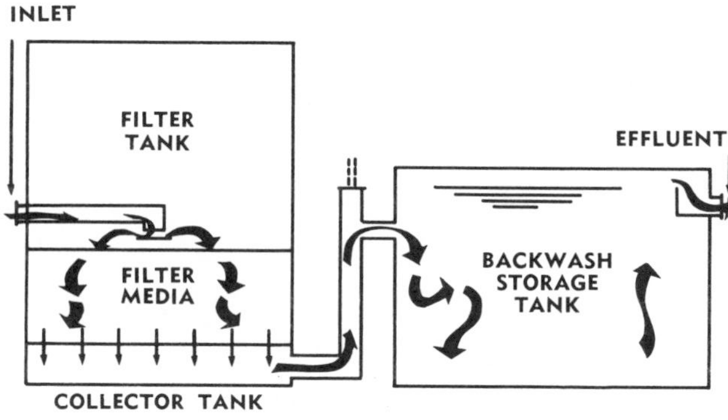

FILTER CYCLE

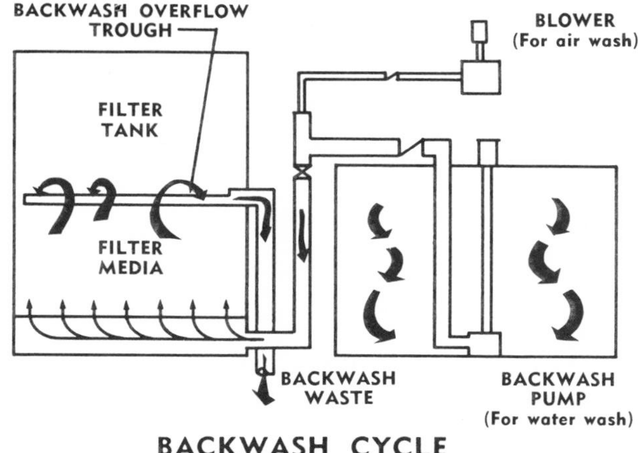

BACKWASH CYCLE

FLOW DIAGRAM

The TF2 Tertiary filter is a gravity type system for third stage treatment. Effluent from a waste treatment plant is directed automatically to one of the two filters and the filtering process continues until the operating filter requires backwashing due to accumulation of suspended solids. (See flow diagram) When this occurs, the system automatically switches the treatment plant effluent to filter No. 2 and the filtering process continues uninterrupted. While filter No. 2 is operating, Filter No. 1 proceeds into a backwash cycle. A controlled air scrubbing pre-wash action of the filter bed occurs for a predetermined period of time. The air wash scrubbing action allows the filter media to expand and the filtered solids to become dislodged in the controlled amount of water remaining in the filter. Upon completion of the air wash, a water backwash automatically and instantaneously removes the entrained solids from the filter media by way of a specially designed and placed overflow trough. The water which contains the filtered suspended solids is directed to the backwash line for disposal.

At no time is a filter bed left "dry." A level of water remains over the filter media between operating cycles, insuring maximum filtration rates. There is no interruption of the flow at any time. A cleaned filter bed is always ready to accept a new filter cycle completely automatically, regardless of the number of times the filtering or backwash cycles occur.

Pre-determined settings of the control system assures a backwash cycle which removes collected suspended solids from the filter media prior to the next filtering cycle. Upon completion of the backwash cycle, the dual deep bed media settles back to its original state in a uniform manner.

TF-2 TERTIARY FILTER TEST DATA

Actual Test Data Results on Domestic Wastes Tests Conducted Continuously Over a Ten Month Period

Unit Flow Rate Applied (gpm/ft.^{2})	Average Filter Cycle (Hours)	Average TSS Applied (PPM)	Average Effluent TSS (PPM)
1	64.0	30	5
2	34.4	18	2
3	19.9	28	4.5
4	12.6	20	3
5	4.8	16	2

Can-Tex Industries

TEX-A-ROBIC CONTACT STABILIZATION WASTE TREATMENT PLANT

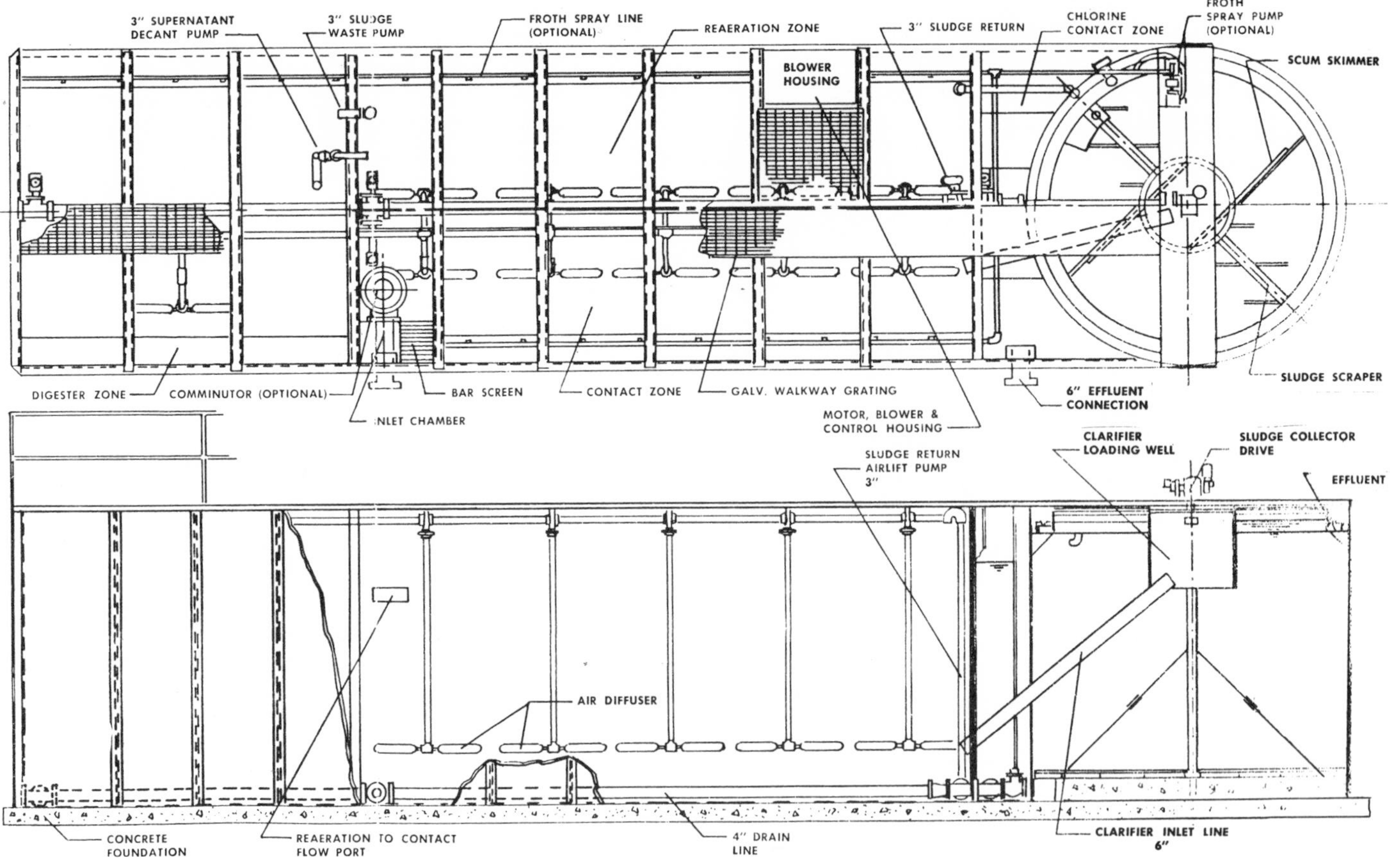

The Contact Stabilization treatment is a modified activated sludge process which separates the aeration volume into two zones.

Sewage flows into the treatment plant through the influent piping, entering an inlet chamber containing a manually cleaned bar screen or, optionally, a comminutor to cut up or shred waste solids. From the inlet chamber, the sewage flows to the "Contact Zone" where it is quickly, thoroughly mixed with the well-conditioned activated sludge from the "Reaeration Zone." The mixture then flows through a pipe to a loading well in the center of the "Clarifier" or "Settling Basin" where the inlet velocity energy is dissipated and the flow is evenly dispersed.

The sludge or solid material settles to the Clarifier bottom where it is collected and pumped back to the Reaeration Zone for re-cycling. In this Zone the sludge is aerated for approximately seven hours to provide a well-conditioned or "healthy" activated sludge to mix with the raw incoming waste in the Contact Zone.

The liquid in the Clarifier, free of settleable solids, overflows the weir and passes through a Chlorine Contact Zone where it is retained for approximately one-half hour and mixed with chlorine solution to disinfect the effluent. This effluent can now be emptied into an intermittent stream or dry water course without creating a nuisance.

Can-Tex Industries

TEX-QUAD RECTANGULAR SEWAGE PUMP STATION

The Tex-Quad Rectangular Station is the first major improvement in packaged sewage pumping equipment to be made in many years. Because of its rectangular shape many benefits are available. It permits the use of horizontal pumps. Additional interior space offers ease of servicing. In addition, the straight-line flow on the suction side eliminates the "L" shaped bend in suction pipes found on vertical pumps. Subsequently, the hydraulic flow in the suction inlet is improved.

Capacity can also be increased because the Tex-Quad's rectangular shape allows more room for the addition of pumps and motors as the need arises. Depending on the size station installed. up to six pumps can be used to handle capacities up to 10,000 GPM in a single factory-built unit.

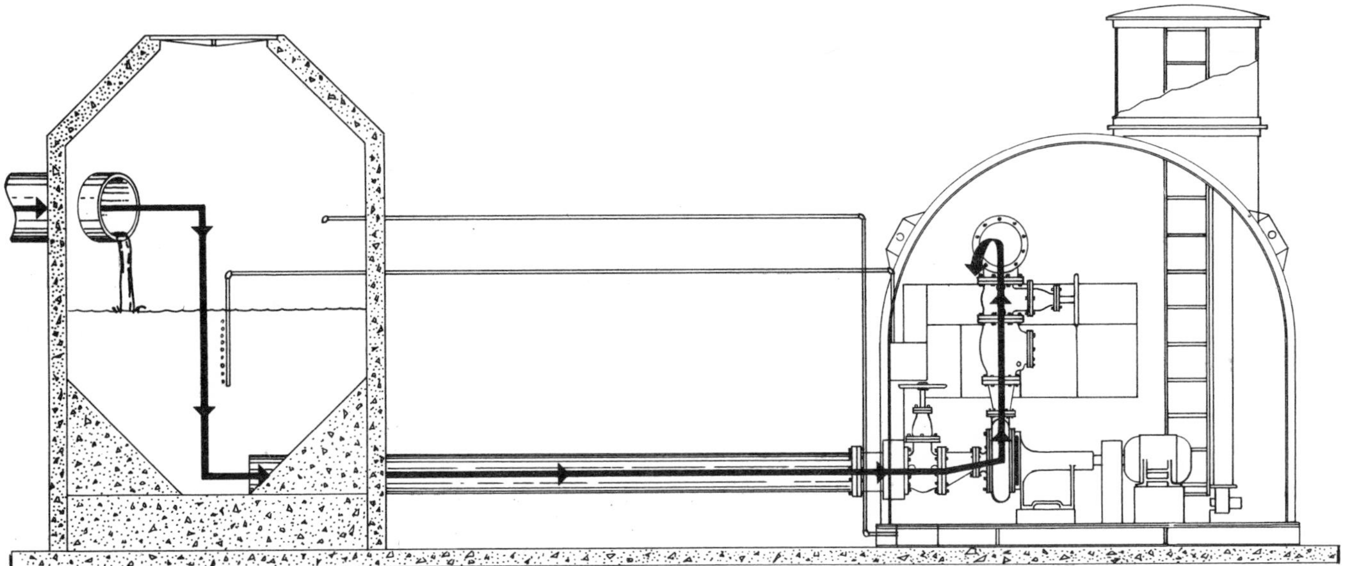

HOW IT WORKS

A compressor supplies a stream of air through the air bubbler pipe to the wet well. As sewage rises, back pressure in the line actuates a "low level" Tex-Trol pressure switch, energizing a magnetic starter and starts one of the pumps.

Sewage flows from the wet well, through the suction pipe, a gate valve, into the pump and then is pumped out through a check valve, discharge gate valve and discharge force main.

If sewage continues to rise above the starting level of the first pump, a "high level" Tex-Trol pressure switch actuates the starters of the other pumps and these pumps operate until the wet well level drops to a pre-determined shut-off point.

WASTEWATER CHLORINATION

Chlorine is used in both municipal and industrial sewage treatment for disinfection purposes, for the control of odors and septicity and for the reduction of biochemical oxygen demand (BOD). It is also used to prevent filter ponding, protect masonry structures, increase efficiency of filter beds and disinfect plant effluent before discharge.

The application of chlorine at specific points in the sewage treatment process is a part of standard treatment practice. The drawing below illustrates a typical ADVANCE wastes treatment installation using recirculated effluent, thereby saving water.

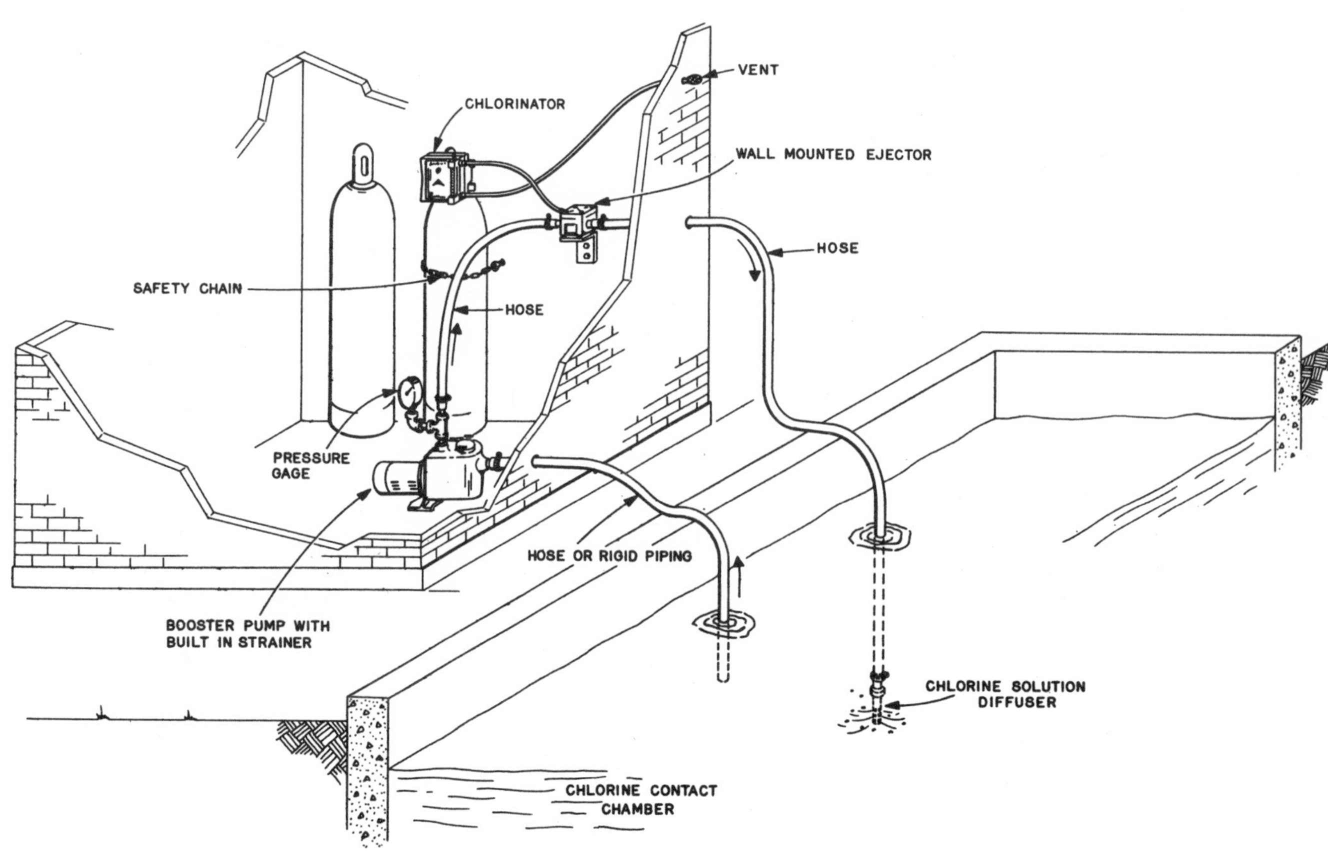

Chlorine is especially adaptable in treating many of the complex and varied wastewaters (liquid wastes) produced by modern industry. ADVANCE chlorination systems, because they are so flexible, with no high pressure piping, remote pressure couplings or flexible connectors, are ideal for industrial treatment applications. ADVANCE chlorination systems can be installed very easily in most existing plants, without expensive shut-downs, alterations and construction. Added benefits are derived from the fact that ADVANCE systems can be moved or expanded with ease.

Soundly engineered Carter Floating Covers and Gasholders are match-marked at the factory to assure swift, trouble-free erection. Designed to receive either mechanical or gas re-circulation type mixing systems for improved digestion and any type of roofing desired, the units are standardized in increments of 5′ from 15′ to 110′ in diameter. This means fast availability of drawings, quick fabrication and shipment and lowest possible cost. For maximum safety, all covers and gasholders are structurally designed for a dead load of 35 lbs./sq. ft. and a live load of 50 lbs./sq. ft. Special covers and numerous shapes and sizes can be produced on request.

CARTER PRE-FABRICATED ROOFING "Flex-Joint"Ⓣ

Carter "Flex-Joint" pre-fabricated roofing, consisting of readily removable sections, aluminum clad and styrofoam insulated, offers high insulating qualities, long life and convenient, easy installation. A special "Z" section joint provides ample adjustment for initial installation, assures vertical rigidity for foot traffic and readily adjusts to thermal expansion and contraction. Stainless steel screws and neoprene washers permit quick dis-assembly for access to the "attic space" in the cover.

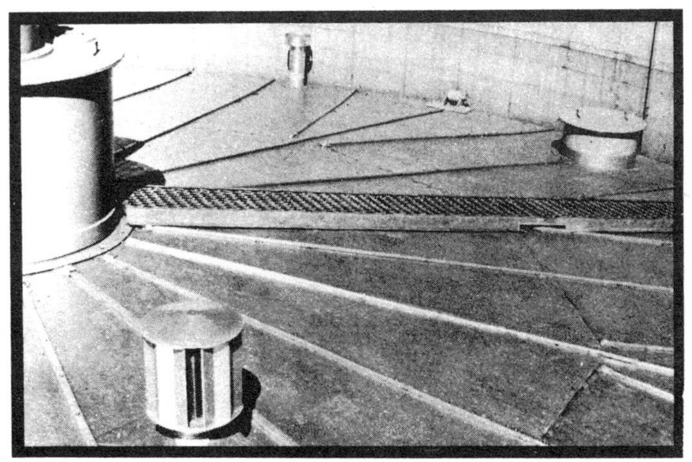

STUART-CARTER WALKING BEAM FLOCCULATION UNITS

The Stuart-Carter Walking Beam is the most advanced equipment available for flocculating and mixing in water, sewage and industrial waste treatment plants. The design permits adaptation into any floc basin regardless of size or configuration. Shape or depth of a basin do not affect its efficiency. Features include:

• No joints, gears, sprockets, or other moving parts underwater. Heavy duty construction • Laboratory results reproducible on a production scale • Fast forming floc • No floc fallout • No dead spots • Fewer drives per plant • Adaptability for covered tanks • Low installation and maintenance cost • Unique 5-year warranty, includes normal wear!

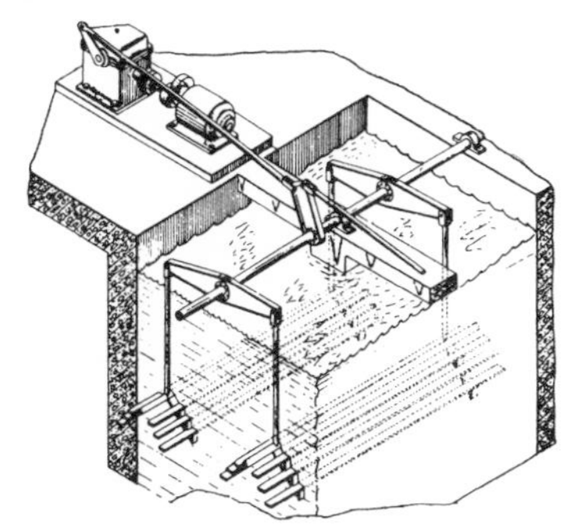

ATARA-CASCADE AERATION MIXERS

The Carter "Atara-Cascade Method" combines process efficiency, economy and maximum design flexibility in aerobic waste treatment systems....aerated lagoons, extended aeration, or activated sludge.

Economics are greatest when the Aerator-Mixers are placed in deep basins (10 to 25 ft.). The "Atara-Cascade" units are installed in unlined earthen basins (only waterline protected), concrete tanks, rubber-lined tanks--any existing or new tankage.

In each Aerator-Mixer, stacked venturi sections produce high pumpage rates. But true mixing rates are two to five times greater than pumpage through the unit, and produce intimate air-liquid contact.

There Are Five Zones of Aeration:

1. Air Discharge Zone - from the orifice to the primary venturi.

2. First Expansion Zone - venturi pressure differential breaks up the bubbles and sucks oxygen depleted mixed liquor through the ports.

3. Second Expansion Zone - same action as first.

4. Induced Mixing Zone - turbulent stream velocity above muzzle of gun induces "entrained" mixing, 2-5 times as great as the primary pumpage.

5. - Surface Aeration Zone - provides a high percentage of the oxygenation.

Multiple units for each application assure uniform distribution of mixing energy and uniform dissolved oxygen levels *throughout* the aeration basins or activated sludge tanks.

Efficient operation with very little maintenance because the "Atara-Cascade" unit *has no moving* parts, and is constructed of corrosion resistant materials.

Flexibility of operation is assured as mixing can be maintained while varying energy and oxygen input to the system.

No icing problems Complete submergence eliminates ice build up. The lack of moving parts protects the system even in the unlikely event the basin should freeze solid.

The 5-zone aeration process produces efficient oxygenation.

Pumpage is greatly augmented by induced mixing, created by turbulent stream discharge of the pumpage vertically into the basin.

Economy is achieved by low first cost, minimum power requirements, and little maintenance.

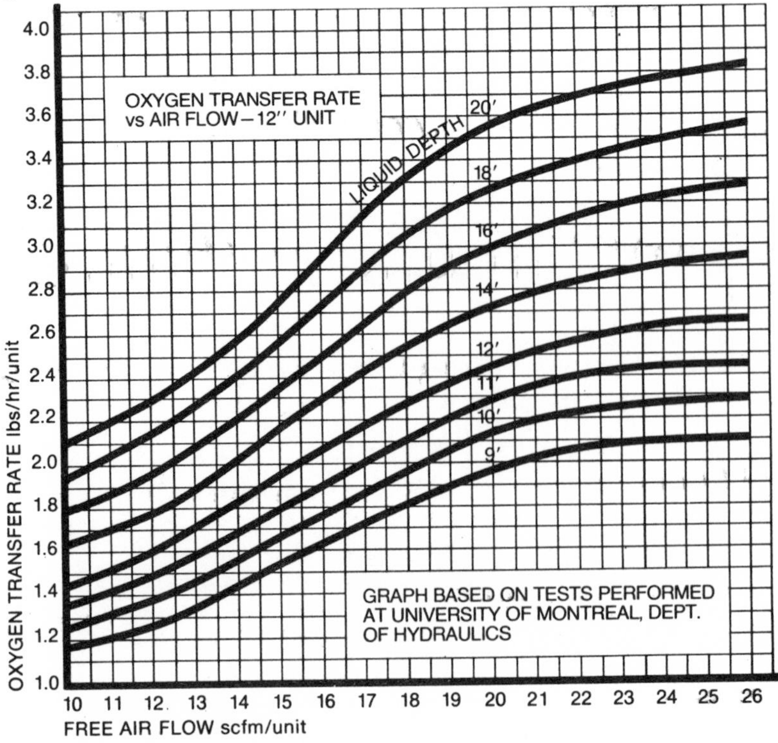

PRIMARY PUMPAGE vs AIR FLOW

Note: Not including entrained mixing of 2 – 5 times primary pumpage.

12″	8 CFM	1150 GPM
12″	12 CFM	1435 GPM
12″	15 CFM	1650 GPM
12″	20 CFM	1850 GPM
18″	20 CFM	2160 GPM
18″	30 CFM	
18″	40 CFM	
18″	50 CFM	

Ralph B. Carter Company

"AERO-HYDRAULIC" DIGESTION METHOD WITH INTEGRAL SYSTEM ANALYZER

FIGURE 1

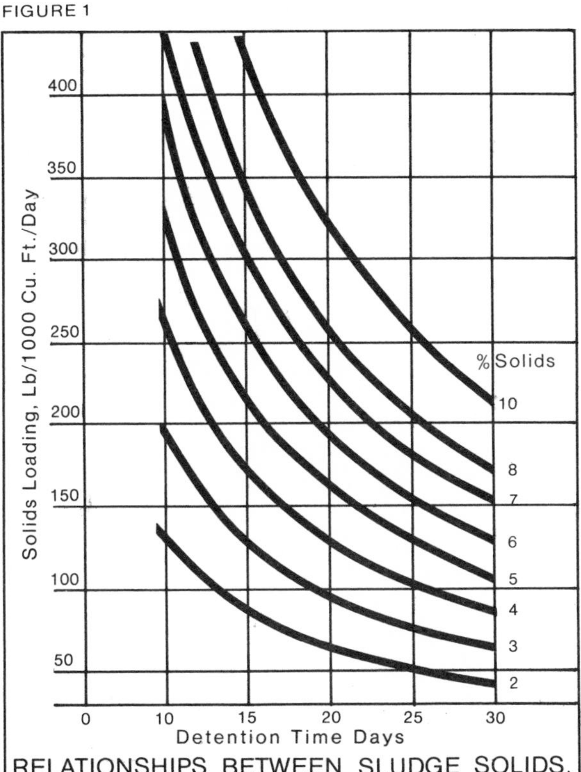

RELATIONSHIPS BETWEEN SLUDGE SOLIDS, DIGESTER LOADINGS, AND DETENTION TIME

With the above five points combined in one system, here are samples of what has happened:

1. In a 110 ft. diameter digester, a conventional gas mixing system required 20 Hp continuous compressor operation. The Carter A-H method used only 4 Hp when not heating and 15 Hp when heating. It *"required"* only half of this. Furthermore, the "conventional system" was one utilizing multiple energy application points, *but it did not utilize true interfering current mixing, or the entrained pumpage characteristic of the "A-H" gun.*

2. In a typical design, a mixed liquor flow of about 5.6 cu. ft./sec. is achieved using approximately 26 cfm of free gas (60°F and 14.7 psia), with a power consumption of less than 2 H.P.

3. In another design, mixed liquor pumpage is about 4.7 cu. ft./sec., using approximately 10 cfm of gas at discharge pressure (19 cfm of free gas).

4. An independent, three-month comparison study of gas production in two adjacent digesters at a plant, showed Carter A-H produced 14% more gas than the companion, conventional method of digester mixing.

5. In a controlled, radioactive tracer test of mixing in two digesters, the Carter A-H method showed: tracer detected after 1.3 hours; tracer concentration uniform after 8 hours; actual tracer withdrawal matched mathematic prediction. Comparison, conventional mixing system failed to produce uniform concentration after 19 hours; experimental rate of tracer removal failed to follow mathematic model.

In late 1964, we introduced the Carter Aero-Hydraulic method emphasizing the following five points of its uniqueness:

1. Heating and mixing should be managed as a single, integrated factor for true, high-rate digestion.

2. Single point mixing in unbaffled tanks produces few interfering currents. It is best described as "pumping" not "mixing." It occurs principally in the immediate vicinity of the mechanism, thus is very inefficient.

3. Three horizontally dispersed points of energy application time sequenced to produce varying flow patterns, are the minimum required for true mixing of organism and substrate in an unbaffled vessel of large diameter. The interfering currents thus produced, contribute even more to *mixing efficiency* than the dispersion of energy application points.

4. Time sequenced schedule of pumpage from opposite Guns optimizes true interfering current mixing within the entire working contents; but we also leave the very bottom, central cone area of the digester relatively undisturbed. This is an important feature in aiding heavier inert solids (such as sand) to pass thru the system as rapidly as possible. It also minimizes withdrawal of sludge which is still in its gas producing stage.

5. The heated, Aero-Hydraulic Gun, with its single large-diameter gas bubble, (D. in Fig. 2), is essentially an "expandable piston" pump. Its efficiency for inducing liquid motion within a vessel far exceeds all previous devices, because the stream discharge at its "muzzle" is designed to entrain a secondary mixing of 2-3 times the flow through the gun.

FIGURE 2

COMPRESSOR EQUIPMENT AND SYSTEM CONTROLS

UTILITY GAS OR OIL

DIGESTER

HEATING & AGITATING GUNS

D C A B F

DIGESTER GAS

BOILER AND CONTROLS

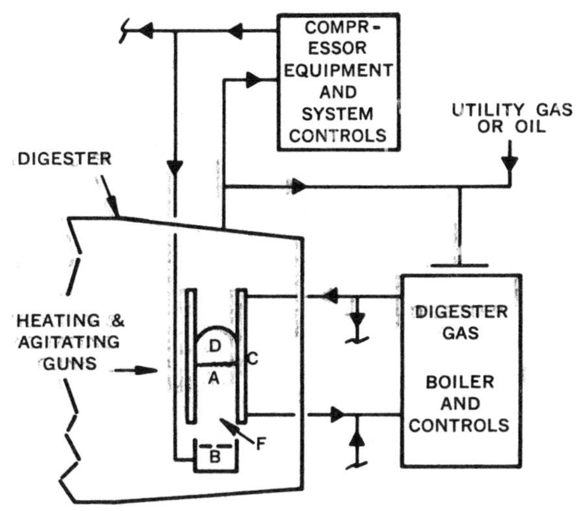

DIGESTER SCHEMATIC
CARTER AERO-HYDRAULIC SYSTEM

System design is simplicity itself—eliminating much conventional equipment, piping, valves and controls, pumps; and considerably reducing control building size and construction costs.

CARTER AERO-HYDRAULIC METHOD FOR ANAEROBIC DIGESTION

60 ft. diameter installation

NOTES:

PIPING SHOWN IN BROKEN LINES ARE OPTIONAL AND SERVE THE FOLLOWING PURPOSES

[A.] FOR TRANSFER FROM SECONDARY TO PRIMARY (THIS MIGHT OCCASIONALLY HELP COUNTERACT THE EFFECTS ON THE PRIMARY OF A SHOCK LOAD OF TOXIC WASTE).

[B.] SUPERNATANT SELECTOR: IF USED SOME OF THE OTHER SUPERNATANT LINES MAY BE OMITTED.

[C.] SUPERNATANT PROVISIONS ON PRIMARY (SHOWN DOTTED) PERMIT EMERGENCY OPERATION WITH INTERMITTENT MIXING, AS A SINGLE STAGE DIGESTER AND PROVIDE FLEXIBILITY IN PROBLEM SITUATIONS.

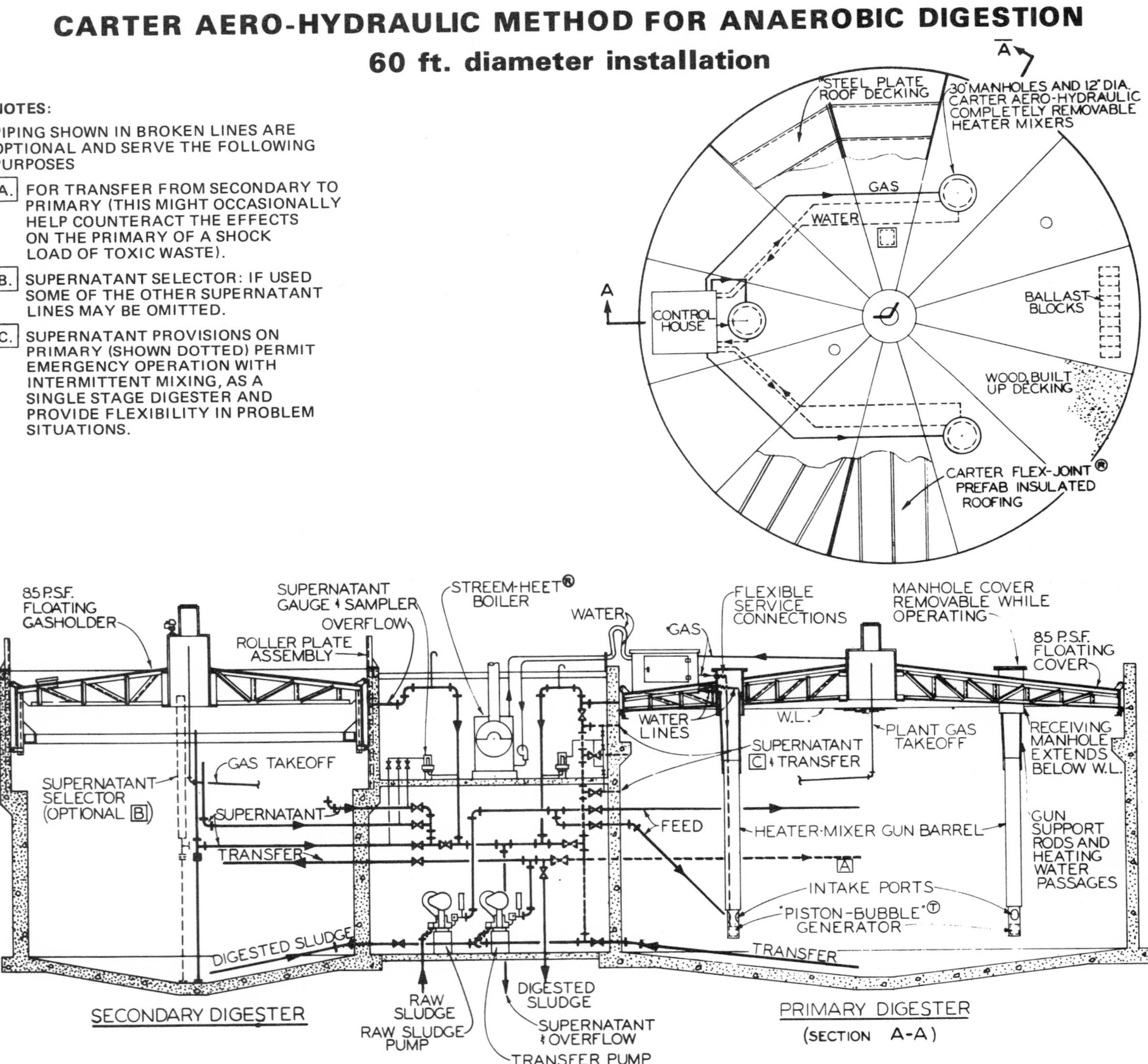

PNEUMATIC EJECTORS

As manufacturers of both centrifugal pumps and pneumatic ejectors, Ralph B. Carter Company has come to recognize that a true need exists for both devices. For relatively high flow rates, in sewage lines large enough not to be clogged by the solids anticipated, the centrifugal pump is often the most logical selection. On the other hand, the pneumatic ejector still is the only good solution for the installation that must dependably handle small flows, entering and leaving through relatively small lines and fittings. It is easily serviced and requires no wet well. The pneumatic ejector has a proven record of efficiently handling sewage and other solids laden liquids without aesthetic nuisance and a minimum of operator attention. Following is a list of the more obvious advantages:

1. When compressors and controls are mounted separate from the ejector pots (and above flood level), the ejector installation can still operate, even if the pot and pit are flooded.
2. If a pneumatic ejector is installed with an emergency air connection, the station can be kept operating during a power failure by attaching a portable compressor unit to the emergency air connection.
3. Equipment life is generally longer with a pneumatic ejector, particularly when cast iron pots are used.
4. The pneumatic ejector station requires no wet well. Fewer structures means lower maintenance cost, and generally lower first cost.
5. With an Eject-O-Pac an inexpensive installation can frequently be made by setting it in a pit constructed from standard 6' or 8' diameter sewer pipe--still no wet well is required.
6. Ejectors are much more easily serviced, particularly when compared with submersible pumps.
7. At the pumping rates usually involved, any centrifugal pump can't pass solids larger than 2". Even the smallest ejector, when equipped with special non-clog check valves, will handle 4" solids.
8. A single air supply system can operate many separate lift stations, thereby eliminating multiple service operations.

How the Ejector Works:

The pneumatic ejector is an extremely simple and workable mechanism. It consists of a receiver or "pot" that allows flows to enter by gravity without obstruction. Once the pot is filled, compressed air displaces the liquid contents up to a higher outlet. The pneumatic ejector is unique as a pumping mechanism inasmuch as no mechanical parts are used to do work directly on the material being pumped and there are no obstructions to restrict full-pipe flow.

Under normal flow conditions, the equipment is designed to operate within a one minute cycle. This cycle consists of two phases. First, the filling of the pot, and then the discharging of its contents. Operation is fully automatic with a choice of electrical or mechanical control systems.

Built-In Flexibility

Since the incoming flow to any pumping station is seldom constant, particularly in a sewage lift station where it may vary from high peak loads to a mere trickle, automatic flow adjustment would provide a distinct advantage. The pneumatic ejector does just that. Should the flow rate exceed the nominal rated capacity of the pot, the pot will fill in less than the normal period, speeding up the cycle. During periods of low flow, where the time required to fill the pot is greater, the system simply awaits the completion of the filling half of the cycle before discharging. There is no wasted work.

Ralph B. Carter Company

EJECT-O-PAC DUPLEX PNEUMATIC SEWAGE EJECTOR

DIMENSIONS																						
CAP GPM	A	B	C	D	E	F	G	H	J	K	L	M	N	P	R	S	T	U	V	W	X	Y	Z
30	6"	4"	2'-8"	3'-1"	3'-4"	8"	2'-1 3/4"	4'-4 7/8"	5/8"	6"	3'-4"	1'-10"	1'-10 3/4"	2'-5 1/2"	6'-6"	3'-2"	3'-4"	2'-4"	3'-6"	7'-0"	3/4"	8'-0"	7'-0"
50	6"	4"	2'-8"	3'-1"	3'-4"	8"	2'-1 3/4"	4'-4 7/8"	5/8"	6"	3'-4"	1'-10"	1'-10 3/4"	2'-5 1/2"	6'-6"	3'-2"	3'-4"	2'-4"	3'-6"	7'-0"	1"	8'-0"	7'-6"
75	6"	4"	2'-8"	3'-1"	3'-4"	8"	2'-1 3/4"	4'-4 7/8"	5/8"	6"	3'-4"	1'-10"	1'-10 3/4"	2'-5 1/2"	6'-6"	3'-2"	3'-4"	2'-4"	3'-6"	7'-0"	1"	8'-0"	8'-0"
100	8"	6"	3'-6 7/8"	3'-11 7/8"	4'-3 7/8"	9"	2'-3 1/2"	5'-0 1/8"	5/8"	7 1/4"	4'-0"	2'-0"	2'-4 3/4"	3'-3 1/8"	9'-0"	4'-4"	4'-8"	2'-9"	4'-0"	8'-0"	1 1/4"	10'-6"	8'-6"
150	8"	6"	3'-6 7/8"	3'-11 7/8"	4'-3 7/8"	9"	2'-3 1/2"	5'-0 1/2"	5/8"	7 1/4"	4'-0"	2'-0"	2'-4 3/4"	3'-3 1/8"	9'-0"	4'-4"	4'-8"	2'-9"	4'-0"	8'-0"	1 1/2"	10'-6"	9'-0"

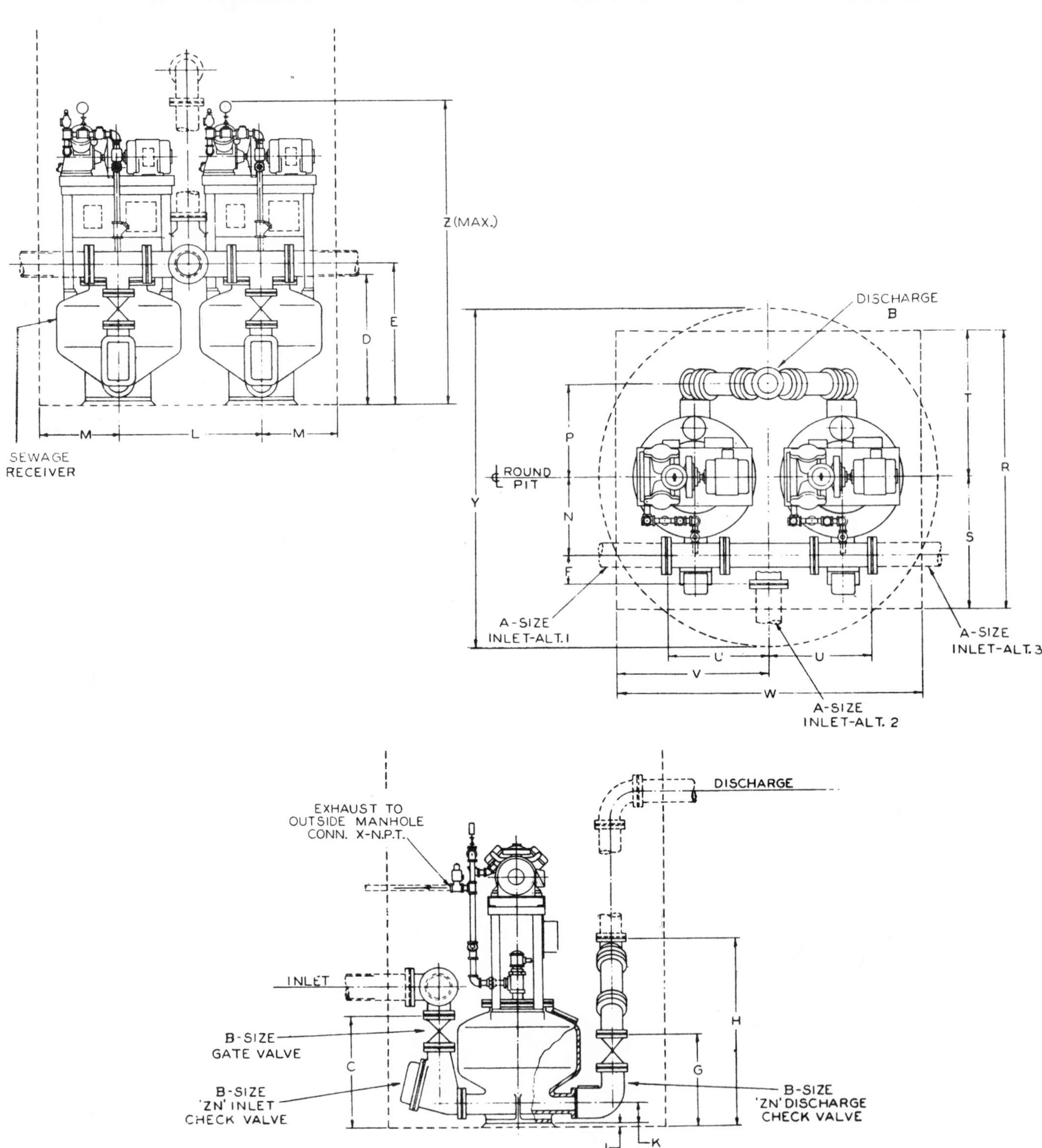

Carver-Greenfield Corporation

DEHYDRATION PROCESS

The Carver-Greenfield Process completely removes all moisture from sludges and other waste products. The unique oil-based evaporative-drying technique is odorless, creates no air or stream pollution, and produces dry, sterile solids of uniform consistency that can be burned as boiler fuel, stored without decomposition or bagged and marketed.

In processing many types of solutions, slurries and sludges, *all* available water can be removed and purified, resulting in complete elimination of waste water disposal problems. In any case water leaving the system is generally of distilled grade, low in biological oxygen demand (B.O.D.), and solids-free.

Depending on composition, processed solids can be marketed as food products, animal feed, fertilizer, reclaimed chemicals, etc. Frequently, sale of valuable by-products will more than offset the cost of operation, in some cases even producing a sizable profit. Since the solids are sterile, dry and stable, disposition is simple and economical. Evaporation of the moisture substantially reduces bulk and consequently lowers handling costs.

The dehydrated solids can also be burned in a boiler furnace to produce steam. Often the high BTU value will deliver enough steam to run the process and supply steam for other uses as well.

SYSTEM COMPONENTS AND OPERATION

There are only three basic components in a Carver-Greenfield system — a fluidizing tank (1), an evaporator (2) and a centrifuge (3).

The sludge is mixed with oil in the fluidizing tank and metered into an evaporator where the water is removed. (Since the water vaporizes at a lower temperature than the oil, carefully controlled temperatures in the evaporator drive off the water but *not* the oil.) In most cases, the remaining dry solids and oil are then separated by a continuous centrifuge.

Mixing oil with the raw feed is the key to the process. The oil keeps the feed in a fluid state *during* and *after* the evaporation of the water. Actually, the moisture-free oil-solids combination — a suspension of dry solids in the oil — is even *more* fluid than the oil-water-solids mixture entering the evaporator.

In addition to the three basic components, additional units are added as necessary for each situation. For example, if the sludge contains large solids — which is often the case — a sludge grinder (4) is usually placed before the fluidizing tank.

Depending on cost factors, it is often desirable to use a multi-stage drying evaporator rather than the single-stage unit.

The solids will contain oil, even after centrifuging; some as little as 3%, others as much as 50%. Depending on the end use of the solids, a further extraction of oil may be indicated. Each case must be viewed in light of the user's requirements.

If the solids are to be burned for the production of steam, a boiler furnace (5) can be included in the system.

FLOW DIAGRAM

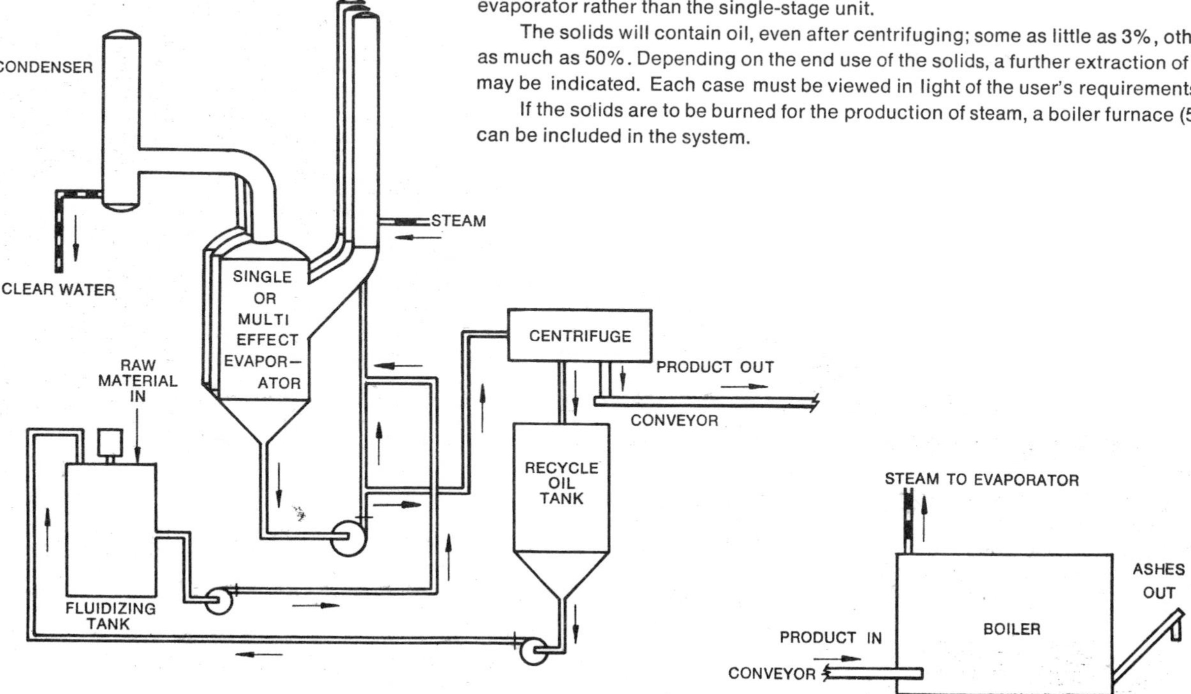

FLOATING DISC TYPE OIL RECOVERY UNIT

NEW FROM CENTRI-SPRAY® Corporation, Centri-Clere® Filter Division — A Floating Oil Skimmer for use in settling ponds, large tanks, rivers, or other bodies of water to remove surface tramp oils efficiently and in a condition that permits reclaiming or resale.

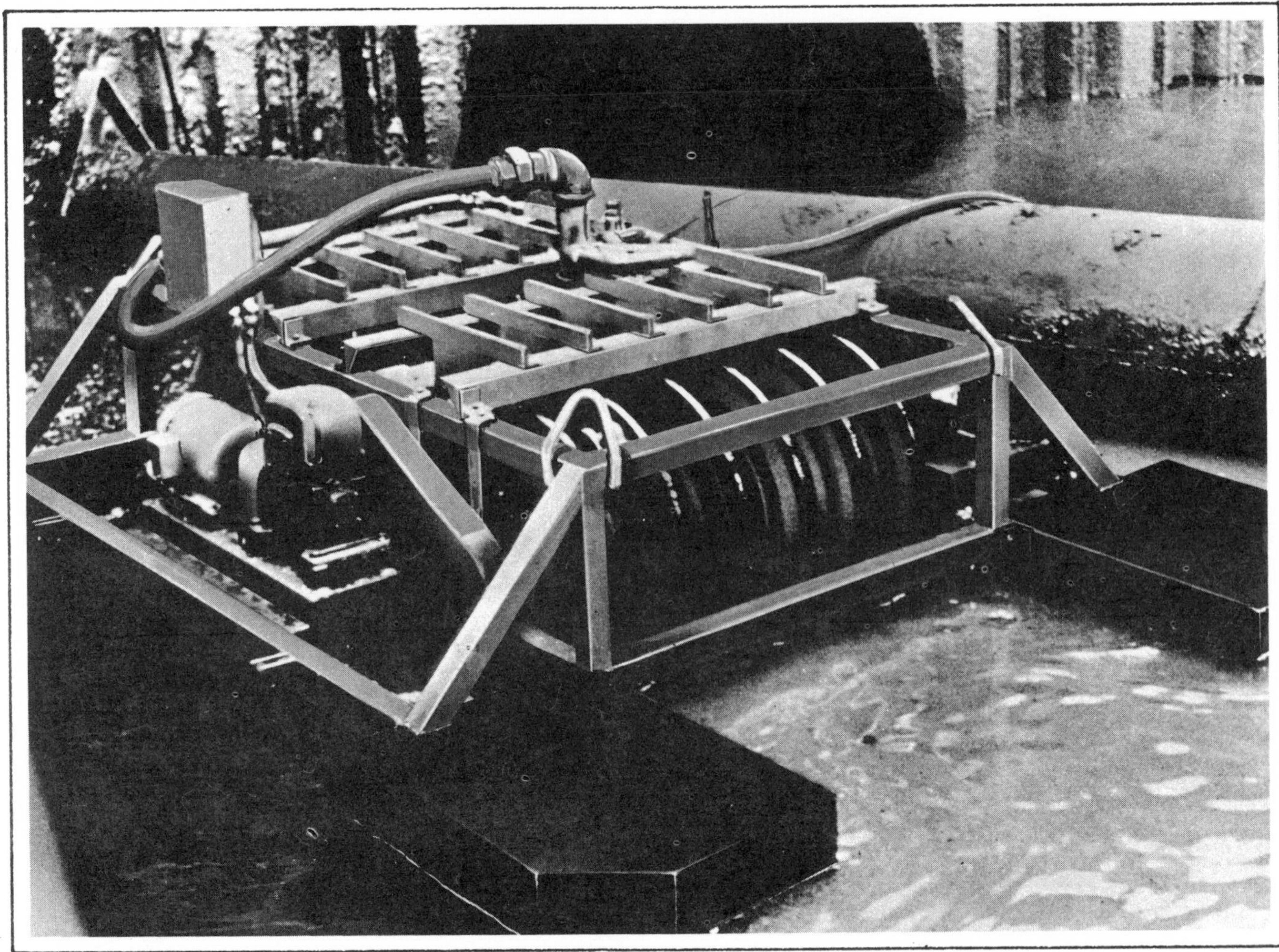

This multi-disc unit is capable of removing 350 GPH of 500 SSU oil at 70° F. or 600 GPH of 2500 SSU oil. Surface floating oil adheres to the continuously rotating multi-disc assembly. Scrapers remove the oil which is deposited in a self contained oil storage tank. An integral sump pump transfers this oil from the skimmer tank to a permanently located storage tank through appropriate piping.

A floating oil baffle completes the installation, thus preventing oil contamination of sewers, lakes, or rivers, and at the same time provides a valuable end product.

The unit shown contains 12, 24″ diameter discs, 6 on each end. Other sizes are availabe on request ranging from a single disc stationary mounted unit through any number of multi-disc units as required for your particular installation.

A barge type self powered unit is also available. Details will be furnished on request.

BELT TYPE OIL RECOVERY UNIT

The Centri-Clere® Belt Type Oil Recovery Unit consists of an endless belt made of selected synthetic materials, and suspended from a special design driven pulley. A low horsepower motor drives the belt through a speed reducer and chain drive at a specific speed. The lower portion of the belt is immersed in an oil-supporting media or a stratified fluid, and is held under tension by a weighted, bottom idler pulley, equipped with guides to maintain belt position.

As the belt descends into the liquid, the floating oil adheres to both surfaces of the belt and is carried downward around the bottom idler and upward over the top pulley. Two horizontally opposed scrapers, mounted in staggered position, wipe the oil or stratified organic liquid from both sides of the belt. The recovered oil drains into a trough and is deposited into a storage container.

The Centri-Clere® Oil Recovery Unit is the most efficient unit of its kind available to industry. It will operate *unattended* in an area of tranquil oil accumulation, and continuously remove all oil that contacts the belt. The recovered oil contains less than 3% water, and is in a condition that permits *resale* or *reclaiming*.

There are many installations where the recovered oil is returned to fuel storage areas and eventually burned in the plant heating equipment, thus reducing operating expenses. In other installations it is sold, *without further refining*, to asphalt companies, road oilers, or reprocessing firms, and thus becomes a profitable by-product that quickly pays for the installation.

In all installations, *water pollution from oily waste is prevented.*

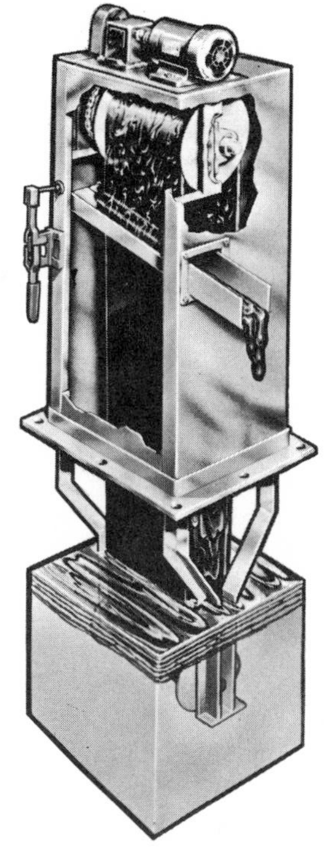

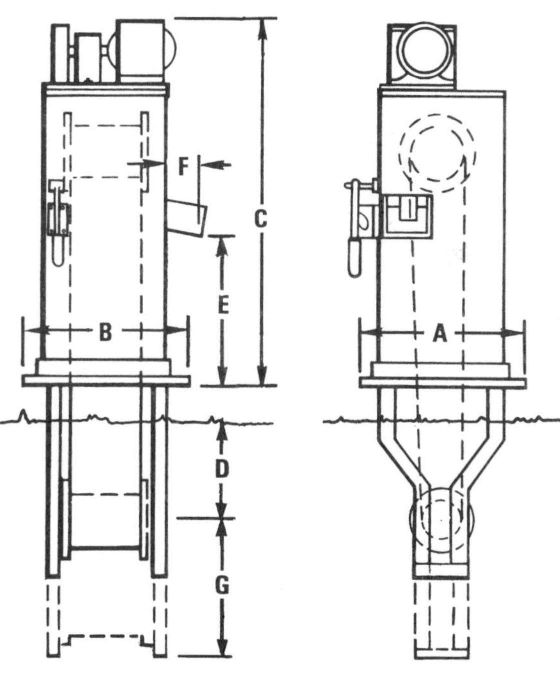

INDUSTRIAL APPLICATIONS

Many different industries have found that the Centri-Clere® Oil Recovery Units have *eliminated water pollution* by reclaiming oily waste, and at the same time provided them with a profitable by-product. Some of these are as follows:

- Oil Refineries
- Steel Mills
- Chemical Processing
- Food Processing
- Metal Working
- Automotive Manufacturing
- Ship loading docks
- Industrial sewage disposal systems

TYPES OF MATERIALS RECOVERED

Many floating oils and stratified organic or synthetic materials can be efficiently recovered from the surface of a supporting liquid, as listed below. It is also possible to recover suspended solids contained in the stratified layer along with the oil.

- Machine oils
- Lubricating oils
- Hydraulic oils
- Mineral seal oil
- Heavy Bunker "C" oil
- Very thin cooking fats
- Other organic synthetic fluids

RECOVERY OF SOLUBLE OR EMULSIFIED OILS

Specific chemicals can be added to a solution to split-out soluble oil and other dispersed fluid components, causing them to rise to the surface. These treated soluble oils can then be removed by the recovery unit.

GENERAL SPECIFICATIONS

Belt	Mounting Dimensions							Motor	Max. Cap.
Width	A	B	C	D	E	F	G	H.P.	GPH
6″	14¼	19¼	33″	10″	9″	5″	7½	¼	30
12″	26¼	29¼	91¼	18″	54″	9″	10	½	75
18″	26¼	35¼	91¼	18″	54″	9″	10	1	112
24″	26¼	41¼	96″	18″	54″	4″	10	1	150

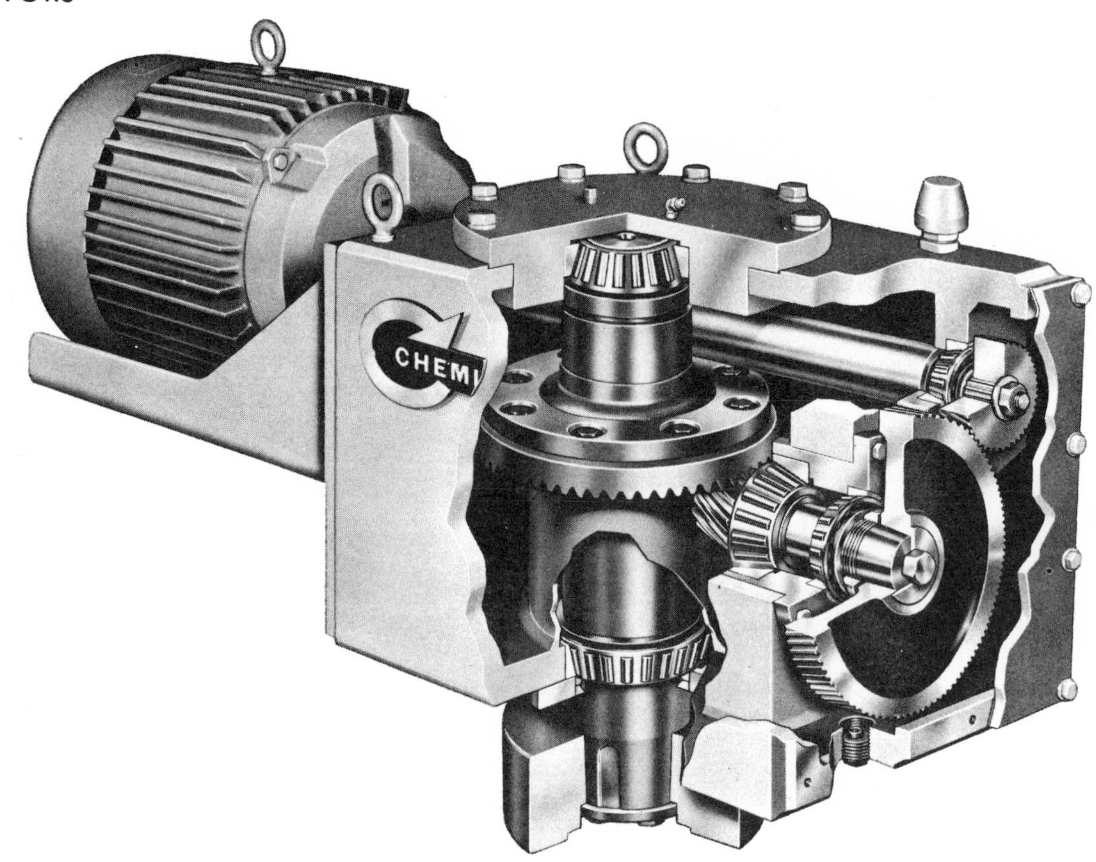

Double Reduction Drive - 1 to 25 horsepower aerators.

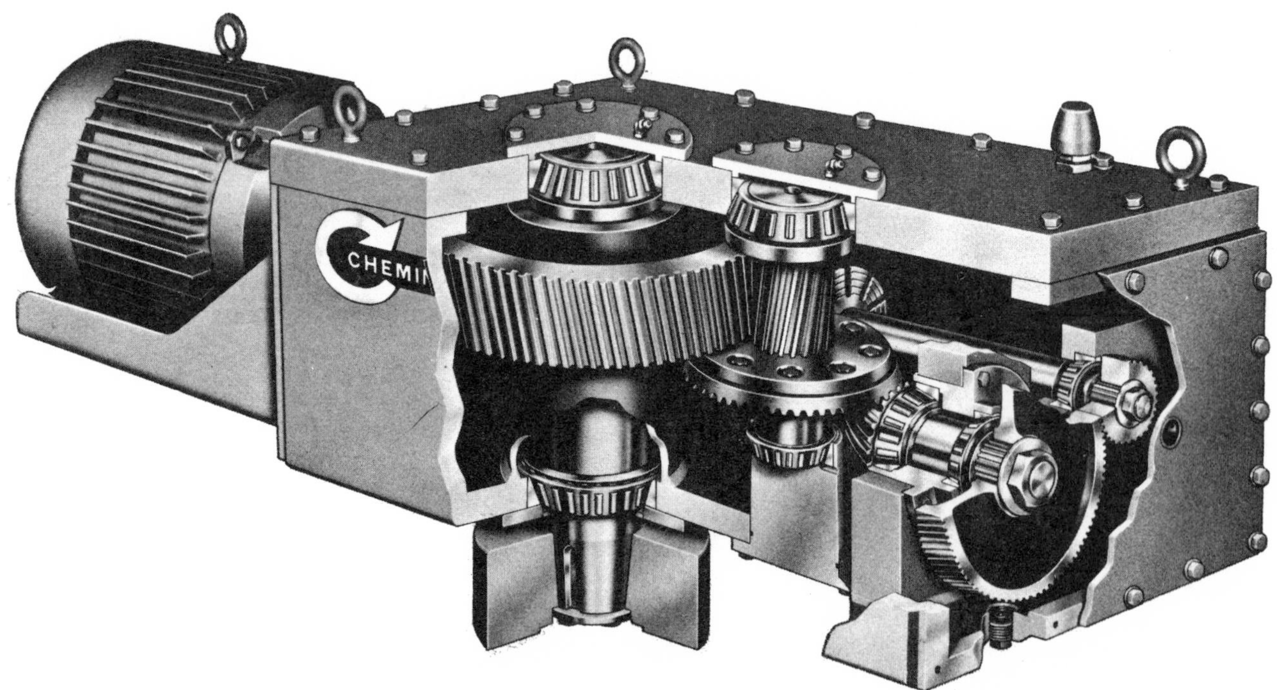

Triple Reduction Drive - 30 to 150 horsepower surface aerators.

Use of the Chemineer Aerator in Common Activated Sludge Systems and Aerated Lagoons

Conventional Activated Sludge System: This is the most common of the activated sludge systems. Sludge and waste water are mixed together and introduced to the aerated basin. As the aerated waste water passes from the basin to a clarifier, biological solids (sludge) settle out. A portion of the sludge is then recycled to the aerated basin in the influent waste stream.

Contact Stabilization System: This design is for systems containing a high portion of colloidal BOD. The waste water is assimilated in the contact basin, then advanced to the clarifier for sedimentation. In the stabilization basin, the sludge is reactivated and then recycled with the influent waste water.

Continuous Extended Aeration System: Influent waste is fed continuously to the basin. The system operates as plug flow zones to cause BOD reduction. The biological system reduces the aerated waste in the waste assimilation zone. Biological solids formed are then oxidized in the aerated endogeneous zone. Sludge entering the settling zone is later recycled to the influent waste stream. The treated effluent is then drawn off.

Aerated Lagoons: Activated sludge systems and aerated lagoons utilize aerobic bacteria for BOD reduction. Lagoons typify systems handling low BOD concentrations in the waste water influent. Biological systems are continuously in contact with the BOD and oxygen. There are three types of aerated lagoons: aerobic mixed, aerobic-facultative mixed, and aerobic-facultative partially mixed. Oxygen is required to assimilate BOD in aerobic mixed lagoons. The facultative lagoons will assimilate the BOD in the presence or absence of oxygen.

Surface Aerator

The installation costs of your surface aerators are reduced by installing the least number of aerators necessary for each basin or lagoon. The aerator horsepower chart, Figure 1, provides data that indicates the size and number of aerators at various standard oxygen transfer rates.

Submerged Aerator

Submerged aerators are best when high rates of oxygen uptake are required. This style of aerator uses an impeller located in the bottom of the basin or lagoon to disperse air and suspend solids. The air is introduced into the impeller from a sparger or other appropriate device. Since submerged aerators require compressors to sparge air into the impeller, the additional power requirement must be considered when selecting a submerged aerator. At various invested energy levels the combined compressor and aerator horsepower yields a minimum or optimum invested horsepower. Figure 2 is a typical invested horsepower curve for a submerged aerator application. This curve is the output data from Chemineer's computer model for selecting submerged aerators. Chemineer can apply your data to the same computer model and optimize your waste treatment application.

Figure 1

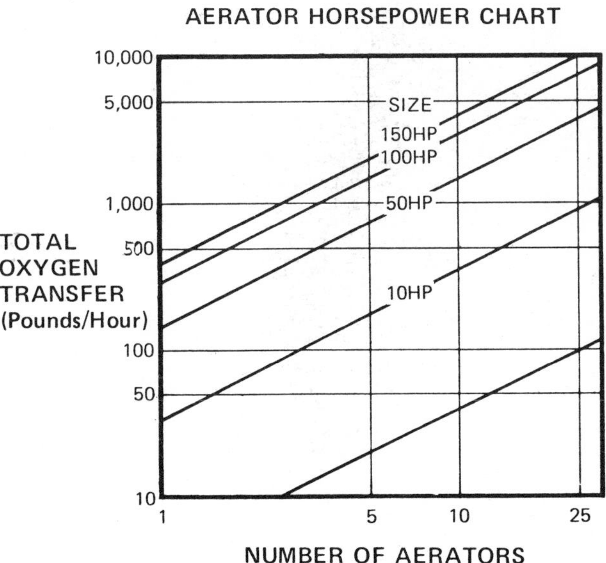

Figure 2

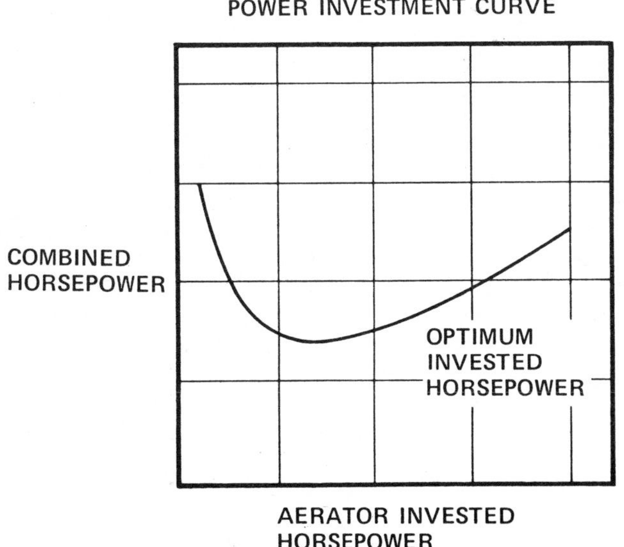

TERTIARY TREATMENT SYSTEM

The general acceptance and increased usage of packaged extended aeration plants as a dependable, economical method of modern sewage treatment has created the need for equally efficient backup systems to guarantee a consistently high quality effluent for discharge into critical water courses and to provide a higher degree of treatment for these critical locations. Aer-O-Flo has answered this need with the first completely automated, safe, dependable Tertiary Treatment System.

Effluent from the extended aeration plant flows through the inlet piping and is introduced into the tertiary pre-settling chamber. This section, designed on the tray settling concept settles out any heavy solids which may reach the tertiary unit during plant upsets, thus preventing blinding or clogging of the filter. The effluent from this section is then introduced on to the filter. The filter, operated at one gallon per minute per square foot of filter surface, provides optimum removal of the remaining solids in the effluent.

Water flows by gravity through the filter into the clear well. The solids in this liquid are retained by the filter. A portion of the filtered water is stored in the clear well — chlorine contact chamber for backwashing the filter. The final effluent is discharged either by gravity or pumping through the outlet piping.

As effluent from the extended aeration plant is filtered, solids will build up on the surface of the sand bed and water will begin to back up. When the water reaches a predetermined level, the Aer-O-Flo backwash system will be automatically started. Air will be introduced into the bed. The turbulence produced will scour the solids from the surface of the bed. Following this the backwash pumps will be automatically started. These pumps, located in the clear well, will force filtered water backward through the filter bed at a rate of 15 gallons per minute per square foot. The water and solids on the filter then flow over the backwash trough and into the backwash sludge chamber. This operation takes approximately five minutes. At the end of this time, backwashing stops and the filter resumes normal operation. Chlorination of the effluent is accomplished by introducing chlorine directly into the clear well.

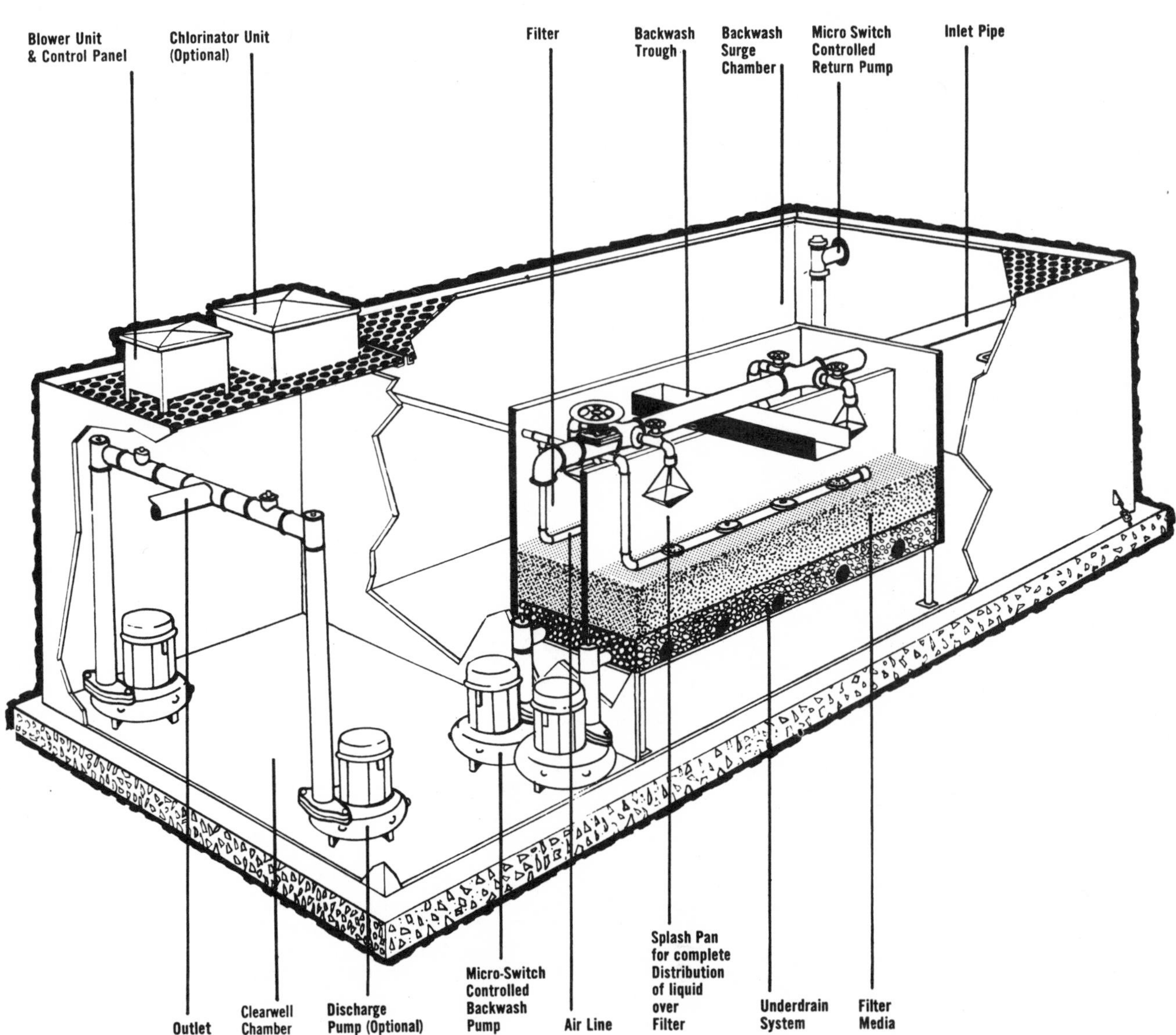

Clow Corporation

FIELD ERECTED CS SEWAGE TREATMENT SYSTEM

The Aer-O-Flo CS System is available in standard sizes to meet a wide range of applications up to 500,000 gallons per day or multiple units can be installed to provide a plant of even greater capacity to accommodate a growing sewage load from any expanding community, subdivision or industrial development. Sound engineering design of any system in this range must provide the flexibility to meet present needs as well as future growth potential; the CS System is designed to accomplish this. For the initial period when the flow is very low the CS System can be designed, if specified, to operate on the basis of the extended aeration process, thus providing the proper ratio of sewage and microorganisms to obtain a high degree of treatment. The plant can be adjusted to utilize a portion or all of the aeration capacity.

As the community or development expands, the CS System can be converted to operate on the contact stabilization process, simply by adjustment of the equipment. This process provides the maximum amount of complete treatment capacity in the minimum amount of space and at the lowest cost per gallon of any aeration process.

In small communities and remote areas away from large cities there is always risk associated with a complex construction project of modest cost. If it is not large enough to attract the skilled companies there is a danger of high costs and possibly poor workmanship. The Aer-O-Flo Field Erected CS System is fabricated and assembled as completely as possible in our factory under quality controlled conditions, then delivered to the job site and erected. The owner receives a top quality, complete, ready to operate plant at the lowest possible cost.

TYPICAL SECTION

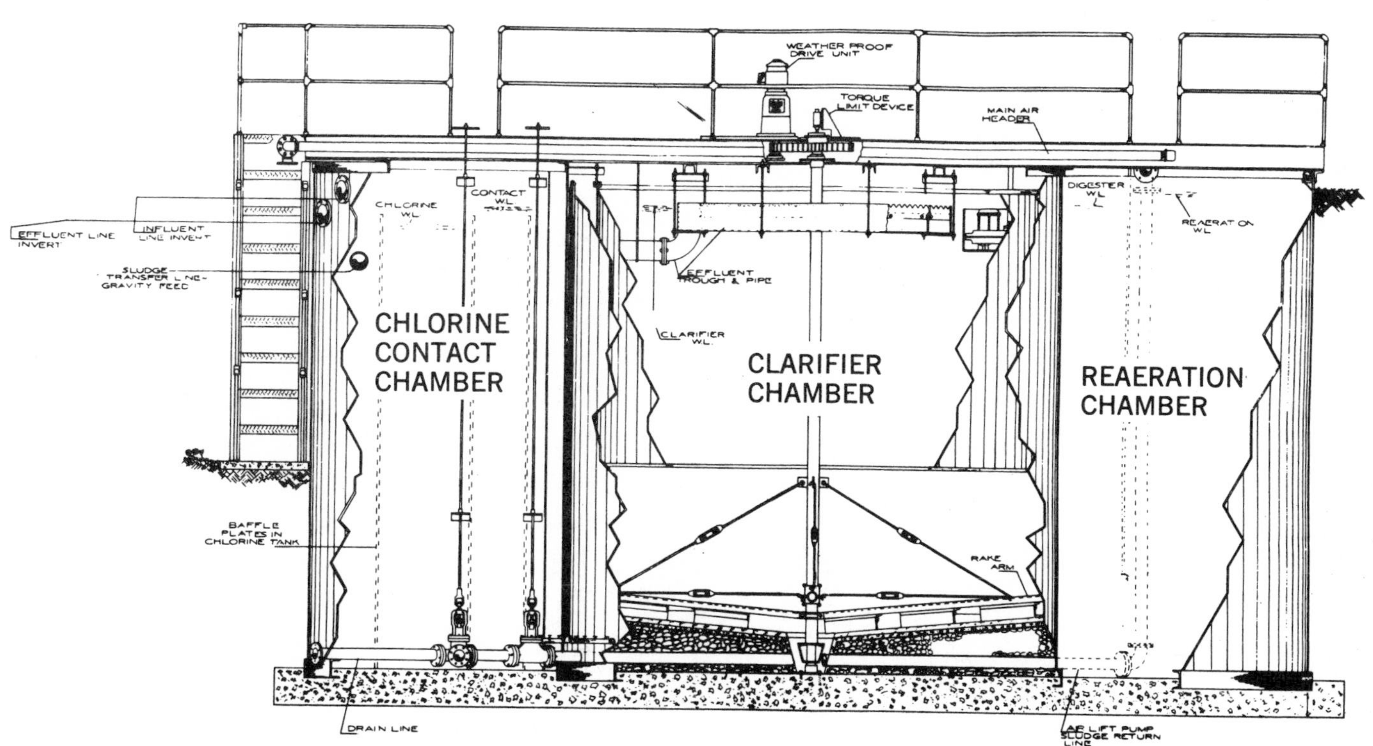

Contact Stabilization

The contact stabilization system is basically the same as other types of aeration processes; microorganisms decompose the sewage in the presence of oxygen or air. These microorganisms, called activated sludge, are the basis of most present day complete sewage treatment systems; the processes differ only in the way which the microorganisms are put to work.

The decomposition or digestion of sewage by microorganisms is much like human eating and digestion processes. In the human process, the first step is eating or consuming the food; this consumption takes place very quickly. The second step is digestion or breakdown of the food consumed. This process is more involved and requires a longer period of time for its completion.

The contact stabilization system is based on this same principle. The raw sewage flows by gravity or is pumped into the plant and first enters either a manually cleaned bar screen or a properly sized Aer-O-Flo comminutor with an automatic emergency bypass and bar screen installation. The comminutor is recommended because of the use of sludge piping and air lifts in the treatment plant.

From the comminutor or bar screen the screened or comminuted raw sewage flows to the "Contact Zone" where the raw sewage is intimately and quickly mixed with the well conditioned activated sludge discharging from the "Reaeration Zone". The "Contact Zone" normally retains the mixture of activated sludge and raw sewage approximately 1½ hours, based on both the raw sewage flow rate and the reaerated sludge flow rate being equal to the plant design flow rate. This mixing time can be varied somewhat when desirable by adjusting the discharge rate of the return sludge air lift.

The flow from the "Contact Zone" enters the "Clarifier" through a submerged transfer pipe and is introduced into the center stilling well. The solids settle slowly downwards; the liquid passes underneath the bottom of the stilling well and enters the main clarification compartment. The sludge or solids settle to the basin bottom where the sludge collector mechanism assists in moving the sludge to the center well from which it is withdrawn through the sludge pipe and sludge return air lift, to the "Reaeration Zone". The clean effluent overflows the adjustable effluent weir trough which carries the plant effluent to the "Chlorine Contact Zone".

The sludge drawn from the bottom of the "Clarifier" through the sludge return air lift is aerated for 5 to 8 hours as it flows through the "Reaeration Zone" toward the "Contact Zone". Here the raw sewage organic material adsorbed by the activated sludge in the "Contact Zone" is digested and assimilated so that at the outlet end of the compartment the activated sludge is ready to be mixed with a fresh load of raw sewage organic material in the "Contact Zone".

Periodically part of the activated sludge from the main units (contact, clarifier, and reaeration) is discharged to the "Aerobic Digestor" compartment. The wasting of sludge is accomplished by operating for a few minutes each week or month the waste sludge air lift discharging to the "Aerobic Digester" compartment. The heavier the plant loading, the more frequently the sludge wasting operation must be performed. The long period of aeration provided in the "Aerobic Digester" permits the wasted solids to be completely stabilized. The outlet from the "Aerobic Digester" is at the opposite end of the compartment from the point of discharge of the waste sludge air lift and the overflow returns to the "Contact Zone" by an overflow weir.

Routinely, possibly as often as once a month when the plant is fully loaded, the operator must drain or pump sludge from the "Aerobic Digester" and dispose of it by hauling to agricultural land, by discharge to sand beds for dewatering, or by other convenient methods.

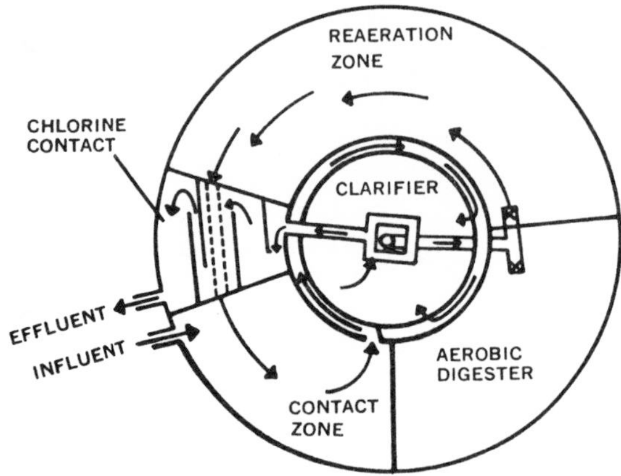

Extended Aeration (Optional)

The contact stabilization treatment system outlined above is capable of producing the greatest degree of sewage purification per unit volume of treatment plant structure. This is the system for which the CS system is designed. However, to provided the flexibility required to meet the needs of a growing community or development, the CS system can be designed to allow it to be operated as an extended aeration system.

When the extended aeration system is used, the incoming raw sewage enters the inlet and passes through the bar screen or comminutor as in the previously described process. The discharge from the inlet chamber enters the head end of the aeration compartment which is the entire annular tank surrounding the clarifier. Here the raw sewage mixes with the sludge drawn from the bottom of the settling basin and is aerated continuously as it flows completely around the plant to the settling basin inlet. The settling basin performs the same operation as in the first process and the plant effluent is discharged in the same way at the same location.

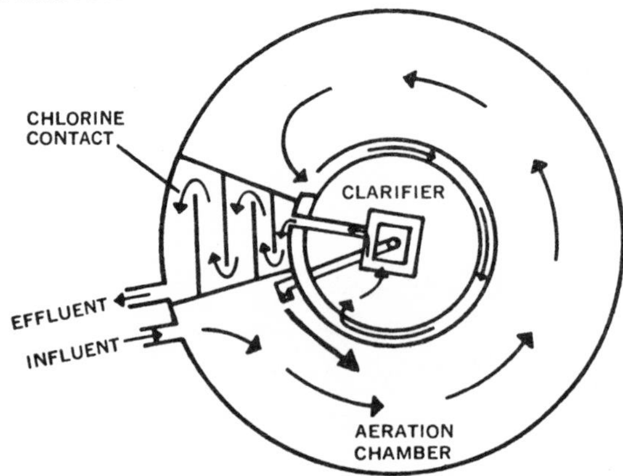

COMMINUTOR

Raw sewage enters the comminutor thru the open top inlet trough, flows into the comminutor housing and then passes through the ¼" slots in the rotating drum. Any sewage solids too large to pass thru the slots are caught up by the projecting teeth on the rotating drum and carried to the fixed cutter comb where the intermeshing of the projecting teeth with the slotted comb cut, shear and shred the solids until they are small enough to be carried thru the slots in the rotating drum by the sewage flow.

Should an exceptionally hard or tough object be encountered jamming the cutting teeth the automatic control system reverses the rotation of the drum approximately one revolution. After a momentary pause forward direction of rotation is resumed. This operation is continued until the jamming object is reduced to small enough size to pass thru the rotating drum. Normal forward rotation is then continued.

In the event of a power outage the comminutor will resume forward rotation as soon as power is restored. In the event of a prolonged power outage in which case sewage solids may cause clogging of the slots in the normally rotating drum, the sewage can bypass the comminutor by overflowing the top of the open top inlet trough.

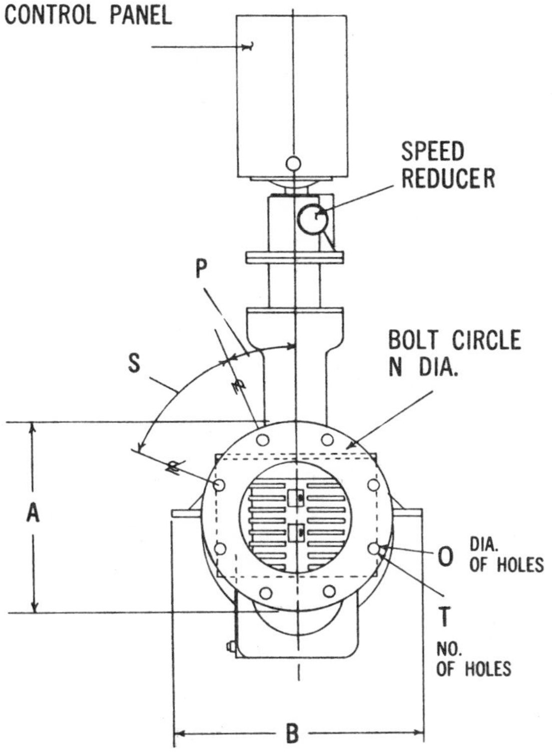

INLET VIEW

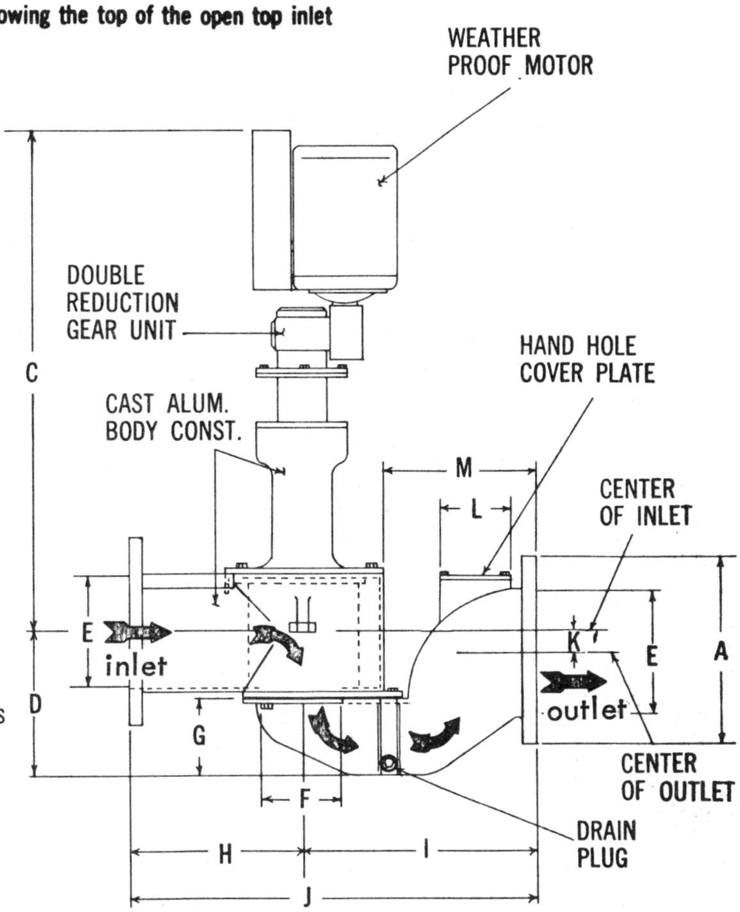

SIDE VIEW

DIMENSIONS

MODEL #	A	B	C	D	E	F	G	H	I	J	K	L	M	N	O	P	S	T
A-5	11"	12½"	36⅞"	8 3/16"	6"	5⅛"	4⅜"	9¼"	12½"	21¾"	1 3/16"	4"	8¼"	9½"	⅞"	22.5°	45°	8
A-8	13½"	15½"	39⅝"	11"	8"	8⅛"	6⅝"	15¾"	16 3/16"	31 15/16"	⅜"	4"	10 7/16"	11¾"	⅞"	22.5°	45°	8
A-12	19"	20½"	45¼"	15⅛"	12"	12⅛"	8¾"	19"	23¼"	42¼"	⅞"	4"	15"	17"	1"	15°	30°	12

Clow Corporation

PREFAB STEEL SEWAGE TREATMENT PLANTS

The Aer-O-Flo Prefab Steel Treatment System provides, within a small amount of space, the necessary environment for nature's processes of sewage decomposition and treatment. Utilizing a modification of the diffused air-aerobic digestion process known as extended aeration, it works by maintaining sufficient oxygen, mixing and detention time to allow the micro-organisms or sludge floc to decompose the treatable wastes into harmless carbon dioxide, water and ash.

The Aer-O-Flo System is shipped to the job site as a complete unit or in modular sections. All internal piping and wiring is complete. This "package" concept provides the flexibility of adding future parallel units, a most economical means of meeting the needs of any growing sewage loads, or for the removal, reuse or resale of the units as municipal sewer facilities extend into development areas. The Aer-O-Flo System is completely factory assembled and factory tested under rigorous conditions for highest standards of performance and durability. The units can be delivered in special Aer-O-Flo trucks, with installation usually requiring less than a day.

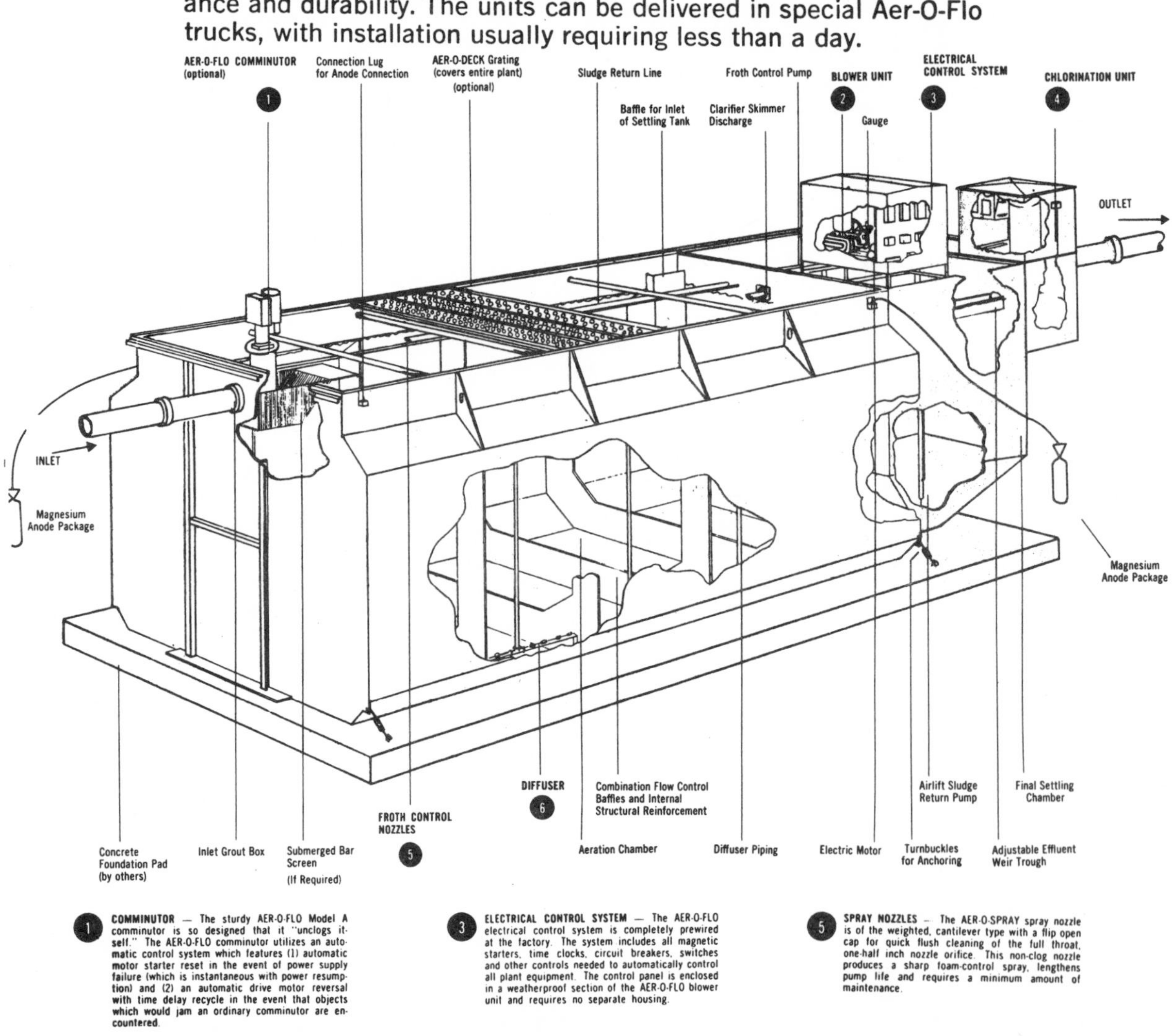

1 **COMMINUTOR** — The sturdy AER-O-FLO Model A comminutor is so designed that it "unclogs itself." The AER-O-FLO comminutor utilizes an automatic control system which features (1) automatic motor starter reset in the event of power supply failure (which is instantaneous with power resumption) and (2) an automatic drive motor reversal with time delay recycle in the event that objects which would jam an ordinary comminutor are encountered.

2 **BLOWER UNITS** — The AER-O-FLO motor-blower unit is delivered complete and ready to operate with motor, "V" belt drive, air blower, controls and a sturdy, attractive weatherproof housing for this equipment. The steel housing is equipped with locking access doors, attractive ventilation louvres and rugged motor-blower supports and is finished with long life, lustrous baked enamel.

3 **ELECTRICAL CONTROL SYSTEM** — The AER-O-FLO electrical control system is completely prewired at the factory. The system includes all magnetic starters, time clocks, circuit breakers, switches and other controls needed to automatically control all plant equipment. The control panel is enclosed in a weatherproof section of the AER-O-FLO blower unit and requires no separate housing.

4 **CHLORINATION UNIT** — The rugged, compact AER-O-FLO chlorination unit provides all equipment necessary for effective chlorination. The unit is delivered complete with chlorine feed pump, solution tank, timer, space heater, and an insulated, weatherproof housing with locking lid.

5 **SPRAY NOZZLES** — The AER-O-SPRAY spray nozzle is of the weighted, cantilever type with a flip open cap for quick flush cleaning of the full throat, one-half inch nozzle orifice. This non-clog nozzle produces a sharp foam-control spray, lengthens pump life and requires a minimum amount of maintenance.

6 **DIFFUSERS** — The AER-O-FLO non-clog Disc Diffuser, constructed of 302 stainless steel, is designed to handle a wide range of air transfer, varying the orifice opening automatically to the air flow and insuring small uniform bubble distribution. Diffusers and diffuser assemblies, with individual control valves, may be easily removed from the tank for minimal maintenance requirements.

Clow Corporation

SIGMA AERATOR

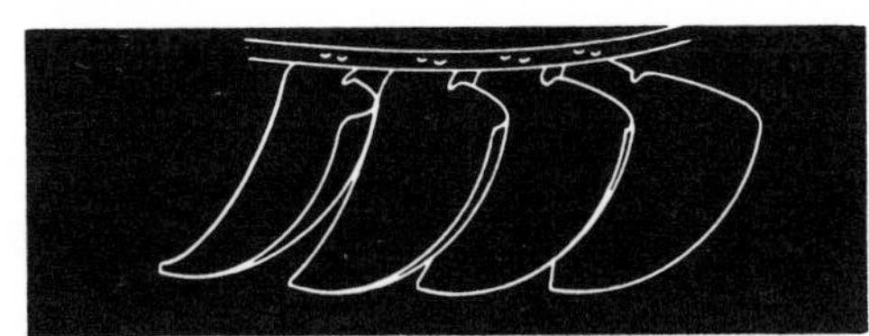

Deceptively simple looking, SIGMA's blades are hydraulically designed with complex characteristics:

1. Low-drag contour on bottom edge squeezes more useful work out of electrical power.
2. Compound curvature gives a fast pick-up of huge volumes of liquid. Resulting "pumping rate" is extremely high which produces complete mixing, fast turnover of tank contents, and bottom-scouring velocities.
3. Concave inner face delivers liquid to top edge of blade with practically no side loss.
4. Top edge feathers out liquid into "acres" of amazingly thin sheets or films for rapid oxygen absorption.

Flexibility for future expansion to higher capacity, or higher horsepower, is another aspect of SIGMA's total performance. Blades are individually mounted so that additional, or larger blades can be substituted as desired.

All season, all weather operation. Drive ring hood prevents ice accumulations from interrupting 24-hour operation. No shutdowns to reduce plant efficiency. From frosty Minnesota to the Rio Grande, you operate the SIGMA when you want to . . . *not* just when the weather says you can.

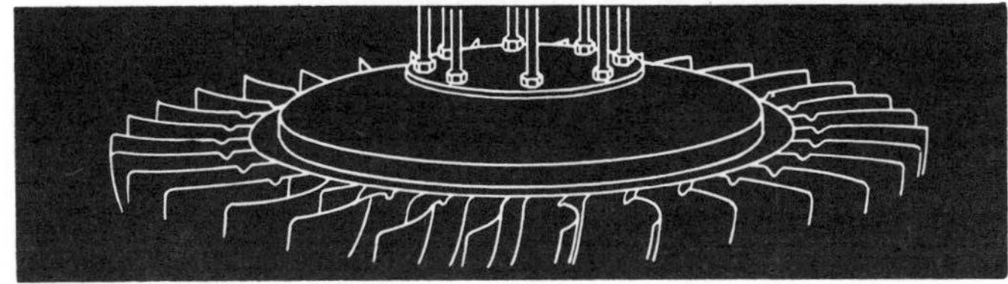

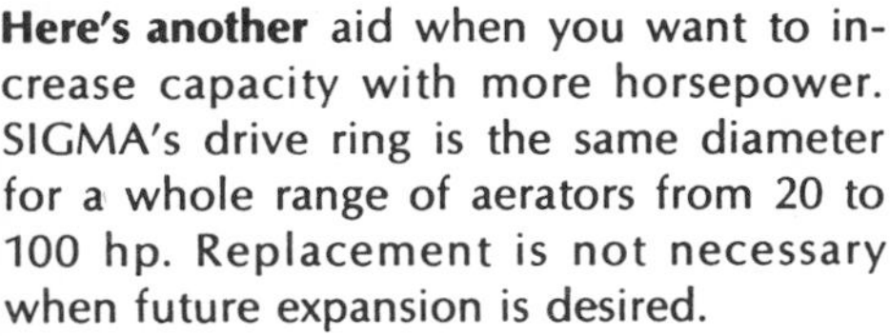

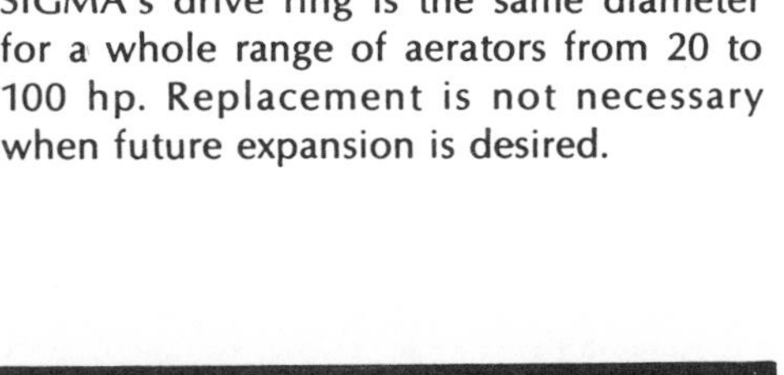

Here's another aid when you want to increase capacity with more horsepower. SIGMA's drive ring is the same diameter for a whole range of aerators from 20 to 100 hp. Replacement is not necessary when future expansion is desired.

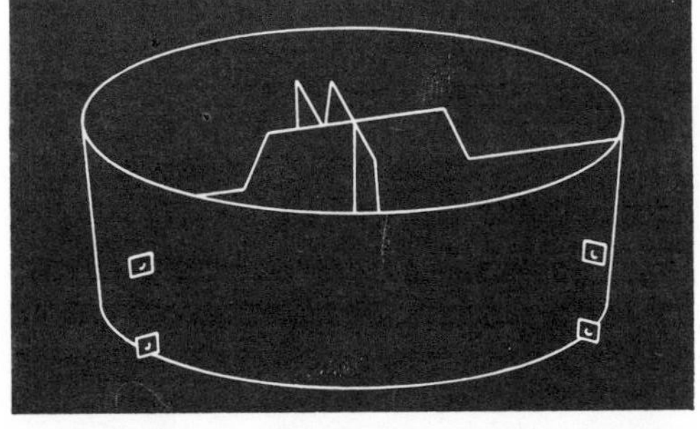

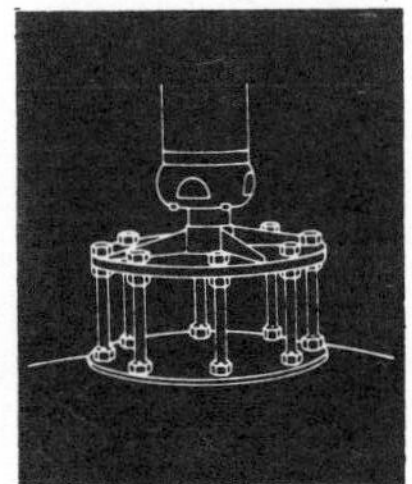

Installation adjustments and operational flexibility are a relatively quick and simple matter for plant personnel. Eight rods (a SIGMA exclusive) are all that need be adjusted. This is especially helpful when the water level of the plant must be changed.

About the only thing that might cause serious power fluctuations and unbalanced conditions with a SIGMA aerator is an earthquake. SIGMA's surge ring prevents wave action being transmitted to the blades which causes head differentials. This also lets motor and gears enjoy a much longer life. The surge ring further makes it possible for the aerator to be independent of tank area, tank depth, and blade submergence.

Depth of the surge ring is also carefully engineered to prevent liquid with a high oxygen content from being short circuited back to the aerator. The entire contents of the tank benefit from high oxygen transfer.

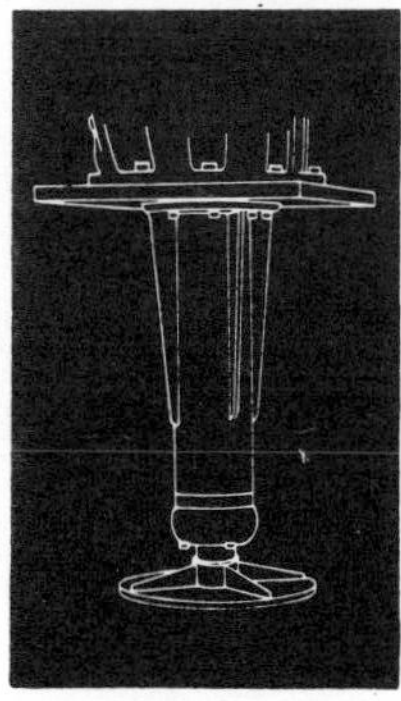

Total engineering by Yeomans includes protection of the most costly single item of the aerator assembly, the gear reduction unit. A lower bearing assembly mounted in a cast iron housing is provided to absorb radial shock loads and to transmit them through the housing, to the base plate, and on to the support bridge or platform. The gears and their bearings do not take these shock loads and consequently the life of the unit is materially extended.

OPERATING FLEXIBILITY

Efficient treatment plant operation demands that the aerators utilized be capable of meeting the oxygen requirements of not only the average organic and hydraulic loads, but also those at maximum and minimum conditions. The oxygenation capacity of the Sigma Aerator can be tailored to the needs of the treatment process by automated or manual adjustment of the liquid level in the aerated basin. Rotation of the hinged weir (available only from Clow) will change the liquid level, thus varying blade submergence, power drawn and total oxygen transferred. Hence minimum power consumption and maximum efficiency is insured at all flow and organic load conditions.

Automated operational control can be provided by means of simple yet sophisticated instrumentation. Manual control is also available if desired.

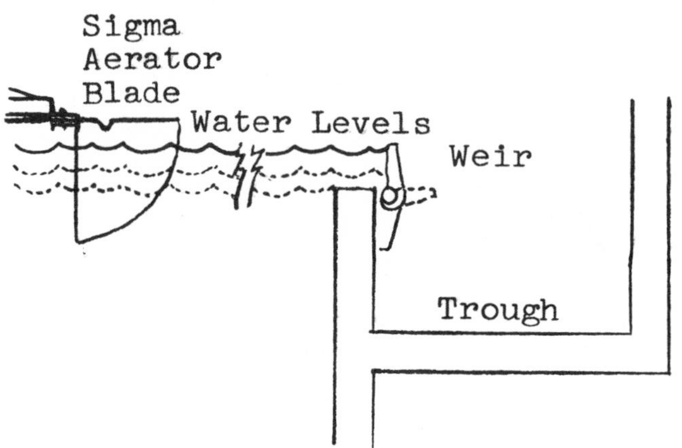

MODELS FOR ANY AERATION APPLICATION

Adaptability is one of the prime characteristics of the Sigma aerator. Standard sizes range from three to one hundred horsepower. They are used in the aeration of domestic as well as industrial wastes. Typical variations include: Tanks, lagoons, ponds, and streams.

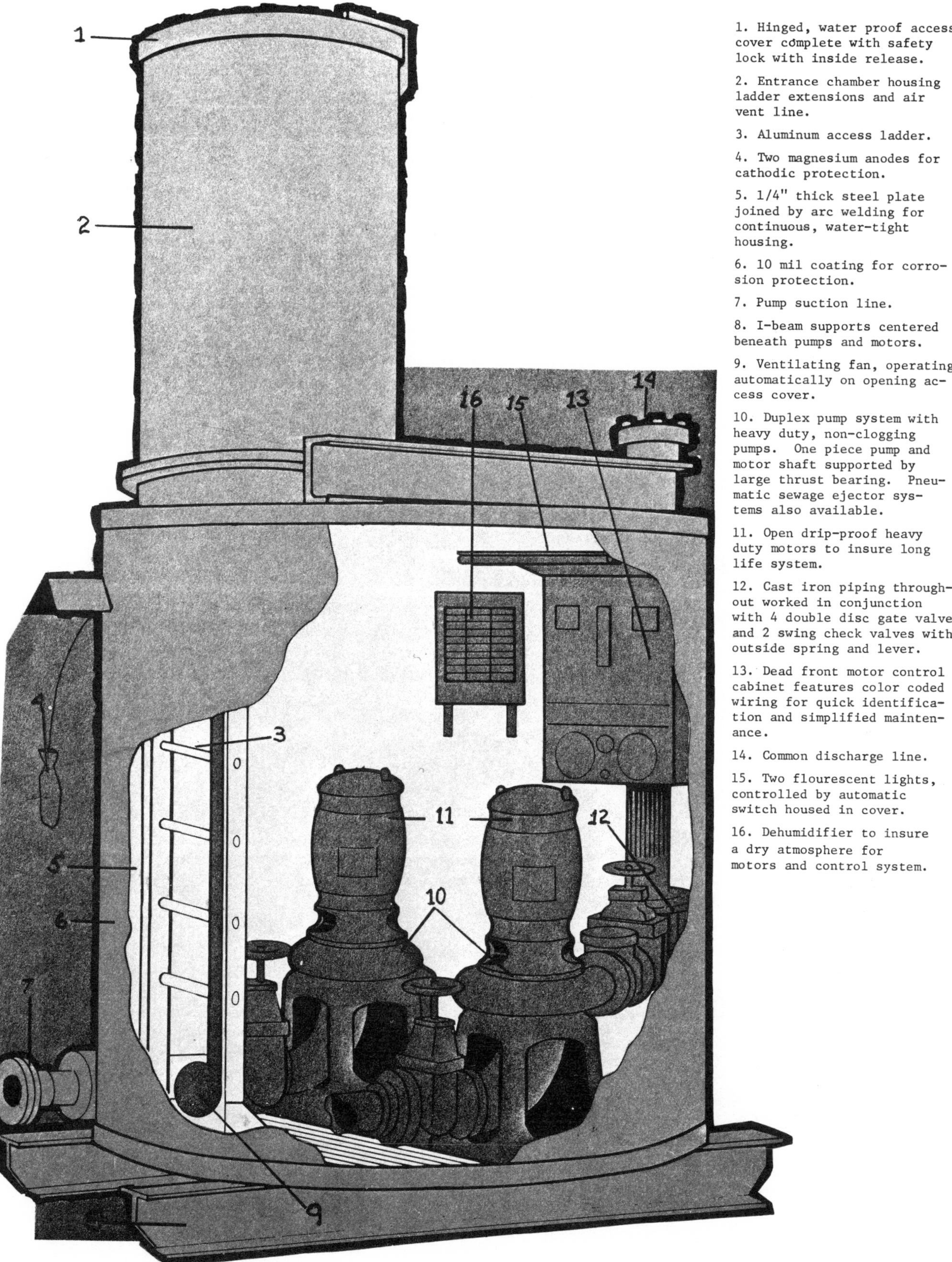

1. Hinged, water proof access cover complete with safety lock with inside release.

2. Entrance chamber housing ladder extensions and air vent line.

3. Aluminum access ladder.

4. Two magnesium anodes for cathodic protection.

5. 1/4" thick steel plate joined by arc welding for continuous, water-tight housing.

6. 10 mil coating for corrosion protection.

7. Pump suction line.

8. I-beam supports centered beneath pumps and motors.

9. Ventilating fan, operating automatically on opening access cover.

10. Duplex pump system with heavy duty, non-clogging pumps. One piece pump and motor shaft supported by large thrust bearing. Pneumatic sewage ejector systems also available.

11. Open drip-proof heavy duty motors to insure long life system.

12. Cast iron piping throughout worked in conjunction with 4 double disc gate valves and 2 swing check valves with outside spring and lever.

13. Dead front motor control cabinet features color coded wiring for quick identification and simplified maintenance.

14. Common discharge line.

15. Two flourescent lights, controlled by automatic switch housed in cover.

16. Dehumidifier to insure a dry atmosphere for motors and control system.

CENTRIFUGAL PUMPS - STANDARD DESIGNS

Vertical Wet Pit Non-Clog

Principal Application - Sewage; liquids having solids.

Discharge Sizes - 4" to 6"

GPM - 50-1600

Head - 3' - 180'

Solids - 2 1/2" & 3"

Submersible Non-Clog

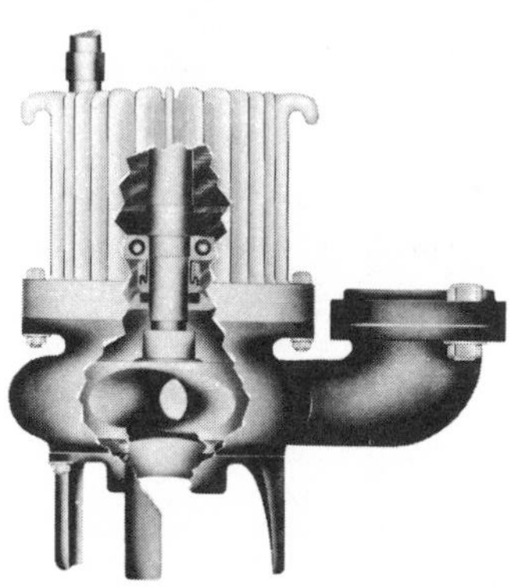

Principal Application - Sewage

Discharge Sizes - 3" & 4"

GPM - 50-500

Head - 6' - 90'

Solids - 2" & 2 1/2"

Pneumatic Ejectors

Mechanically Controlled

Principal Application - Sewage

Discharge Sizes - 4" - 10"

GPM - 30 - 600

Head - 0 - 50'

Solids - 4" to 10"

Electrode Controlled

Principal Application - Sewage

Discharge Sizes - 3" - 10"

GPM - 20 - 600

Head - 0 - 50'

Solids - 3" to 10"

The Fairbanks Morse Figure 5420BP self-priming solids handling pump features superior hydraulic characteristics of the Fairbanks Morse non-clog impeller making this pump especially efficient for handling raw sewage and industrial wastes.

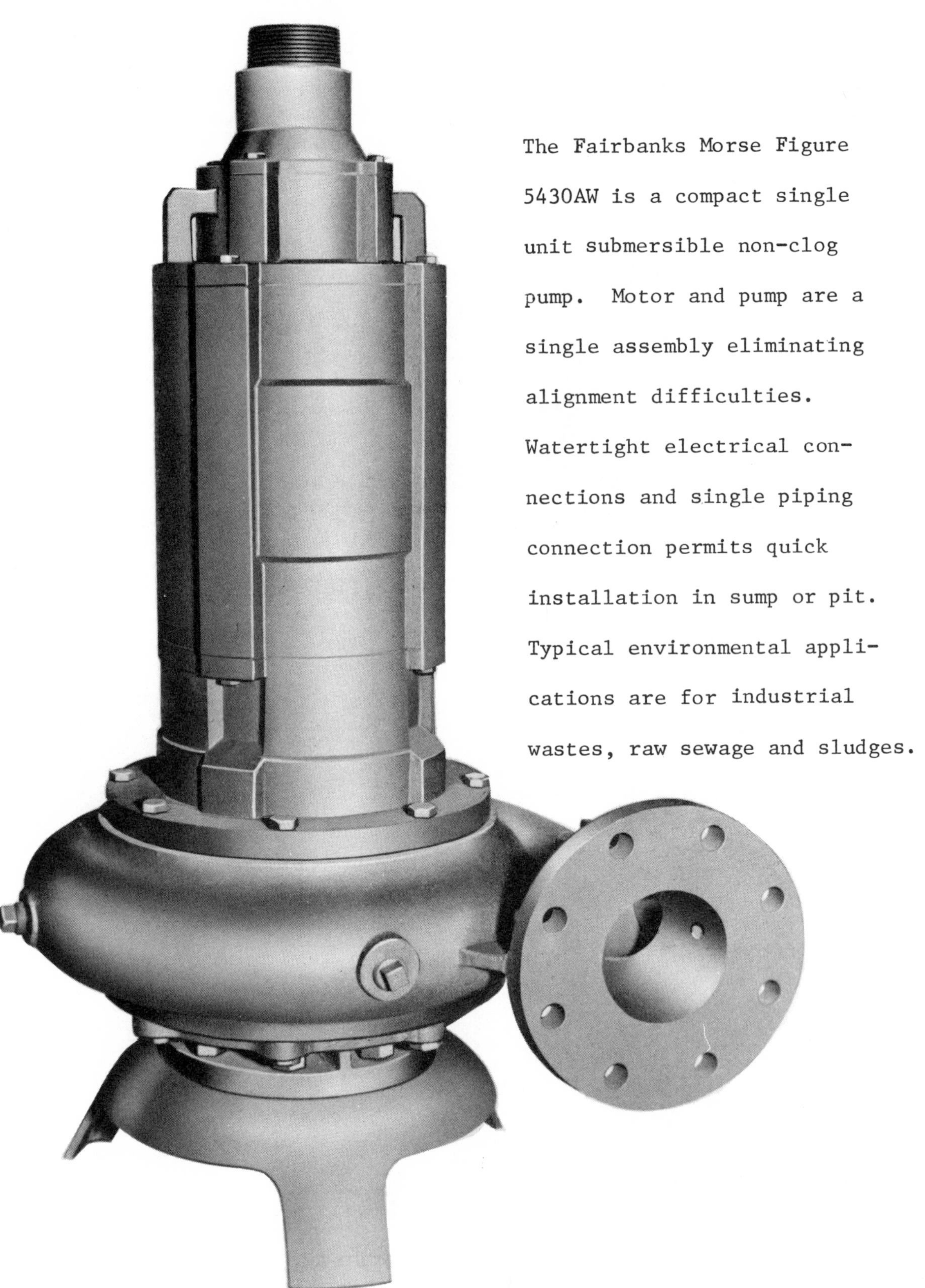

The Fairbanks Morse Figure 5430AW is a compact single unit submersible non-clog pump. Motor and pump are a single assembly eliminating alignment difficulties. Watertight electrical connections and single piping connection permits quick installation in sump or pit. Typical environmental applications are for industrial wastes, raw sewage and sludges.

Fairbanks Morse Figure 5430B Vertical Biltogether non-clog pumps have fluid passages which are designed to handle large diameter solids in suspension without clogging at the highest possible efficiency consistent with large solids passing ability.

Capacities to 3,000 GPM in sizes from 2" to 6".

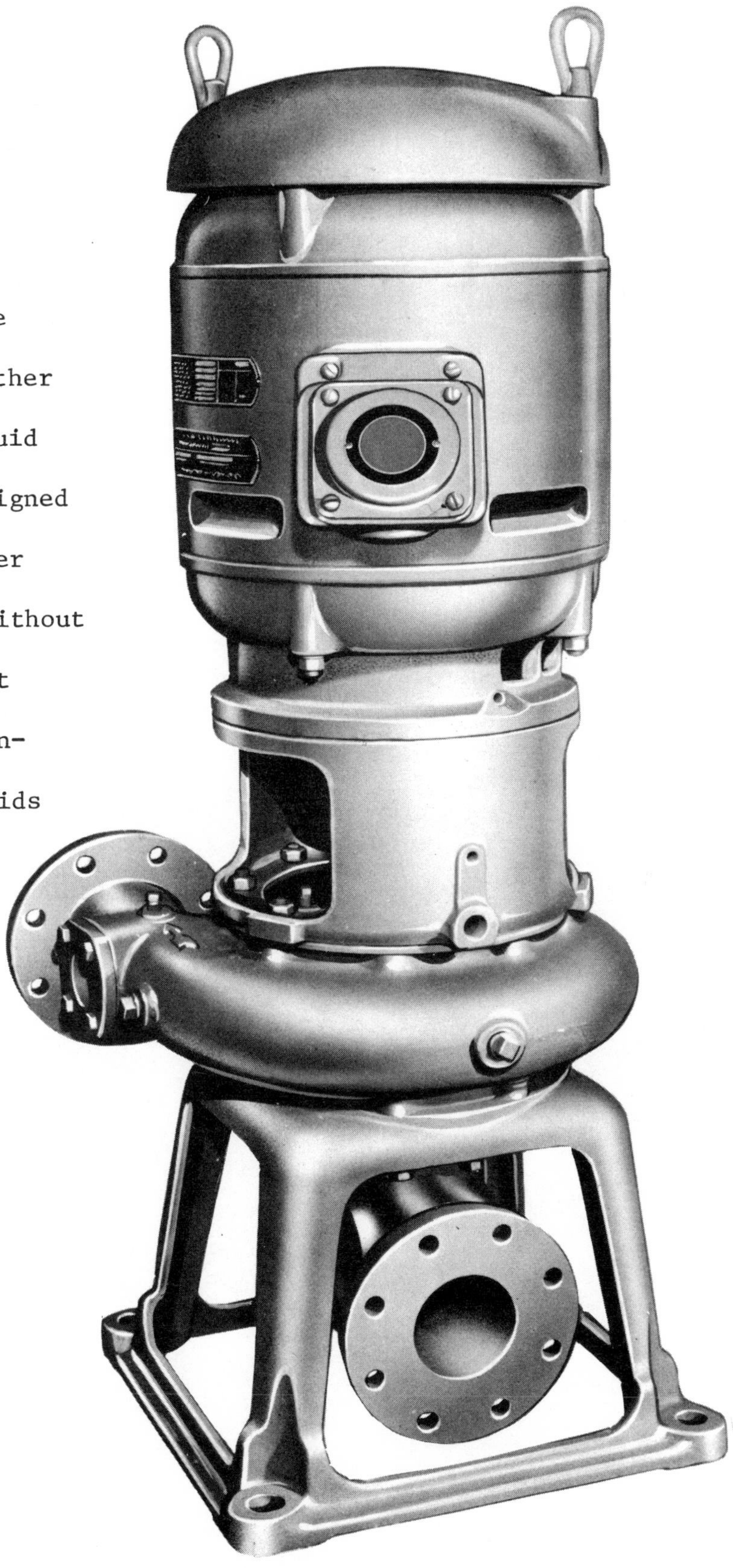

Fairbanks Morse Figure 5700 Angleflow pumps are available in either vertical or horizontal types, and in sizes from 8" to 54", capacities up to 80,000 GPM and pumping heads to 250 feet. Because of wide, unobstructed passages through the impeller and volute, these pumps are ideally adaptable to sewage and wastewater service.

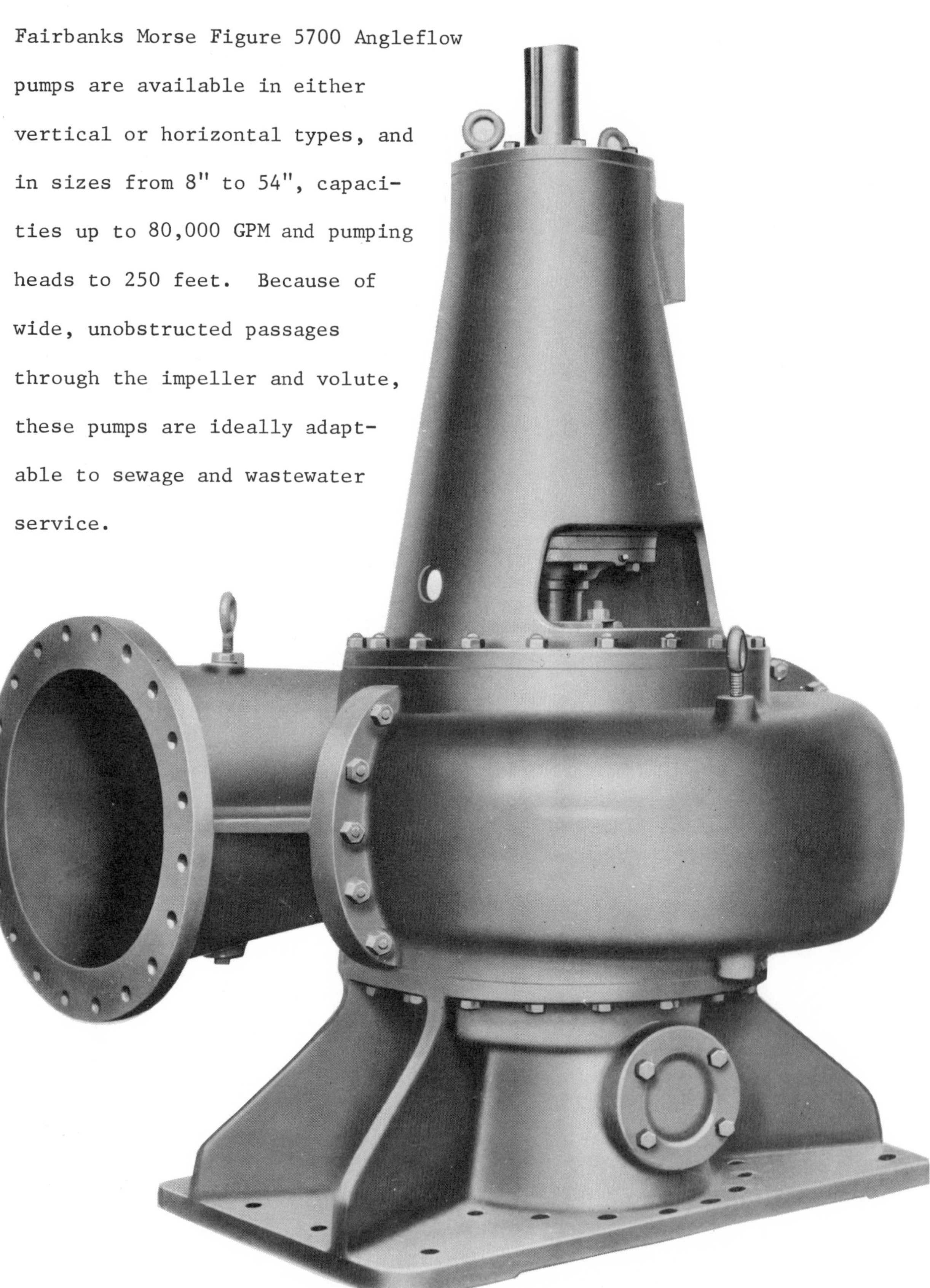

Continental Water Conditioning Corporation

REVERSE OSMOSIS WATER SYSTEMS AND SERVICES

The use of Continental Systems as a substitute for or an adjunct to conventional on-site ion exchange equipment makes a significant contribution to the reduction of dissolved solids in waste streams. Combined Continental Reverse Osmosis and Deionization Service can result in a dramatic 100% on-site reduction. Feeding your deionizer with Continental Reverse Osmosis Systems reduces dissolved solids in waters as much as 95% as shown in the diagram below. This means substantially less chemicals added to the environment, plus reduced regenerant waste. Using Continental Systems and Services rather than distillation equipment reduces water waste of 10-100 gallons for each gallon produced. We are pleased to offer equipment which brings ecological benefit to all.

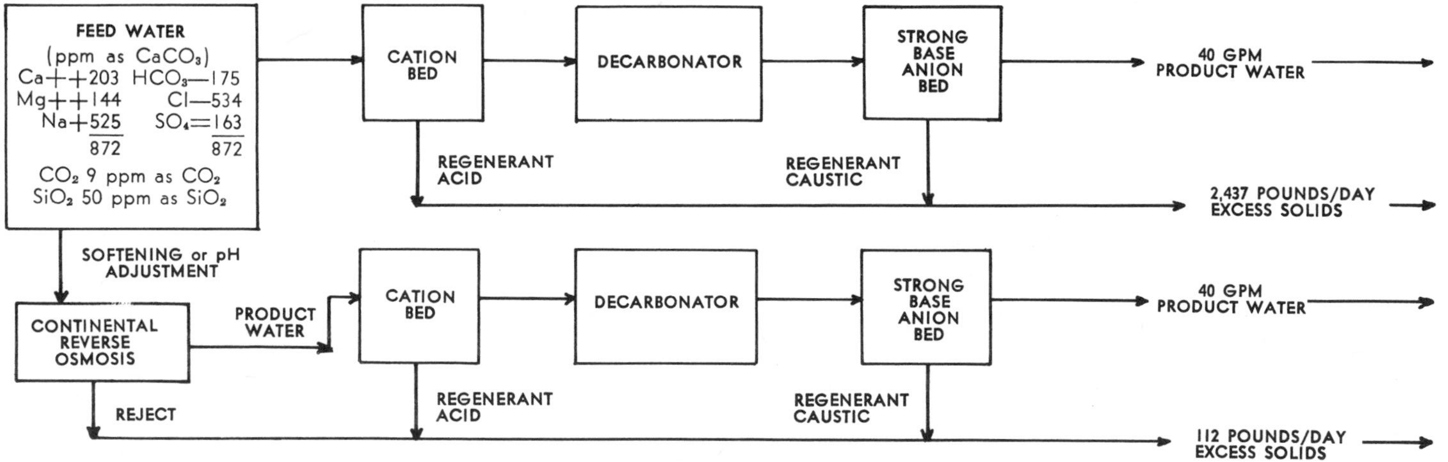

This PERMEATOR is the heart of Reverse Osmosis by Continental. A durable shell and hollow fiber membranes assure long useful life.

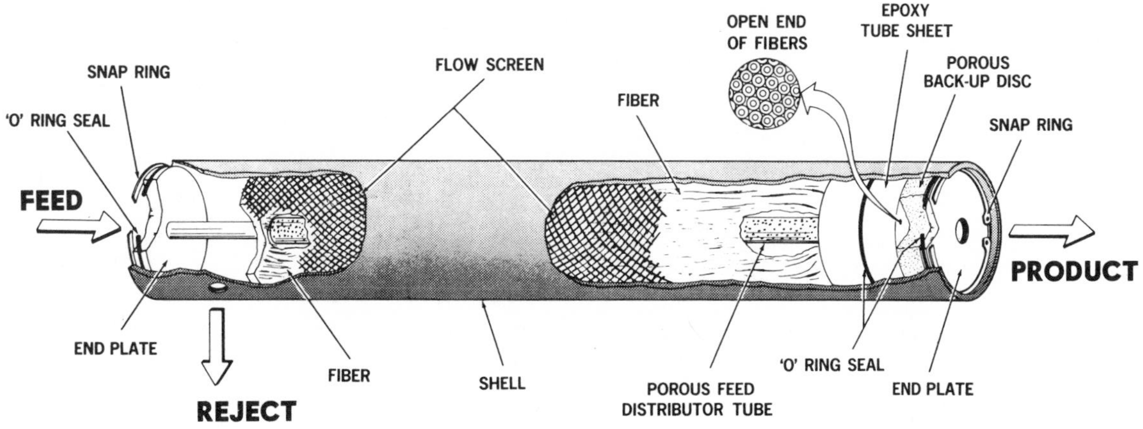

WATER RECLAMATION SYSTEMS

Crane-Cochrane equipment is used in both municipal and industrial waste treatment. Crane-Cochrane components can be used as single units or as components of systems. Basic components include:

FILTERS — Pressure and gravity-types with multi-media beds for removing suspended solids, turbidity, color and bacteria.

MICROSTRAINERS for treatment of potable and waste waters.

ION EXCHANGE SYSTEMS — In treatment of wastes, they are used for removal of dissolved minerals.

ACTIVATED CARBON removes soluble organic contaminants. Recent studies show supplemental chemical feed also effects a reduction in nitrate content of waste waters. These columns come with either fixed or moving beds.

DE-GASIFIERS are efficient in the removal of dissolved gases such as hydrogen, krypton, xenon, etc.

OZONE — Oxidation by ozone is the most efficient means of removing bacteria, color and odor from effluents almost instantaneously.

CLARIFIERS — These units are used to separate and concentrate suspended matter from effluent.

Some representative systems illustrated in the flow diagrams indicate solutions to particular specific waste treatment problems.

TERTIARY TREATMENT OF MUNICIPAL EFFLUENTS —REMOVAL OF SUSPENDED SOLIDS, BOD

Microstraining is used to remove suspended solids and BOD, followed by ozone for color reduction and disinfection. If complete turbidity removal is required for water reuse, high-rate filtration can be added. Foam fractionation is applied when detergent residue is a problem.

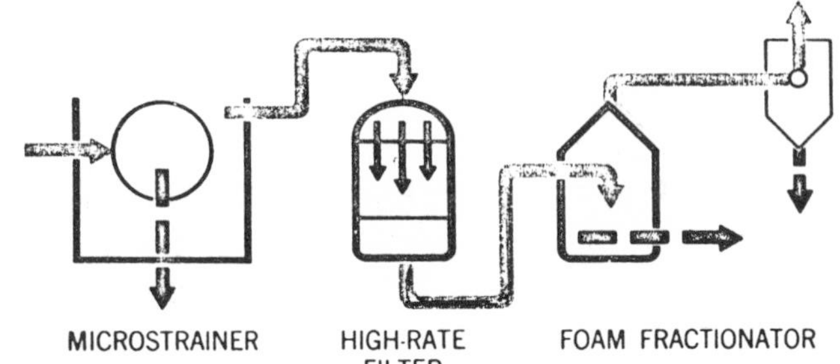

TERTIARY TREATMENT OF MUNICIPAL EFFLUENTS —REMOVAL OF PHOSPHATES, NITRATES

Chemical coagulants are added in a clarifier to precipitate phosphates. This is followed by a deep-bed filtration system to remove remaining suspended matter. Activated carbon absorbs soluble organic matter and, with supplemental chemical feed, reduces the nitrate content.

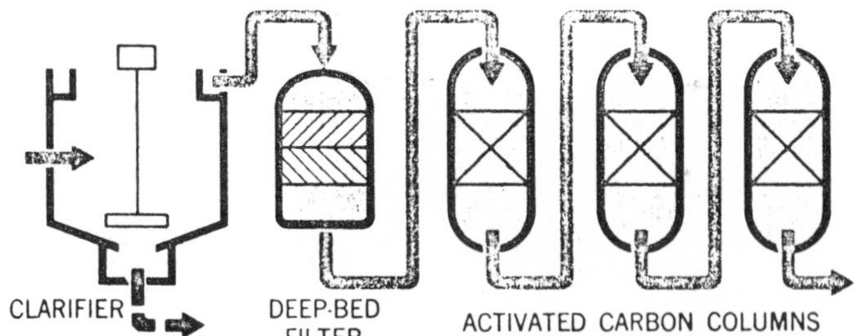

METAL AND METAL FINISHING INDUSTRIAL WASTES

These wastes are extremely diversified in character. Examples might be: suspended matter, removed by high-rate filtration; soluble oils absorbed by a carbon column, or valuable metallic ions, recovered by ion exchange.

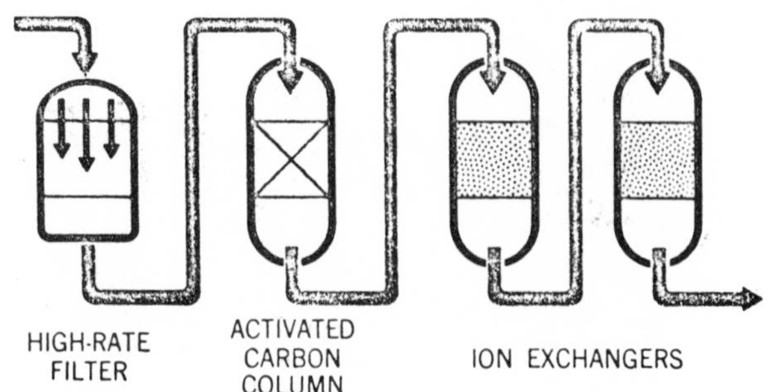

Crane Company, Cochrane Division

RECLAMATION SYSTEM

Cochrane's reclamation system combines coagulation with newly-developed techniques of filtration and foam removal, a combination which economically reclaims sewage plant effluent without settling. The system has only four main components and works in the manner described below.

FLOCCULATION

Sewage-plant effluent first enters the flocculation vessel, where alum, chlorine, and coagulant aid are added. (About 3 ppm alum is added for each ppm phosphate in the sewage.) No lime is used. The alum reacts with the phosphate to form a precipitate that coagulates suspended solids and colloids in the water. The coagulant aid and retention time in the flocculator cause the precipitates to agglomerate and grow in size.

After floc is formed, it is removed from the water by a unique pressure filter specially-developed by Cochrane for this application.

FILTRATION

Cochrane's new "Penetrex" filter contains several layers of filter media, graduated in size and density to give greatest porosity at the top of the bed. The filter acts as a graded sieve, trapping coarse solids at the top and finer ones deep in the bed – providing true "in-depth" filtration. As a result, "Penetrex" filters have 5 to 10 times the capacity of ordinary filters between backwashings.

Following filtration, the clean effluent is passed into the "Frothair" foaming unit for detergent removal.

FOAMING

In Cochrane's "Frothair" unit, water is contacted by a counter-current flow of fine air bubbles to produce a rich foam that concentrates and entraps detergent. The foam-detergent mixture then rises to the top and separates from the water.

De-foamed water is withdrawn by pumping from the bottom of the unit. The original treated sewage influent is now free of suspended solids, dissolved phosphates, organics, and detergents, and is suitable for a variety of industrial uses.

FOAM REMOVAL

The rich foam from the top of the "Frothair" unit is forced through a cyclone foam breaker where centrifugal forces collapse it into a highly-concentrated liquid stream. This stream may be discarded or partially recycled as desired.

The cyclone foam breaker is another recent Cochrane development. It provides efficient foam separation with no moving parts. No extra water spray or air supply is required; the same air that creates the foam conveys it to the cyclone.

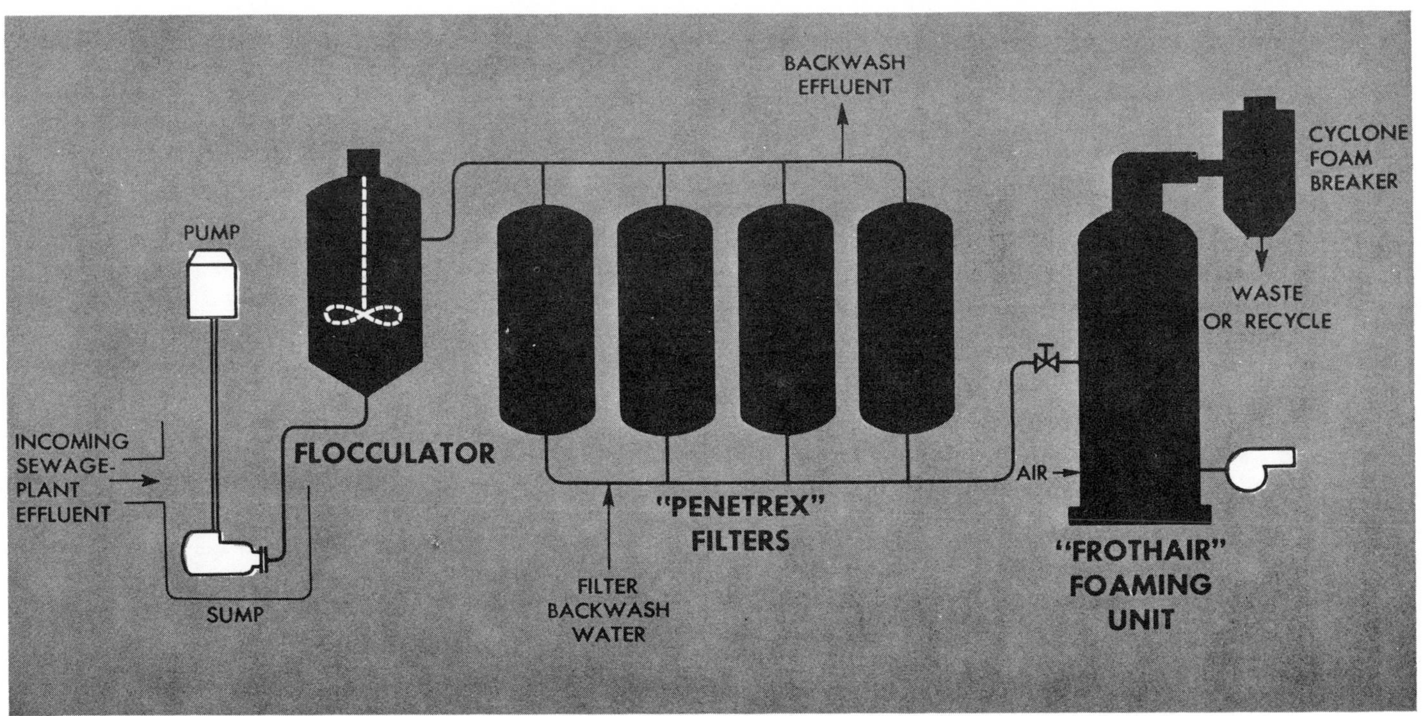

ATM MARK III CENTRIFUGAL

The Mark III is a fully automatic batch basket centrifuge which is especially qualified for efficient solids concentration of waste streams. It is a link-suspended machine with hydraulic top drive, providing the maximum removal of solids through its unobstructed bottom discharge.

In most applications, expensive polyelectrolytes are not required in the Mark III since the solids are compacted in the basket over a longer period of time with less agitation or turbulance. For example, digested mixed sludge, in one instance having 4.7% dry solids (DS) in the feed, was concentrated through a range of 15-18% DS with a recovery of over 98% of the solids.

The basket centrifuge is not affected by grit or erosive materials in the feed slurry and is equipped with skimmer for removing free water from the surface of the cake for optimum solids compaction.

Operation

A feed pipe on top of the Mark III cover provides a passage into the spinning basket for the slurry to follow. Centrifugal force separates the slurry, distributing solids onto the basket wall as an easily removable mass. The perforate Mark III basket wall allows liquid to pass through into the outer chamber. (On models with imperforate baskets, the liquid overflows the basket cap.)

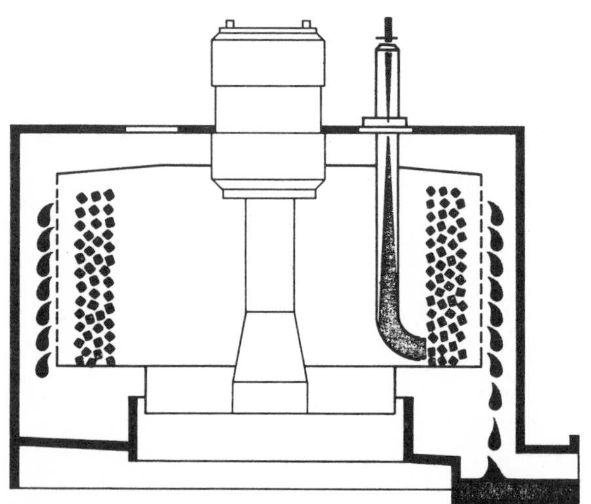

Solids collected on the Mark III wall during separation are discharged by the unique plow, which loosens the mass, pushing the solids out. An opening in the curb of the basket bottom allows solids to exit completely unobstructed.

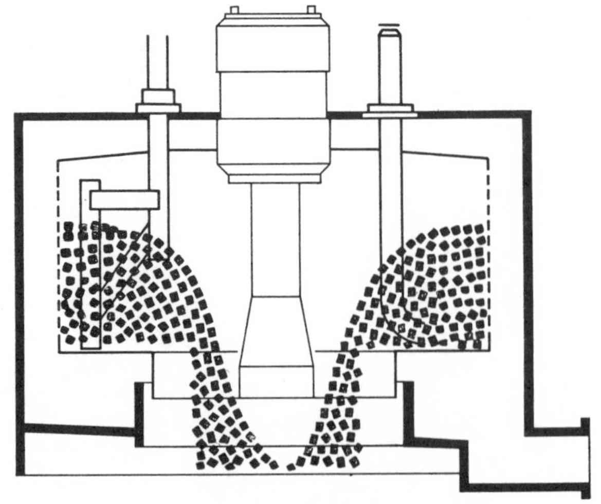

The ATM Mark III integral curb-basket design allows the plow and basket to oscillate together providing a constant clearance for a balanced, clean discharge.

BRPX-217 SEPARATOR

This high-G effect clarifier centrifuge continuously dewaters waste activated sludge while periodically and automatically ejecting concentrated solids. The BRPX can be utilized for dewatering without the use of flocculating agents and has the ability to concentrate waste activated sludge from an initial .75-1% dry solids (DS) up to as high as 12-15% DS dependent upon the S.V.I. (Sludge Volume Index).

The BRPX is self-cleaning which means it does not have to be taken apart at frequent intervals for cleaning. It is so designed that the sludge is discharged automatically while the liquid phase is continuously pumped away. The BRPX is of corrosion resistant stainless steel and is best applied to grit-free sludges such as activated or alum sludge.

How the BRPX Clarifies

The unclarified liquid enters the BRPX through the inlet connection and flows downward into the rotating bowl. Centrifugal force created by the bowl's high speed of rotation causes the heavier particles in the liquid to move outward and away from the center of rotation. This material accumulates at the sludge space around the outer periphery of the bowl wall until the bowl discharges, ejecting the accumulated material from the machine. The lighter clarified liquid flows inward and up to the paring device. The paring device, a stationary pump impeller, utilizes the energy of the revolving liquid to pump it away under pressure, free of foreign material.

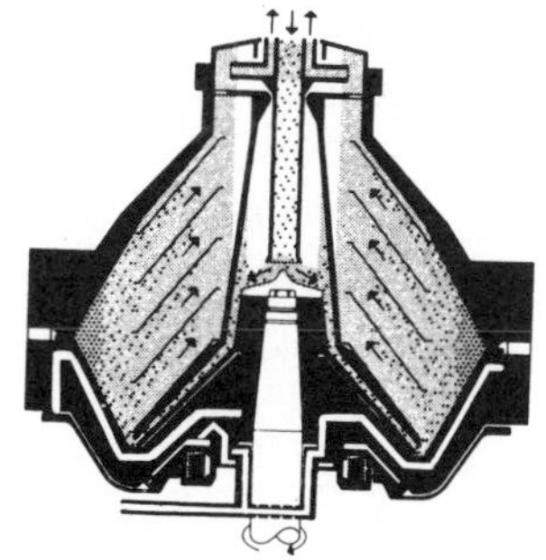

To empty the bowl of sludge the bowl bottom ,hydraulically sealed to the bowl top, is lowered for an instant by relieving the hydraulic pressure. The high centrifugal force created by the bowl's rotation shoots the accumulated solids out elongated discharge ports to the sludge collecting cover. In a fraction of a second the bowl is closed again, allowing no interruption of the efficient clarifying process. With partial solids discharge the sludge concentration is at a maximum, with minimum loss of mother liquor. Partial solids discharge may be combined with self-triggering for peak efficiency.

THE IMMEDIUMFILTER

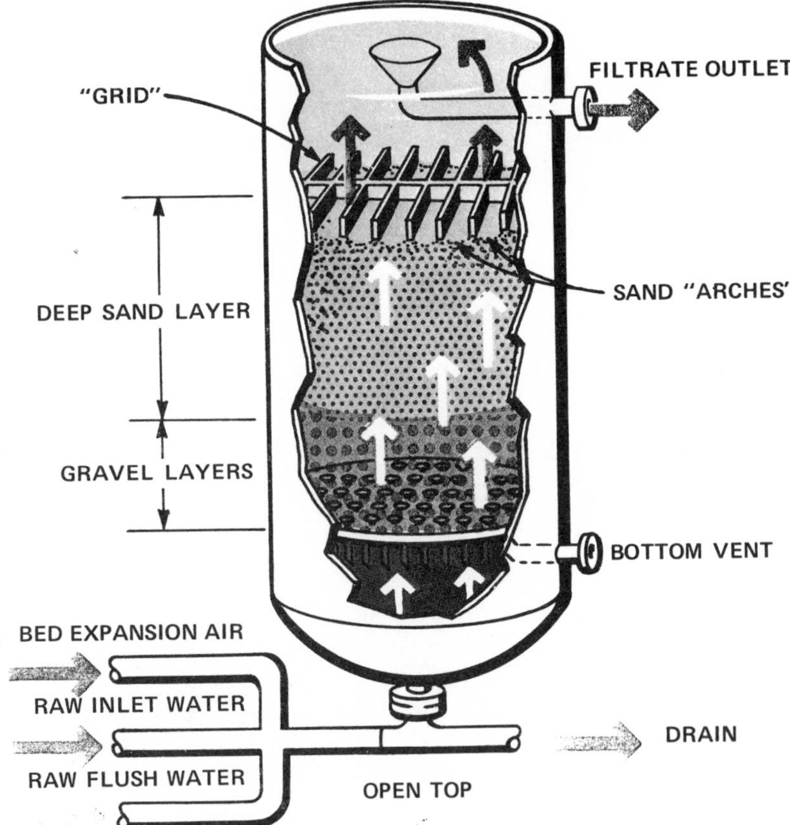

The De Laval Immediumfilter is a high capacity, high efficiency graded-bed sand filter extremely well qualified for removing suspended solids that might have carried over from primary and secondary treatment. Unlike conventional sand filters, the Immediumfilter directs the flow upward through a deep bed of over 7 feet of gravel and sand. Thus, the whole sand bed provides dirt retention with the top 5 feet of fine-grain area serving as the polishing zone. Filter rates are from 4 to 12 gpm/sq.ft.

In one application involving tertiary treatment, the feed to the filter containing 15-20 mg/l suspended solids (SS) was reduced to the level of 4-6 mg/l SS at flow rates of 5 gpm/sq.ft. of filter area.

Settling or retention tanks are seldom necessary even when filtering raw water from streams or lakes. If chlorine or coagulating chemicals are used, they are injected ahead of the filter. All chemical contact with the water is carried out in the sand bed. With a sand depth of 5 feet, the Immediumfilter's dirt-holding capacity is about 5 to 10 pounds per square foot of surface area. The conventional sand filters are capable of holding only one-half pound of dirt per square foot of surface area because they don't filter in depth as the De Laval Immediumfilter.

The Immediumfilter is easily cleaned by injecting air into the vessel to expand and fluidize the filter bed. In this way the sand particles rub against one another cleansing themselves of all dirt. The lower coarse gravel layer is not disturbed and the gravel layer only slightly.

Water and air are all that is necessary to perform the cleaning cycle, flowing through the filter media at about 14 gpm per square foot. The water/air combination effectively and efficiently dislodges the dirt from the sand layers. The dirt is carried up and out of the filter discharging at the top of the vessel. No sand is lost during the cleaning cycle.

De Laval Immediumfilters use raw water for forward flush cleaning. The conventional downflow sand filter requires as much as 10% of the filtered water to do the same job by backflushing.

The Immediumfilter's sand bed is held in position by a grid fixed into the top of the vessel. The sand arches, concavely, between the patented spaced grid bars and keeps the bed stationary. This makes the rapid up-flow process possible thereby taking advantage of the sand's depth and cake holding capacity.

Other De Laval equipment suitable for water, waste and sewage treatment:

AVNX SLUDGE DEWATERING DECANTER CENTRIFUGE

The AVNX Decanter is a continuous dual-discharge conveyor separator especially designed for dewatering or concentrating solid suspensions in municipal and industrial wastes. It may be applied on raw primary sludge and excess activated sludge, by the addition of suitable quantities of polyelectrolytes. In one example, using a polyelectrolyte on a digested mixed sludge with a feed of 3 - 4% dry solids, a concentrated sludge suitable for landfill of 20 - 22% dry solids was obtained with 90% recovery.

With the AVNX, liquid clarity and solids dryness can be controlled, and maximum dewatering is obtained with flocculating agents introduced into the conveyor bowl through a special floc tube.

NOZZLE-TYPE SOLIDS DISCHARGING CENTRIFUGE

The nozzle machine continuously discharges solids as a concentrated slurry and is particularly suited to soft, gelatinous concentrations such as activated and alum sludges. A nozzle type centrifuge can concentrate excess activated sludge from an initial 0.5 to 1.0% dry solids to the range of 5 to 7% dry solids with +90% recovery.

In cases where pumpable sludges are desired the nozzle machine has a high capacity. It is best applied to grit-free sludges.

SPIRAL HEAT EXCHANGER

The Spiral Heat Exchanger is a unique and compact heat exchanger which is used to maintain proper temperatures for maximum anaerobic digestion. It is especially well suited to this application because of its very low fouling tendency and high heat-transfer efficiency. Two sheets of metal are wound into a pair of concentric spirals which produce passages through which media and product can flow counter to each other.

Spiral heat exchangers can be furnished for design pressures of up to 150 psig or more in sizes ranging from 5 square feet to 2,000 square feet of heat transfer surface.

15 x 10 MANUAL CLARIFIER

Designed for removal of insoluble metallic hydroxides from ECM electrolytes. Clarification is semi-continuous, centrifuge requires periodic cleaning of accumulated sludge by manual activation of a skimmer tube; sludge removal by skimming is accomplished while centrifuge is operating at full speed with feed off. Can be utilized for clarification of other liquids whose solids are gelatinous and capable of removal from basket by skimming.

Nominal flow rate 10 gpm. Centrifuge designed to handle up to 2500 amp capacity ECM machine at a utilization factor of 75%. Requires manual skimming every 15 minutes at these capacities.

Del-Pak Corporation

ABF — The Activated Biofilter System that combines activated sludge performance with the stability and simplicity of trickling filters.

ACTIVATED SLUDGE

- 90% + removals.
- Lowest first cost.
- Requires less area.
- Odor free.

TRICKLING FILTERS

- Simplicity.
- Low sludge volume.
- Low operating costs.
- Absorbs shock loads.

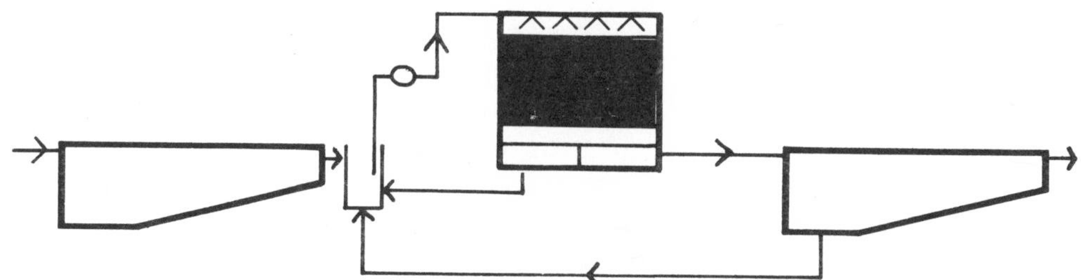

The raw sewage flows to the primary settling tank, then is pumped over the Del-Pak tower with a portion of the tower underflow being recycled. The remainder is sent to the final tanks. In addition, sludge from the final tanks is recycled over the tower to produce a true activated sludge mixed liquor. This mixed liquor is continuously aerated as it splashes from layer to layer of Del-Pak media. In effect, Del-Pak is putting the mixed liquor into the air rather than forcing the air into the mixed liquor.

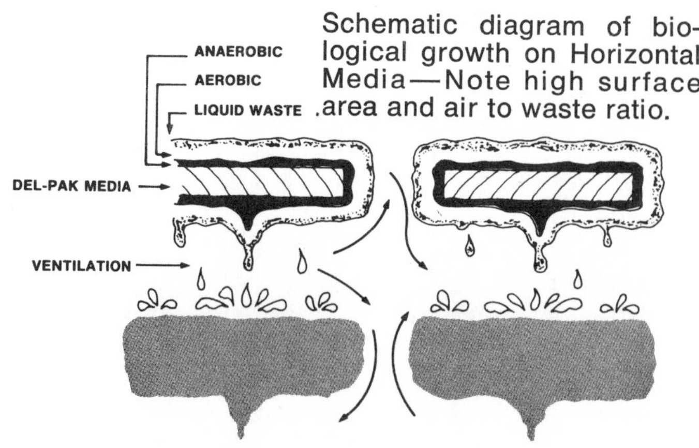

Schematic diagram of biological growth on Horizontal Media—Note high surface area and air to waste ratio.

Desalinations Systems, Inc.

WATER PURIFICATION SYSTEMS
SPIRAL DESALINATORS
PRESSURE OSMOSIS MEMBRANES

The Research and Development programs of Desalination Systems, Inc. have produced a family of proprietary water purification semi-permeable membranes, each designed to perform molecular separation processes with high efficiency for a variety of specific desalination requirements and systems conditions. These membranes are known by the trademark Braylon. Braylon pressure osmosis membranes have various salt rejection ratios up to 99% of all dissolved salts and various quality water production rates up to one half gallon per hour per square foot of membrane. The selection of the right membrane for each specific systems application is an important cost and efficiency decision. Braylon 10 membrane is available in twenty six inch wide rolls.

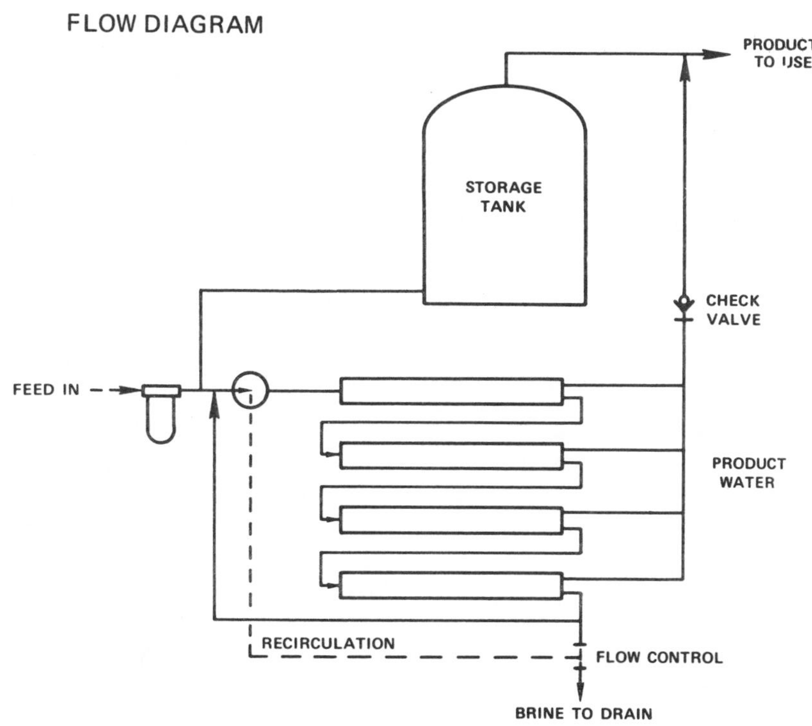

DESAL reverse osmosis water purification units produce low mineral content, high quality water. These systems store and deliver processed water at existing line pressure and, during periods of excess demand, will automatically deliver line water to the point of use.

The units use reverse osmosis cartridges containing Braylon 10 membrane. They remove from 70 to 95% of most dissolved salts and organic compounds and essentially 100% of the particulate matter. They will process feedwater salinity up to 3,000 ppm and with decreasing efficiency to about 5,000 ppm.

Detroit Filter Corp.

DETROIT AUTOMATIC LAGOON

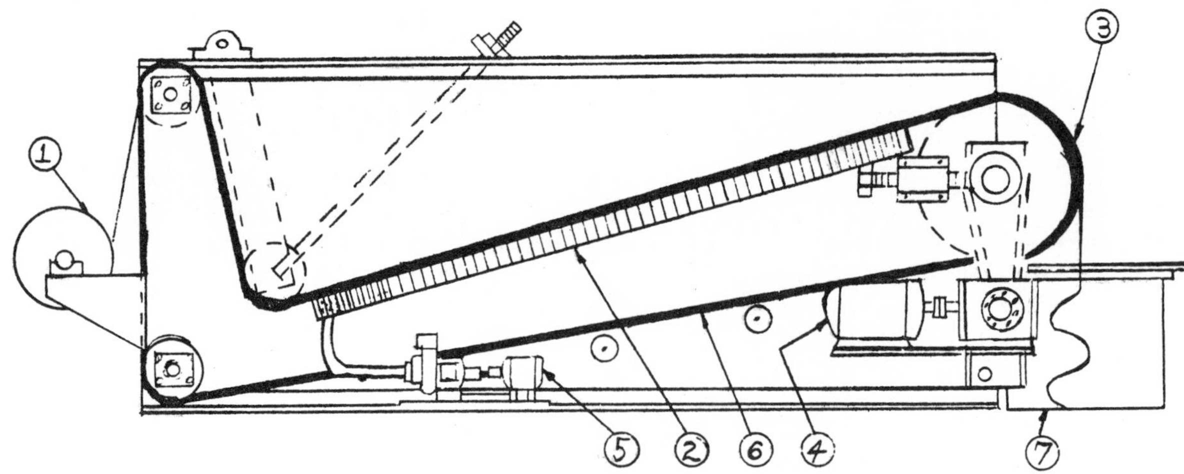

HERES HOW IT WORKS-AUTOMATICALLY

Belt 6 travels at a continuous speed over Vacuum box 2, pump 5, draws on the Vacuum box, media 1 rides on top of belt as solution is drawn through media along grooves in belt which is being pulled by roller 3, driven by drive motor 4, media and waste is collected in box 7, and removed when full.

The Detroit Automatic Lagoon separates solids from liquids automatically allowing clean liquids to flow to tanks or sewers. It will stop or reduce pollution, eliminating need for large acreage lagoons. It can be installed inside, outside, in the basement, or suspended from a ceiling or wall.

The Lagoon is self-cleaning. Operation of plant facilities does not have to be stopped to clean the machine as is required with settling lagoons. As it works it cleans itself. Contamination is collected in a box which is removed when full. No draining is required.

SIZES: This unit is constructed in sizes starting with one square foot and larger to suit the requirements of the solid/liquid separation to take place. By laboratory testing of a sample of the liquid we develop the flow in gallons per square foot per minute. The total gallons to be worked divided by the gallons per square foot per minute, gives us the size of the lagoon.

LOW OPERATING AND MAINTENANCE COSTS: Low horsepower motors do the whole job. A continuous roll of inexpensive paper or clot does the work of bulldozers or power shovels. This cuts costs to a fraction of a dollar per hour in comparison to large settling lagoons. Simply designed, it works only with motors, belts, pumps, rollers and ball bearings. It requires only simple loading with media when necessary and only simple lubrication of the bearings at regular intervals. The Automatic Lagoon can be constructed to withstand acid or alkaline solutions, thus eliminating costly wall repair, as is often required with regular cement basins.

FILTER TOWERS - ACTIVATED CARBON COUNTERFLOW

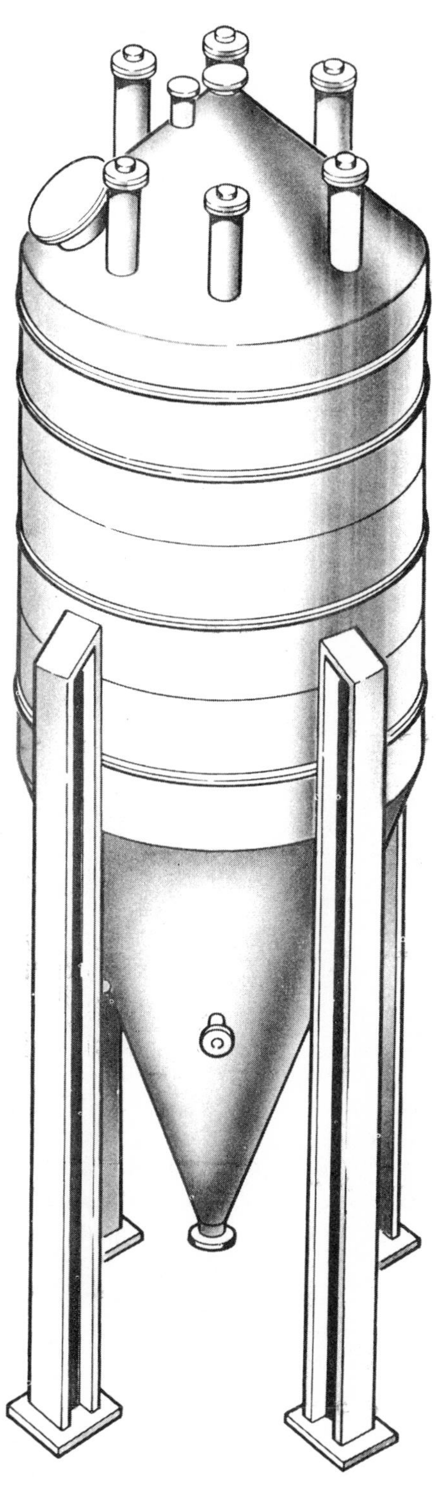

FEATURES

VINYL LINED
MILD STEEL CONSTRUCTION
EXPANDABLE
ECONOMICAL

PURPOSE

Detroit Filter Carbon Towers are designed for the purification of waste water by removing organic contaiminants whether measured as BOD, COD, TOC, color, odor, optical density, or by other analytical techniques.

Granular carbon treatment has built-in flexibility since the waste water can be either upgraded partially to meet effluent treatment and receiving stream standards or upgraded completely for direct re-use in high quality water applications.

Advanced waste water treatments--A process utilizing the Detroit Filter Tower can be incorporated as a final polishing step. It can be used to complement industrial and municipal waste water treatment systems whose sedimentation, clarification, and bacteriological techniques are inadequate for meeting higher standards of effluent quality.

Industrial process and waste water treatment---Detroit Filter Tower can be incorporated into manufacturing process systems to remove particular types of pollution. It can be used in by-product recovery operations and in industrial waste treatment programs for the removal of phenols, color, insecticides, TNT, detergents, polyols, dyes and other waste.

Detroit Filter Corp.

DETROIT AUTOMATIC FILTER - DELUXE MODEL

The Detroit Automatic Filter is especially designed for the electroplating industry, is a self-sufficient pressure unit automatically performing all the operations of a conventional (leaf) filter, and more. It filters, drains, cleans and repacks itself, all automatically. It improves plated finishes, eliminates dirt and down-time headaches, boosts production and profit dollars.

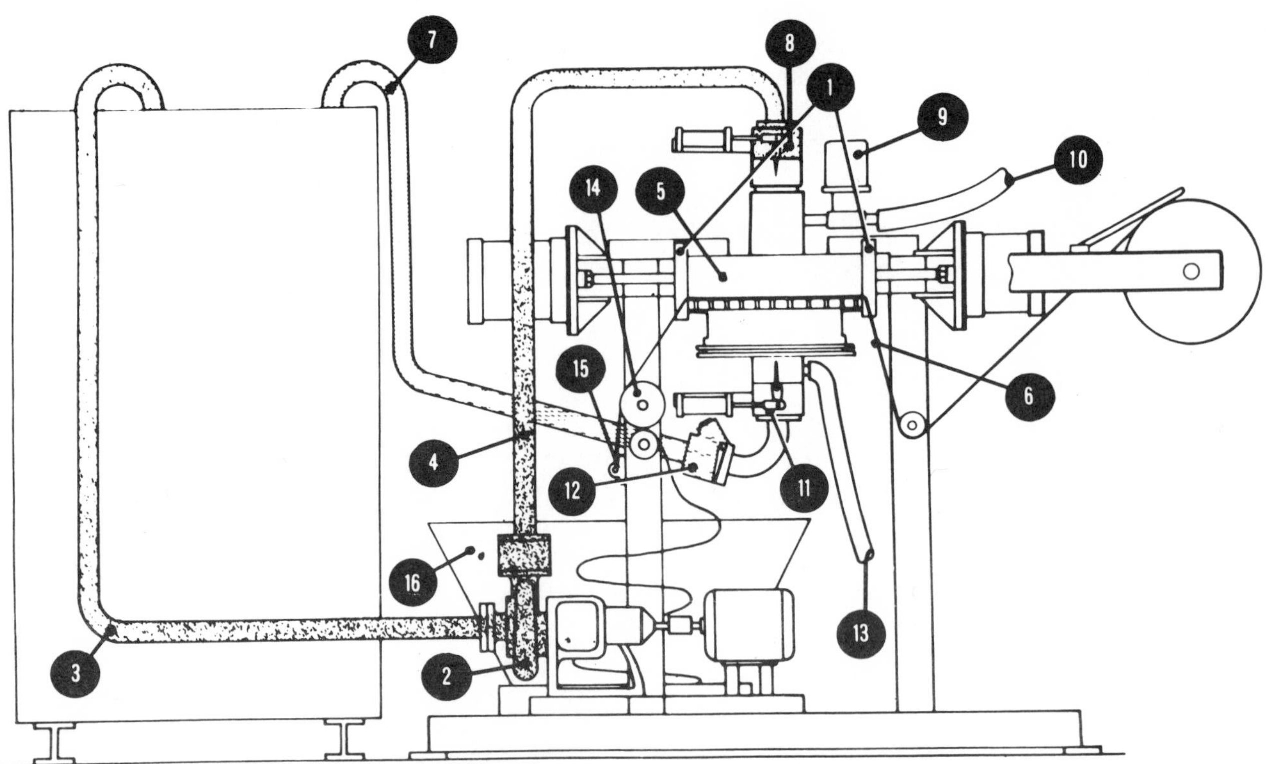

Here's How it Works . . . Automatically

1. Doors (1) are closed. Pump (2) draws dirty solution from work tank, via suction line (3). Inlet line (4) feeds contaminated solution into filter chamber (5) where it is forced downward, through pre-pack filter media (6) under continuous circulating (pump) pressure. Clean solution is returned to work tank, under pressure, via return line (7).

2. Pump (2) stops on automatic signal from pre-set timer. Butterfly valve (8) closes, automatically. Air valve (9) opens and air pressure, via line (10) expells remaining solution from filter chamber.

3. Timer closes valve (9), cutting off downward air pressure. Doors (1) open, automatically. Butterfly valve (11) also closes on automatic cue, as does swing-check valve (12). Upward air pressure is applied, via line (13), lifting media sheet clear of filter chamber platen. Powered roller (14) pulls used media over platen on a cushion of air. Excess media cake is removed by "doctor blade" (15). Used media and cake drop into toté tub (16).

4. Upward air pressure is cut off. Butterfly valve (11) opens. Doors (1) close. Machine is ready to repeat filtration cycle. Swing-check valve (12) prevents back-up of clean solution into filter chamber.

Diebold Inc., Lamson Division

LAMSON MULTI-STAGE CENTRIFUGAL BLOWERS FOR SEWAGE AERATION

Lamson multi-stage centrifugal blowers are ideally suited for the round-the-clock service demanded of sewage aeration equipment.

Lamson heavy-duty cast iron blowers are engineered with but one moving component--the shaft and impeller assembly. This assembly is the heart of the blower, rotating (usually at 3500 rpm) inside a cast iron housing with ample clearances throughout. These housings have heavy cast iron inlet and outlet heads with interlocking cast iron intermediate sections and integral annular diffusers.

The inlet head includes a diffuser to direct air to inlet of first impeller. The outlet head is of vortex design to eliminate friction. Cast iron bearing housings are bolted to outside of head sections to assure cool bearing operation. Labyrinth seals prevent air loss.

Lamson centrifugal blowers provide a smooth, uniform increase in pressure as air passes through the several stages of the machine. They are self-governing, and power consumption is determined by load factor.

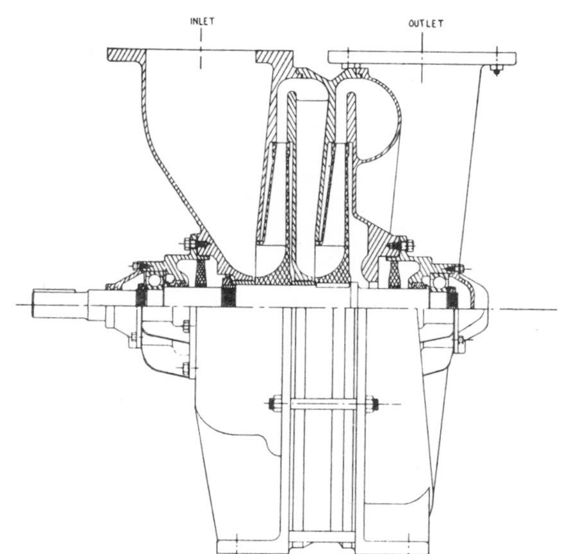

Lamson centrifugal blowers are virtually maintenance-free, requiring only routine lubrication of the two outboard-mounted bearings, which are anti-friction ball type and grease-packed.

The impeller assembly consists of a precisely machined heavy steel shaft, and one or more aluminum alloy shrouded-type impellers. This assembly is machine balanced, securely keyed to shaft and held in position with lockwashers and locknuts.

Power units consist of electric motors of standard manufacture, steam turbines or gas engines with suitable controls. Power may be connected directly with flexible coupling or through belt drive. Standard units are designed to rotate at 3500 rpm.

Dorr-Oliver Inc.

ABC SYSTEM FOR PRIMARY AND SECONDARY WASTE TREATMENT

The ABC System is the complete liquid treatment system for liquid waste, combining primary clarification, complete mix aeration and final clarification.

Primary clarification efficiency greatly affects the design and operating characteristics of subsequent liquid and solids treatment units. To obtain high primary treatment efficiency, proper tank configurations must be provided. Statistical analysis of the data indicated that the side water depth was the most important factor affecting efficiency. The new improved Dorr-Oliver Clarifier with its deep tank walls and sloping floor is endurance-rated for dependable operation and is a vast improvement over other types.

The ultimate in primary clarification is the Dorr-Oliver Clariflocculator, which combines flocculation and clarification. An additional 20% capture of suspended solids can be achieved with the Clariflocculator, resulting in substantial savings of down stream aeration facilities. It also reduces the expensive handling of waste sludge from secondary treatment.

The ABC System approach to aeration is homogeneity of tank content—complete mix. As feed enters the tank, it is instantaneously distributed throughout the entire tank. The aeration volume smooths out variations in the feed and maximizes the effectiveness of the mixed liquor in reducing the load. With a complete mix system an increase in loading of 20% is possible over a conventional aeration system.

Grand Island, Neb.: Combined Municipal and Industrial Wastes - This plant with a design flow of 5.8 MGD contains two 90' diameter Dorr-Oliver primary clarifiers two 75' diameter by 12' SWD aeration tanks and two 100' diameter Dorr-Oliver RSR final clarifiers. Each aeration tank is equipped with four 30 HP dual impeller D-O Aerators with air addition to the bottom impellers.

DORR-OLIVER AERATORS:
The Dorr-Oliver Aerator is a high capacity mechanical aeration unit designed to increase both the effectiveness and the rate of oxygen absorption over conventional methods. The unit can be adapted to a variety of tank sizes and is readily incorporated into existing tanks. Motor and the shock-protected drive and shaft are secured to cross beams spanning the tank. Air is introduced into the system by means of a sparge ring located beneath the lower impeller. The Aerator is easily adaptable to various aeration requirements.

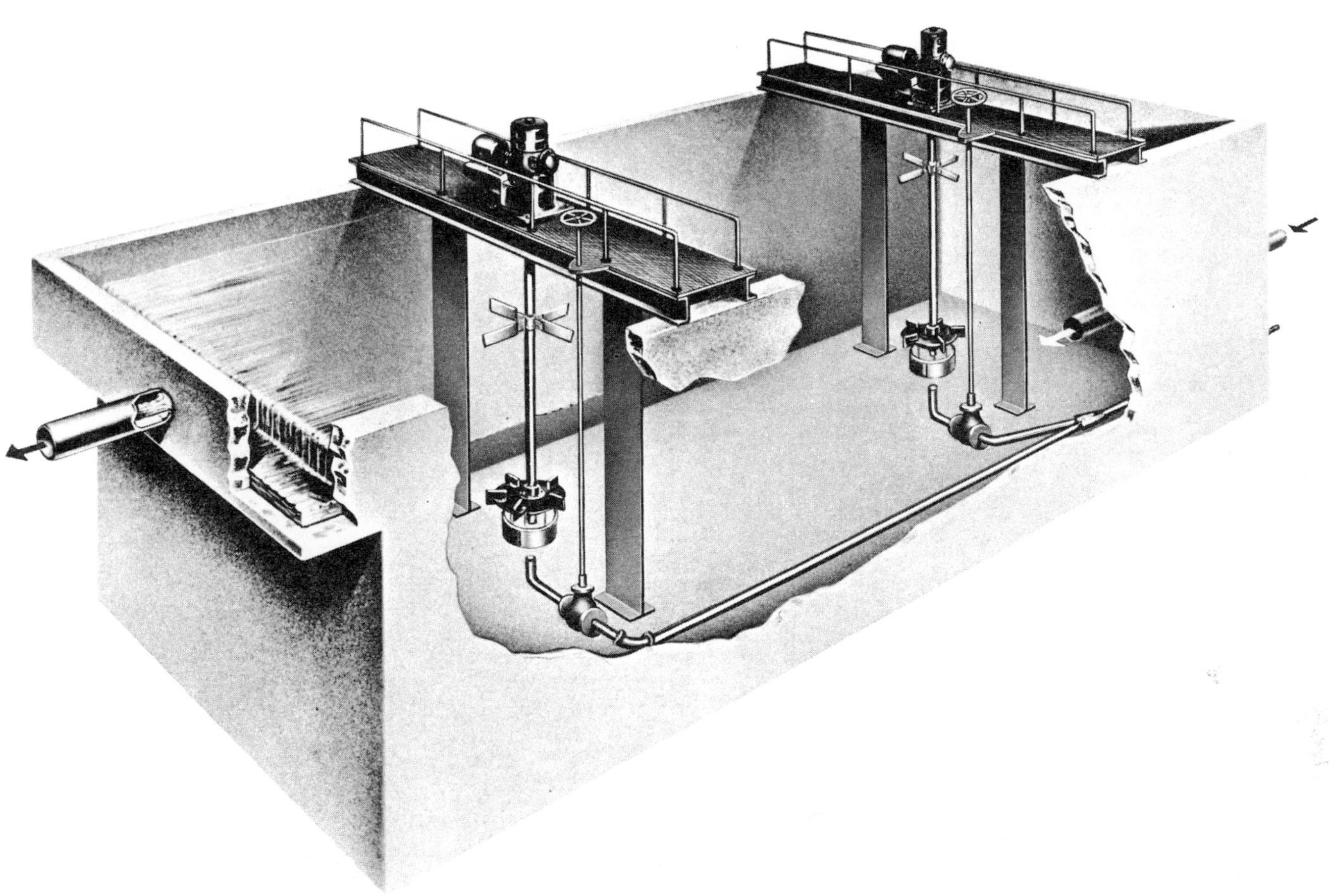

Complete mix aeration, the most effective kind, is best accomplished in square or rectangular basins with a length to width ratio up to 2 to 1. Common walls can be eliminated in rectangular multi-bay tanks. The complete-mix system is adaptable to single point feeding and withdrawal greatly reducing tank construction costs.

Mechanical aerators are ideally suited for complete mix aeration due to their radial flow pattern. Mechanical aerators mix in all directions at high velocities, affecting the entire tank volume. Mechanical aerators can be serviced without dewatering the tank, reducing need for multiplicity of tankage.

RSR CLARIFIER:

The Dorr-Oliver RSR Clarifier is designed for optimum operating efficiency as a final clarifier. It provides for continuous rapid removal of fresh solids and is equipped to make a three product separation:

1. An improved effluent with higher dissolved oxygen and lower suspended solids.
2. Rapid removal of fresh sludge.
3. Positive removal of low volatile solids.

Larger feedwells provide 100% recirculation. High volatile solids removed from the tank floor can be visually observed. The rate of solids removal is easily controlled. Low volatile solids are continuously raked inward over a sloping floor to a separate discharge sump. The high overflow rate is a direct result of maximum tank depth.

The design incorporates:

1. Rapid solids recirculation for fresher recycle and elimination of denitrification.
2. High rate recirculation for more effective aeration.
3. Adequate floor slope for high area loading capacity and less solids detention.
4. Large and deeper feedwells to improve entrance hydraulics and increase tank detention efficiency.
5. Deeper tank standards which will effectively handle variations normal to waste treatment.

The following table represents the depth ratios that are recommended and considered minimal for efficient final clarification

Diameter (feet)	Recommended SWD (feet)	Flow Rate MGD
30 - 65	12	0.7 - 3.3
75 - 100	13	4.4 - 7.8
105- 140	14	4.4 - 7.8
145- 190	15	8.6 -15.3
200+	16	16.4 -28.3

RSR Clarifier

Dorr-Oliver Inc.

THE DORR-OLIVER FARRER SYSTEM

The Dorr-Oliver Farrer System for continuous heat treatment brings a new dimension in the economics of sludge conditioning and sludge dewatering. Fully automatic in operation, the Farrer System continuously heats the sludge and maintains it at a temperature range of 360°-380°F. for the desired period of reaction. The result is rapid sludge coagulation, complete breakdown of the sludge gel structure and reduction of the hydration and water affinity of the solids. The bulk of the water can subsequently be separated in a continuous decantation tank. Thickened sludge residue can be easily dewatered by filtration or centrifugation.

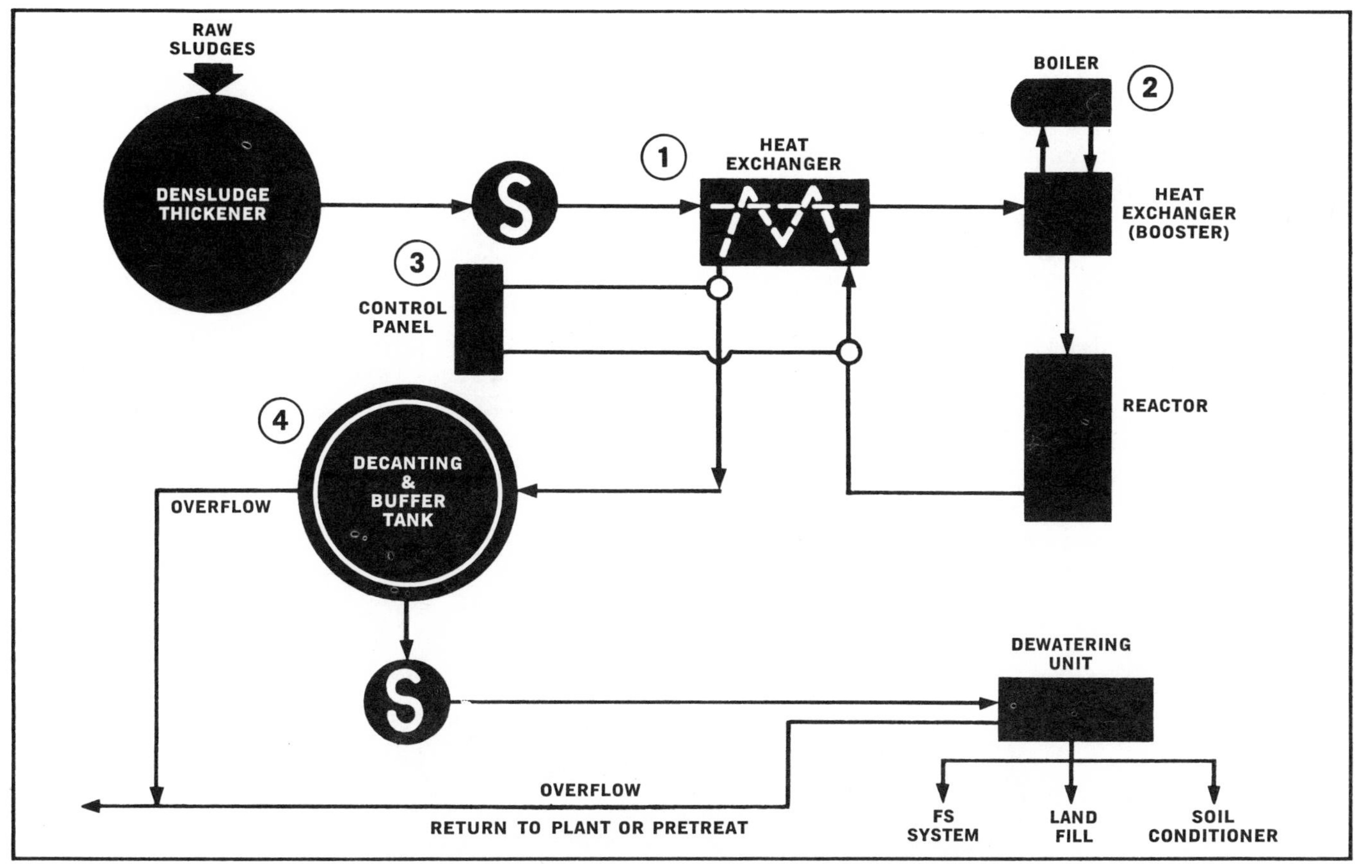

1 The heat treatment process is comprised of a two stage heat exchanger followed by an economizer. The economizer section of the heat exchanger preheats incoming sludge to approximately 300°F by controlling the outgoing treated sludge temperature at 85°F. A booster heat exchanger raises the preheated sludge to the desired final temperature of 360°-380°F by means of heat from the boiler. The heated sludge then flows through a specially designed reactor, which eliminates short circuiting and insures the desired retention time for full sludge conditioning.

2 The boiler for the booster heat exchanger can be fired by any of the conventional fuels including digester gas. However, when used in conjunction with an FS System, a waste heat boiler can be utilized for additional overall economy in operating costs.

3 Operation is fully automatic at every phase. A central, automated control panel with full instrumentation minimizes operation attention.

4 Following heat treatment the sludge moves to a continuous flow decanting tank for separation. The separated sludge is then dewatered with a final moisture content of 50%-60%.

FS DISPOSAL SYSTEM

The Dorr-Oliver FS Disposal System is an efficient, low cost sludge disposal system for both primary and secondary sludge.

Combining advanced processing techniques with thermal oxidation, the FS System reduces waste organic sludges from wastewater treatment plants to an inert, low volume ash.

FLUOSOLIDS REACTOR - Dewatered sludge is introduced directly into a fluidized sand bed. The violent agitation of the bed insures immediate and uniform distribution of fresh sludge. Combustion is rapid and absolutely complete. Exit gas temperature is 1400 - 1500°F., destroying all organics and odor. The Reactor produces completely oxidized off-gases using a maximum of only 20% excess air, for true operating economy.

FLUIDIZING BLOWER - Air for fluidization of the sand bed is supplied by this multi-stage, low maintenance blower. Air pressure is a low 3.5 to 4.5 psig and excess air requirements less than 20% reducing blower hp requirements.

AIR PREHEATER - The Preheater, an integral part of the FS System when sludge cakes with a high moisture content are being combusted, increases the efficiency of burning by 30%.

DORRCLONE DEGRITTER - Removes 95% of 150 mesh grit from clarifier underflow before introduction into the Densludge Thickener. Efficient grit removal increases sludge volatile content and minimizes wear on ensuing equipment.

GAS SCRUBBER - Ash produced by combustion in the Reactor is carried out by the flow of hot gases into the wet gas scrubber. All solids present in the gases are removed. Exhaust gases from the scrubber are about 160°F. and are free from fly ash and odor, meeting strict air pollution codes.

DEWATERING - Dewatering of sludge prior to combustion takes place in the MercoBowl Centrifuge. Featuring a new long bowl design, the MercoBowl discharges sludge containing only 65 to 75% moisture depending on sludge source. This new design of centrifuge can produce a separating force of 3700 g's.

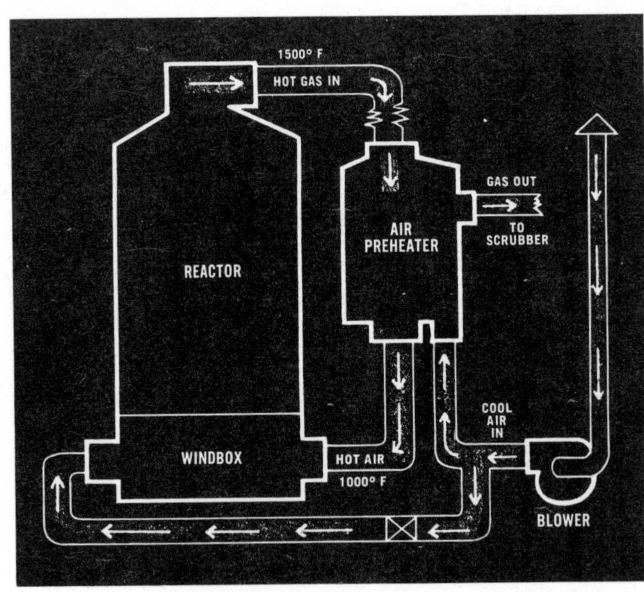

FLOW SHEET

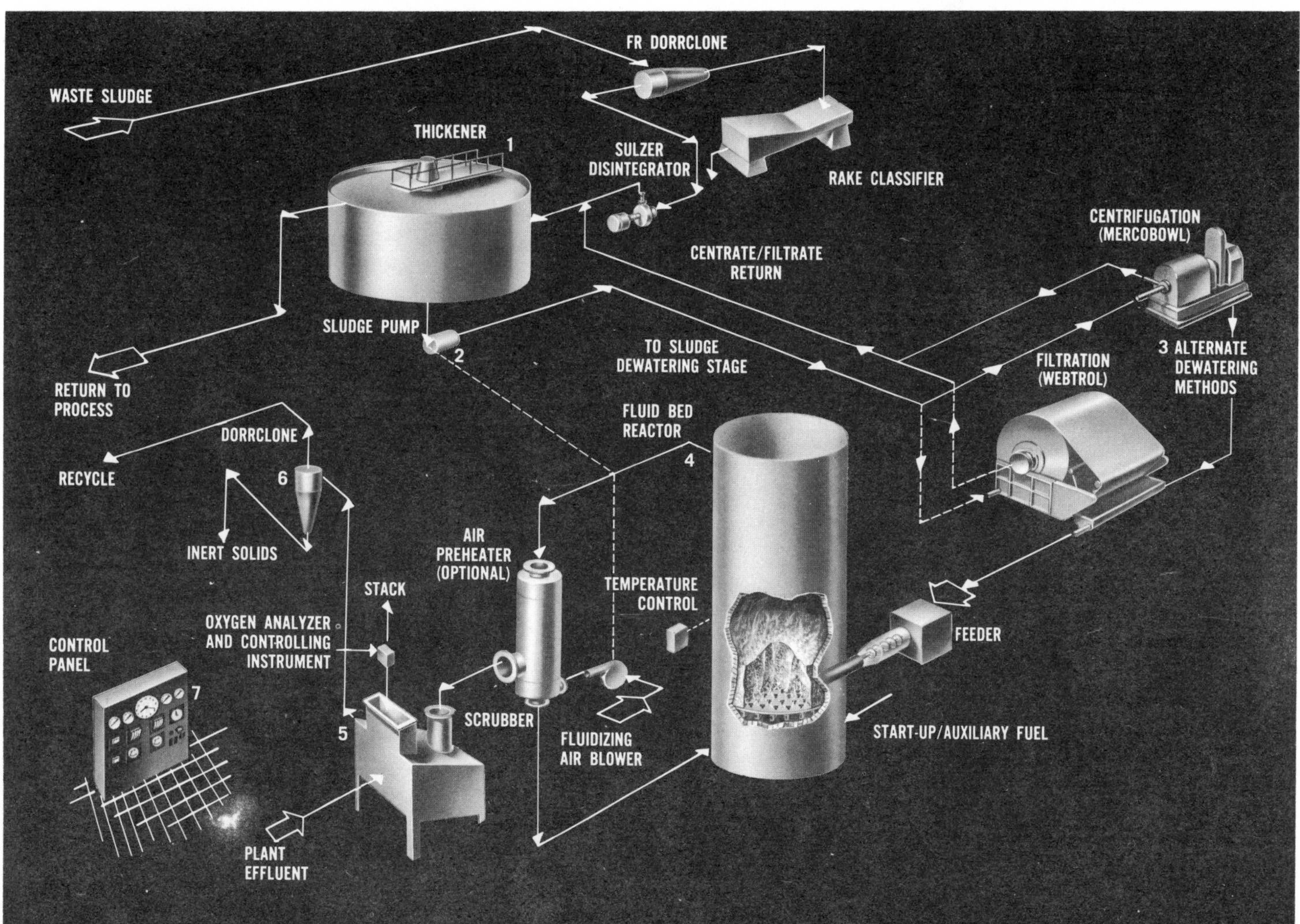

1 **Concentration stage**...the dilute degritted waste sludge is concentrated in a Densludge Thickener. Clarified overflow from the thickener is returned to the main raw waste stream.

2 **Sludge concentration**...the concentrated sludge is pumped from the thickener by a metering pump to the dewatering step.

3 **Sludge dewatering**...high hydraulic capacity produces extremely dry sludge cake. Sludge dewatering is accomplished either by centrifugation (MercoBowl) or by filtration (Webtrol Filter) depending on conditions. Filtrate/centrate is returned to the raw waste stream while dewatered cake is introduced directly into the fluid bed for combustion.

4 **Thermal oxidation**...dewatered solids are extruded directly into the reactor...the fluid bed within the reactor (at a pressure of approximately 2 psi) operates at a temperature of 1400–1500°F. Rapid combustion occurs and the solids are retained in the bed until reduced to an inert ash. Inert solids are removed from the fluidized bed by the upflowing combustion gases.

5 **Combustion gas scrubbing**...combustion gases are subjected to wet gas scrubbing using main plant effluent as a scrubbing medium. This action cools the gases and entrains the inert solids in the scrubber water.

6 **Inert solids removal**...the inert solids, contained in the scrubber water are separated in a DorrClone...the solids fraction removed as a by-product for disposal, and the liquid portion returned to the raw waste stream or recycled to scrubber.

7 **Control system**...combustion air rate to the thermal oxidation reactor is automatically controlled by the oxygen content of the exit combustion gases. The oxygen content is continually measured; instrumentation adjusts the air rate to maintain a predetermined oxygen excess. Remote variable control of the sludge pumping rate is an integral feature. Automatic fuel control results in minimum fuel consumption. These controls provide for maximum utilization of reactor capacity while automating the operation of the system.

PURIFLOC® polymer flocculants are synthetic, water-soluble, high-molecular-weight organic polymers tailored specifically as flocculants for water and waste-water treatment. Their use raises to unprecedented levels the efficiency of solid-liquid separations in clarification, thickening, and filtration processes. The flocculants agglomerate a wide variety of inorganic and organic solids, including colloids, which are present in waste waters. They operate efficiently in waters of widely varying pH or chemical content.

Tests indicate that in the flocculation reaction, the products are absorbed to a great extent on the solids present and therefore will not be discharged to a stream with the plant effluent.

This bulletin discusses in detail the evaluation and use of PURIFLOC flocculants as sludge-conditioning agents. Further information concerning the specific flocculants used in this type of application is available upon request from the nearest Dow sales office.

ADVANTAGES:

PURIFLOC flocculants offer numerous advantages for conditioning municipal and industrial sludges for dewatering. Their use can:

1. Eliminate the need for inorganic chemicals
2. Produce higher filtration rates
3. Permit reductions in chemical storage and metering facilities
4. Minimize maintenance of equipment
5. Increase filter cloth life
6. Improve cake incinerator operations
7. Make chemical handling safer and cleaner

Thickening and elutriation processes can be greatly improved by the use of the flocculants. Increased solids capture efficiencies, solids loadings, and solids concentrations are likewise all possible.

APPLICATIONS:

Low dosages of PURIFLOC flocculants may be utilized in the following solid-liquid separation processes:

1. **Sludge Dewatering**
   a. Vacuum filters
   b. Sand beds
   c. Centrifuges
   d. Other mechanical devices
2. **Sludge Thickening**
   a. Gravity settlers
   b. Air flotation units
   c. Centrifuges
3. **Elutriation**

Centrifuges

Outstanding centrifuge performance is possible due to the extremely rapid flocculation rate attainable with PURIFLOC flocculants. The use of these flocculants in centrifuges dewatering or concentrating sludge provides the following advantages: increased through-put, clear centrate resulting from high solids capture, and increased sludge concentration. Maximum efficiency is attained by the use of a dilute flocculant solution (.1%) and by adding it to the slurry at the proper point. Equipment manufacturers' recommendations concerning pool depth, speed of rotation, slurry flow rate and exact flocculant addition point, or points, should be followed.

Sludge Drying Beds

More rapid drainage is possible when sludge is conditioned with PURIFLOC. The flocculant should be added to the sludge at a point where agitation is sufficient to provide adequate mixing but not severe enough as to cause floc degradation. These criteria may be met by adding the flocculant several feet ahead of the discharge end of the sludge feed line or at the drying bed splash plate. A small mixing device could be attached to the end of the sludge feed line if the above points of addition do not produce sufficient flocculation.

Sludge Concentration and Elutriation

Thickening processes by gravity settling, flotation, or centrifuging can be greatly improved with the use of PURIFLOC flocculants. Increased solids capture efficiencies, increased solids loadings, and increased solids concentration are all possible with the use of these polyelectrolytes.

Surfpac biological oxidation media

SURFPAC® biological oxidation media is a plastic trickling filter media used in the treatment of municipal and industrial wastewaters.

Trickling filters purify wastewater in essentially the same way as rivers and streams do. A trickling filter, however, concentrates into a small space the same biological process that would occur naturally only over many miles of waterway. The capacity of a stream to purify itself, however, is limited and can be overloaded.

SURFPAC media is the result of continuous research by The Dow Chemical Company in the area of biological oxidation of liquid wastes. It provides:

A uniform surface area, completely available for the development of healthy biological slimes.

A large specific surface area—27 square feet per cubic foot of media—for greater slime surface per unit volume.

A void space of 94%, which gives excellent aeration of waste and trouble-free solids release.

Excellent redistribution of wastewater with virtually no free-fall.

Considerable savings in land area and over-all costs, because of its design flexibility and light weight.

Self-supporting strength for heights up to 21.5 feet, so expensive containing structures are not required.

High hydraulic loading capacity which, when uniformly distributed, will prevent the forming of larvae, thus eliminating trickling filter flies.

Applications

Dual System

In this application, SURFPAC is used to "clean up" strong, varying or difficult to treat wastes before they are passed on to a second biological system for polishing. The dependability and ruggedness of SURFPAC media makes it ideal for Dual System treatment processes, particularly ahead of the more sensitive activated sludge system. SURFPAC can protect the activated sludge process from variations in waste composition or strength which can cause it to be seriously upset for long periods of time.

For the treatment of strong wastes, the reliability and economics of the dual system can be particularly attractive.

Plant Expansion

An organically or hydraulically overloaded secondary treatment plant can be expanded rapidly by building a tower with SURFPAC media following primary settling. This type of expansion is both efficient and economical. It is an effective means to increase plant capacity where space is limited.

Moreover, this method of plant expansion incorporates all of the advantages of a dual system.

Chemical-Biological Treatment

In some waste treatment plants, a substantial fraction of BOD can be removed by using chemical coagulants in the primary settling step.

When this is done, SURFPAC can provide an exceptionally economical and simple biological process for polishing the chemically treated effluent. This type of treatment is often used following the Dow phosphorous removal system.

Pretreatment

Many industrial wastes require pretreatment before they are discharged to the sewer in order to:

meet municipal ordinance requirements
reduce sewer charges that are based on waste strength
treat difficult wastes, (e.g., phenol).

SURFPAC media has been used successfully to perform all of these functions.

Complete Treatment

In a small or medium sized plant, SURFPAC will serve economically as a complete secondary biological facility. Its use is dictated by a strong concern for dependable, trouble-free operation, the desire for an advanced trickling filter system, or because land is limited and costly.

CENTRIFUGALS

ROOTS Single Stage Centrifugal Compressors are designed for high efficiency, high reliability, low noise level, and oil-free delivery. Because of the centrifugal characteristic (constant head, variable volume) this type of compressor lends itself well to automatic monitoring and control of air flow. Suitable for gas or steam turbine, engine/gear, electric motor or motor-gear drive. Heavy duty cast iron construction for years of rugged service.

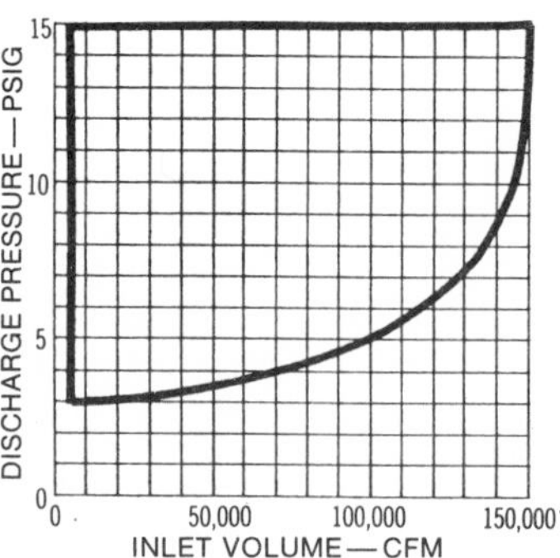

ROOTS MV Compressor is a vertically split multi-stage centrifugal unit designed in modules to meet various flow and pressure requirements to 15,000 CFM and to 10 PSIG. It is furnished with sufficient stages to meet the necessary head (pressure) specification.

Its operating reliability, efficiency and low noise level make it well suited for use in aeration, conveying or any other application where air or gas must be compressed within the range covered.

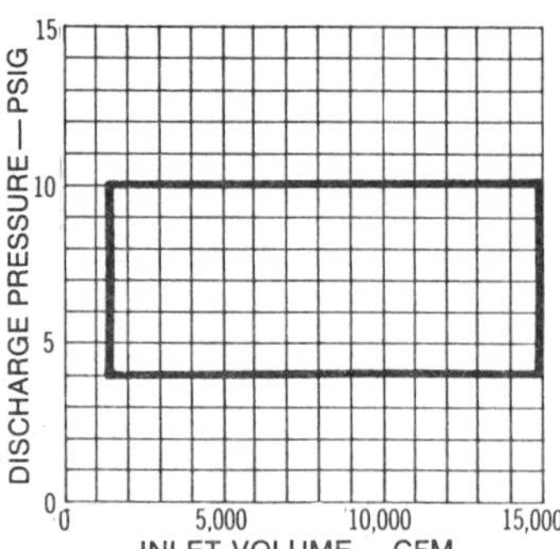

The type H, horizontally split multi-stage centrifugal compressor is a heavy duty unit of cast iron construction suitable for handling air or other gases to pressure levels requiring as many as seven stages of compression. Like the single stage, its constant head/variable volume characteristic makes it ideal for applications where flow control is a requirement. Flows up to 90,000 CFM are possible, using a variety of driver options.

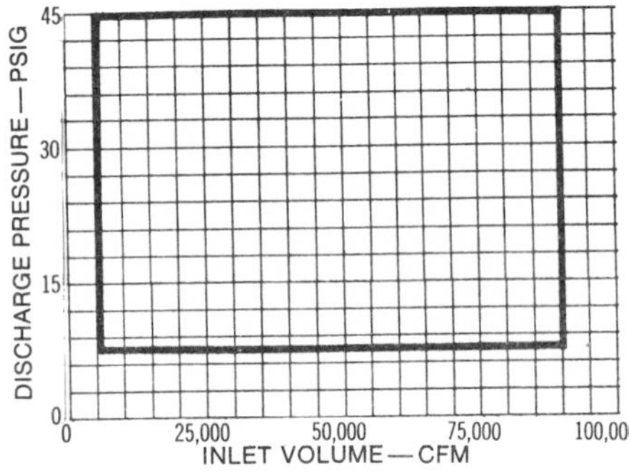

BLOWERS

These oil-free blowers operate with no internal contact of moving parts. High efficiency and positive air displacement make this rotary air blower ideal for systems where system resistance may vary and where power savings are important. Roots exclusive **Whispair Blower** provides all the advantages of the conventional design plus significant noise level reduction. The chart shows our standard size range . . . Roots also builds units to 40,000 CFM or higher.

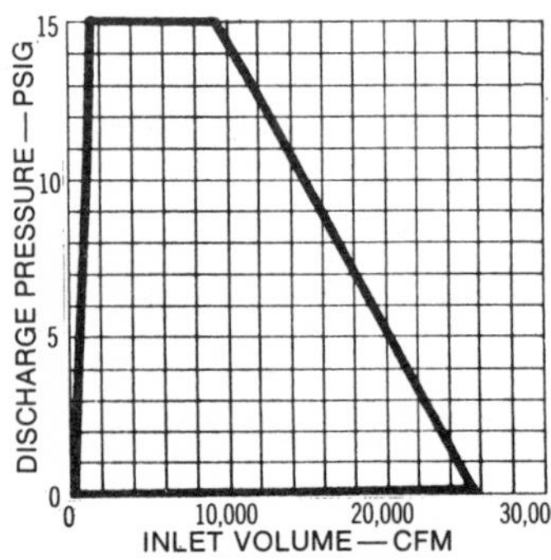

We manufacture an extensive line of Roots small positive displacement blowers, vacuum pumps and gas pumps. These sturdy, dependable machines are ideally suited to OEM applications. Typical tasks include aeration of fluids or solids, agitation of liquids, pneumatic conveying, cooling, drying, vacuum processing, exhausting vapors, pressurizing, air blasting, spraying, blending, suction pickup, and dozens of other jobs.

Small air blowers range from 1 to 13 PSIG and 2½ to 2800 CFM. On vacuum service, the range is 4″ to 22″ Hg and up to 2300 CFM.

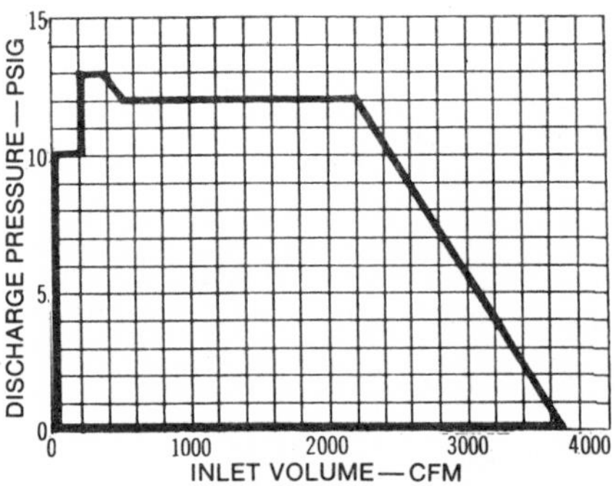

Roots Rotary Gas Pumps are similar to Roots Rotary Air Blowers in operation. They are used where it is necessary to compress gas with minimal gas leakage. Seals on the larger machines are the mechanical face type; the small frame sizes use a split lip-type seal to minimize gas leakage. The larger pump sizes are available as **Whispair Gas Pumps** and operate at a significantly reduced noise level.

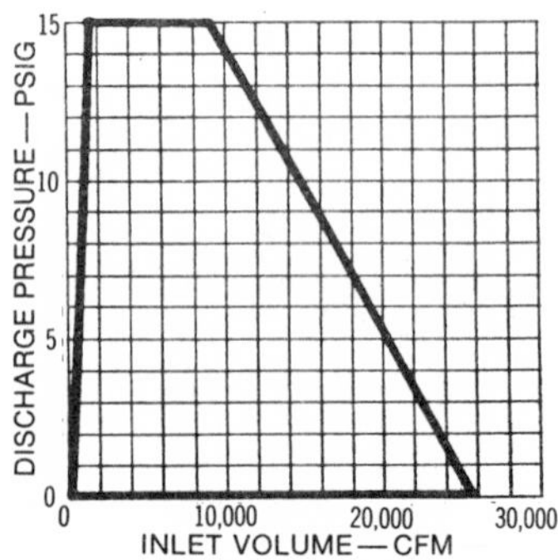

PERMASEP® PERMEATOR MODEL NO. 044 4" DIAMETER B-9 PERMEATORS PRODUCT SPECIFICATIONS

Membrane Type	B-9
Membrane Configuration	Hollow Fiber
Shell Dimensions	5¼" OD x 4-5/8" ID x 47" long
Shell Material	*Filament-wound Fiberglass Epoxy
End Plates	**Aluminum A356-T6
Snap Rings	15-4 PH-Mo Stainless Steel
Connections	Feed and Permeate, ¾" female, NPT Concentrate, 3/8" female, NPT
Permeator Weight, filled with water	50 pounds
Product Water Capacity[1]	2,500 gpd
Salt Passage[2]	$<$ 10%[1]
Rated operating Pressure	400 psig
Temperature Range	32°F–95°F
pH Range[3], continuous exposure	4-11
Conversion Range[3]	25-90% (for soluble salts)
Cl_2 Tolerance[3], continuous exposure	
pH <8	0.1 ppm max
pH $\geqslant 8$	0.25 ppm max
Operating Position	Horizontal or vertical
Permeate Back Pressure[3]	50 psig max
Reverse Flush Pressure Drop (Flushing through concentrate port)	50 psi max

Notes:

(1) Based on operation with a feed of 1500 ppm NaCl at 400 psi, 68°F, and 75% conversion. For operation at other conditions, consult Permasep® Products.
(2) Dependent on water analysis and conversion.
(3) For operation outside this range, consult Permasep® Products.

*Model 043 Available 6063–T6 Aluminum
**Available with Fiberglass epoxy end plates on special order

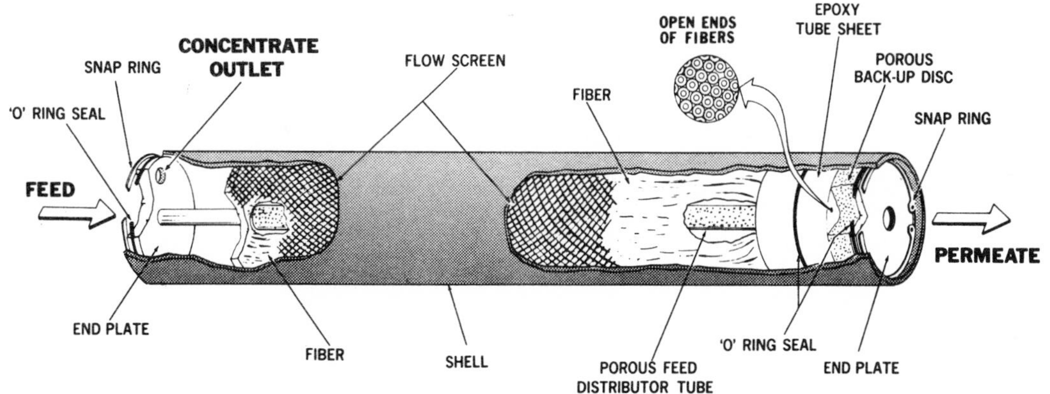

CUT AWAY DRAWING OF PERMASEP® PERMEATOR

Like all reverse-osmosis devices, Du Pont's permeators fractionate a salts solution — brackish water, for instance — into a relatively saltfree effluent and a salts-concentrated effluent, by subjecting the feed to a pressure higher than its osmotic pressure and bringing it into contact with a semipermeable membrane. The pressure drives water, but relatively few salt ions, through the membrane, leaving a concentrated solution.

A basic difference among reverse-osmosis designs is the physical configuration of the membrane. Whereas some units use flat membrane plates and some others use tubular or spiral-wound membranes, Du Pont has been the pacesetter in developing a new configuration wherein the membranes consist of very fine hollow fibers. These are assembled into an apparatus similar to a shell-and-tube heat exchanger, with the fibers representing the tubes. The feed solution goes to the shell, and the applied pressure in excess of the osmotic pressure forces water inwardly through the individual fiber walls. Fresh water thus emerges from the tube side, while the salts-concentrated stream leaves the shell side.

The hollow-fiber approach's big advantage over other reverse-osmosis configurations is its high productivity per unit of volume, due to effective packing of much membrane area into a small space. Another advantage is that membrane-support is relatively easy and a given membrane is much more resistant to mechanical damage when it consists of hollow fiber than when it is a flat plate.

In Du Pont's version of hollow-fiber reverse osmosis, the fibers are of an aromatic polyamide. The wall is asymmetric; most of it is porous, but an outer layer about 0.1 micron thick is dense. Each fiber is about 85 microns outside diameter and 42 microns inside diameter. The company calculated its inside diameter as the optimum balance among (1) fiber-bore pressure drop, (2) the amount of permeation area made available, and (3) expected time-trends in fiber permeability and cost. The ratio of inside to outside diameter was fixed by strength considerations.

The present version of the permeator, designated the B-9, contains a bundle of about 900,000 fibers. The open ends are embedded in an epoxy tubesheet; the closed ends are likewise sealed in epoxy, at the opposite end of the permeator. The fiber bundle, having a membrane area around 1,900 sq. ft., is mounted in an aluminum shell 4 ft. long and 4 1/2 in. dia.

A porous feed-distributor tube passes along the bundle axis, dead-ending just short of the tubesheet for the hollow fibers. Feed solution is introduced through this tube and passes radially outward through the fiber bundle, which is interspersed with cloth layers to maintain the bundle configuration and promote orderly flow. Most of the water passes radially inward through the individual fiber walls, then flows axially through the fibers toward the open or tubesheet end. Leaving the tubes, it is collected by a porous sintered-alumina plate that is in contact with the epoxy tubesheet and acts as a mechanical support and as a header that directs the water to the permeator outlet.

Meanwhile, the concentrated solution that stays on the shell-side flows through an annulus between the fiber bundle and the shell proper, and is discarded through an outlet in the shell wall. When operating on a 20°C. feedwater containing 1,500 ppm of sodium chloride, a permeator of this size can produce 2,000 gal./day of water containing 150 ppm of salt. Operating pressure for such a unit is 400 psig. Full of water, it weighs only 65 lbs.

CHEMICAL SOLUTION FEEDERS

Sewage treatment requirements are met by a variety of Mec-O-Matic products, each with a flexibility in feed rate and pressure. Odor control, for example, is provided by blanketing storage tanks with a citrus oil base sprayed through a series of nozzles. This highly aggressive chemical is metered with a Mec-O-Matic pump. The feeders are also used to introduce enzymes into the sewage lines, thus increasing the handling capacity of undersized treatment plants. Filter aids and bacteria control chemicals, pH control materials and algicides are all commonly metered and fed with Mec-O-Matic equipment.

SERIES 50 is a positive, fully adjustable cam type drive system powered by a heavy duty motor to prevent burn-out even under the most strenuous duty cycles. Feed rate is adjusted by loosening a thumb screw on the cam and rotating an arrow indicator to the required setting. The head is mounted on a heavy die cast frame which also serves as the mounting bracket for the pump.

BARRACUDA SERIES features a unique control mechanism which allows feed rate adjustment while the unit is in operation; at the lowest feed rate, the diaphragm moves completely forward in the head to exhaust the head cavity providing fine control.

BARRACUDA TWIN SERIES has two motors included, each of which operates as an indendent system; ideal for treatment systems where two chemicals are used.

H-J SERIES features a concealed diaphragm drive system with a time proven dual cam arrangement.

MARLIN NO. 60J1, the largest of Mec-O-Matic's diaphragm pumps, is ideal for installations where higher volume feed rates are required; a quality instrument with a 200 pound pressure rating. Feed rate can be controlled easily from the low to high range because of a unique drive system which doubles the output speed of the motor shaft to provide diaphragm movement of 120 strokes per minute. This rapid diaphragm movement substantially reduces the possibility of solids buildup in the valve system. The unit is self priming, even at the lower feed rate settings.

G-D SERIES uses a PVC piston in place of the diaphragm used on smaller chemical feeders. The result is a high pressure capability, more reliable performance — particularly in the low feed rate ranges — and greater accuracy throughout both feed rate and operating pressure range.

ECODYNE GP SERIES, a major development in chemical pump design, incorporates the best features of both diaphragm and piston type chemical feeders by combining a piston and a diaphragm in one part. The diaphragm segment is used to prime the pump; since it moves at 75 strokes per minute and is operating at atmospheric pressure, reliable priming is assured every time. The piston samples the required quantity of chemical from the priming chamber and pumps only that portion required through the discharge line to the point of injection. The balance of the chemical in the priming chamber is recirculated back to the chemical tank. This method of piston feed results in accurate control, even at the lowest feed rates, against comparatively high pressures. Because of the high pressure capability, this pump is ideal for remote installations, well away from the point of injection.

Ecodyne Corp., Smith & Loveless Division

OXIGEST SEWAGE TREATMENT PLANT WITH AUTOMATIC SURFACE SKIMMING

The factory-built Oxigest sewage treatment plant is designed specifically for small subdivisions, mobile home courts, motels, shopping centers, apartments, resorts, hospitals, schools and factories in outlying areas without municipal sewer facilities. With the new Automatic Surface Skimmer, the Oxigest requires less maintenance and much less operating attention than accepted conventional treatment systems.

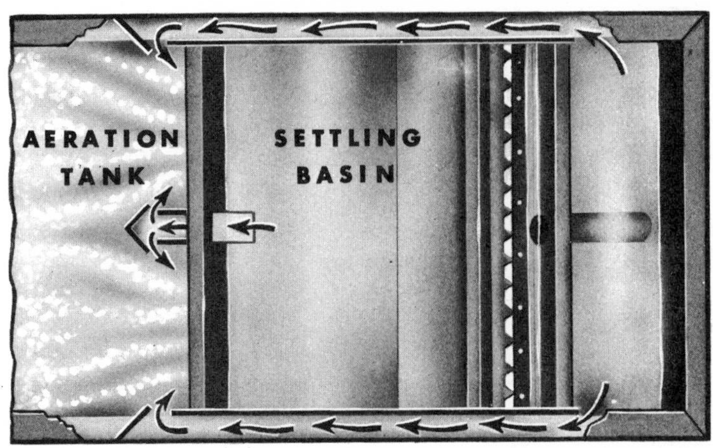

A new non-mechanical surface skimmer on all standard factory-built models provides automatic removal of floating material from the surface of the settling basin by a new hydraulic system—an exclusive with Smith & Loveless. The system reduces maintenance by removing floating particles (such as grease, garbage and denitrified sludge) from the settling basin.

The circulating liquid in the aeration tank (at left), flowing past the strategically located eductors, sets up a "return flow" which skims the surface of the settling-basin compartments by drawing the surface liquid through the skimming troughs to the aeration tank. The recirculation effect eliminates operation problems, reduces maintenance expense.

Aerobic Digestion Process:

This process is best described as an "extended aeration" or "aerobic digestion" treatment system. It provides simple, dependable treatment for domestic sewage by introducing an abundant supply of air to the sewage to keep the sewage solids in suspension for a sufficient period of time to permit adsorption and digestion to take place.

Microbes and other living organisms live off the organic matter and consume it. They are stimulated to activity by the abundant oxygen and thrive on the rich food source of high-energy organic wastes.

The turbulence in the aeration tank aids the digestion process by rapidly mixing the fresh sewage solids with the activated sludge, by breaking up the sewage solids and by bringing the contents of the aeration tank in contact with the atmosphere where additional oxygen may be absorbed.

Thus the aerobic bacteria and microbes reduce the organic matter and waste to a stable form, which is odor and nuisance free.

SEWAGE LIFT STATIONS

The VAC-O-JECT pneumatic ejector provides dependable two-stage lifting action, incorporating the principle of vacuum intake and air pressure ejection. It is designed for installation on a standard four-foot wet well or receiving manhole.

As sewage rises in the wet well, it actuates a level switch. The control system starts a reversible air pump which evacuates air from the receiver. Atmospheric pressure forces sewage from the wet well up to the receiver. The sewage fills the receiver and touches a "probe" which stops the air pump, reverses it and ejects the sewage by forcing air into the receiver.

At the end of each ejection cycle, the air pumps are automatically alternated to evenly distribute wear between the pumps. The discharge cycle can be controlled by an adjustable timer. The system is fully protected against voltage surges and overloads.

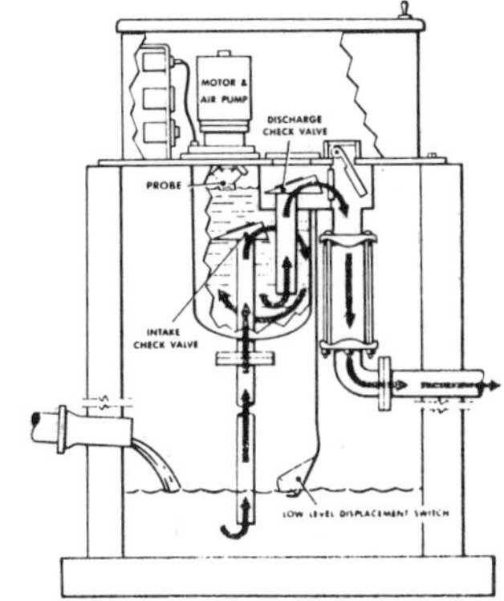

The MON-O-JECT simplex pneumatic ejector lift station is a cylindrical steel chamber with three individual compartments. The pneumatic ejector lift station is recommended where the rated capacity is less than 100 gpm. A four-inch centrifugal pump, designed for three-inch solids, is the smallest which will operate with reasonable freedom from clogging on raw sanitary sewage. Such a pump is not available with a rated capacity less than 100 gpm, therefore, to lift smaller flows, the pneumatic ejector is the only satisfactory answer.

The DU-O-JECT duplex pneumatic ejector lift station equals two MON-O-JECT stations, combined into the same structure with dual piping and receivers. It provides the added dependability of complete stand-by equipment throughout the extra capacity for peak loads.

The WAY-O-MATIC pneumatic ejector lift station is constructed similar to the pump station but with ejector pots (receivers) replacing the pumps. A superior control system "weighs" the sewage to provide dependable, trouble-free operation with minimum maintenance.

PRES-O-JECT is a compact, factory built pneumatic ejector. It can be specified with the conventional electrode system or the exclusive Smith & Loveless "No-Fail" Electrode System. Designed for capacities up to 200 gpm, it can be installed in simplex or duplex arrangements. Also available for stored-air applications and remote installation of compressors.

Factory built sewage pumping stations are built to provide a lifetime of dependable service — available for capacities from 100 to 4,500 gpm per pump with two, three and four pumps per station with a wide variety of control systems including variable speed controls and flow-matching systems, ozone generators, elevators, and alarm systems. Even larger capacity stations are built to order.

A compressor (1) supplies a stream of air through the air-bubbler pipe (2) to the wet well. As sewage rises, back pressure in the line actuates a "low-level" mercury pressure switch, energizing a magnetic starter and starting one of the pumps (3)

Sewage flows from the wet well, through the suction pipe (4), a gate valve (5), into the pump (6), out through a check valve (7), discharge gate valve (8) and discharge force main (9).

If sewage continues to rise above the starting level of the first pump, a "high-level" mercury pressure switch actuates the starter of the second pump and both pumps operate until the wet well level drops to a predetermined shut-off point.

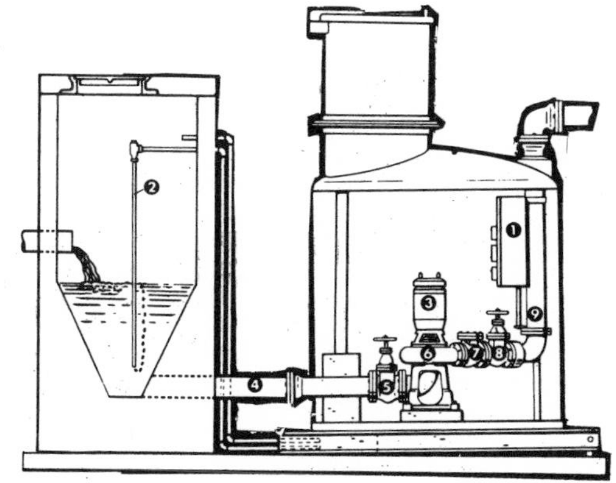

PRIMARY WASTEWATER SEPARATORS

The "Standard" machines are used for the mechanical processing of industrial wastewater at the primary level to remove free (surface) waste oils, floating oil/water emulsions, buoyant wastes and settled solids.

Floating oils and emulsions are retrieved from wastewater by partially immersed, 2-speed primary and secondary rotary cylinders. The intake compartment is also equipped with a transversely positioned slotted pipe for emergency purposes in the event of power failure to the primary oil-retrieval cylinder, or waste oil surges in volumes beyond the capacity of this cylinder.

Settled solid wastes are removed from the machines by an integral, floor-mounted solids ejector with vertical discharge ramp. Buoyant wastes can be retrieved from the water surface in the intake compartment by means of a partially immersed surface conveyor.

The discharged wastes are highly de-watered due to the vertical ramp and the (patented) obtuse angle of the flights; the water content of the retrieved waste oils is less than 5%, on average, by volume except for included emulsion content. All driven assemblies are cycled by program timers to minimize wear and eventual maintenance. Performance logs are available on request.

Basic design parameters are: (1) average detention time of 30 minutes, and (2) maximum water velocity under the primary and secondary baffles of not more than 3 fpm. Water discharge from the machines is optional between float-switch controlled pump or gravity flow from a weir to exterior pipe connection and bolting flange.

The "Standard" Primary Separators are steel plate, pre-assembled, pre-tested, integral machines which are ready for immediate operation at site when connected to power and piping.

Composite (reduced) drawing showing general arrangement and inside dimensions of EDENS steel plate Primary Waste Water Separators for influent flow rates in range of 1,000 gpm through 2,000 gpm. (Discharge port of the influent diffuser is shown in cross-hatch.) The integral, pre-assembled, pre-tested structures with gravity discharge are sized for shipment by railroad flat car.

(Detention times approximate 30-minutes. Water travel ranges from 3 fpm to 4.84 fpm under the primary baffle.)

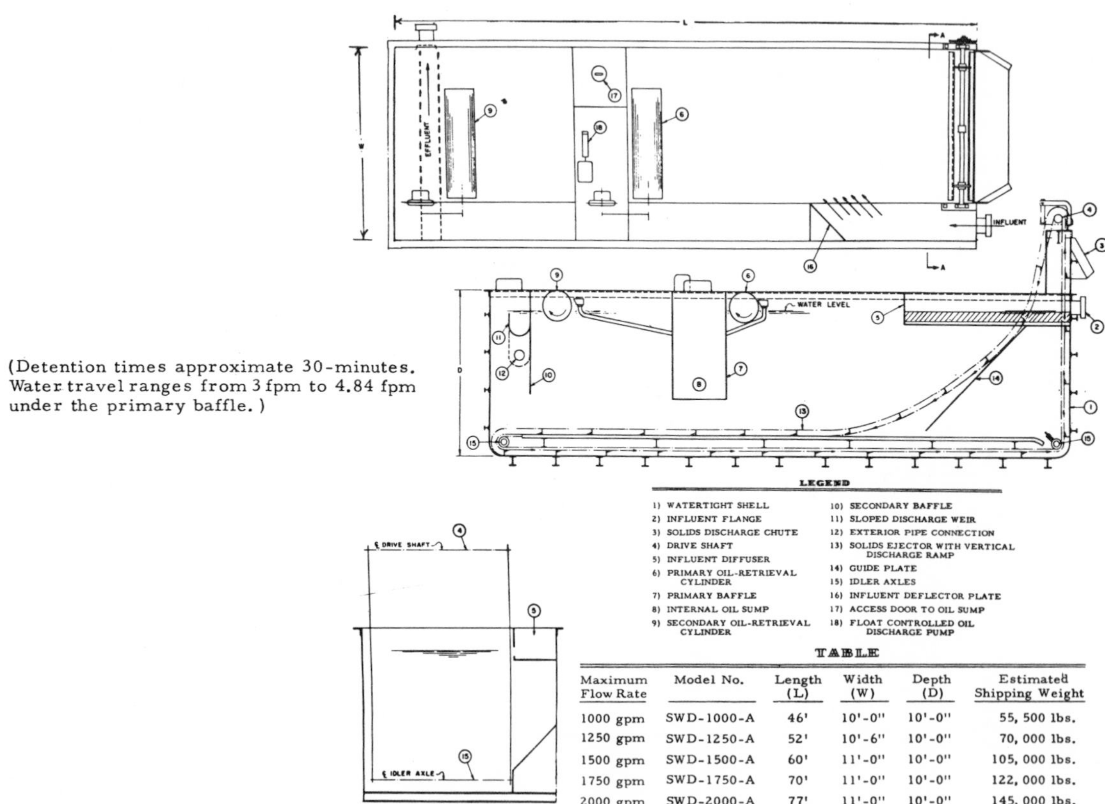

TABLE

Maximum Flow Rate	Model No.	Length (L)	Width (W)	Depth (D)	Estimated Shipping Weight
1000 gpm	SWD-1000-A	46'	10'-0"	10'-0"	55, 500 lbs.
1250 gpm	SWD-1250-A	52'	10'-6"	10'-0"	70, 000 lbs.
1500 gpm	SWD-1500-A	60'	11'-0"	10'-0"	105, 000 lbs.
1750 gpm	SWD-1750-A	70'	11'-0"	10'-0"	122, 000 lbs.
2000 gpm	SWD-2000-A	77'	11'-0"	10'-0"	145, 000 lbs.

Economy primary waste water separators are also offered for processing industrial wastewater at low flow rates and reduced machine investment. Standard models are available for influent flow rates to 30, 50, 60, 75, 100, 150 and 200 gpm.

VERTICAL WASTE EJECTORS

EDENS VERTICAL WASTE EJECTORS are used to separate settled solids from industrial waste water. The integral machines are custom built, angle flight conveyors with vertical discharge ramps designed for installation in existing concrete or steel plate water-holding structures with vertical walls and rectangular cross section. For enclosures with sloping ends, the discharge ramp can be designed accordingly.

The vertically positioned ramp abuts the vertical endwall precisely over which the lifted wastes are to be discharged for accumulation and disposal. The exclusive ramp configuration of the **EDENS VERTICAL EJECTORS** eliminates the need for modifying the endwall to accept the inclined discharge ramps of conventional angle flight conveyors. The effect of the vertical ramp on installation costs is significant and may offset the cost of the machine.

EDENS EJECTORS are in use in waste water sumps, quench tanks, gravity separators, scale pits, coolant reclamation sumps, large metal washers, and refinery and railroad oil-water separators. The machines may also be used to remove settled solids from waste water in settling basins, steel mill clarifiers, water curtain dust collectors, fly ash and cinder sumps in power houses, phenol tanks, railroad sanding pits, slag quenching sumps in cupola foundries, etc.

In the removal of compressible wastes from water, the machines will retrieve such settled materials as sludges, asphalt, tar, fly ash, latex, as well as silts from rotoclones, wheelabrators, and conveyor exhausts. The ejectors will also discharge incompressible wastes from the vertical ramps with equal facility such as steel mill scale, quenched slag, machining chips, quenching scale, cinders.

In contrast to about 2% solids content by weight of pumped sludges on average **EDENS EJECTORS** discharge sludges from locomotive truck washers with measured solids content of 40%. Consequently, for the mechanical handling of setteable sludges, including oil refinery sludges, the economics of this ratio are extremely favorable in terms of eliminating the customary investment in, and maintenance of, pumps and controls, valves, piping, and drying basins.

Height, length and width of the ejectors may be as specified by the user. Single channel ejectors usually are confined to 12′ widths; multiple channel machines (with common power pack) may be sized to 24′ or 36′ widths. **EDENS WASTE EJECTORS** have been built to lengths of 99′, widths of 16′, and ramp heights of 32′. Drives are electric or hydraulic; heavy duty, variable speed, mechanical transmissions are also available.

In order to minimize wear and eventual maintenance, the motor controller in standard enclosures is connected to an adjustable 24-hour clock timer for intermittent conveyor movement. The machines are designed with excess discharge capacity in order to obtain the most favorable ratio of "off" time to "running" time.

Each machine is equipped with a pivoted, automatic stripper bar assembly around and above the drive shaft at the top of the discharge ramp. This assembly strips the wastes mechanically from each ascending angle flight to prevent return to the water.

All **EDENS WASTE EJECTORS** are equipped with full-length and full-width "false" bottoms which are flat plates seam-welded to the sidewall constructions with ground transverse welds to eliminate interferences and over-stresses. The angle flights do not move over rough concrete floors, or uneven steel plates in waste water sumps, gravity separators, etc.

Engineered Products

CONTINUOUS OPERATION GRINDER

VARIED APPLICATIONS

Applicable in all areas where it is important that sewage be ground to acceptable minimum standards. Particularly suitable for installation preceding lift stations and in packaged wastewater treatment plants, flumes, and other places where effective grind is important. Can handle up to 500,000 gpd of average domestic raw sewage. Has standard ASA flanged inlet connections.

CONTINUOUS OPERATION

The Engineered Products grinder will operate continuously to grind sewage common to wastewater. Its heavy-duty, hardened die steel rotor disc automatically changes rotation direction whenever an unusually hard object is encountered. Rotor direction continues to change until the object is ground small enough to pass through the rotor opening.

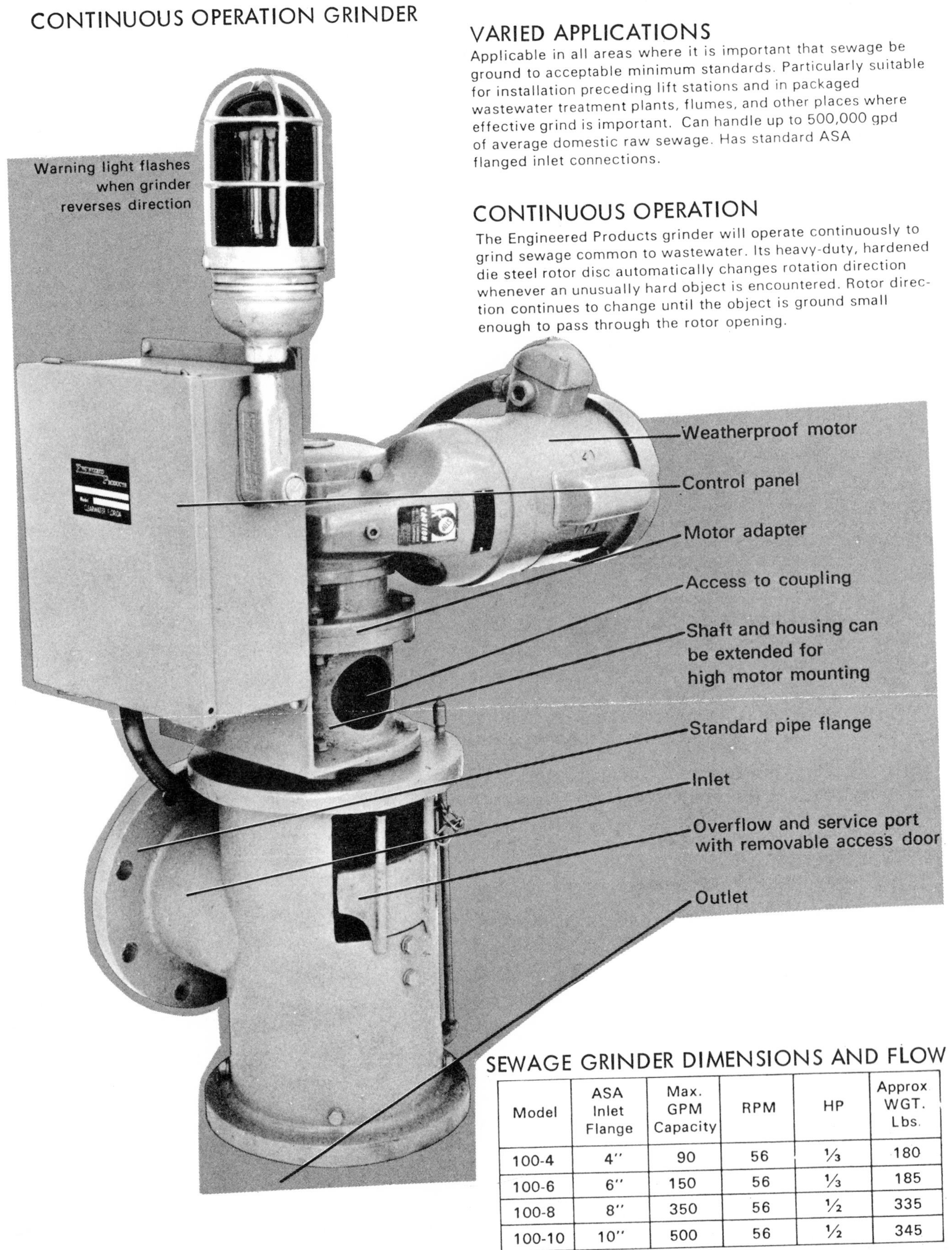

SEWAGE GRINDER DIMENSIONS AND FLOW

Model	ASA Inlet Flange	Max. GPM Capacity	RPM	HP	Approx WGT. Lbs.
100-4	4″	90	56	⅓	180
100-6	6″	150	56	⅓	185
100-8	8″	350	56	½	335
100-10	10″	500	56	½	345

ENRO SYSTEM

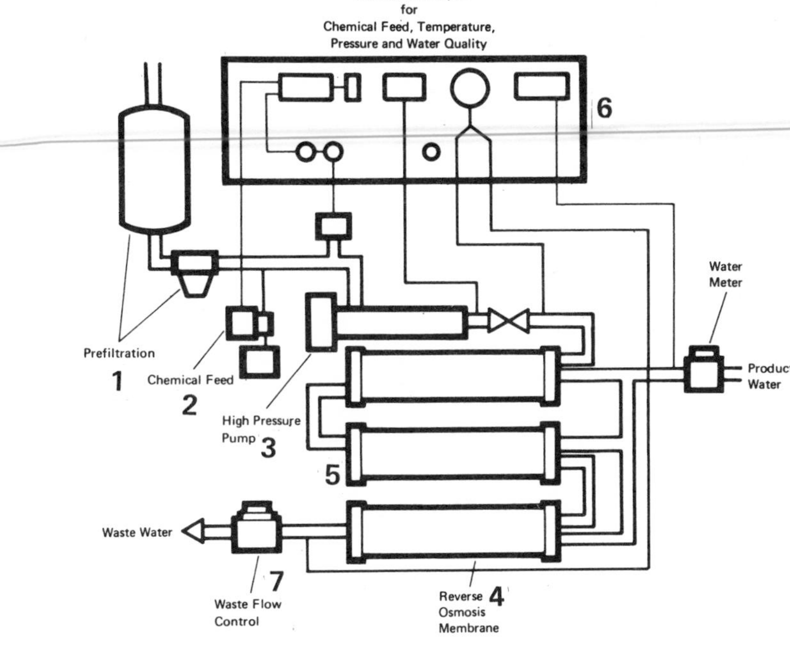

The schematic, at left, shows the major parts and subsystems of an **ENRO** spiral wrap reverse osmosis system:

1. Prefiltration removes particles of suspended matter that could cause damage to the pump and membranes.

2. Chemical Feed – used to control pH, scale formation and bacteria. pH control is necessary to prevent the build-up of calcium carbonate scale. Membrane life has a large effect on the cost of operating an RO system, thus pH control is often advisable. Systems can be alarmed to shut off when pH or water quality is off specification.

3. High Pressure Pump – provides driving force needed for the reverse osmosis process to take place efficiently. Operating pressures are typically between 400 and 600 psi, regulated by a pressure control valve.

4. Reverse Osmosis Membrane Modules– the heart of the system which separates water into two streams — purified product water and concentrated brine. Product water is usually directed to storage tank, for distribution later.The concentrated brine stream contains all matter rejected by the RO membrane. It has a much higher concentration of solids than the feed water. The concentrated brine may be sent to drain, if convenient, or to evaporation ponds.

5. Pressure Vessels – Epoxy coated steel, PVC end caps, Victaulic type pressure seals. Pressure rating of modules and system to 800 psi.

6. Instrument Cluster - Pressure gauge, water quality meter, flowmeters for product, brine, pH meter and control provide at a glance, monitoring of system operation and product quality.

7. Waste Flow Control – provides precise control over the amount of water being diverted to drain and thereby the recovery.

SPIRAL WRAP MEMBRANE

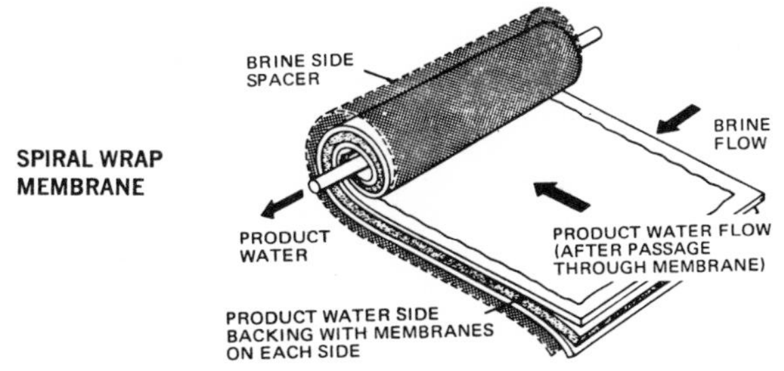

MODULE	DIMENSIONS	FEED WATER	SODIUM CHLORIDE REJECTION	SULFATES, HARDNESS, SOLIDS REJECTION	GALLONS PER DAY WATER PRODUCTION PER MODULE AT 600 PSI, 75°F.	TYPICAL REQUIREMENT
UHP-90	4" diameter x 22"	Tap Water, to 1000 ppm	90%	95%	up to 1400	Ultra-pure water
UHP-95	4" diameter x 22"	Process Wastes, natural waters up to 5,000 ppm	95%	98%	up to 900	Drinking water
UHP-98*	4" diameter x 22"	Brackish Water, Process Wastes, up to 10,000 ppm dissolved solids	98%	99.5+	up to 500	Industrial process

*UHP-98 is unique in its field because it can clean such dirty water in a single pass.

Environment One Corporation

GRINDER PUMPS

The GRINDER PUMP is a compact, dependable, long-life grinding lift station for individual residences, multi-family dwellings and/or multiple unit developments, industrial and marine applications. It grinds solids to yield a finely-divided, easily pumped slurry. Unlike conventional sewer pumps, the GP can handle even those things which shouldn't be in sewage, but are . . . plastic, rubber, cloth, wood, even metal . . . without clogging or jamming.

The GP can pump this slurry great distances at heads up to 81 feet through lines only 1¼ inch in diameter! This makes possible connection into shallow gravity or pressure sewers at higher elevations, even if they are far away by previous standards.

The GRINDER PUMP makes it easy. The problem of "difficult lands" is now obsolete. The GP solves most problems commonly encountered in bringing gravity sewers to hilly terrain, very flat land, high-water-table areas, soils with poor "percolation," waterfront property, rocky areas, seasonal occupancies, remote or low-density areas. So, you have complete freedom in site layouts; locate roads, streets, houses, open spaces where you want them. Not where gravity sewers force you to put them.

MODEL FARRELL 210

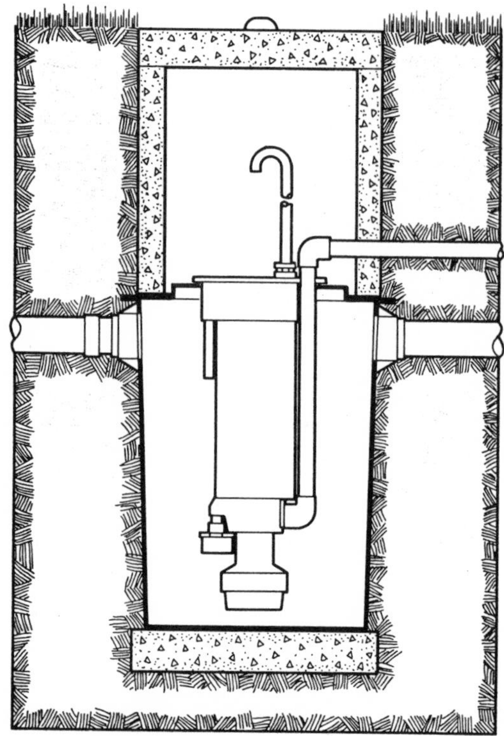

Its size, efficiency and operating economy make the GP 210 ideal for single dwellings, waterfront property, subdivision developments, motels and marinas. It can be used to handle the single dwellings and more remote structures which contribute to a pressure sewer system network.

MODEL FARRELL 211

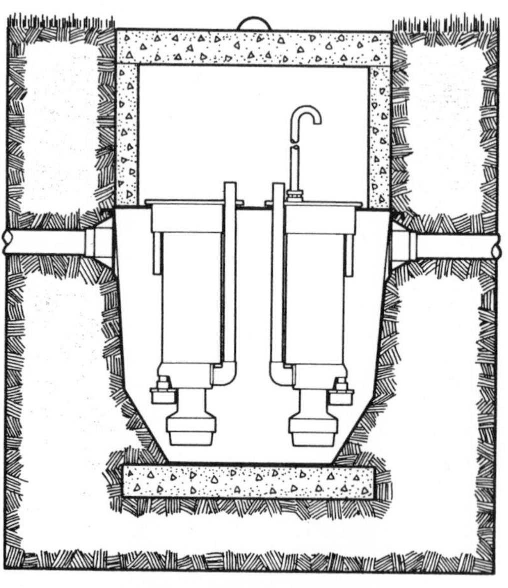

The GP 211 finds a broader range of applications as a result of increased capacity. For instance, one GP 211 can effectively handle the normal volume of wastewater from as many as four single-family dwellings, six apartments or eight motel units.

The GP 211 offers tremendous advantages in pressure sewer systems for existing areas, as well as new developments. Its ability to handle multiple dwellings reduces the number of units needed in a system, and thereby reduces the cost of the system.

Environmental Sciences, Inc.

CHEMICAL SLUDGE SOLIDIFICATION

Chemfix is a realistic ultimate disposal process which treats liquid, semi-liquid, and solid wastes. It converts these wastes into a non-toxic, non-polluting material suitable for use as landfill.

Economical conversion of liquids, sludges, powders, and unstable "solids" into easily handled, chemically and physically stable solids. Through "chemical fixation", the resulting solid provides a material which can be utilized in any location where earth is used.

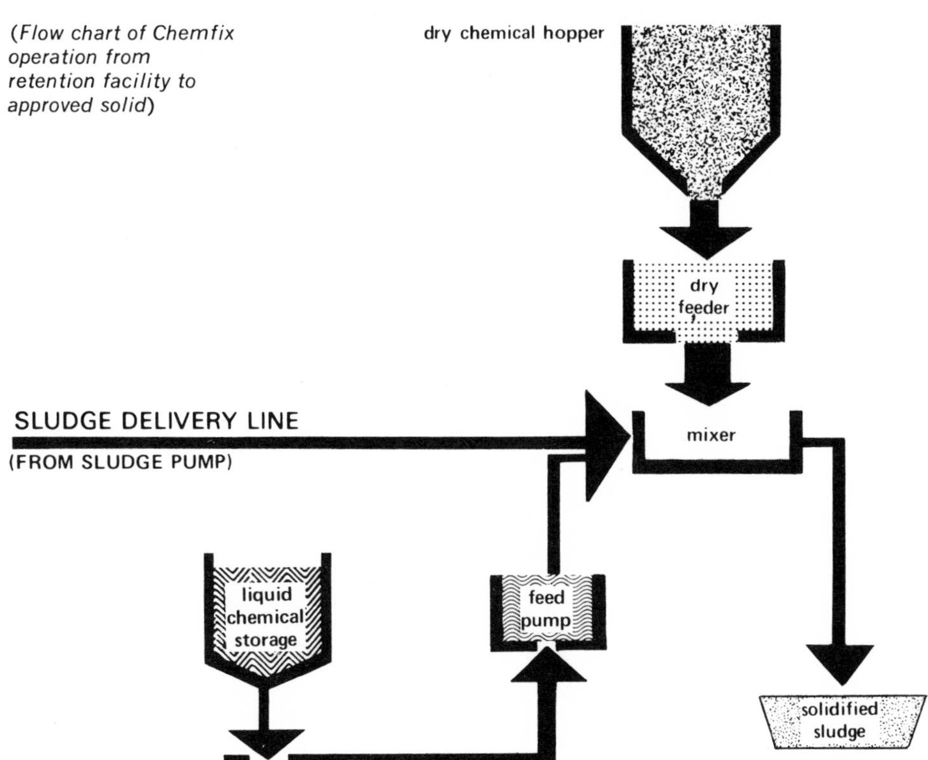

(Flow chart of Chemfix operation from retention facility to approved solid)

The CHEMFIX process involves the reaction of at least two chemical components with the waste material to form chemically and mechanically stable solids. The reaction occurs at normal, ambient temperatures and is not affected by temperature variations. No heating or other special reaction conditions such as pressure, catalysis, or long contact times are necessary.

We employ the Chemfix process through the means of special technical equipment to turn liquid wastes into non-polluting landfill. The Chemfix mobile plant will work with customers' on-site lagoons or retention facilities to convert wastes to landfill. This plant is operated by Chemfix personnel. Also, the necessary technical assistance can be provided to qualified waste haulers . . . *In either case, eliminating the need for large capital equipment expenditures.* The Chemfix mobile treatment plant can treat up to 100,000 gallons per eight-hour shift!

ESI-FLX

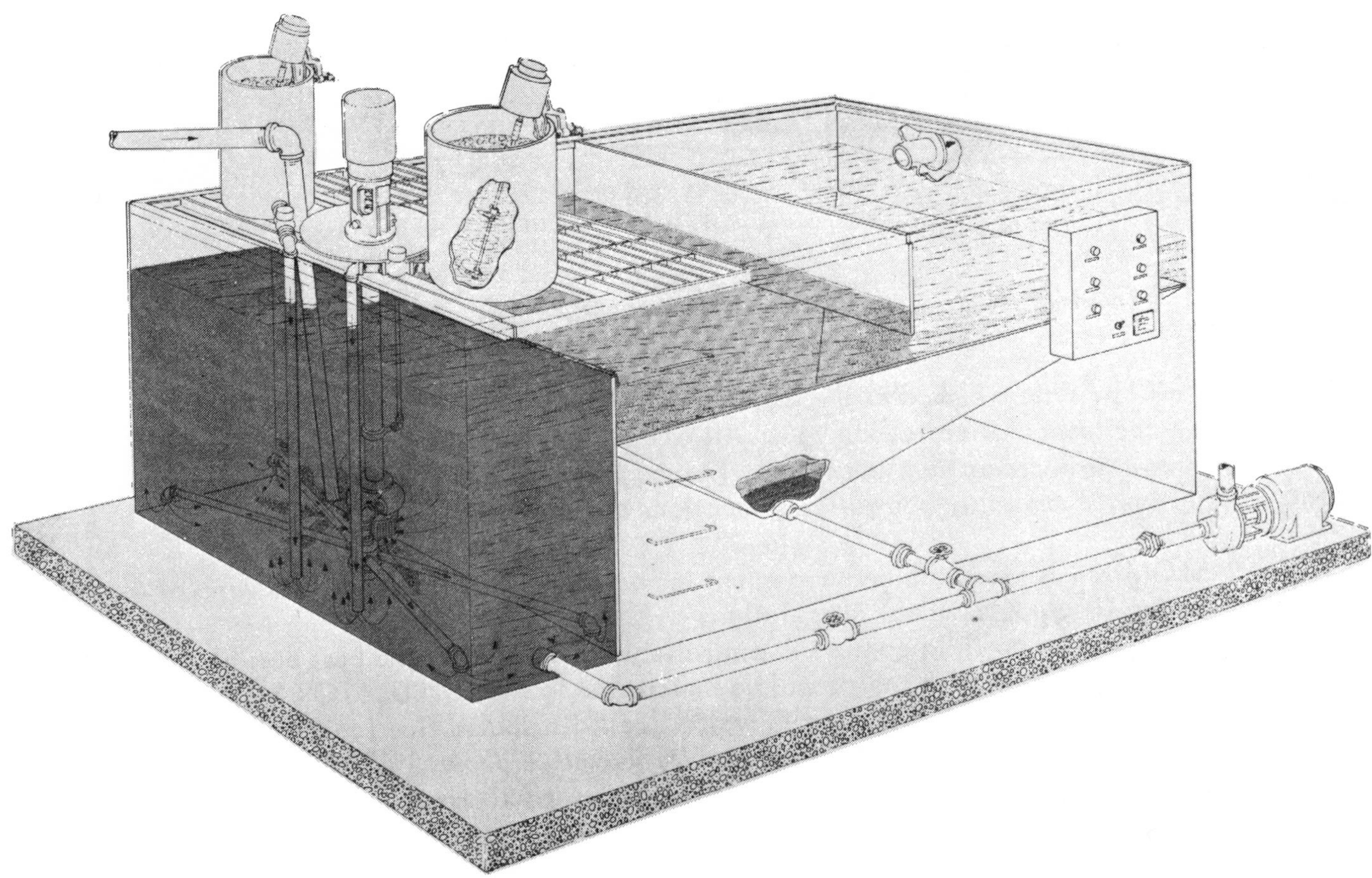

This ESI-FLX unit is being used successfully by several of the largest paper companies in plants printing containers. The ESI-FLX units come as package plants sized to treat the waste from one or any number of printing machines. All units can be built to be completely automatic or semi-automatic.

The first compartment serves as a chemical dosing and mixing chamber. Located on the floor of the mixing chamber is a pump which picks up the washdown influent and the chemical additives, forces the mixture through arms attached to the pump that extend to the four corners of the chamber, and expels the liquid downward toward the floor of the chamber. This patented device provides extremely thorough mixing without creating surface turbulence.

The second, or settling, chamber is equipped with a baffle, crosscurrent to the flow, extending downward 15 inches from the top of the chamber. The walls of the chamber are tapered to collect the sludge, created by the reaction of chemicals with the polluted waste, in the center of the chamber at floor level.

Chemical treatment consists of the addition and mixing of two combinations of chemicals, in slurry form. The principles involved are adsorption, absorption, flocculation and settling. The first mixer contains a chemical additive called ESI-FLX-R1 and the other mixer contains ESI-FLX-R2.

"ALL IN ONE"

THE ABOVE PHOTO SHOWS THE "ALL-IN-ONE" T-50-T UNIT ON STREAM AT A TEXTILE MILL

The collected influent is pumped by the influent pump to the mixing chamber of the unit, where a predetermined amount of special chemical reagents are added and thoroughly blended and mixed with the waste by the ESI patented mixing unit. The chemically treated waste water flows into the settling chamber, where the sludge collects in the inverted cone for periodic removal by the sludge pump. After settling, the treated water flows into the aeration portion of the unit for further reduction of suspended solids, BOD and color. pH is adjusted to the neutral range and the cleaned waste water then is ready for disposal into the municipal sewer or stream, pond or lake.

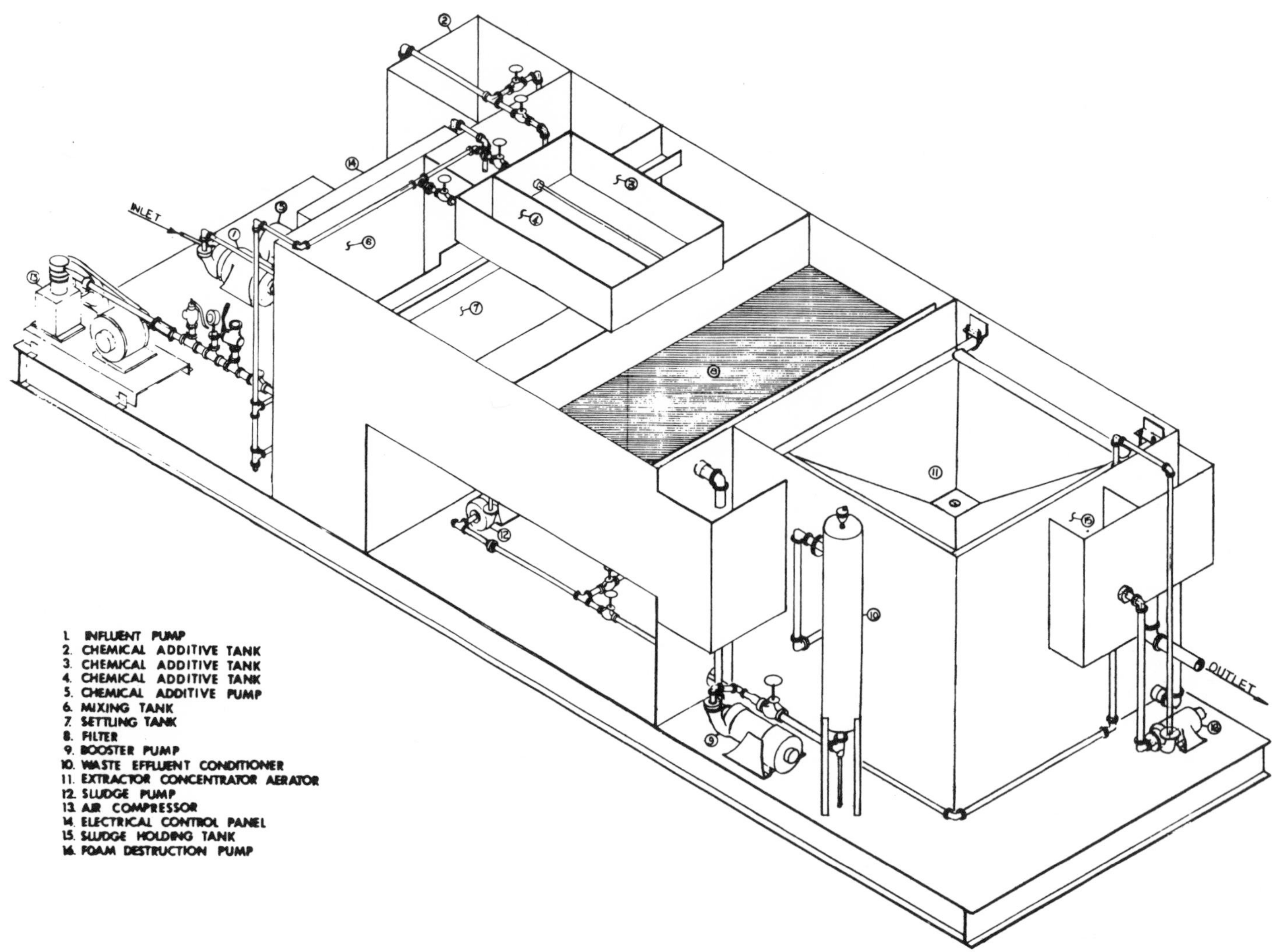

WASTE TAMER

Treatment begins immediately when liquid waste from one or more dwellings is discharged into the system's underground tank. Two pneumatic pumps provide a continuous recirculation and pressuring of the waste while putting dissolved oxygen into the system. Addition of oxygen to the waste solution causes a rapid reduction in the BOD (biological oxygen demand), dissolved solids, total solids and nitrogen and eliminates odoriferous gases. A compact control unit provides the air which operates the pneumatic pumps and supplies the dissolved oxygen.

Designed as an aerobic digestion system in which bacteria grow only where oxygen is present, the Waste Tamer provides both primary and secondary treatment of home sanitary waste. It produces a clear, odorless effluent and does not require the extensive tile drains of a septic tank system. The ESI unit consists of two packages connected by pipes: (1) a tank manufactured of corrosion resistant materials and containing 4 chambers but no moving parts; and (2) a mechanical assembly with controls for rapid installation in a protected and accessible location.

Environmental Services, Inc.

MODEL 600-8RC WASTE TAMER SPECIFICATIONS

The 600-8RC Waste Tamer is a complete Sewage Treatment Plant, packaged to serve a single dwelling, with a maximum of eight people. The plant is designed for aerobic action to completely break down domestic waste by a patented process, and its function is to produce an odor-free effluent with up to 95% BOD reduction. The design capacity of the plant is 600 gallons per day with a total BOD load of 1.36 lbs. per day, and tanks are sized to provide at least a 48-hour retention period. The complete plant consists of a multi-sectioned underground tank, two pneumatic pumps and a console control cabinet.

The Control Console Model 600-8RC is a completely factory assembled and wired cabinet containing an air compressor (fused) with relief valve and intake filter, safety alarm bell with transformer, pressure switch, sequence timer and solenoid valve.

The Pneumatic Circulator Model 600-8RC (2 supplied per Waste Tamer) is completely shop manufactured, including pump barrel and tank mounting plate, bottom sewage inlet connection complete with check valve, sewage outlet pipe complete with discharge orifice, air inlet, sparger and air vent with check valve. Air inlet and air vent on each circulator are manifolded so that only one pipe is required from each circulator to the control console. The Pneumatic Circulators (pumps) have bitumastic rust resistant coating.

The Waste Tamer Tank is a 1200 gallon multi-chambered underground tank furnished by the dealer. The tank shall conform with specification and design as furnished by Environmental Services, Inc.

The Air Piping from the console to the pneumatic circulators; the Sewage Influent Line between the house and the tank; the Sewage Effluent Line and Tile Field are furnished by the dealer and installed in accordance with Environmental Services, Inc. as well as with local requirements.

Location of Equipment

The control console shall be installed in the location selected by the owner. The factory assembled and wired cabinet shall be placed in a garage, basement or other protected clean, dry area within 50 feet of the pneumatic pumps and at an elevation no lower than the pneumatic pump mounting plate (top of tank). In selecting the location, judgment shall be exercised relative to the following:

1. Ease of running air lines to pump from console.
2. Ease of making vent line connection.
3. Ease of running electrical power line from source (house panel box) to console.
4. Ease of electrical power connection to console.
5. The console is designed to operate on a level concrete base without anchoring to the floor. If floor location is not level, it will be necessary to pour a 30" x 24" x 3" high concrete pedestal for the console.
6. The control console contains an air compressor requiring circulation of ambient air to aid in cooling. This is accomplished with the especially designed cabinet. In locating the console, leave a minimum 6" space between the rear face and any adjacent wall or obstruction.
7. Minimum clearance around the console should be as follows:
   Front 24", Left Side 24", Right Side 6", Rear 6".

Electrical

The control console is designed to operate on single phase, 115 volt, 60 cycle current. Operating amperage draw of units is 6 amps. A separate 15 amp fused circuit should be installed to the control console. Electrical connections are through left side panel of cabinet. All electrical work must be installed in accordance with local requirements.

Two other models of Waste Tamer are available, the largest of which will handle a total BOD load of 3.3 pounds per day. Tanks are sized to provide at least a 48-hour retention period.

Z-M PROCESS PHYSICAL-CHEMICAL PILOT PLANTS

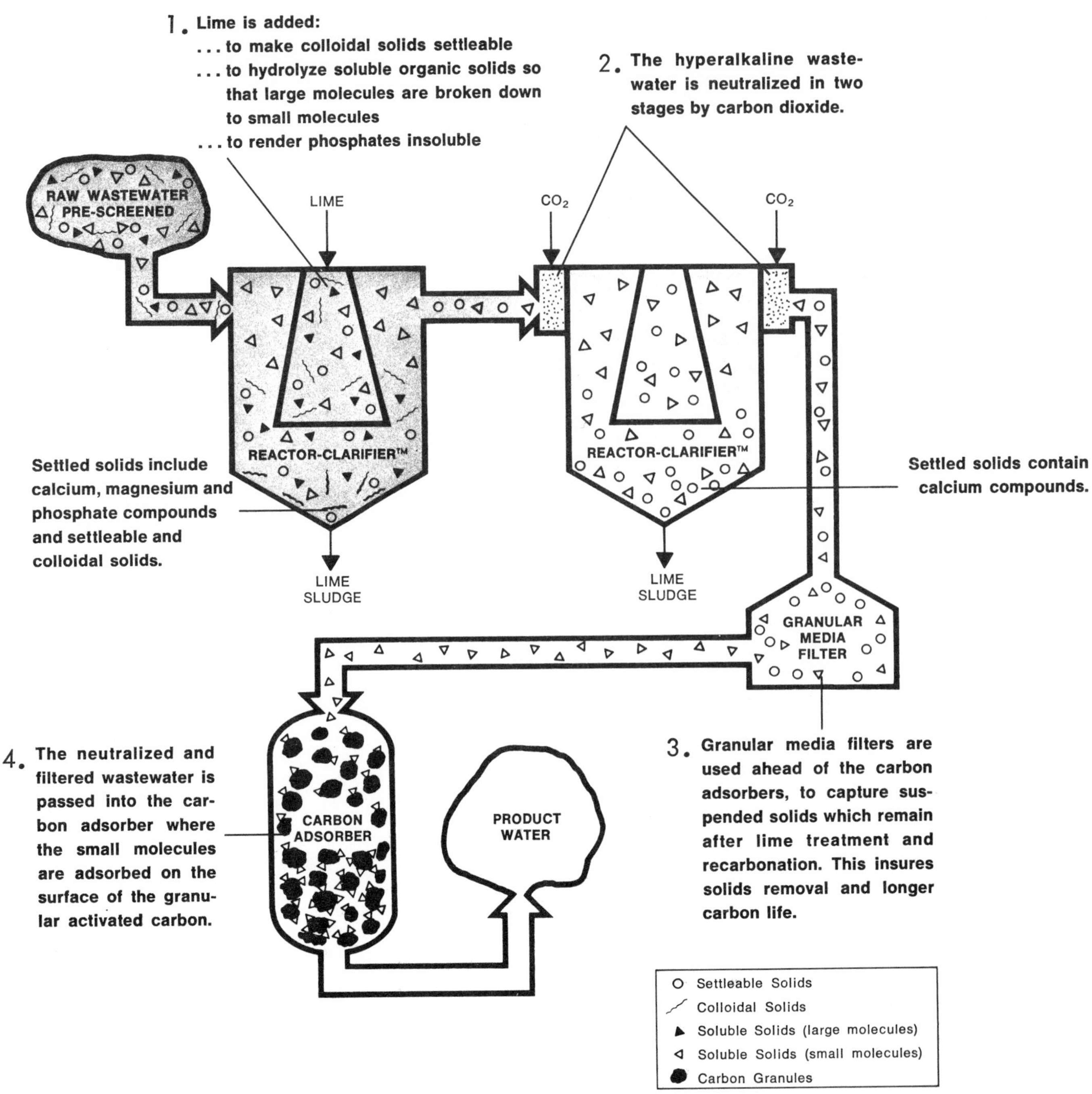

Complete treatment mobile units with laboratory.
Complete treatment on-site erected units.
Unit process equipment for specialized studies.

Envirotech Corporation

Eimco QR7 Floating Mechanical Motor Speed Surface Aerator

The Eimco QR7 mechanical floating surface aerator is a motor speed unit that will provide efficient transfer of oxygen from the atmosphere and excellent mixing action for municipal and industrial waste treatment plants.

The QR7 is recommended for the aerating of large stabilization ponds and reservoirs.

An exclusive tri-pontoon mounting method and lower center of gravity provides a unit that is extremely stable and not subject to overturning in foaming wastewater applications.

A mechanical method of controlling discharge to a low profile reduces fogging, cuts down on possible icing and improves the oxygen transfer efficiency.

Installation is simple. The QR7 is self-contained and can be placed into the water, floated to the desired location and secured by mooring cables.

A direct drive motor-to-impeller construction eliminates the expense of a gear reducer. The motor is positioned above the high intensity spray zone for better environmental protection, although all motors are weatherproofed and protected for chemical service and are totally enclosed, fan cooled. A housing drain is provided to remove internal condensation.

The Eimco QR7 floats keep the unit level during all operating conditions and maintain an optimum impeller depth level at all times. Water level variations in the operating area do not affect the unit's efficiency.

The unit is available in standard sizes up to 75 hp with larger sizes on request.

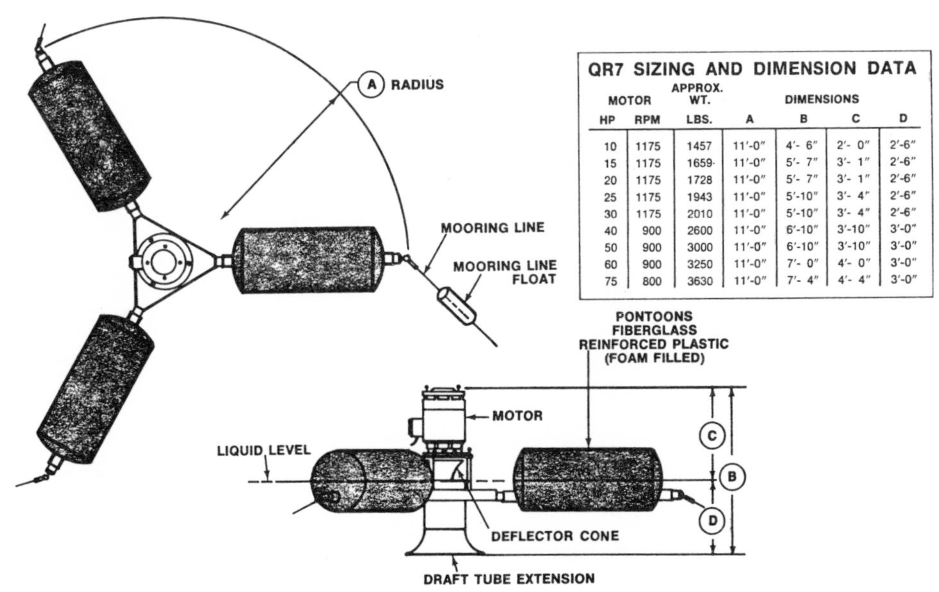

QR7 SIZING AND DIMENSION DATA

MOTOR		APPROX. WT.	DIMENSIONS			
HP	RPM	LBS.	A	B	C	D
10	1175	1457	11′-0″	4′- 6″	2′- 0″	2′-6″
15	1175	1659	11′-0″	5′- 7″	3′- 1″	2′-6″
20	1175	1728	11′-0″	5′- 7″	3′- 1″	2′-6″
25	1175	1943	11′-0″	5′-10″	3′- 4″	2′-6″
30	1175	2010	11′-0″	5′-10″	3′- 4″	2′-6″
40	900	2600	11′-0″	6′-10″	3′-10″	3′-0″
50	900	3000	11′-0″	6′-10″	3′-10″	3′-0″
60	900	3250	11′-0″	7′- 0″	4′- 0″	3′-0″
75	800	3630	11′-0″	7′- 4″	4′- 4″	3′-0″

EIMCO AERATORS

The Eimco aerator is a highly efficient mechanical surface aerator for transferring oxygen from the atmosphere into water or wastewater. At the same time it provides vigorous agitation and mixing in the water to disperse the oxygen and keep solids in suspension.

Oxygen transfer capacity of the aerator is high, ranging from 3.5 to 4.0 pounds per impeller horsepower hour, depending on size and conditions. High bottom velocities are maintained, accurately measured in a 700,000 gallon test tank at 1.3 to 1.7 feet per second at a point 30 feet from the aerator center line and 1 foot above the bottom.

The aerator is easy and economical to install and maintain. In large basins or ponds, aerators can be platform-mounted, supported either by columns or pontoons. Smaller installations can be supported on bridges.

The entire unit is manufactured to keep maintenance time and costs at the very minimum. There are no submerged bearings. A major maintenance advantage is that rags or other solids cannot catch on the self-cleaning impeller vanes.

The drive is sized and selected from available reducer or gearmotor equipment to provide a generous service factor. This enables dependable and durable operation around-the-clock, without maintenance or parts replacement. Eimco aerators are furnished with reliable lubrication facilities for 24-hour continuous-duty operation. Eimco's wide experience with aerators is applied to each project to insure reliability and suitability to the particular treatment problem.

Avoidance of overload is vital to reducer life and freedom from maintenance. This must be accomplished by the designer, who must incorporate facilities to limit water level variation in the aerated basin.

An indicating ammeter should be provided for each aerator, to enable operating personnel to observe at any time the load condition on gearing and motor. This enables the operator to estimate and control the oxygen input to the waste.

The uniquely efficient design of the aerator's impeller is the real difference between the Eimco unit and other mechanical surface aerators.

The impeller is an inverted cone with blades radiating outward from a boss at the center. As the impeller turns, it draws the liquid upwardly toward the cone which then propels it upwardly and outwardly in a low but free trajectory, with a minimum loss of energy. The liquid is broken into a fine spray which creates a large surface area and thus entrains more air. As the spray rejoins the tank contents, large volumes of air in the form of small bubbles are entrained and circulated below the surface.

The inverted cone to which the blades are attached assures hydraulic stability — an action which provides the efficient pumping force which keeps high solids concentrations in suspension. There is minimum deposition on the tank bottom or at the perimeter of the tank or basin.

The driving motor and gear reducer can be either totally enclosed or otherwise suitable for outdoor service. The reducer bearings and gears are oil-lubricated and weather-protected.

The true cost of an aerator is the combination of power cost plus depreciation per pound of oxygen transferred to the liquid. The Eimco aerator is the low-cost way to provide aeration for all biological and chemical oxidation applications and for the release of gases or for gas absorption.

The information in this bulletin is provided to help you with the proper application of Eimco aerators. Municipal Equipment Division sales engineers will be pleased to provide additional assistance if required.

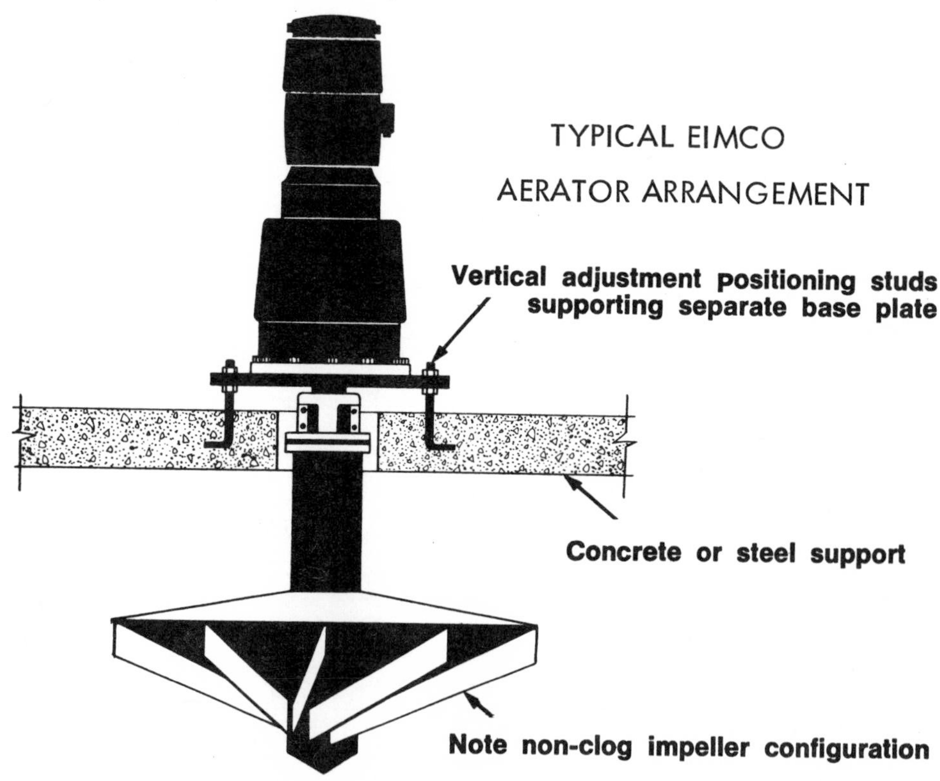

REACTOR-CLARIFIER TREATMENT UNITS

Solids-Contact Types HRC and HRB . . . use the proven, highly-efficient upflow solids contact action. Large-diameter turbines internally recirculate large quantities of previously formed flocs or precipitates at low peripheral turbine speeds. In softening operations, this recirculation can be up to 15 times the feed rate with slurry density up to 5 percent by weight. Excellent overflow qualities and dense underflows are obtained in this simple, stable operation. There is no unstable sludge blanket to upset the operation. These units are excellent physical-chemical treatment clarifiers.

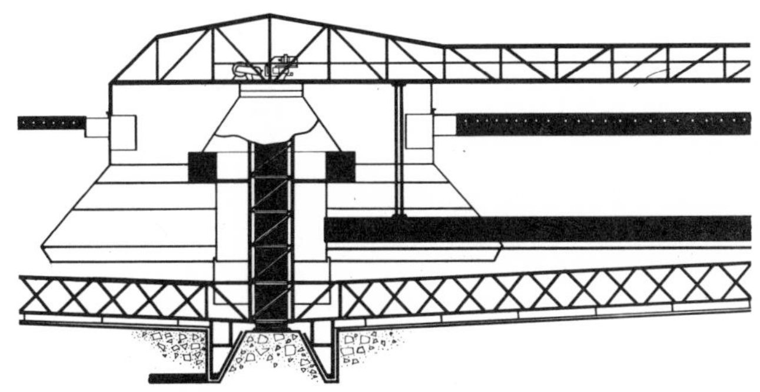

Type HRC from 50 to 200 ft. diameter

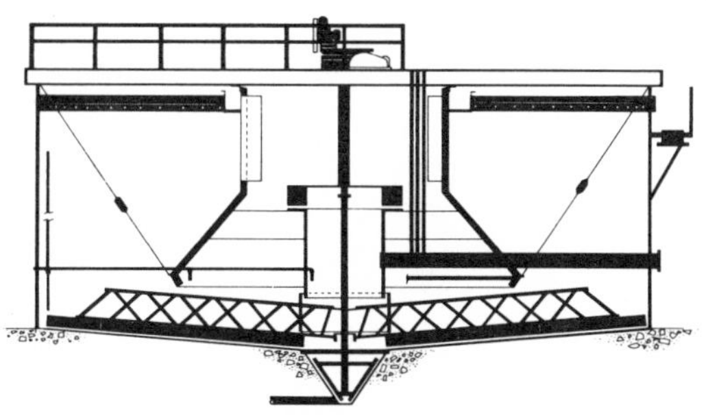

Type HRB from 10 to 75 ft. diameter

The **Eimco Reactor-Clarifier** water treatment unit is a highly versatile machine that combines flocculation, coagulation, clarification and positive sludge removal in a single tank. For municipal use, these are the most compact and economical-to-operate units available today. They will remove turbidity, color, algae and other contaminants in water treatment. Wastewater treatment with chemical treatment removes suspended solids, BOD and COD. They accomplish lime or lime-soda softening, magnesium precipitation, brine softening or clarification as well as wastewater clarification.

Type CF — Standard rate 30 to 200 ft. diameter center-column-supported. This is a standard rate unit which combines vertical paddle flocculation with clarification. Center column units in sizes from 30 to 200 ft. diameter are standard. Recommended for turbid water clarification, algae and color removal. The standard unit has a uniform influent distribution system.

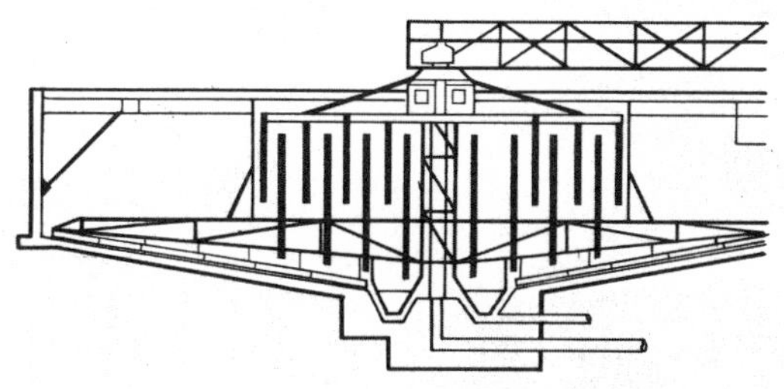

Type CF from 30 to 200 ft. Diameter

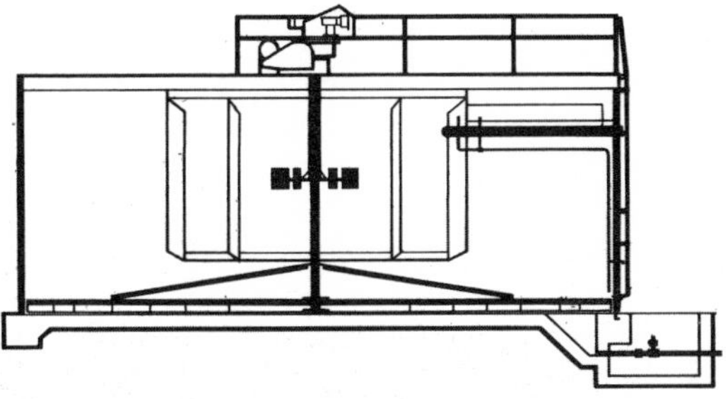

Type BFR from 20 to 70 ft. diameter

Type BFR — Standard rate type up to 70 ft. diameter, beam-supported. The BFR is a standard rate type that combines slow-speed turbine flocculation with clarification. Beam-supported units only for diameters 20 to 70 ft. For turbid water clarification and treatment where gentle flocculation by turnover is beneficial.

EIMCO CLARIFLOTATOR CLARIFIER

The sewage treatment plant installed at Milan, Illinois is a good example of an Eimco equipped system. This plant, designed by Missman, Stanley, Farmer & Assoc., Rock Island, Ill., treats the sanitary sewage from Milan, which is adjacent to Moline and Rock Island, Illinois. The Davenport Packing Plant wastes, containing a large amount of blood, grease and paunch manure, are discharged to the Milan sewers.

Primary Treatment

Because of the packing house wastes, dissolved air flotation, using an Eimco ClariFlotator clarifier, was selected to remove floating and settleable solids before biological treatment. The ClariFlotator clarifier is 70 ft. in diameter with a 44 ft. diameter flotation compartment. The plant average design flow is 1.7 mgd. The flow averages 1.0 mgd, ranging from about 0.85 to 1.20 mgd. The system is designed for recycle pressurization, with about 10 SCFM of air applied to a recycle flow of 745 GPM @ 142′ TH (40 HP). The skimmers remove approximately 30 gallons of float every 6 minutes. The float averages 5.35% dry solids and 1.47% hexane solubles. The float is further thickened in two sludge holding tanks.

The coarse solids, mostly paunch manure, are removed by sedimentation in the ClariFlotator clarifier. The underflow concentration averages 6.06% dry solids and 0.71% hexane solubles. The underflow and the float are mixed and filtered together.

Biological Treatment

The ClariFlotator clarifier effluent is treated by two-stage biological treatment, using a trickling filter followed by activated sludge treatment.

Trickling Filter

A 90 ft. diameter trickling filter was installed to handle the high surge loadings from the packing house wastes. It is equipped with an Eimco Type HR-AW-NP rotary distributor having a capacity of 600 to 3200 GPM. Also, the engineers did not believe that the required degree of treatment could be consistently obtained without the trickling filter followed by activated sludge treatment.

Activated Sludge Treatment

The trickling filter effluent flows by gravity to a two-compartment aeration tank each compartment being 20 ft. x 110 ft. x 13 ft. The two tanks operate in parallel. Each tank has an air distributor with 60 Eimco non-clog diffusers. Mixed liquor suspended solids are maintained at about 6,000 mg/l. The aeration basin effluent flows to a 70 ft. diameter x 8 ft. SWD Eimco Type C2 final Clarifier. The underflow contains 1.47% dry solids and 0.14% hexane solubles. Approximately 80% of the underflow is recycled to the aeration tanks, with the waste sludge being pumped back to the wet well ahead of the ClariFlotator clarifier.

Sludge Filtration

There are two 6 ft. diameter x 6 ft. face EimcoBelt® filters operating 7 hours per day, 5 days per week. Conditioning chemicals applied are 127 lbs. CaOH and 72 lbs. $FeCl_3$ per ton of dry solids in the cake. The filter cake averages about 73% moisture. The filtrate containing about 550 mg/l suspended solids is returned to the wet well ahead of the ClariFlotator™ clarifier.

A comparison of plant influent and final clarifier effluent are listed below:

	DAY		NIGHT	
	Influent mg/l	Effluent mg/l	Influent mg/l	Effluent mg/l
Suspended solids	338	2.4	158	3.2
Hexane Solubles	212	8	46	2
BOD_5	281	7	203	8
COD	276	37	365	37

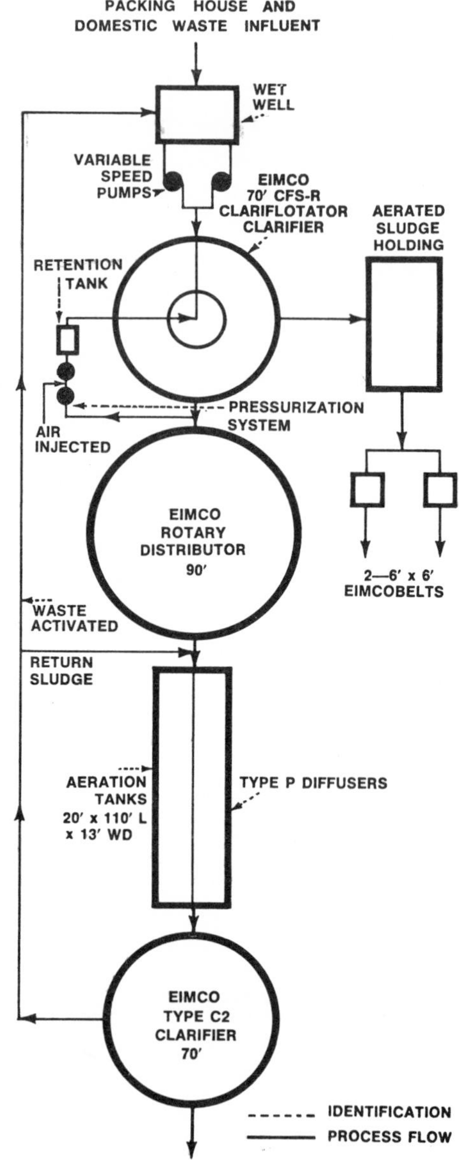

Clarifiers — For pre-sedimentation of silt and sand or for settling after flocculation. Beam supported units are available for tanks 20 to 45 ft. diameter; center column and traction types for tanks 30 to 325 ft. diameter. Types available for rounded corner square tanks with cross-flow arrangement, side feed or center siphon feed with conventional overflow. Driveheads are of efficient design with quality construction throughout using high grade materials. Adequate drive gear to machine size ratio equals trouble-free, low maintenance operation.

Eimco offers a package combination to thicken and dewater the sludge which results from softening. Sludges can be dewatered to approximately 65 percent by weight of dry solids on an **EimcoBelt®** continuous belt filter. The blinding characteristic of the softening sludge presents no filtering problem for an EimcoBelt filter, which operates at all times with a washed, clean medium. Simplified **EdgeTrack™** belt filter operates with minimum of operator attention and provides long belt life. The Eimco thickener used in conjunction with the EimcoBelt filter can be equipped with an automatic lifting device to prevent damage to raking arms by the heavy sludge.

Dual media double gravity filters which utilize **Eimco FlexKleen™** nozzles and the **Eimco Air Wash System** have several major advantages for polishing treated water. The FlexKleen nozzles, which are manufactured of non-corrodible plastic with flexible stainless steel screens, are threaded into precast concrete blocks to form a filter bottom. In dual media applications they are covered with a layer of sand and a layer of anthracite. The filter underdrain blocks are 24 in. square, 3 in. thick. They can be supported by circular or square piers on 24 in. centers or on longitudinal beams placed across the filter floor. Threaded plastic inserts cast into the blocks receive the threaded nozzles. In the airwash system, the nozzles are supplied with plastic tubes which extend below the blocks. The use of air with the backwash provides an especially vigorous wash, with thorough scrubbing of the filter medium. The amount of wash water required is reduced by as much as one half. The agitator eliminates the possibility of "mud balling" tendencies and requires no surface washers or the additional water they use. There is no drifting of gravel or breakthrough caused by surges in backwash. The FlexKleen nozzle is highly resistant to clogging.

Envirotech Corporation offers equipment for all kinds of water and wastewater treatment. Equipment which has not been described includes:

Eimco-Extractor Horizontal Belt Filters
Wemco Static Screens
Wemco Hydro Cyclones
Rectangular Flotator Clarifiers
Wemco Pumps

and a complete line of water treatment plants using the Reactor-Clarifier, SVG automatic backwashing sand filters and including activated carbon systems, if further treatment is desired.

SVG FILTERS

The Eimco SVG Filter is a simplified, automatic gravity filter with a self-contained backwashing system designed to provide efficient and economical operation of municipal plants. The filters can be installed in multiples to meet capacity requirements.

The SVG Filter has these major advantages: (1) It is completely automatic in operation, (2) All piping, pumps and control and regulating valves required for a conventional separate backwash system are eliminated, (3) No raw water is wasted during the backwash cycle, and (4) Complete control--both hand and fully automatic--is provided.

A single three-way butterfly type valve provides complete hydraulic control. A valve can be operated either with an electric, hydraulic or pneumatic actuator. While filtering, the valve opens the inlet line and closes the waste line. During backwashing the valve reverses, closes the inlet line and opens the waste line. The backwash cycle is automatically initiated by an adjustable loss-of-head switch in the inlet line (normally set at 5 ft. head loss), or manually by pushbutton. The backwash cycle terminates when a low level probe in the storage compartment is actuated. The backwash rate is adjustable and varies from approximately 24 gpm/sq.ft. with an average of 15 gpm/sq.ft. The backwash period lasts for approximately 4 1/2 minutes. The complete control package includes a selector switch which permits manual pushbutton control and control over-ride as well as the normal automatic control.

FLOCCULATORS

The Eimco Flocsillator horizontal oscillating flocculating mechanism provides a highly efficient way to gently mix and agitate water for developing floc. All wearing components are above water assuring long life. The paddles of the Flocsillator mechanism travel through approximately 80% of the tank volume and influence 100% of the volume. The action of the lower vertical arms effectively prevents precipitation in the corners of the tank, although there is no high-velocity turbulence to destroy previously formed floc. The Eimco Flocsillator mechanism has many mechanical advantages not found in any other type of flocculator.

The drive shaft and bearings are located entirely out of water and are easily accessible for servicing. There are no dry wells, submerged bearings or stuffing boxes to construct or maintain. Mechanisms can be custom designed to fit other basin configurations. Standard Flocsillator mechanisms are available in sizes designed for operation in basins 20 ft. wide by 12 ft. deep and from 16 to 96 ft. or more in length.

Eimco Package Plants — Available for gravity operation using S V G automatic sand filters. Eimco offers a complete line of these pre-designed water plants for boiler feed. Plants are available in sizes from 50 to 1,000 gpm as standard and can be custom designed for larger volumes on request. Many operating efficiencies contribute to lower operating costs. Automatic units are fully dependable.

Ferro Corporation, Electro Refractories and Abrasives Division, Filtros Plant

FILTROS DIFFUSER PLATES

Whether used in the construction of new water treatment plants or the expansion of existing facilities, Filtros porous ceramic underdrain plates offer inherent efficiency and economy unmatched by graded gravel systems or other porous plate bottoms.

ceramic bonded FAO

Filtros underdrain plates are manufactured of fused alumina grain carefully proportioned and ceramic bonded at high temperatures to produce a strong, rigid body. Emphasis is placed on controlling uniformity in composition, structure, and the physical and chemical properties essential to optimum filtration through a truly porous material.

uniform filtration and backwash

Controlled Filtros plate permeability assures a more uniform flow of filtered water through properly sized filter sand, and the support of filter media without penetration. Rapid backwash discharge through the entire plate area also assures complete suspension of all filter media, eliminates objectionable mud balling and bed upsets, and reduces backwashing time and wash water requirements.

easily installed

With the use of flexible polyvinyl gaskets to seal joints between plates, Filtros underdrains are quickly and easily installed. Standard plates, 11-7/8" square by 1" or 1-1/4" thick, are notched for either 3 or 4 point suspension. Irregular shapes, circular plates, and special sizes also are available.

lower initial and maintenance costs

By replacing graded gravel beds with Filtros underdrains supporting filter media of lower than normal depth, considerable savings in combined construction, labor, and maintenance costs are realized. Elimination of gravel and stones not only prevents the loss of filter sand but also avoids plant downtime and expense of rebuilding upset beds. In addition, the more shallow beds possible with Filtros underdrains reduce excavation and construction costs of the filtration gallery. Long plate life also is assured since Filtros underdrains are resistant to corrosion by water or water treatment chemicals, and withstand various solvents used for underdrain cleaning.

PROPERTY DATA

Material:	Fused Aluminum Oxide.
Standard Size:	11⅞" x 11⅞" x 1" or 1¼" thick. Notched for 3 or 4 point suspension. Special sizes and shapes can be provided for irregular or circular filter beds.
Grade:	Underdrain "standard" to effectively support standard filter sands. Special grades can be manufactured to meet unusual requirements.
Weight:	12 lbs. per plate — 1" thick. 14½ lbs. per plate — 1¼" thick.
Flow Rates:	With standard filter sand, the generally accepted flow is 2 gal./min./ft.^{2}. With proper floc and media maintenance, flow rates of 4 gal./min./ft.^{2} have been used.
Backwash Rate:	24 inches rise per minute.
Strength:	Modulus of rupture — 2000 psi.
Pore Diameter:	800 micron—as determined by ASTM E-128-61.
Air Permeability: (dry)	70-90 SCFM air @ 2 inches H_2O per 12" x 12" plate.
Recommended Cleaning:	As required with solvents, sandblasting or heating.

PROPERTY DATA

	Filtros	Kellundite	Electroflo
Materials:	Glass-bonded silica	Ceramic-bonded fused alumina	Resin-bonded silica
Color:	White	Blue-Grey	Brown
Strength: (Std. Permeability) (Modulus of Rupture)	1000 psi	3000 psi	1500 psi
Maximum Reheat Temp.:	800°F	2000°F	400°F
Standard Sizes:	**Plates**—12" x 12" x 1" thick —12" x 12" x 1½" thick **Tubes**—2½" O.D. x 1¾" I.D. x 12", 18" & 24" Long 3" O.D. x 2" I.D. x 12", 18" & 24" Long 4¼" O.D. x 3" I.D. x 12", 18" & 24" Long Plates up to 12" x 24", special shapes and additional tube sizes are available.		
Permeability:	Standard Rating 60 (ft.^{3} air/ft. 2/2" H_2O pressure) Special permeabilities also available from 1 to 100		

FLOW DIAGRAM OF BACKWASH CYCLES

Note the more uniform distribution and shorter travel of the dirt laden water in the UNICEL from filter media to waste. This results in a cleaner filter in less backwash time and a considerable saving in water.

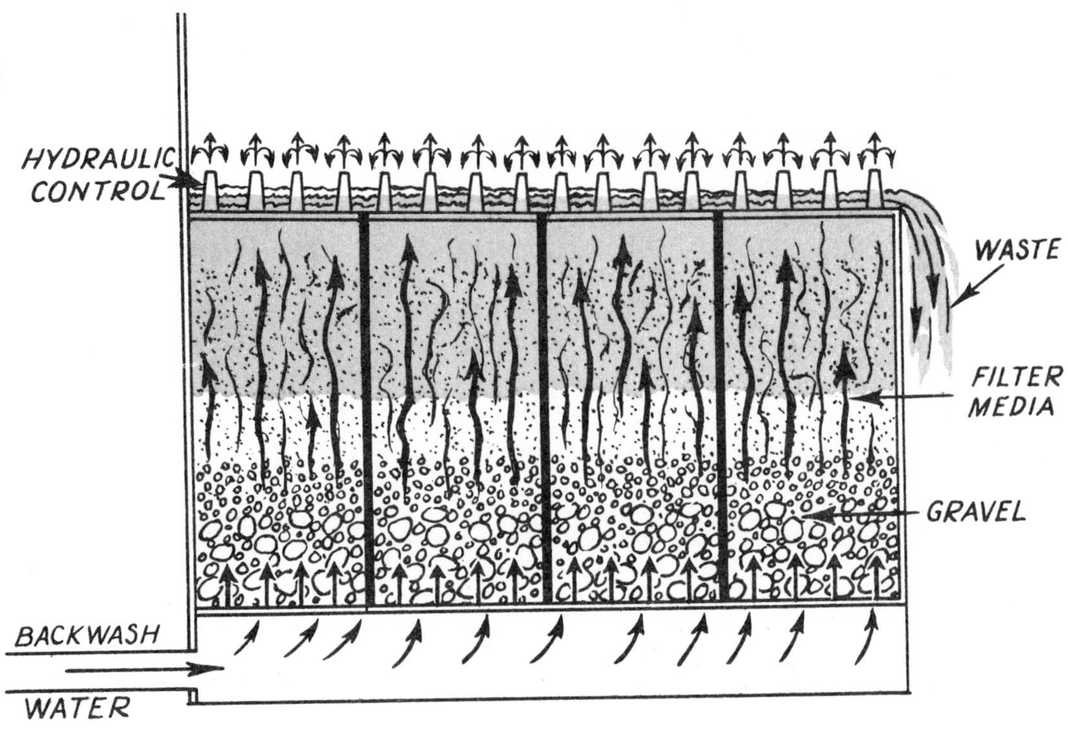

UNICEL FILTER

As the thickness of the filter bed remains constant and uniform it is not necessary to allow a factor of safety for the varying thickness of the filter media, as in conventional type filters. This fact, together with the ability to clean the filter in a very short time, makes a filter rate of six (6) gallons per square foot per min. not only possible but practical and economical.

On a test filter consisting of eight (8) two-foot square cells which we have in operation at a rate of 6 gallons per sq. ft. per minute we are getting runs of 20 hours and longer with the backwash water used less than .70%.

The top plates provide an area where solids can be precipitated during the filter cycle. This reduces the load on the filter and produces longer filter runs especially where the water being filtered has a high turbidity. During the backwash this surface will automatically be cleaned, no extra time, equipment or personnel are required.

Perforated plates on top of the cells provide positive control and uniform flow of backwash water. These plates also distribute water to be filtered uniformly over the surface of the filter media. Uniform fluidity of the filter media and the short travel of the backwash water to the top plate provides positive cleaning of all areas in less than one-half the time required for conventional filters.

Visual observation of the uniform velocity and turbidity of the water thru the holes in the top plate during the backwash cycle is positive assurance of the efficient operation of the filters. The top control plates are fastened to the top of the cells with corrosion resistant bolts and wing nuts and can easily be removed for inspection of the filter bed.

Gardner-Denver Company

CycloBlower

for 40 to 6900 CFM
Pressures to 18 PSIG
Dry Vacuum to 16" HG
Wet Vacuum to 24" HG
(continuous service ratings)

3 CDL-5 CyloBlower
Max. 289 CFM at 5 PSIG
Weight 165 pounds.

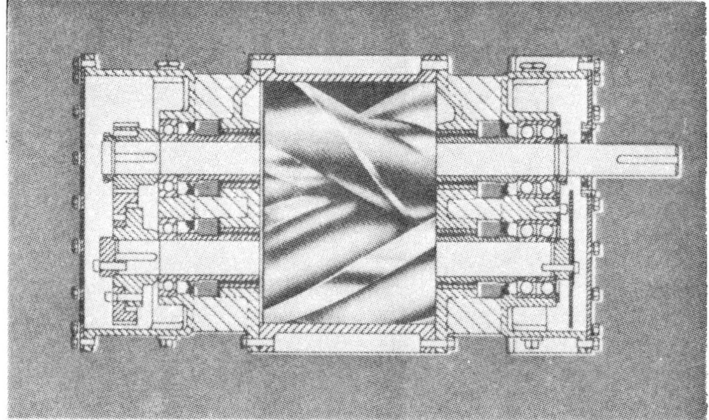

11 CDL-31 CycloBlower
Max. 6940 CFM at 5 PSIG
Weight 3530 pounds.

CycloBlower is a compact, rotary lobe-type, axial flow, positive displacement blower/compressor. It combines the most efficient rotor form with other unique features which result in user benefits not found in similar equipment. The meshing of two screw-type rotors that are synchronized by timing gears provides controlled compression of air for maximum efficiency and shock-free discharge.

CONSTRUCTION FEATURES

Rotors –Helical four-flute gate rotor and two-lobe main rotor are milled from high tensile strength cast iron, and stress relieved. Dynamic balancing assures vibration-free performance.

Bearings –Anti-friction bearings carry the shaft loads in all models. Low capacity and low pressure models use standard single and double-row ball bearings. Larger models use pairs of angular-contact ball bearings and cylindrical roller bearings.

Timing Gears –Synchronization of rotors is through a pair of helical timing gears. Precision alloy steel gears insure quiet, accurate operation.

Housing –One piece, high strength housing resists deflection to retain accurate running tolerances between rotors. Porting is properly contoured for smooth air flow.

Seals –Labyrinth-type shaft seals provide a minimum of controlled leakage of air or gas. Purged labyrinth-type seals or mechanical-type seals are available with units handling gas, where leakage cannot be tolerated. Lip type seals retain lubricants.

Lubrication –Basic design requires no lubrication of rotors. Gears and bearings are lubricated by a splash oil system, in combination with grease lubrication of bearings in some smaller models. An oil pump is not required.

A CONFIGURATION OF GRAVITY FILTERS OF CONVENTIONAL DESIGN EMPLOYING A CENTRAL CONTROL PRINCIPLE WITH AUTOMATIC FLOW DISTRIBUTION AND WITHOUT MECHANICAL CONTROLLERS.

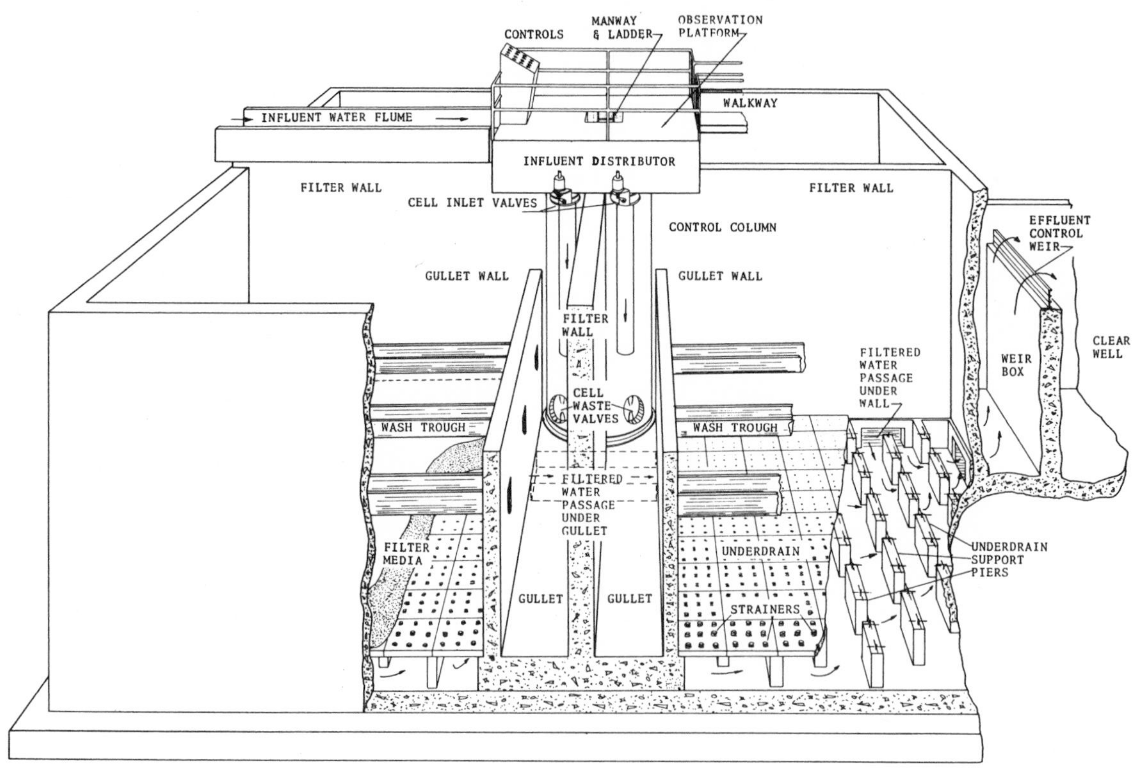

CAPACITY

A single CenTROL center can handle capacities up to approximately 12 MGD. Multiple centers are used for higher capacities. A single control center normally serves four cells but for special applications it may be adapted to serve more. Multiple centers may be used in all sizes to provide higher backwash rates.

WATER FLOW

Water which has received the pretreatment necessary for filtration flows through the inlet flume into the center inlet distributor, under stilling baffles and over filter cell inlet weirs that divide the flow equally and automatically among the filters in operation. From the filter cell inlet weirs water flows into each filter cell gullet through valved inlet lines and then onto and through the filters.

Filtered water is collected through an underdrain chamber common to all cells. This is the principal long employed throughout our MULTICELL line. Underdrain design with cell isolating valves for any unit may be employed if required in specific plant design.

The filtered water is discharged over the adjustable outlet weir to the clear well. No flow controllers are required.

Waste Water GFC AUTOMATIC GRAVITY Filter

THIS SELF-CONTAINED GRAVITY FILTER IS COMPLETE WITH DUAL MEDIA FOR THE DESIGNED FILTER APPLICATION, HAS AUTOMATIC AIR WASH AND WATER BACKWASH CONTROL AND STORES ITS OWN BACKWASH WATER SUPPLY. IT IS USED IN THE INDUSTRIAL AND MUNICIPAL FIELDS OF TREATMENT OF PROCESS WATER AND WASTE WATER.

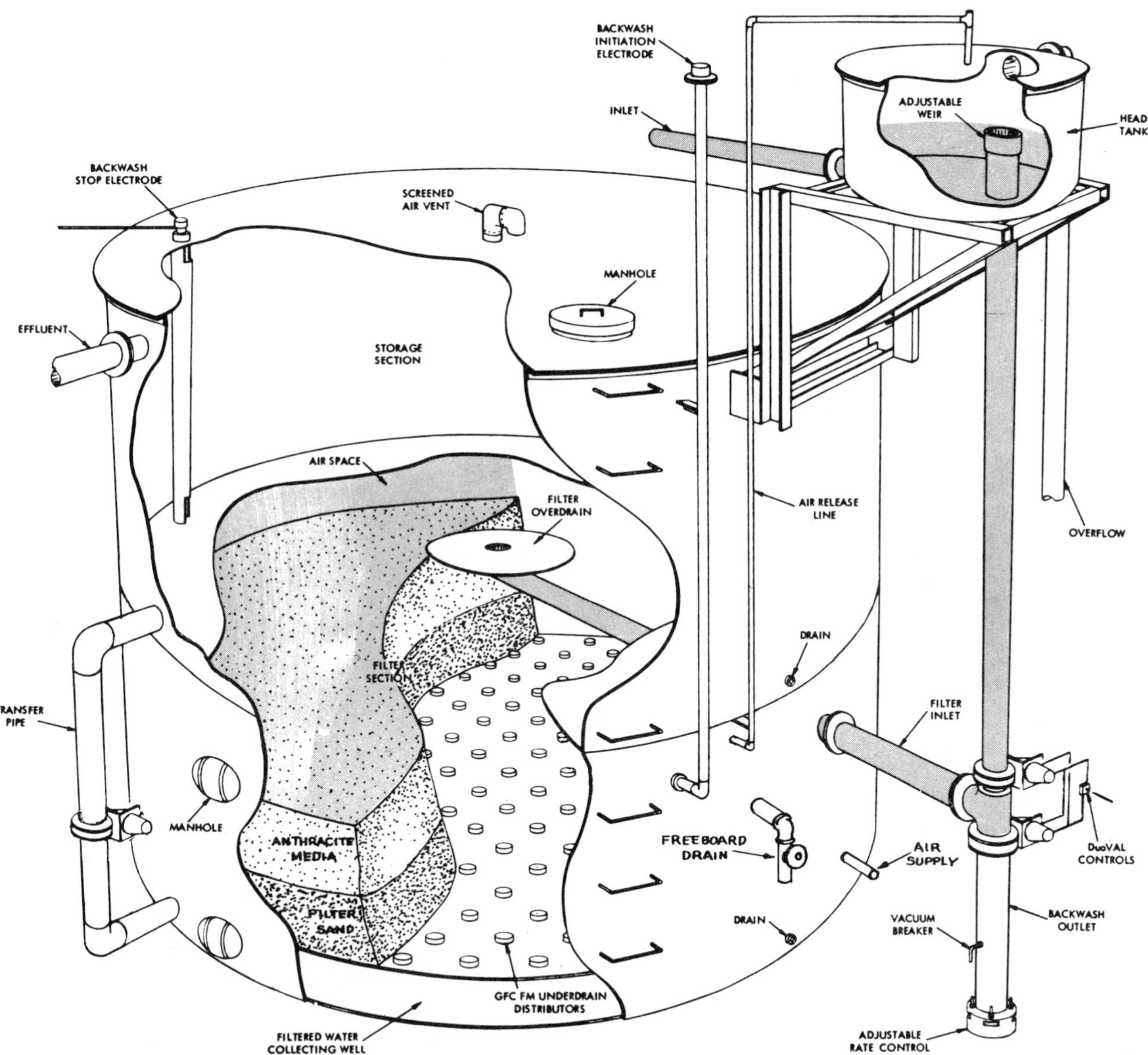

TYPICAL APPLICATIONS:

TYPE II - TYPE IV - TYPE V

Side-stream filtration for cooling towers.
Polishing water for reuse.
Filtering waste water in advance waste treatment processes after required pretreatments.

THREE DuoVAL TYPES FOR WASTE WATER

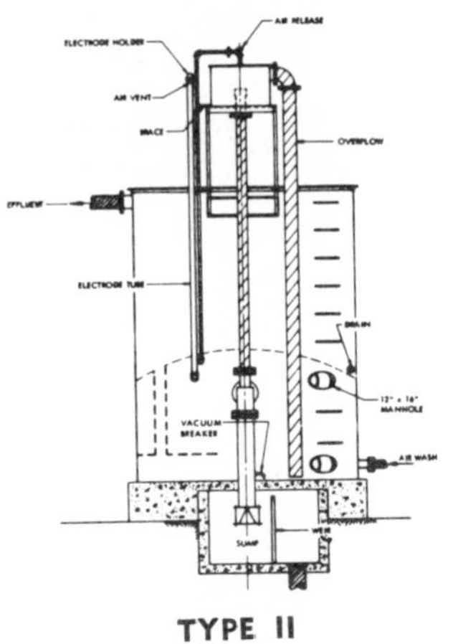

TYPE II

TYPE II DuoVAL has common walls between filtered water and unfiltered water compartments. Air wash is easily provided by increasing filter shell height, by adding air wash extensions to the underdrain distributors and by providing isolation line with valve and air supply as called for in the air wash table. Used in industrial process water treatment.

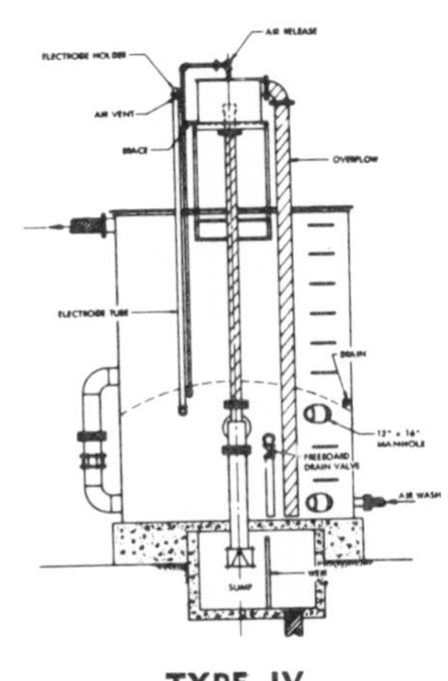

TYPE IV

TYPE IV DuoVAL is used primarily in advanced waste treatment processes and waste water reclamation. It has common walls between compartments, it has a greater backwash water storage, and the backwash water supply tank is not usually covered. Dual media with air wash preceding backwash is used. For large plants multiple units up to 20′ dia. are used.

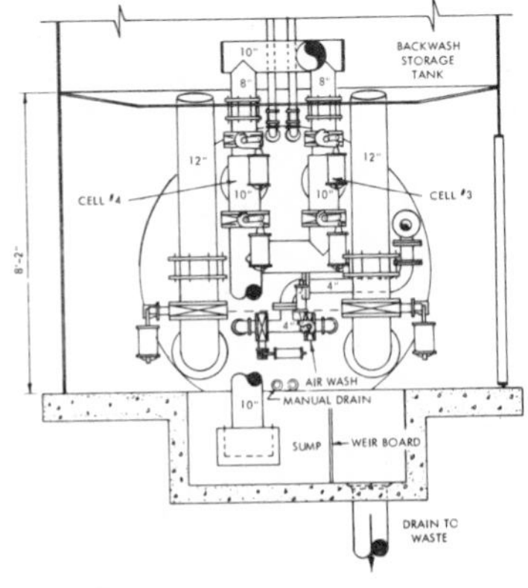

TYPE V

TYPE V DuoVAL is built in larger sizes utilizing one or two horizontal filters each with one or two cell arrangements. One common backwash storage tank is provided. The unit is self-housing with an integral operating room. Dual media with air wash preceding backwash is used.

GARD® ROTARY DISTRIBUTORS FOR TRICKLING FILTERS:

GARD® rotary distributors are offered for all sizes of trickling filters. They are furnished with the center column constructed of properly coated fabricated steel with aluminum distributor arms and reactor assemblies. Gravity flow principal eliminates troublesome mercury, air or mechanical seals. Reactors are non-clogging.

DESIGN DETAILS

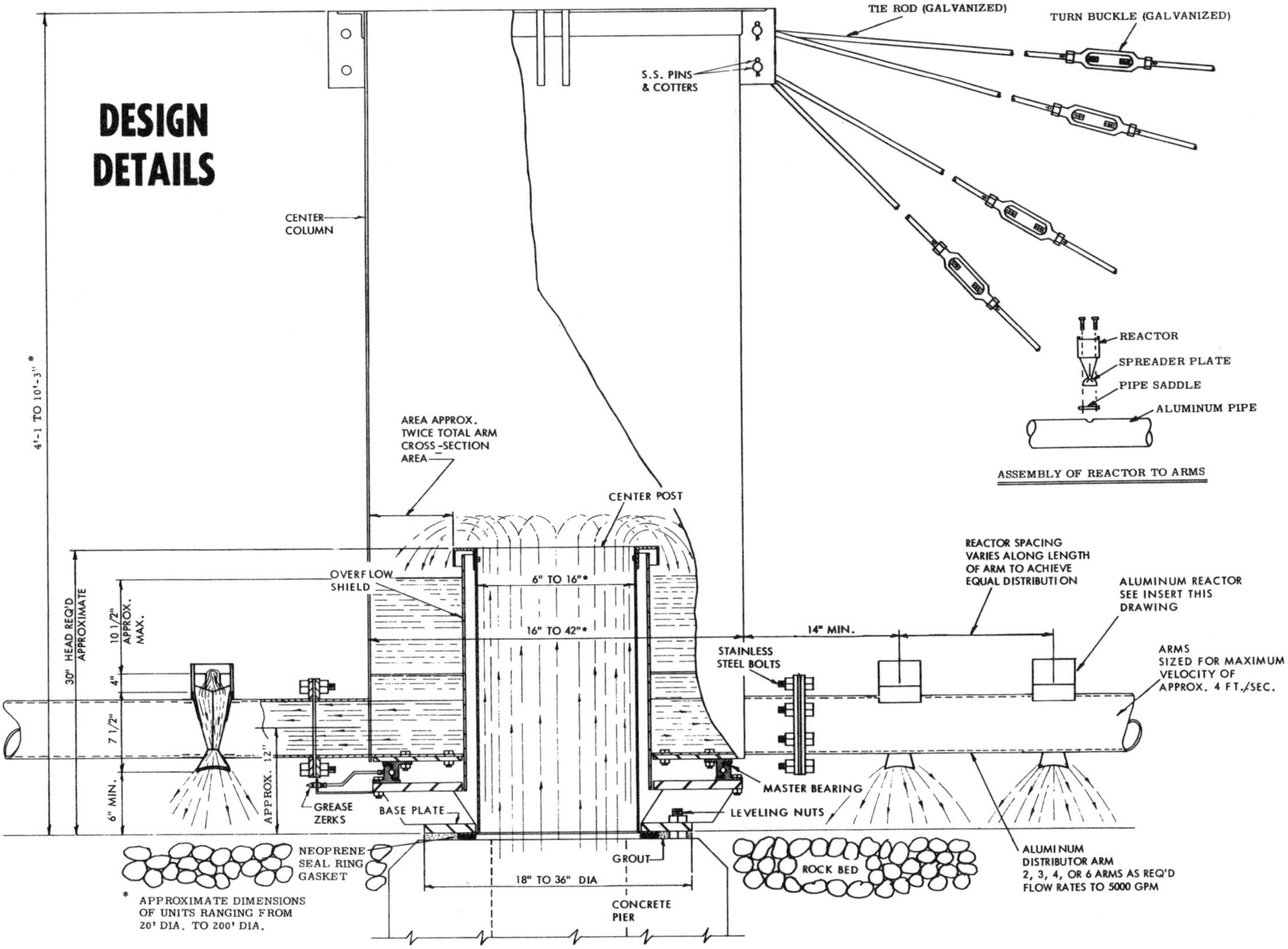

TIE ROD (GALVANIZED)
TURN BUCKLE (GALVANIZED)
S.S. PINS & COTTERS
CENTER COLUMN
REACTOR
SPREADER PLATE
PIPE SADDLE
ALUMINUM PIPE
ASSEMBLY OF REACTOR TO ARMS
4'-1 TO 10'-3" *
AREA APPROX. TWICE TOTAL ARM CROSS-SECTION AREA
CENTER POST
REACTOR SPACING VARIES ALONG LENGTH OF ARM TO ACHIEVE EQUAL DISTRIBUTION
ALUMINUM REACTOR SEE INSERT THIS DRAWING
30" HEAD REQ'D APPROXIMATE
10 1/2" APPROX. MAX.
OVERFLOW SHIELD
6" TO 16"*
16" TO 42"*
14" MIN.
STAINLESS STEEL BOLTS
ARMS SIZED FOR MAXIMUM VELOCITY OF APPROX. 4 FT./SEC.
4"
7 1/2"
APPROX. 12"
6" MIN.
GREASE ZERKS
BASE PLATE
MASTER BEARING
LEVELING NUTS
NEOPRENE SEAL RING GASKET
GROUT
18" TO 36" DIA
ROCK BED
CONCRETE PIER
ALUMINUM DISTRIBUTOR ARM 2, 3, 4, OR 6 ARMS AS REQ'D FLOW RATES TO 5000 GPM

* APPROXIMATE DIMENSIONS OF UNITS RANGING FROM 20' DIA. TO 200' DIA.

THE GFC AERALATER

The GFC Aeralater is a new, completely self-contained filter plant. It combines all the functions of aeration, detention and filtration in one package unit. It is equipped with fully Automatic or Manual Control.

Optimum reduction of CO_2, H_2S, and entrained gases is achieved with controlled induced draft aeration. Oxidation of dissolved metals such as ferrous iron is effected. Application of any chemical which may be necessary for the treatment involved is made at the aerator discharge.

Reaction time is provided following aeration. The length of this time depends upon the character of the water to be treated.

Filtration is through four gravity filter cells furnished with anthrafilt media, operating under a low and always positive head at nominal rates. Other filter media such as greensand are used for specific applications. The media is supported on 16" of graded filter gravel which rests on the underdrain plate.

The filtered water flows into the collecting well from the filter bed through GFC stainless steel Multiplate baffled orifices in the underdrain plate. The collecting well is common to all filter cells. The filtered water from the collecting well is pumped to storage. If desired, the pump may be eliminated and the collecting well outlet furnished with a suitable control valve which will permit gravity discharge from the collecting well into a ground level storage reservoir.

The service pump is controlled by an end point limit control which insures a positive head at the bottom of the filter. The end point limit control will initiate the backwashing procedure when the filter is dirty.

Employing the GFC Multicell principle, the Aeralater furnishes its own source of filtered backwash water at a uniform rate from in-service cells.

During the backwash operation the total plant flow remains at its constant operating rate. The in-service cells filter this total flow, discharging into the collecting well common to all cells. All of this filtered water flows upwardly through the cell being backwashed at the constant designed rate and is discharged to waste through the backwash outlet until the cell is clean. This step is repeated for all cells and the plant is then returned to service.

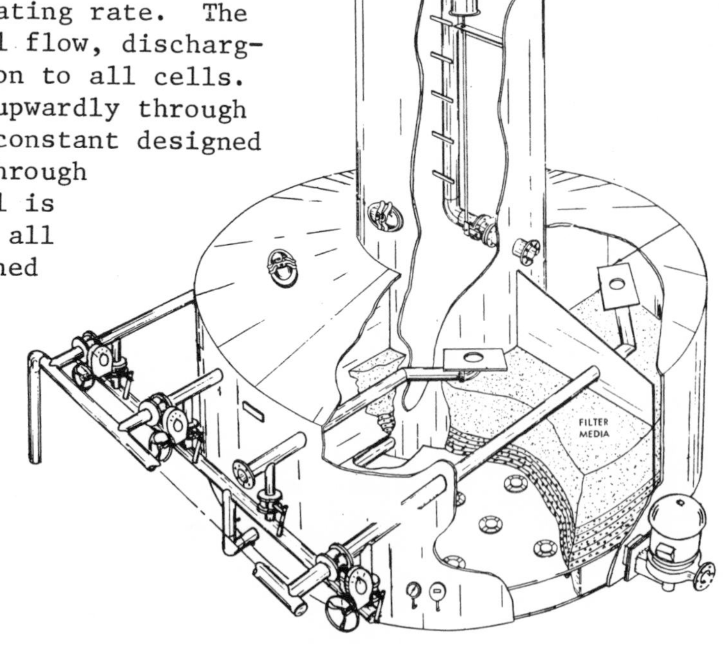

GFC AERATORS

Aeration of water is a mechanical method of removing gases such as carbon dioxide, hydrogen sulfide, methane and odors which may be dissolved or entrained therein. It is also a most important method of oxidizing dissolved metals such as ferrous iron and manganous manganese.

POSITIVE DRAFT aeration is the most efficient type of aeration for dissolved gas removal and oxidizing iron and manganese. High efficiency gas removal is accomplished by providing for each specific application a predetermined number of surfaces or trays against which free falling droplets and thin films of water impinge again. A counter-current flow of air continuously sweeps through the aerating water, carrying away released gases and supplying oxygen for oxidation processes that may be involved.

In the GFC POSITIVE DRAFT AERATORS water discharges through the top center inlet over a distributor with orifices to insure even droplet distribution into the aeration zone immediately under the distributor.

This zone contains the slats or screens and constitutes the actual aeration chamber in which the object is to obtain the smallest droplets to expose the greatest mass of water particles directly to the air.

The INDUCED DRAFT AERATOR uses a top mounted blower system which induces from the top an evenly distributed air flow through the chamber at slightly below atmospheric pressure. Air is evenly distributed passing at high velocity through the mist zone, oxidizing the iron and removing gases. The air is then discharged through the internal air stacks into the moisture separator and out the top.

The FORCED DRAFT AERATOR uses forced air introduced from the blower at the base. The uniform air flow through the chamber effects the desired oxidation reactions and together with the dissolved gases removed from the water is discharged out the top through moisture separators.

GFC aerators are designed to operate at full blower output for greatest efficiency at the lowest power consumption.

ALUMINUM AND STEEL INDUCED DRAFT AERATORS

ALUMINUM TYPE 14.10 STEEL TYPE 14.30

CAP. GPM	CROSS SECTION SQUARE INCHES A	CROSS SECTION AREA SQUARE FOOT	LIQUID LOADING gpm/sq.ft.	MATERIAL THICKNESS HOUSING	MATERIAL THICKNESS COVER	AIR REQ'D CFM (1)	AERATOR HEIGHT B—FAN HP, CO_2 REDUCTION TO 10[2] FROM 40 PPM HGT.	40 PPM HP	100 PPM HGT.	100 PPM HP.	150 PPM HGT.	150 PPM HP	WATER CONNECTION SIZES (INCHES) INLET C	OUTLET D	DIMENSIONS E	F	G	H	OPERATING WEIGHT (LBS.) STEEL Std. 10' High Unit	STEEL Each Add. Ft. of Body	OPERATING WEIGHT (LBS.) ALUMINUM Std. 10' High Unit	ALUMINUM Each Add. Ft. of Body	
40	18" x 18"	2.2	18.2	3/16"	3/16"	200	10'	1/2	12	1/2	14'	1/2	2"	2 1/2"	12"	15"	24"	13'-0"	1,121	60	870	32	SQUARE
50	21" x 21"	3.0	16.7	3/16"	3/16"	250	10'	1/2	12	1/2	14'	1/2	2"	3"	12"	18"	27"	13'-0"	1,355	80	1,010	36	
70	24" x 24"	4.0	17.5	3/16"	3/16"	350	10'	1/2	12	1/2	14'	1/2	2 1/2"	4"	12"	21"	30"	13'-2"	1,620	90	1,180	38	
110	30" x 30"	6.25	17.6	3/16"	3/16"	390	10'	1/2	12	1/2	14'	1/2	3"	4"	12"	27"	38"	13'-2"	2,258	116	1,520	42	
160	36" x 36"	9.0	17.8	3/16"	3/16"	560	10'	1/2	12	1/2	14'	1/2	4"	5"	12"	33"	52"	13'-2"	2,911	140	1,980	50	
215	42" x 42"	12.3	17.5	3/16"	3/16"	760	10'	1/2	12	1/2	16'	1/2	5"	6"	12"	39"	58"	13'-2"	3,789	164	2,490	56	
280	48" x 48"	16.0	17.5	3/16"	3/16"	990	10'	1/2	12	1/2	16'	1/2	5"	6"	12"	42"	64"	13'-2"	4,615	208	3,090	64	
355	54" x 54"	20.2	17.5	3/16"	3/16"	1250	10'	1/2	12	1/2	16'	1/2	6"	8"	12"	48"	70"	13'-2"	5,600	243	3,760	73	
440	60" x 60"	25.0	17.5	3/16"	3/16"	1550	10'	1/2	12	1/2	16'	1/2	6"	8"	12"	54"	76"	13'-2"	6,666	270	4,540	85	
530	66" x 66"	30.2	17.5	3/16"	3/16"	1850	10'	1/2	12	1/2	16'	1/2	8"	8"	12"	60"	82"	13'-2"	7,764	309	5,380	96	
630	72" x 72"	36.0	17.5	3/16"	3/16"	2200	10'	1/2	12	1/2	18'	3/4	8"	10"	12"	66"	88"	13'-2"	10,013	421	6,290	108	
740	78" x 78"	42.2	17.5	1/4"	3/16"	2580	12'	1/2	14	1/2	18'	3/4	8"	10"	12"	72"	94"	15'-2"	11,450	453	7,290	123	
860	84" x 84"	49.0	17.5	1/4"	3/16"	3000	12'	1/2	14	3/4	18'	3/4	8"	10"	12"	78"	104"	15'-2"	12,940	500	8,410	138	
985	90" x 90"	56.3	17.5	1/4"	3/16"	3450	12'	1/2	14	3/4	18'	1 1/2	10"	12"	12"	84"	110"	15'-2"	14,475	542	9,530	153	
1120	96" x 96"	64.0	17.5	1/4"	3/16"	3900	12'	3/4	14	1 1/2	18'	1 1/2	10"	12"	15"	90"	116"	15'-5"	16,413	621	10,820	172	
1260	102"x102"	72.2	17.5	1/4"	3/16"	4400	12'	1 1/2	14	1 1/2	18'	1 1/2	10"	12"	15"	96"	122"	15'-5"	18,164	669	12,080	190	
1420	108"x108"	81.0	17.5	1/4"	3/16"	4950	12'	1 1/2	14	1 1/2	18'	1 1/2	10"	14"	15"	102"	128"	15'-5"	19,976	712	13,420	209	
1750	120"x120"	100.0	17.5	1/4"	3/16"	6100	14'	2- 3/4	16	2-1 1/2	18'	2-1 1/2	12"	14"	15"	114"	140"	17'-5"	24,351	855	16,380	253	
2120	132"x132"	121.0	17.5	1/4"	3/16"	7400	14'	2-1 1/2	16	2-1 1/2	18'	2-1 1/2	12"	16"	15"	126"	152"	17'-5"	28,607	955	19,580	298	
2520	144"x144"	144.0	17.5	1/4"	3/16"	8800	14'	2-1 1/2	16	2-1 1/2	18'	2-1 1/2	14"	18"	15"	138"	164"	17'-5"	33,236	1057	23,020	350	
	(A)																						RECTANGULAR
2000	126" x 132"	115.5	17.5	1/4"	3/16"	7000	12'	2- 3/4	Aerator height B limited to 12' by railroad height limitation. If greater height is required use multiple square units.				12"	16"	15"	10'-6"	11'-6"	15'-5"	27,500	1052	18,700	290	
2200	126" x 144"	126.0	17.5	1/4"	3/16"	7700	12'	2-1 1/2					12"	16"	15"	11'-6"	12'-6"	15'-5"	29,700	1116	20,300	310	
2400	126" x 156"	136.5	17.5	1/4"	3/16"	8400	12'	2-1 1/2					14"	18"	15"	12'-6"	13'-6"	15'-5"	31,700	1183	21,900	323	
2580	126" x 168"	147.0	17.5	1/4"	3/16"	9000	12'	2-1 1/2					14"	18"	15"	13'-6"	14'-6"	15'-5"	33,900	1269	23,500	355	
2760	126" x 180"	157.5	17.5	1/4"	3/16"	9600	12'	2-1 1/2					2-10"	18"	15"	14'-6"	15'-6"	15'-5"	36,200	1340	25,100	370	
2940	126" x 192"	168.0	17.5	1/4"	3/16"	10200	12'	2-1 1/2					2-12"	20"	15"	15'-6"	16'-6"	15'-5"	38,600	1406	26,700	390	
3125	126" x 204"	178.5	17.5	1/4"	3/16"	10900	12'	3-1 1/2					2-12"	20"	15"	16'-6"	17'-6"	15'-5"	41,200	1466	28,500	398	
3320	126" x 216"	189.0	17.5	1/4"	3/16"	11600	12'	3-1 1/2					2-12"	20"	15"	17'-6"	18'-6"	15'-5"	43,400	1535	30,100	419	
3500	126" x 228"	199.5	17.5	1/4"	3/16"	12200	12'	3-1 1/2					2-12"	20"	15"	18'-6"	19'-6"	15'-5"	45,400	1607	31,600	434	

CONTRAFLO SOLIDS CONTACT CLARIFIERS

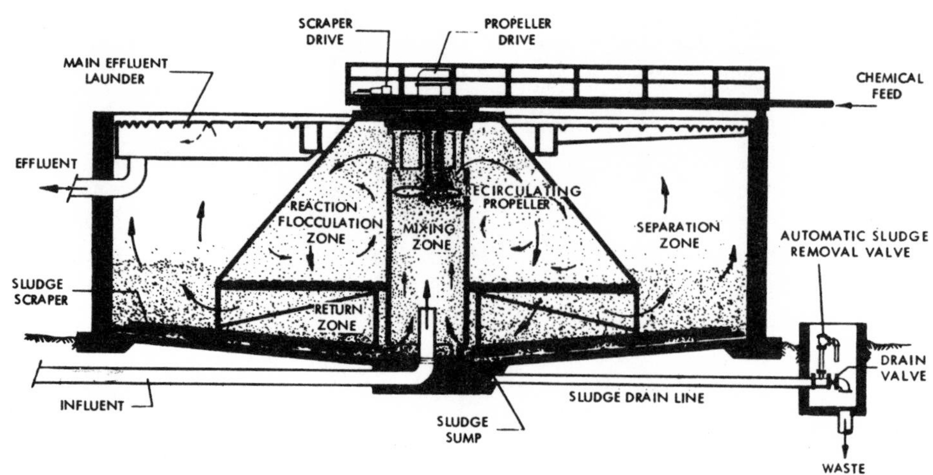

THE CONTRAFLO PROCESS

Mixing Zone: Raw water and chemicals are intimately mixed in the presence of large amounts of previously formed slurry recirculated from the reaction-flocculation zone and large amounts of precipitates recirculated from the return zone at the bottom of the basin. A slow speed, high efficiency propeller recirculates this flow into the reaction-flocculation zone. Propeller speed is adjusted to attain the optimum recirculation rate and slurry concentration for the treatment involved.

Reaction-Flocculation Zone: The ample space under the cone receives the total flow from the mixing zone. Here flocculation is greatly accelerated by the intimate contact between the reacting chemicals and the existing precipitates on which the newly forming material deposits. One part of the flow is then discharged into the separation zone and the remaining flow recirculated into the mixing zone.

Separation Zone: The large area under the edge of the cone insures even distribution and low velocity entrance to the separation or upward filtration zone. The upward velocity of the water maintains a zone of suspended reacted slurry which acts as a filter and catalyst collecting small particles of sludge and forcing the chemical reactions to completion.

Sludge Removal and Recirculation: A portion of the sludge in the separation zone becomes too heavy to maintain in suspension and settles to the bottom of the tank. This material is moved to the center of the tank by the sludge collector. Some of this material falls into the sludge hopper and is removed, and some is recirculated and used as a flocculation aid. The sludge discharge lines may be provided with a connection for back flushing, should this prove necessary.

Clarification Zone: The water velocity constantly decreases as it rises toward the effluent launders. At some level, the velocity is no longer great enough to carry the fine slurry particles and the clarified water escapes upwardly toward the effluent launders. The flow of water in the clarification zone must not be allowed to develop localized high velocity; therefore, design and placement of launders within the tank is an important aspect of clarifier design.

General Filter Company

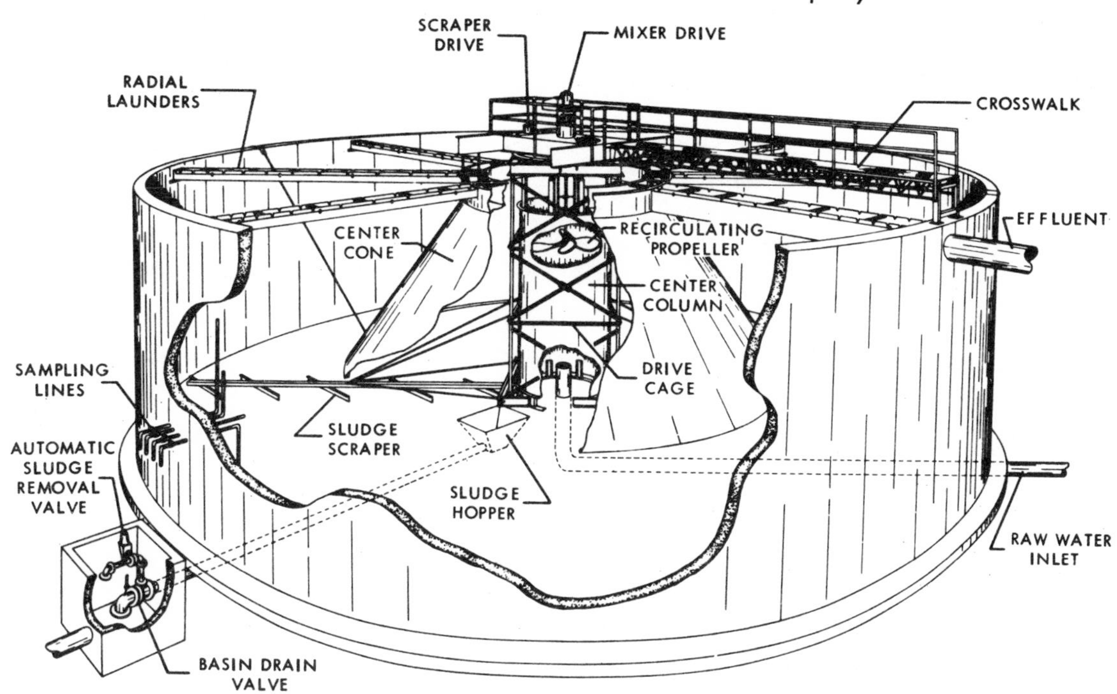

INLET
The raw water is introduced into the center column through a bottom center inlet. Alternately it may be introduced into the side of the center column through a basin side inlet.

CONTRAFLO® "C"

CENTER COLUMN

The CONTRAFLO® mechanism and one end of the crosswalk are supported on a steel center column. The center column serves as a mixing chamber and is equipped at top and bottom with openings for recirculation of previously formed precipitates. Openings are sized for velocities necessary to insure sludge suspension. The center column is designed to withstand the full torque developed by the mechanism.

MIXER AND SCRAPER DRIVES

The mixing propeller has a ball bearing equipped drive shaft connected to a variable speed drive. No bearings are submerged. The slow speed, high efficiency propeller assures thorough mixing and eliminates floc breakup.

The scraper drive consists of a motor and speed reducer driving a turntable, which is mounted on a large diameter locked race pressure grease lubricated ball bearing. A torque limiting device is provided to protect all parts of the mechanism.

CENTER CONE

The area under the center cone forms the flocculation zone. The center cone may be constructed to rotate the scraper arms or it may be fixed and the scraper arms rotated by a drive cage or the center column.

ROTATING SCRAPER ASSEMBLY

Two rotating rake arms with scraper blades are arranged to move settled sludge to the sludge hopper and to the recirculation ports in the center column. The rotating radial arms are adaptable to any desired bottom slope. The scrapers normally are operated at a tip speed of approximately five feet per minute.

EFFLUENT LAUNDER

The design of the effluent launder is based on the type of operation desired and on the overall dimensions of the treating basin. If the unit is to be operated at variable flow rates, as determined by the level in the clear well or a surge tank, it is usually desirable to equip the effluent launders with V-notch weir collectors. If the raw water rate of flow is to be controlled from the water level in the CONTRAFLO® it is desirable to use orifice hole collectors in the vertical sides of the launders. The positioning of launders within the tank is an important aspect of upflow clarifier design. The flow of water in the clear water area must not be allowed to develop localized high velocity. Sufficient linear feet of launder must be provided to limit weir loadings to approximately seven and one-half gpm per lineal foot. The launder arrangement may be peripheral, radial, concentric, or a combination thereof, depending on the shape and size of the basin.

TYPE C

The Type C is designed for medium and large capacity plants. It is adaptable to round or square basins and utilizes a bottom center raw water inlet.

Round basins may be constructed of steel or concrete. Corner fillets are used in square basins.

In large units a structural steel center support is utilized inside the rotating center column when a side raw water inlet is used.

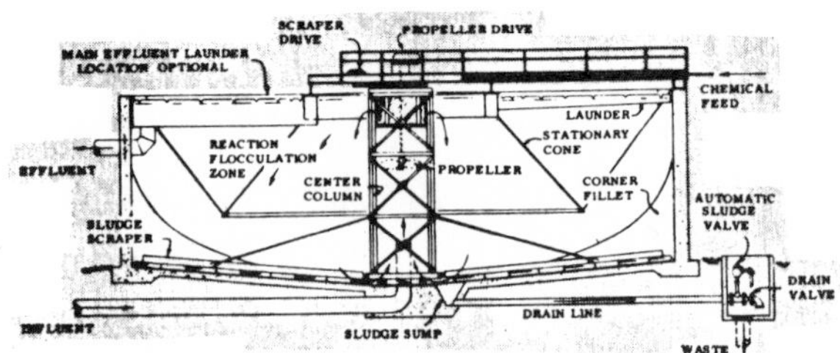

A structural steel cage drives the scraper arms in Type C CONTRAFLOS with 60′ and larger total span.

TYPE CT

Type CT is for small capacity plants up to 20′ total span when treating water where sludge scrapers are not required.

The mechanism is suspended from a bridge across the basin and a side raw water inlet is used. Sludge is automatically withdrawn from the sludge concentrator pockets which are provided for its collection.

Type CT may be installed in concrete or steel tanks. The bottom of the basin may be flat and furnished with fillets or a sloping concrete bottom may be used.

In the smaller sizes the all-steel basins can be shipped completely fabricated with the mechanism installed therein.

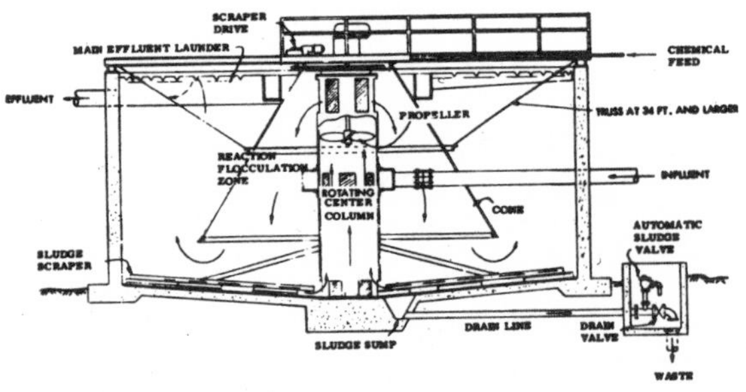

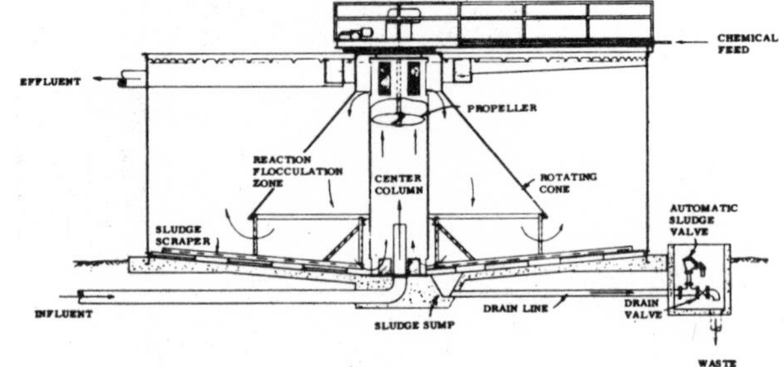

In Type C CONTRAFLOS from 26′ to 60′ total span the rotating cone drives the scraper arms. Peripheral launders are used in basins up to 30′ total span if loading does not exceed 7.5 gpm per lineal foot of weir.

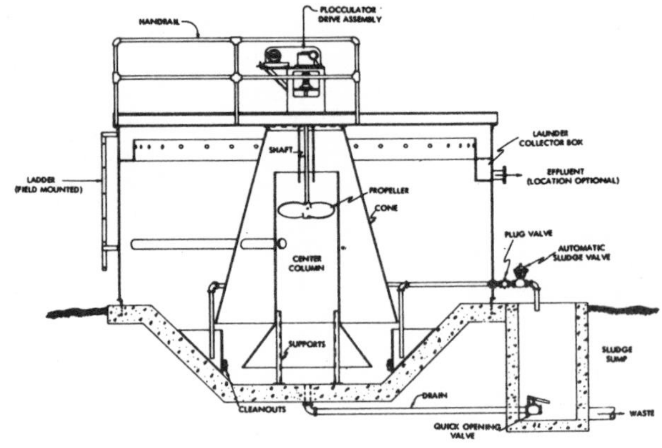

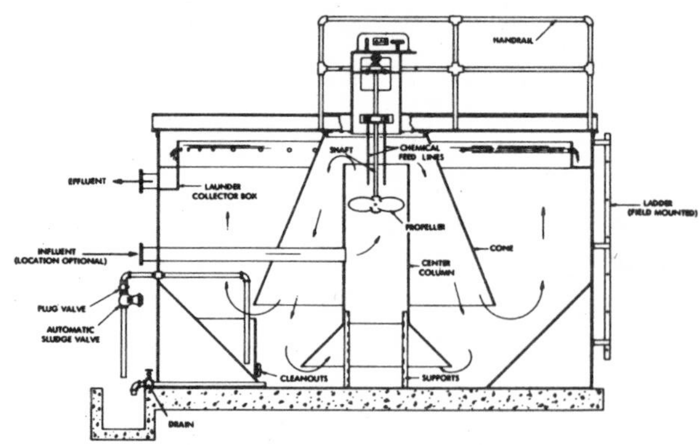

TYPE CSS

Type CSS suspends the entire mechanism from a bridge across the basin and uses a side raw water inlet. It is designed for medium capacities in sizes up to 60′ total span. Design criteria are the same as for Type C.

SEWAGE LIFT STATION Pre-engineered, factory assembled, above ground

A package lift station delivered complete.

Totally factory built, ready to set in place and to operate, once sewage lines and power lines are connected. Complete with **all** required equipment, including two self-priming, horizontal, centrifugal, V-belt motor driven sewage pumps; valves; internal piping; motor control center; liquid level control system; ventilator; internal wiring.

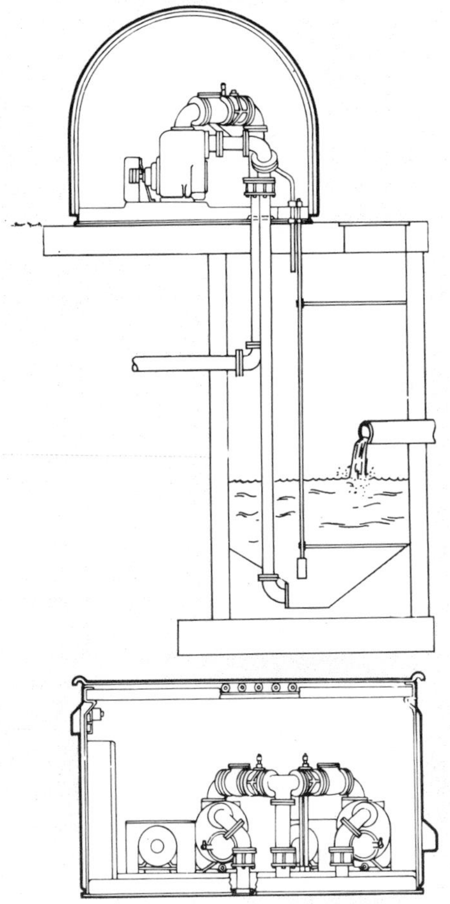

Pumps.

Self-priming centrifugal, "T" Series, designed by Gorman-Rupp specifically for handling raw, unscreened sanitary domestic sewage.

Self-Priming. Pumps retain sufficient liquid in casing to prime and reprime automatically, unattended, in an open system. They mount high and dry above the wet well, with only the suction pipe in the liquid.

Solids Handling Ability. Model T3A3-B handles spherical solids up to 2½" in diameter. Models T4A3-B and T6A3-B handle 3" spheres.

Accessibility. Removable cover plate permits easy access to pump interior (wear plate, impeller and seal). No special tools needed. No need to disturb suction or discharge piping or pump power source.

Seal. A Gorman-Rupp original, the double-floating seal is exclusive and its design and concept are protected by application for patents. It is standard in "T" Series.

This new seal is a double-floating, self-aligning, mechanical type with both rotating and stationary members made of tungsten-titanium carbide, oil lubricated, with mating surfaces lapped to a flatness tolerance of 0 to 5.8 millionths of an inch. Tungsten-titanium carbide properties provide a sealing surface impervious to abrasives, yet will not shatter or fracture, as do carbons and ceramics, under extreme shock loads. Removable through cover plate opening. Exclusive and unsurpassed for sewage service.

Station Enclosure.

Two major sections; base and movable cover. Cover slides in either direction, providing adequate access. Cover can be removed entirely. Made of molded fiberglass reinforced polyester resin, which has structural stability. Corrosive resistant. Impervious to microbes, mold and fungus. Gel coated interior resistant to abrasion, impervious to sewage, greases, oils, gasoline and other chemicals. Maintenance free. Excellent insulating qualities.

Portability. Ideal for the housing developments just beyond municipal collection systems. Once lines are extended, developer may transport lift station to his next development, realizing a substantial cost saving.

Low Silhouette. Designed to provide a low silhouette. Does not look like a typical pump station. Color blends with surrounding shrubbery, grass, other landscaping.

Optional Equipment. High water alarm, discharge pressure gauges, vacuum gauges, space heater, immersion heater, lights, spare parts.

T-SERIES SELF-PRIMING CENTRIFUGAL PUMPS
specifically designed for handling sewage

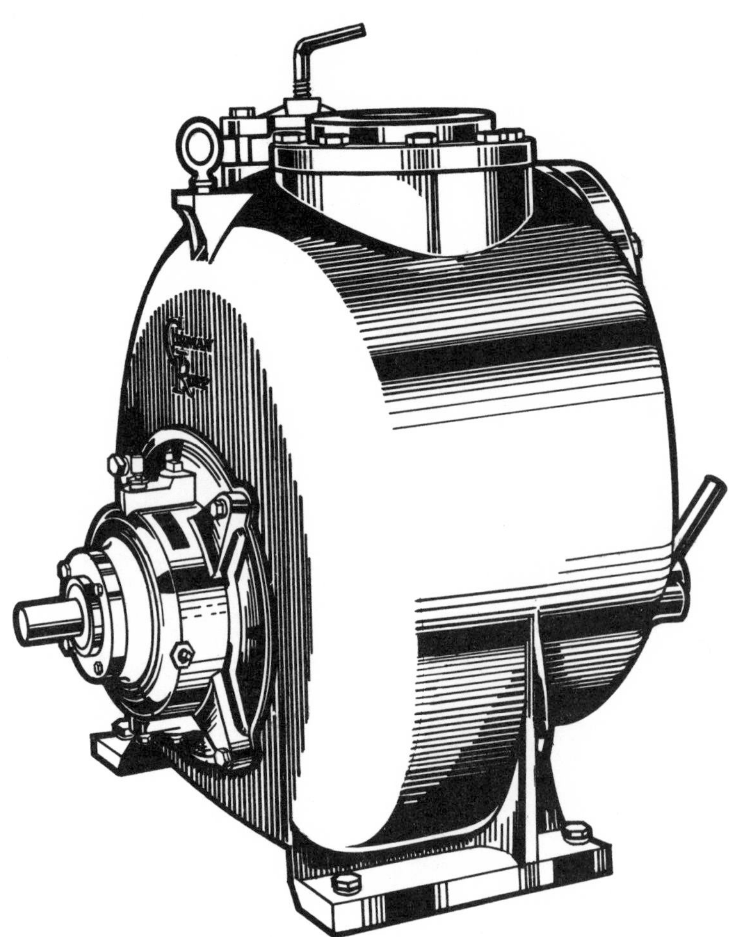

With rags and other stringy material regularly finding its way into waste water, all pumps will occasionally clog. With the T-Series removable coverplate, unclogging a pump takes just minutes . . . not hours. With the coverplate removed, you have complete access to the pump case and impeller. No pipe disconnections are required. And the entire rotating assembly is also removable and replaceable without disturbing pump casing or piping. Regular maintenance employees can completely service T-Series pumps with common tools.

"T" SERIES PUMPS		
Model	Size	Solids Handling Ability*
T3A3-B	3-inch	2½"
T4A3-B	4-inch	3"
T6A3-B	6-inch	3"
T8A3-B	8-inch	3"

*Indicating the diameter of spherical solids which the pump will pass.

PRIMING

Only Gorman-Rupp "T" Series sewage pumps are designed with the capability of priming and automatically repriming in a completely open system, without discharge or suction check valves, where the pump casing is only partially filled with water and a completely dry suction line. All other self-priming pumps must operate at 1750 r.p.m. or above, equipped with a full diameter impeller, and then only on limited priming lifts, no greater than 5′ to 6′.

DOUBLE FLOATING SEAL

As the pump is the heart of a lift station, the seal is the heart of the pump. Development of the double floating self-aligning Gorman-Rupp mechanical seal, standard on "T" Series pumps, is a break-through in mechanical seal design comparable to the first Gorman-Rupp Grease Seal, in 1935, still tops in contractors pumps. This double floating seal was born of compelling necessity to neutralize effects of shock load stress and abrasives in sewage pumping. Structural principles and seal face materials proven in perpetually difficult pumping situations were combined in this exclusive design. Proof of its success is a national pattern of satisfactory replacement on jobs too tough for conventional seals. (Not one failure in the 800 seals already installed.)

A Gorman-Rupp original, the double floating seal is exclusive and its design and concept are protected by application for patents. It is standard in "T" Series. Following is a description:

This new seal is a double-floating, self-aligning, mechanical type with both rotating and stationary members made of tungsten-titanium carbide, oil lubricated, with mating surfaces lapped to a flatness tolerance of 0 to 5.8 millionths of an inch. Tungsten-titanium carbide properties provide a sealing surface impervious to abrasives, yet will not shatter or fracture, as do carbons and ceramics, under extreme shock loads. Removable through coverplate opening.

This seal is exclusive and unsurpassed for sewage service.

FILTRATION WITH THE NCE-MIURA REVERSIBLE FLOW FILTER

A complete filter unit comprises a pump, a specially designed multiport valve, a number of filter elements and cloths and a feeding device for filter aid. A varying number of these elements are clamped together to form a filter case and between each element a filter cloth is located. The standard cloths are of polyester fiber , but other materials such as nylon or cotton duck may be used if necessary.

Changes from one step in the filtration process to the next are accomplished by a single movement of the multiport valve. This operation may be carried out electrically in larger units.

The process steps are shown in the illustration.

(1) Filtration - Liquid to be filtered flows from right to left through a layer of filter aid and the filter cloth. The impurities are removed and accumulate within and upon the filter aid layer.

(2) Backwashing and Precoating - The flow is reversed and raw liquid mixed with fresh filter aid flows from left to right through the filter cloth. A layer of fresh filter aid is formed in the left side of the cloth and its formation, initially in patches, assists the washing away of the used filter aid with the impurities from the right side of the cloth. During this step the wastewater and impurities are led away to a drain. By successively removing patches of dirty aid from the right side and building up patches of fresh filter aid on the left side the cloth is quickly covered with a complete layer of fresh filter aid. This operation is usually completed in less than five minutes.

(3) Filtration - Raw liquid now flows through the precoat and cloth from left to right.

(4) Backwash and Precoating - The flow is again reversed and the dirty filter aid from the left side of the cloth is washed away and a fresh layer built up on the right side. One complete filtration cycle is performed with a single rotation of the multiport valve. If necessary, operation can be made completely automatic. Cycle time between backwashing depends only upon the liquid being filtered.

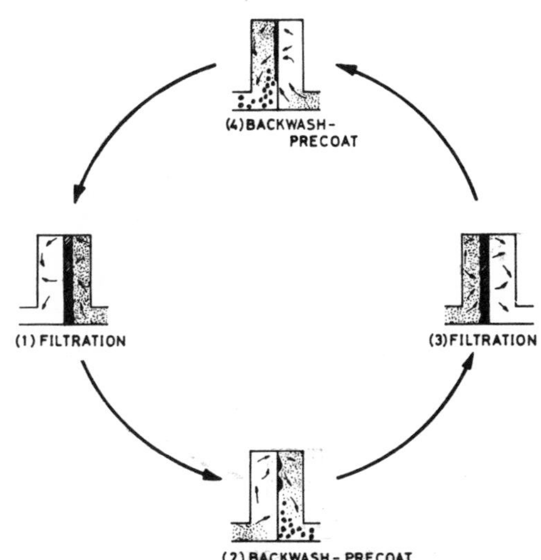

CONTINUOUS ADSORPTION SYSTEM

The Continuous Adsorption System is a fully automated countercurrent system employing adsorbents in granular form. It is used for virtually complete removal of organics, particularly refractory materials.

Waste liquid is introduced continuously at the bottom of the C.A. unit and flows upward through the bed of activated carbon. A carbon retainer and upper collector allow regenerated adsorbent to be fed into the top of the unit, while treated water is discharged.

This operation permits the most regenerated adsorbent to be in contact with the waste stream just prior to its discharge. Acting as a "polishing" step, this last contact cleans the waste stream to a greater degree than is possible with fixed-bed adsorption systems.

The contaminant-saturated carbon is dewatered and then regenerated in a multiple hearth furnace. In this way, the carbon is re-used approximately 30 times before new carbon is required. Colored wastes can be cleared economically using activated carbon in a C.A. System. This is applicable to dye wastes and other clear but colored waste streams, as well as to tertiary treatment. In addition, there appears to be a definite place for the C.A. System following biological treatment, to assure that the final effluent will meet the stringent requirements for discharge set by regulatory agencies.

ION EXCHANGE

Effective removal of essentially all dissolved minerals from fluid industrial waste is accomplished with Graver Demineralizers. Demineralization, either fixed bed or by the Graver CI Process (continuous countercurrent ion exchange), is used to convert industrial waste water into the highest quality process water and boiler feedwater. As a by-product of this water purification, valuable materials are occasionally recovered. In some cases, the cost of the demineralization equipment will be more than offset by this material savings. Waste treatment applications include chromate and other plating metals recovery, concentration of rare metals and rare earths, and recovery of copper and zinc.

In addition to demineralizers, Graver also offers other ion exchange equipment including softeners, dealkalizers and disilicizers. A complete line of packaged demineralizers using the unique Partilok Strainer subfill-less underdrain system and the Monotrol® valve is also available.

GRAVER FLOC BARRIER

CLARIFIER	Rated Capacity, MGD	Capacity with Floc Barrier, MGD	% Increase *
86′ Diam. Reactivator	9.0	16.8	87
27′-6″ Square ** (solids contact)	2.2	4.0	85
16′ Square ** (sludge blanket)	0.35	0.7	100
31′ Square Reactivator	1.5	3.0	100

* Maximum flow increase may be governed by the hydraulic limitations of the existing effluent system.

** These are competitive clarifiers converted to Graver's Floc Barrier System.

The Floc Barrier will increase the capacity of the Graver Reactivator® Clarifier as well as competitive solids-contact units. This table describes four plants, before and after Floc Barrier installation.

The Floc Barrier consists of sloped shallow clarifier ducts, combined into modules. Entering floc particles rapidly settle to the bottom of each duct where they come into contact with previously settled particles. These additional contacts facilitate the formation of denser, more readily settleable particles.

The agglomerated particles eventually reach the lower portion of the Floc Barrier, from which they fall rapidly to the floor of the clarifier. In effect, the Floc Barrier creates a multiplicity of very shallow settling units, reducing the surface area and detention time necessary to produce clarified water.

GRAVER ROTA-RAKE CLARIFIER

The Rota-Rake® provides for the gravity separation of suspended solids from liquids. It offers maximum separation, positive sludge removal, and complete overload protection. Available in a wide range of designs and sizes, the Rota-Rake is a simply designed, ruggedly built unit, economical to install, operate and maintain.

A Rota-Rake installation consists of a round or square, conical bottom, steel or concrete tank with a quiescent flow feed arrangement and an overflow effluent system. Water is fed to a central inlet well and its velocity is reduced. Heavy solids are deposited on the bottom of the tank. As the water flows radially across the unit, it settles out the other solids, the finest particles being deposited near the periphery where the velocity is lowest. The water then flows into collecting launders.

For positive sludge collection and discharge with a minimum amount of turbulence, a motor-driven, heavy-duty, box-truss scraper continuously moves settled solids to a central discharge sump in the bottom of the tank. A rotating paddle in the sump keeps the sludge moving and thickens it further.

The Rota-Rake is used as a primary clarifier for treating paper mill effluents. With the addition of appropriate skimming equipment, it is applicable for the removal of floating solids or free oil.

Graver Water Conditioning Co., Division of Ecodyne Corporation

GRAVER MUNICIPAL WATER TREATMENT EQUIPMENT

Equipment	Graver Trademark	Where Used
Solids contact clarifier	Reactivator ®	High rate coagulation Turbidity removal Softening Alkalinity reduction Iron removal Fluoride reduction
Gravity separator/ thickener	Rota-Rake®	Gravity settling Backwash water reclamation Blowdown sludge thickening
Aerators.		Dissolved oxygen addition Hydrogen sulfide removal Iron & manganese oxidation Carbon dioxide removal Methane removal
Filters	Monovalve ® Mono-Scour ® Mono-Floc ®	Remove suspended solids Oxidize iron and manganese
Ion Exchange	Gravex C. I. ®	Hardness removal T.D.S. reduction (brackish water) Alkalinity reduction Demineralization

REACTIVATOR CLARIFIER

The Graver Reactivator is a high-rate solids contact sludge recirculation clarifier ideally suited for chemical coagulation and clarification of waste water. It combines flash mixing, flocculation, clarification and sludge thickening into one operation.

When raw water enters the Reactivator, it is mixed with previously formed precipitates and treatment chemicals. The benefit of intimate contact with solids is obtained by the full retention time provided in the mixing zone, under the conical hood, so that by the time the water enters the outer settling zone, the bulk of the precipitated particles is large and dense.

In the Graver Reactivator, the precipitated solids are moved mechanically by means of scrapers across the entire bottom of the unit to a central sump, thereby providing positive sludge removal. The ability of the Reactivator to recirculate, collect, thicken and remove sludge, makes this machine particularly applicable to sewage tertiary treatment, metal finishing, paper and steel mills.

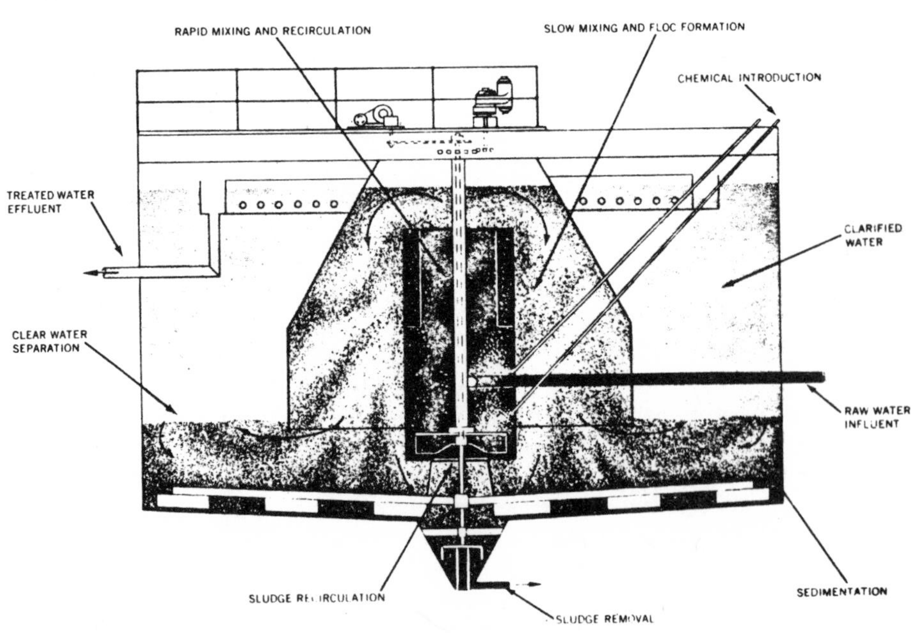

Schematic view of Graver Reactivator. Raw water entering the Reactivator is mixed with previously formed precipitates and treatment chemicals. Intimate solids contact is achieved in the mixing zone under the retention hood. By the time the water enters the outer settling zone, the precipitated particles are large, dense and settleable.

C 21 M – REACTIVATOR

Graver Water Conditioning Co., Division of Ecodyne Corporation

THE MONOVALVE FILTER SERIES

MONOVALVE Standard Filtration Service	Low concentration of filterable turbidity	For any service in which conventional sand filtration is satisfactory.
MONO-SCOUR® Heavy-Duty Filtration Service	High concentration of filterable turbidity	A heavy-duty, high-rate Monovalve filter, designed to remove more suspended solids per run, and to clean itself more vigorously. Excellent for high turbidity loads and sticky particles.
MONO-FLOC Coagulation and Filtration Service	Low or high concentrations of non-filterable turbidity	Where all or part of the inlet turbidity would pass through a filter to make the outlet water unsatisfactory.
MONO-FERM® Iron Removal Service	Iron, manganese and sulfide removal	All Monovalves remove these water contaminants in varying degrees. Mono-Ferm has been specifically designed to remove these impurities to prescribed values of raw water conditions.

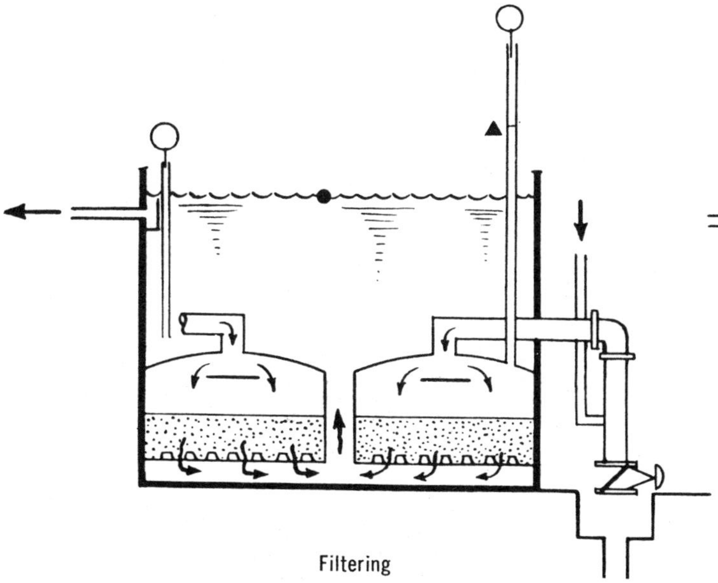

Filtering

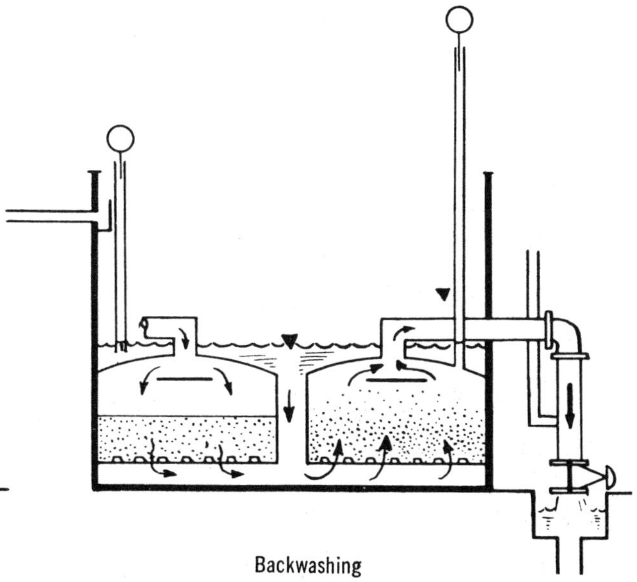

Backwashing

FILTERING. Influent water percolates through the filter bed to the false bottom, where it is collected by Partilok Strainers. From the underdrain compartment, filtered water flows to the backwash compartment and out to service. The water level in the backwash storage compartment remains fixed, while water in the level control pipe rises slowly as filtering continues and the beds collect dirt. This continues until the level differential reaches a predetermined point, usually a head-loss of about five feet.

BACKWASHING. When the level differential reaches its pre-determined point, the high-level control energizes the pilot valve, opening the backwash-initiating control valve. As the water from the backwash compartment flows up through the filter bed, expanding and backwashing the sand and proceeding down the tail pipe and into the sump, the level in the backwash storage compartment drops. Water filtering through the other sections is used as backwash water. As this process continues, the water level in the backwash compartment falls until the low level control deenergizes the pilot valve. This closes the control valve, completing the backwash cycle.

RINSING. Rinse water then begins filtering through the bed, passing from collection chamber up to the backwash storage compartment, where it remains until required for backwash. The filling continues until the previously fixed level is achieved, and water again passes out to service.

low concentrations of filterable turbidity

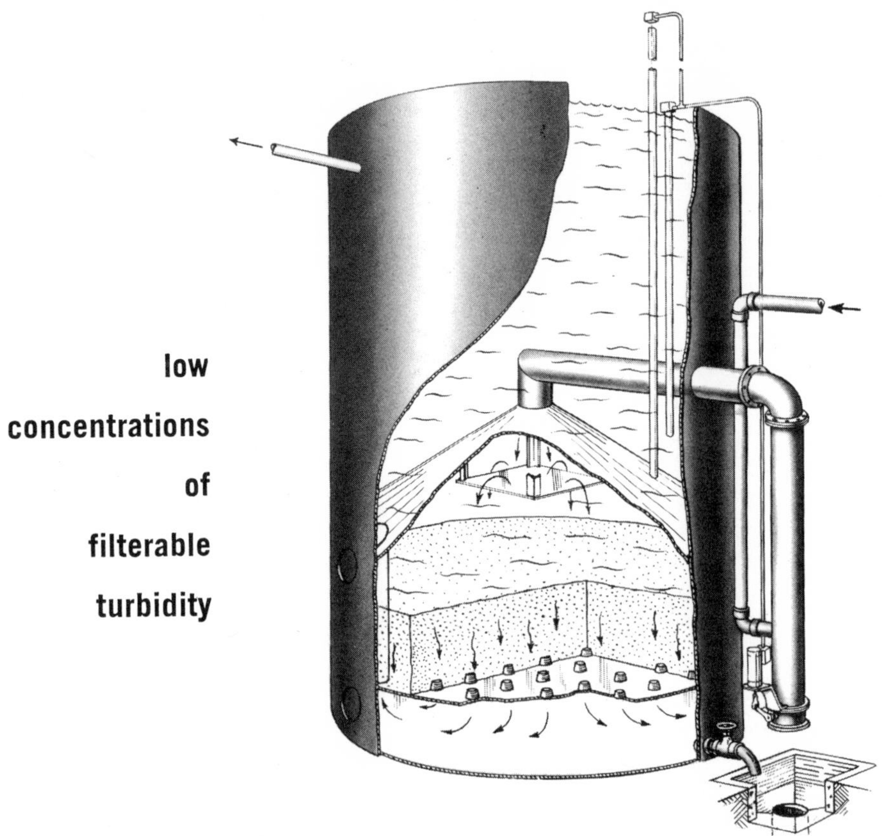

The Monovalve Filter is a fully-automatic gravity filter of advanced functional design. It utilizes conventional filter media, and can be used wherever bed-type filtration is applicable. The Monovalve Filter stores its own backwash water, and has simplified controls for fully automatic operation.

Storing its own backwash water eliminates the need for external backwash facilities like pumps, pipings, storage, etc. The installation costs, as well as the initial equipment costs are, therefore, lower than conventional filter systems.

The fully automatic nature of the Monovalve Filter eliminates the possibility of human error. Backwashing starts automatically, dependent solely upon the actual condition of the bed, so that the effluent will consistently be of high quality.

Multiple compartment design, an exclusive feature of the Monovalve Series of Automatic Gravity Filters, allows reduced headroom, smaller waste pipe size, and lower inlet head requirements.

MONO-SCOUR FILTER

The Mono-Scour filter is a heavy-duty, high-rate Monovalve filter, designed to remove more suspended solids per run. Vigorous self-cleaning, including air scour, is used to dependably restore the filter media. Efficient for high turbidity loads and sticky particles, the MONO-SCOUR filter successfully combines the economic advantages of roughing filtration followed by polishing, European air wash practice, and other proven features of the standard Monovalve filter. This equipment is designed to filter waters containing substantially higher levels of filterable turbidity than "standard" filters, or operate at considerably higher flow rates on low-to-moderate turbidity waters.

Years of experience has shown that when conventional filter media is preceded by roughing filters, much longer operating cycles and higher rate capabilities are obtained. The Graver MONO-SCOUR filter offers two-stage filtration; roughing, followed by fine polishing. This is accomplished by the use of a layer of relatively coarse, low-density material at the top, and a fine-particled polishing layer of high density material at the bottom of the filter bed. The depth filtration obtained with combination media permits high solids accumulations within the depth of the filter bed to a given headloss end point.

MONO-SCOUR units apply high rate filtration to low turbidity waters or moderate rate filtration to high turbidity waters normally uneconomical to treat by conventional filtration.

Certain high solids accumulations within the filter bed may not be effectively removed by conventional backwash. Efficient air scour in conjunction with backwash, is used in the MONO-SCOUR filter to insure positive bed cleaning.

MONO-FLOC AND MONO-FERM FILTER SYSTEMS

The Monovalve and Mono-Scour Filters are used for the removal of filterable turbidity from water supplies. The MONO-FLOC system is suggested for use where all or part of the inlet turbidity is not filterable, and enough turbidity would pass through a filter to make the outlet water unsatisfactory. In such cases, it is customary to use chemical coagulation and clarification equipment to render the particulate matter settleable. This incorporates the use of flash mixing, flocculation and sedimentation to reduce the suspended solids load applied to the polishing filters.

Experience indicates that many water supplies lend themselves to production of high quality effluent by the modification of the turbidity prior to filtration. Our experience in this field goes back four decades to the use of "pot type" alum feeds ahead of pressure sand filters, and more recently, through the use of polyelectrolytes, as coagulant aids.

Depth Filtration, air scour, and other features of the Mono-Scour Filter, combined with Graver's extensive chemical treatment experience, are utilized in the MONO-FLOC System.

An adaptation of the Mono-Floc system is the MONO-FERM® System which incorporates the unique oxidation potential of a catalytic mineral bed, the Depth Filtration and Air Scour features of the MONO-SCOUR Filter, and long term experience in iron, manganese and sulfide removal.

CONCRETE MONO-PAK FILTER

Central Control for Economy and Efficiency

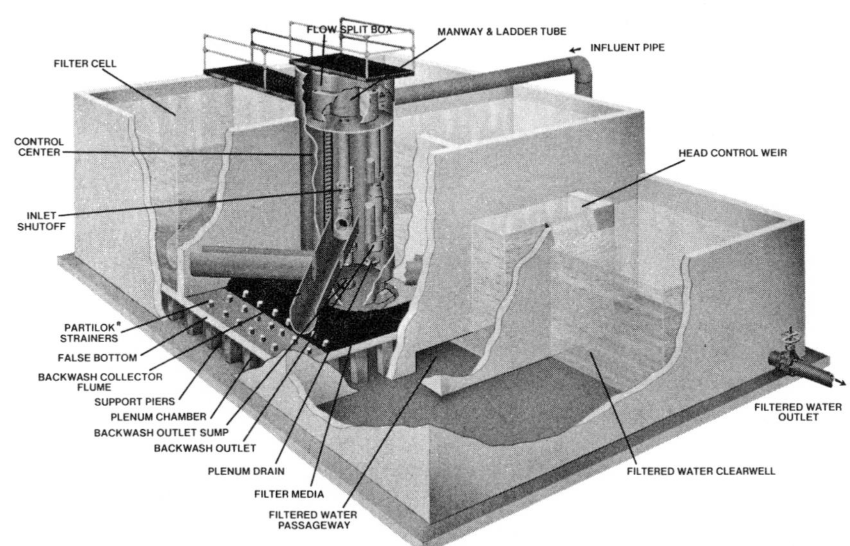

Graver's centrally controlled Mono-Pak Filter conserves building space, and simplifies the operation and maintenance of conventional or high rate rapid sand filters. The filter's factory-built central control includes all flow control and filter operating elements, accomplishing the work of individual filter control systems.

The control center can be installed in square or rectangular filter compartments, and can be operated automatically or by manual control. Substantial cost savings are possible because pipe galleries, large valves and complicated pneumatic flow controllers are eliminated.

BACKWASHING. Backwashing can be initiated manually by the operator, or automatically as a function of head loss, turbidity or time. The backwashing of filter compartments is accomplished with filtered water.

FLOW DISTRIBUTION. A simple hydraulic design for flow distribution is utilized to insure that every filter in the battery receives an equivalent load, helping to prevent mud balls and/or channeling.

New concepts, such as high rate filtration in depth, can be accomplished with minimum modification of the standard Concrete Mono-Pak Filter. Various types of media can be used, individually or in combination. If desired, the Concrete Mono-Pak Filter can be modified to incorporate Graver's unique Mono-Scour air washing system.

REACTIVATOR

Most solids-contact clarifiers rely on a mechanical drive, drive shaft, bearings and an impeller to achieve the solids recirculation necessary to obtain effective contact between incoming water, previously formed precipitates and treatment chemicals.

Graver's Hydro-Circ System achieves the same recirculation hydraulically, and guarantees the same effluent qualities as the Reactivator with mechanical recirculation.

The Hydro-Circ System takes available energy from the influent water and converts it into pumping energy. The resultant "inverted whirlpool" action pumps the previously formed precipitates up through the draft tube, and thoroughly mixes them with incoming water and treatment chemicals.

The recirculation rate is fully adjustable to permit optimum treatment results. Once set, generally at three to five times the influent flow, the Hydro-Circ System automatically adjusts to changes in loads from 20% to 110% of the design capacity in normal operation. The recirculation system is trouble-free, having no motor driven parts to malfunction, and thus no equipment maintenance problems.

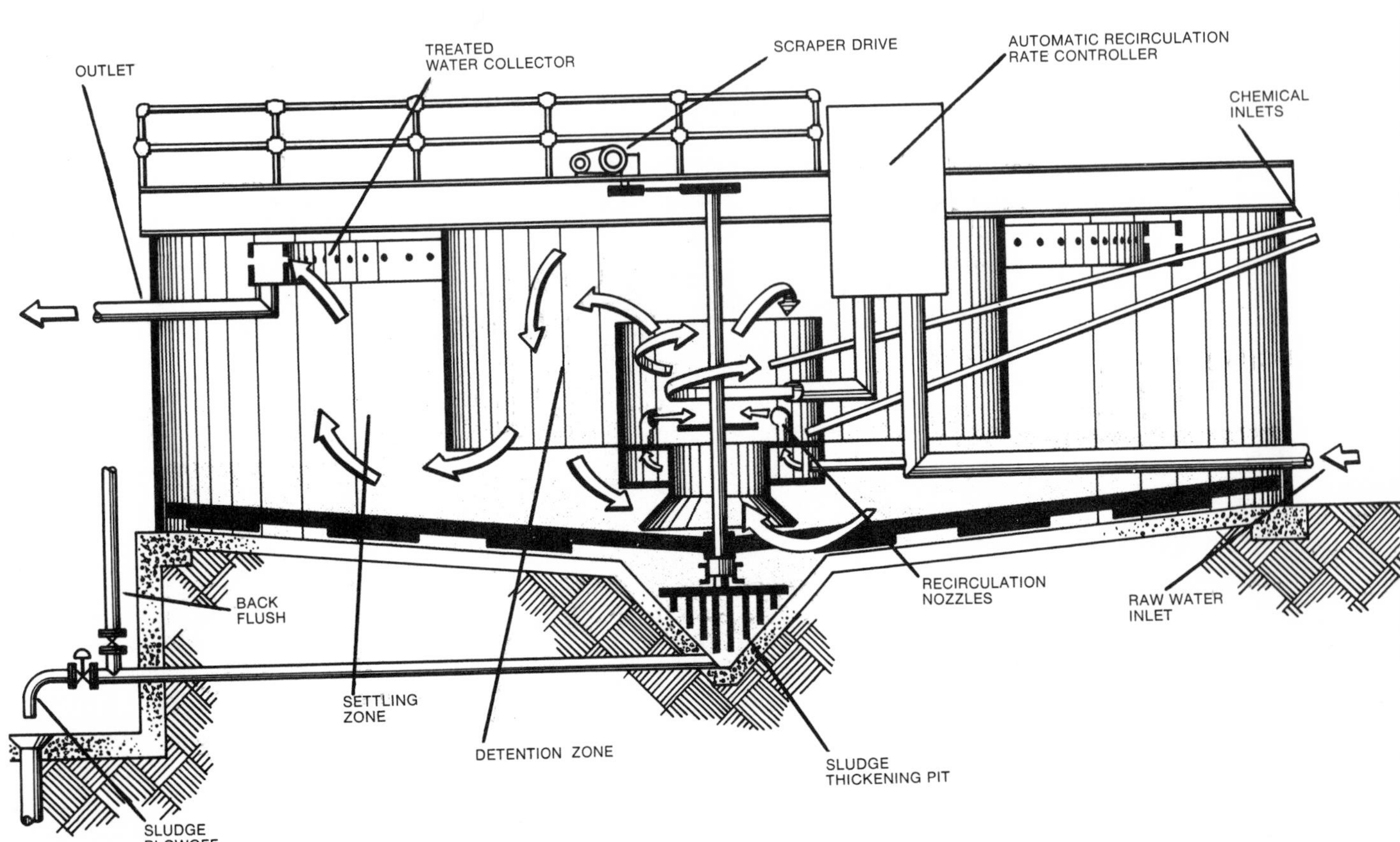

NORTH WATER FILTERS

"North" Filters are rotary, self-cleaning, gravity type screening machines that have been serving industry for over 40 years in more than 50 different applications. They are capable of coarse screening that ranges from 3000 microns up to fine screening of 25 microns. Their primary function is the separation of solids no matter how fine in combination with liquids. The removal of these solids results in clarified water that is easily disposed of, or polished water for use as process water in the plant. These machines are used for filtration purposes in both the influent and effluent end of industry.

The design and operating principle of "North" Filters is simple and efficient. Based on the principle of screening from inside the cylinder to outside, a wire mesh of any size covers the revolving cylinder. The cylinder itself rests and rotates on water lubricated bearings and is driven from a main drive shaft with precision cut pinions mated to segmental cut gearteeth. Within the rotating drum is a stationary and solid inclined trough around which the cylinder revolves. Lift paddles that extend the length of the cylinder, in its interior, pick up solids from the water and gravitationally drops them into an inclined pan. These solids then flow out the rear of the machine through an 8″ flange opening. The water within the cylinder filters through and around the fine wire mesh and is collected in a wooden or steel tank attached to the machine. This water is then used or disposed of. In all cases the influent of liquid and solid material can be gravitationally flowed or pumped into the interior of the machine. This is accomplished through a flanged aperture up to 20″ diameter inclusive. The entire machine is driven by a 3 H.P. gear motor, or separate motor and reducer, and either chain and sprocket driven, or direct driven with a coupling. The drive can be located on any of the four corners of the machine and the cylinder rotated in either direction. This allows for versatility in installation.

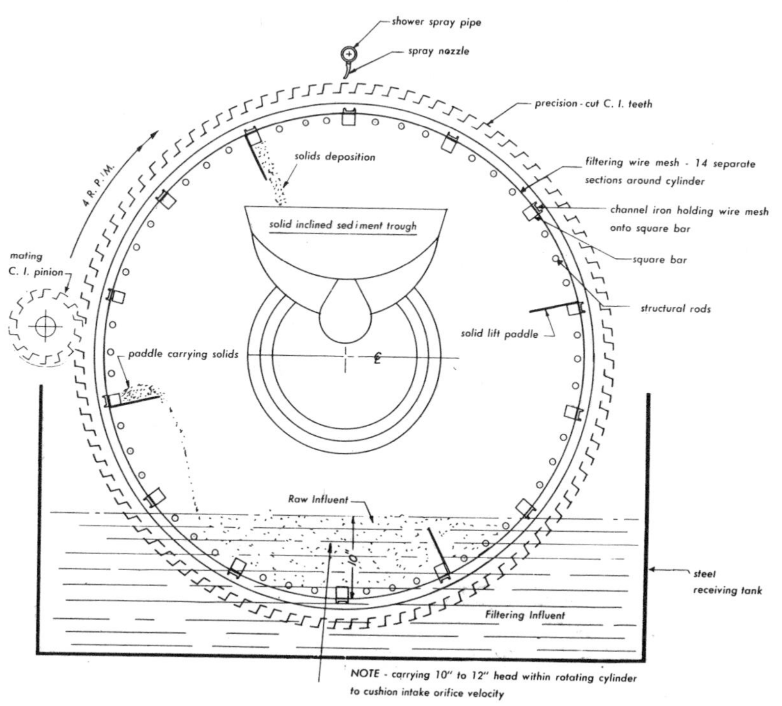

Cross-section of filter unit showing construction and operation. Solids are shown being picked up by paddles and deposited into solid inclined trough.

WIRE MESH DATA

SCREEN MESH NO. Square Weave	EQUIVALENT Corduroy Weave	SIZE OPENING In Inches	MICRON SIZE	% OPEN AREA
6		.126	3327	57.2
8		.090	2362	51.8
10		.068	1651	46.2
12		.060	1397	51.8
14		.051	1168	51.0
16		.045	991	53.0
20		.033	833	43.6
24		.0287	701	47.4
28		.0227	589	40.4
30		.0203	495	37.1
35		.0176	417	37.9
40	12 x 64	.0150	380	36.0
42		.0138	351	33.6
50		.0105	280	27.6
60	14 x 88	.0077	246	21.3
65		.0084	208	29.8
70		.0068	190	22.7
80	24 x 110	.0070	175	31.4

SCREEN MESH NO. Square Weave	EQUIVALENT Corduroy Weave	SIZE OPENING In Inches	MICRON SIZE	% OPEN AREA
90	20 x 120	.0056	159	25.4
100	20 x 150	.0055	147	30.3
115		.0051	124	30.0
120		.0046	120	30.7
130		.0043	115	31.1
140		.0042	109	34.9
150		.0041	104	37.4
160	20 x 200	.0038	96	36.4
170	20 x 250	.0035	88	35.1
180	20 x 300	.0033	82	34.7
200	20 x 350	.0029	74	33.6
120 x 330			70	
120 x 400			40	
120 x 600			30	
200 x 600			25	

Peripheral Screening Area of Machines

4′ machine	Cylinder 39″ dia. x 4′ long	47.33 sq. ft.
6′ machine	Cylinder 60″ dia. x 6′ long	108 sq. ft.
8′ machine	Cylinder 60″ dia. x 8′ long	143 sq. ft.
10′ machine	Cylinder 60″ dia. x 10′ long	178.50 sq. ft.
12′ machine	Cylinder 60″ dia. x 12′ long	214 sq. ft.

Weight of Screen Units

4′ machine	2400 lbs.
6′ machine	8400 lbs.
8′ machine	8700 lbs.
10′ machine	8900 lbs.
12′ machine	9200 lbs.

Horsepower

4′ machine	1½ H.P.
6′ machine	3 H.P.
8′ machine	3 H.P.
10′ machine	3 H.P.
12′ machine	3 H.P.

Examples of "North" Filter Capacities in Other Types of Industrial Applications

A 6′ filter with 60 mesh wire will handle 650 g.p.m. of water from Ross Economizers—depositing water directly into warm, fresh water chests for use in clear water chests, white water surge chests, head box and breast roll showers, wet felt showers, etc.

A 4′ filter covered with 30 mesh wire handles 600 g.p.m. of black liquor, removing wood chips that normally cause damage in liquor burners and Venturi scrubber nozzles. The filtered chips are returned to the pulp system prior to the brown stock washer. The clarified liquor from the "North" Filter is pumped to the weak liquor storage tank.

A 10′ filter covered with 60 mesh wire effectively screens 500 g.p.m. of back water from Fourdrinier machines containing 2 to 3 lbs. per thousand gallons of fibers. The clarified effluent can be used directly on the machine showers of paper machines.

A 6′ screen covered with 100 mesh wire handles 300 g.p.m. of concentrated slurry bark particles having a 3 to 4% consistency from the hydraulic barkers. These bark particles, retained by the "North" Filter, are sent to the power house for use as fuel. The clarified effluent is clear enough to discharge directly into the ocean.

A 10′ "North" Filter is used to screen spent liquor from chemical pulp washing. Flow varies according to particular conditions. Solids filtered out are returned to the head of the pulp washers, and clarified liquor goes into multiple effect evaporators.

"NORTH" SEWAGE DISPOSAL SCREENS

The "North" Sewage Disposal Screens separate solid and liquid constituents from waste materials at a location where they gravitationally flow or can be pumped into the screened cylinder. The machine's large drum rotates at 4 rpm. The lift paddles within the drum pick up the solid material from the water and deposit it into a stationary, perforated hopper within the cylinder. The hopper holds a spiral screw conveyor that moves the solids to the rear end of the machine, then out the discharge spout. In the process, it compresses the wastes and squeezes out more liquid which drains through the perforated hopper, back into the cylinder. The water in the cylinder drains through the wire mesh and collects in a steel or wooden tank which is part of the machine. The fine wire mesh is at all times kept free from clogging by a continuous spray pipe with jet nozzles, located above the rotating cylinder.

Applications: At-source installations for separation of solids from waste waters — meat processing, canning, grain washing, tanning, malting, dairy, woolen and paper mills, etc.

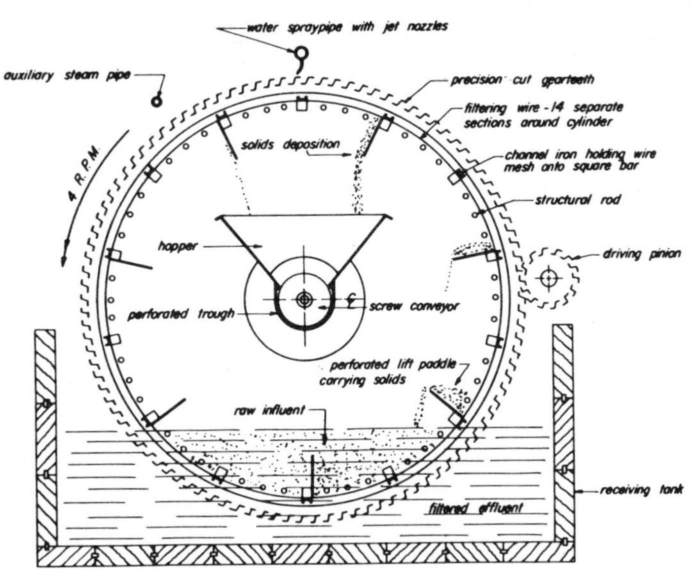

CROSS-SECTION OF SEWAGE SCREEN SHOWING CONSTRUCTION & OPERATION

Peripheral Screening Area of Machines

4′ machine	Cylinder 39″ dia. x 4′ long	47.33 sq. ft.
6′ machine	Cylinder 60″ dia. x 6′ long	108 sq. ft.
8′ machine	Cylinder 60″ dia. x 8′ long	143 sq. ft.
10′ machine	Cylinder 60″ dia. x 10′ long	178.50 sq. ft.
12′ machine	Cylinder 60″ dia. x 12′ long	214 sq. ft.

Weight of Screen Units

4′ machine	2400 lbs.
6′ machine	8400 lbs.
8′ machine	8700 lbs.
10′ machine	8900 lbs.
12′ machine	9200 lbs.

Horsepower

4′ machine	1½ H.P.
6′ machine	3 H.P.
8′ machine	3 H.P.
10′ machine	3 H.P.
12′ machine	3 H.P.

RAW INTAKE AND COOLING TOWER WATER TABLE

APPROXIMATE CAPACITIES OF NORTH ROTARY SCREENS WITH SQUARE WEAVE WIRES

SCREEN	40 Mesh 380 Microns		60 Mesh 246 Microns		80 Mesh 175 Microns		100 Mesh 147 Microns		120 Mesh 120 Microns		150 Mesh 104 Microns		200 Mesh 74 Microns	
SIZE	G.P.M.	G.P. Day	G.P.M.	G.P. Day	G.P.M.	G.P. Day	G.P.M.	G.P. Day	G.P.M.	G.P. Day	G.P.M.	G.P. Day	G.P.M.	G.P. Day
4′	1,000	1,440,000	800	1,152,000	700	1,008,000	500	720,300	350	540,000	200	288,000	150	216,000
6′	4,500	6,480,000	3,500	5,040,000	2,777	4,000,000	2,000	3,280,000	1,200	1,728,000	750	1,080,000	500	720,000
8′	5,900	8,496,000	4,900	7,056,000	4,166	6,000,000	3,500	5,040,000	2,900	4,176,000	2,200	3,168,000	1,500	2,160,000
10′	8,000	11,520,000	7,000	10,080,000	6,250	9,000,000	5,300	7,632,000	4,700	6,768,000	4,000	5,760,000	2,500	3,600,000
12′	10,000	14,400,000	9,000	12,960,000	8,000	11,520,000	7,000	10,080,000	6,000	8,640,000	5,200	7,488,000	3,500	5,040,000

APPROXIMATE CAPACITIES OF NORTH ROTARY SCREENS WITH CORDUROY WEAVE WIRES

SCREEN	14 x 88 Mesh 246 Microns		24 x 110 Mesh 175 Microns		20 x 150 Mesh 147 Microns		20 x 200 Mesh 95 Microns		20 x 250 Mesh 88 Microns		20 x 300 Mesh 82 Microns		20 x 350 Mesh 74 Microns	
SIZE	G.P.M.	G.P. Day	G.P.M.	G.P. Day	G.P.M.	G.P. Day	G.P.M.	G.P. Day	G.P.M.	G.P. Day	G.P.M.	G.P. Day	G.P.M.	G.P. Day
4′	800	1,152,000	700	1,008,000	500	720,300	190	273,600	175	252,000	160	230,400	150	216,000
6′	3,500	5,040,000	2,777	4,000,000	2,000	3,280,000	700	1,008,000	640	921,600	580	835,200	500	720,000
8′	4,900	7,056,000	4,166	6,000,000	3,500	5,040,000	2,000	2,880,000	1,850	2,664,000	1,675	2,412,000	1,500	2,160,000
10′	7,000	10,080,000	6,250	9,000,000	5,300	7,632,000	3,800	5,472,000	3,400	4,896,000	2,900	4,176,000	2,500	3,600,000
12′	9,000	12,960,000	8,000	11,520,000	7,000	10,080,000	5,000	7,200,000	4,500	6,480,000	4,000	5,760,000	3,500	5,040,000

APPROXIMATE CAPACITIES OF NORTH ROTARY SCREENS WITH MICRONIC WEAVE WIRES

SCREEN	120 x 330 Mesh 70 Microns		120 x 400 Mesh 40 Microns		120 x 600 Mesh 30 Microns		200 x 600 Mesh 25 Microns	
SIZE	G.P.M.	G.P. Day	G.P.M.	G.P. Day	G.P.M.	G.P. Day	G.P.M.	G.P. Day
4′	140	201,600	80	115,200	60	86,400	55	79,200
6′	470	676,800	400	556,000	340	489,600	300	432,000
8′	1,300	1,872,000	800	1,152,000	600	864,000	500	720,000
10′	2,200	3,168,000	1,400	2,016,000	1,000	1,440,000	750	1,080,000
12′	3,100	4,464,000	2,000	2,880,000	1,500	2,160,000	1,000	1,440,000

Linear Equivalents

1 inch = 25.4 millimeters = 25,400 microns

1 millimeter = .0394 inches = 1000 microns

1 micron $= \frac{1}{25,400}$ *of an inch = .001 millimeter*

1 micron $= 3.94 \times 10^{-5}$ *= .000039 inches*

OVERHEAD SHOWERPIPE GALLONAGE

"SPRAYCO" NOZZLES—No. ¼ P5010—5/64" ORIFICE

Pressure	15#	20#	30#	40#	60#	80#	100#
	G.P.M.	G.P.M.	G.P.M.	G.P.M.	G.P.M.	G.P.M.	G.P.M.
4′ screen 13 nozzles	7.9	9.2	11.3	13.0	15.6	18.2	20.8
6′ screen 20 nozzles	12.2	14.2	17.4	20.0	24.0	28.0	32.0
8′ screen 27 nozzles	16.5	19.2	23.5	27.0	32.4	37.8	43.2
10′ screen 33 nozzles	20.1	23.4	28.7	33.0	39.6	46.2	52.8
12′ screen 40 nozzles	24.4	28.4	34.8	40.0	48.0	56.0	64.0

Grefco, Inc., Dicalite Division

DIATOMITE FILTERAIDS

WIDE PERFORMANCE RANGE

Diatomite filteraids are practically pure amorphous silica, containing a small amount of microcrystalline material. They are processed in the temperature range of 1500° to 2000° F, which eliminates organic matter. Both calcined and flux-calcined diatomite are non-adsorptive. Dicalite diatomite filteraids have extremely low solubilities in mineral and organic acids at both low and high temperatures. Solubilities in strong alkalis are relatively low and will vary according to temperature and time of contact. Calcined diatomites indicate a pH range of 6.7 to 7.5 and flux-calcined products show a pH range of 9.0 to 10.5 in a 10% slurry. The three kinds of Dicalite diatomite filteraids produced are natural, calcined and flux-calcined.

OPTIMUM FLOWRATES

All Dicalite diatomite filteraids afford maximum filter throughput with consistent clarity at minimum filteraid dosage. This high performance holds down operating costs and maintains finished product quality. A further mark of Dicalite filteraid quality is the ability to cope with the occasional "tough batch" of hard-to-filter liquor with which nearly every plant must contend from time to time.

How Filteraids are Used:

In general there are three different modes of filtration employing filteraids. These are: 1) Fixed Bed Filtration; 2) Body Feed Filtration; 3) Rotary Precoat Filtration. In all three modes of operation either pressure or vacuum may be employed.

I. FIXED BED FILTRATION. A layer of filteraid is applied to the filter septum and used as the filtering medium. As the word "fixed" indicates, no filteraid is added nor is any removed during the cycle. Principal applications are those where the amount of suspended solids is very low.

The fixed bed is applied by recirculating a thin slurry of filteraid and clean liquor through the system until all the filteraid is deposited on the filter septum. Relatively thick layers are used, varying from ⅛″ to ⅜″. Filteraid quantities are as follows: 20 lbs. to 50 lbs. of diatomite, perlite or Diabestos, or 10 lbs. to 35 lbs. of Solka-Floc filteraid per 100 sq. ft. of filtering area.

Selection of the filteraid is based upon using the coarser, more porous grades since the desired effect is penetration of the filterable solids into the filteraid layer beyond its surface. In this way filtration in depth is secured, utilizing the pores in the full thickness of the bed to trap the solids.

Sometimes a thin first coat is applied using a fine filteraid grade next to the septum, followed by a thicker layer of a coarser grade. Thus, colloidal type solids do not penetrate through the cake, but are stopped by the finer filteraid layer before reaching the septum to plug it. Another useful variation is a partial coat of Dicalite diatomite or perlite, or Solka-Floc or Diabestos filteraid, topped by a layer of activated carbon. In this way, both clarification and bleaching of a liquid are secured at the same time.

Principal applications of the fixed bed mode are in trap filtration of aqueous and other low viscosity liquids where the proportion of filterable solids is very low – usually 100 p.p.m. or less. Filters for the many thousands of small home swimming pools (not large commercial pools) usually are operated in this manner. Another example is "insurance" filtration to make sure of removing any unwanted solids which may have entered into or precipitated from a process liquor left standing for some time. Fixed bed filtration eliminates filteraid feeders and high flowrates with satisfactory clarity are secured.

II. BODY FEED FILTRATION, the most widely used method, is so called because filteraid is added to the main body of the liquid.

It is good filtration practice to apply a precoat to the filter septum rather than to recirculate from the main batch of dirty liquor. Precoats insure early clarity and prevent plugging of metal or cloth septums.

Precoats acting as filter septums in body feed filtration are applied using 5 lbs. to 15 lbs. per 100 sq. ft. of filtering area of Dicalite diatomite, perlite or Diabestos filteraids. Using Solka-Floc, filteraid is applied in the ratio of 2 lbs. to 10 lbs. per 100 sq. ft. of filtering area. Thickness of these precoats ranges from 1/32″ to 3/32″. Usually the same grade of filteraid is used for both the precoat and the body feed, except when a Solka-Floc or Diabestos precoat is used.

The use of the cellulose filteraid Solka-Floc for a precoat is the most positive way to coat a metal screen or loosely woven cloth septum. This will eliminate any subsequent bleed-through of diatomite or perlite filteraid used as a body feed. In effect, Solka-Floc "papers" the septum. Even on closely woven cloth septums, an initial coat of Solka-Floc is beneficial to eliminate cloth plugging and blinding, and will extend the filter cloth life. Cleaning is easier and faster. A Solka-Floc precoat, in many instances, can result in cost savings of ⅓ to ½ that of diatomite or perlite, plus faster precoating time and elimination of bleed-through.

Body feed filtration is used to increase the filterability of a liquid to afford maximum length cycles with uniformly high clarity. Choice of the best type and grade as well as the quantity of filteraid is vitally important.

After the precoat application is finished the main batch of liquid to be filtered is pumped into the filter. This process liquid containing both filteraid particles and suspended solids reaches the filter septum (or precoat surface) and the filter cake begins building up.

For a batch process, body feed addition of filteraid either dry or in a slurry is satisfactory, with dry addition being most common. Slurry addition is generally used when batch tanks are under pressure. Continuous or long cycle operations require proportioning equipment capable of maintaining the same percentage of addition throughout the operation. Slurry addition is most often used here.

Whatever the feeding method or equipment, the chief requirement is that filteraid be added uniformly, both as to quantity and rate. Agitation should be provided in any tank where the filteraid is suspended in the liquid, whether this be a slurry feeder or the tank for the main batch of liquid being fed to the filter. All types of filteraids will settle out in time without agitation, and it is required for uniform feeding. Over-agitation should be avoided—only enough to keep the filteraid in suspension is required.

A major advantage of body feed filtration is that the permeability of the cake is maintained over a longer period than is possible with fixed bed filtration. This is because the "working" surface of the filter cake is constantly being renewed. The suspended solids and the filteraid particles collect together, so that the solids are separated and prevented from forming a layer impervious to flow of the liquid. Thus cake resistance to flow is greatly reduced. Another advantage of body feed filtration is that, if desired, the rate of filteraid feed may be changed at any time during the cycle. It can be increased to increase flow, or decreased for economy if clarity considerations permit.

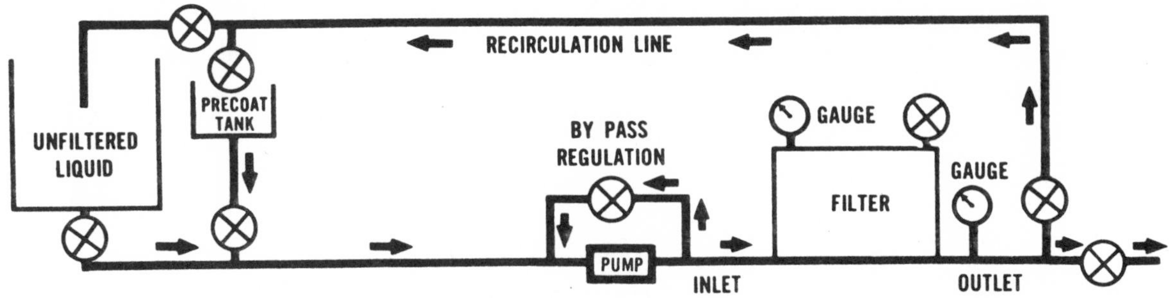

FIGURE 4 — *Schematic sketch of a typical filtration system such as employed in fixed bed and body feed filtration.*

III. ROTARY PRECOAT FILTRATON employs equipment utilizing a rotating drum to which external pressure or internal vacuum is applied. Figures 9 and 10 illustrate and diagram a typical system. The precoat is in reality a cake pre-formed on the drum using a slurry of clean liquid and filteraid. This slurry is recirculated through the system until the precoat cake is 2" to 4" thick, depending on conditions.

Efficient operation requires certain precautions when applying the precoat. A slurry containing 2% to 5% of filteraid applying an average precoat 2½ inches thick in 45 minutes to one hour will normally give best results. Too rapid precoating rate and too high filteraid concentration in the slurry can cause excessive cake shrinkage and cracking, although small hairline cracks are normal. Too rapid flow, causing a washing action by currents in the slurry tank against parts of the drum area, can result in uneven cake thickness. Dirty cloths or screen septums have the same effect.

During the operating cycle following application of the precoat, passage of the process liquid through this precoat leaves the filterable solids in a relatively thin film which just penetrates the precoat surface. A mechanically operated knife blade continuously shaves off this film, cutting just deep enough to remove the filtered-out solids and leave a clean, fresh filtering surface. The knife is usually set to cut between .001" and .010" deep, with the average setting for .002" to .005". Selection of the proper depth of cut depends mostly on the nature and quantity of the filterable solids, and the degree of penetration into the cake. The selection of the grade of filteraid is most important as discussed on the next page.

The most economical drum speed is determined largely from the characteristics of the filterable solids. Lower drum speeds usually produce higher gallons of filtrate per filteraid dollar, but also produce lower flowrates. As a general rule, the slowest drum speed compatible with production requirements is recommended. Speeds commonly range from 1 to 1/5th revolution per minute.

Flowrate in gallons/sq. ft./hour is a most important index of filteraid performance in rotary vacuum precoat filtration. For any given operation the flowrate depends on the knife cut, drum speed, grade and type of filteraid used, the filtration resistance of the suspended solids, and the viscosity of the process liquid. In cases where the flowrate is dictated by production requirements the drum speed and the knife cut must be regulated to give the needed throughput. This may result in higher filteraid consumption, but may be justified by peak demands, lack of sufficient filter capacity, or a temporary run of hard-filtering liquor. Another way to increase flowrate (again at the expense of filteraid economy) is to add filteraid to the charge liquor. In many cases the purchase of additional filter equipment proves most economical on a long-term basis, since advantage is gained in economy of filteraid consumption through the use of slower drum speeds.

Selection of the best grade and type of filteraid is based upon determining the one which will minimize penetration of the suspended solids into the precoat cake so as to make possible the thinnest cuts to maintain desired flow.

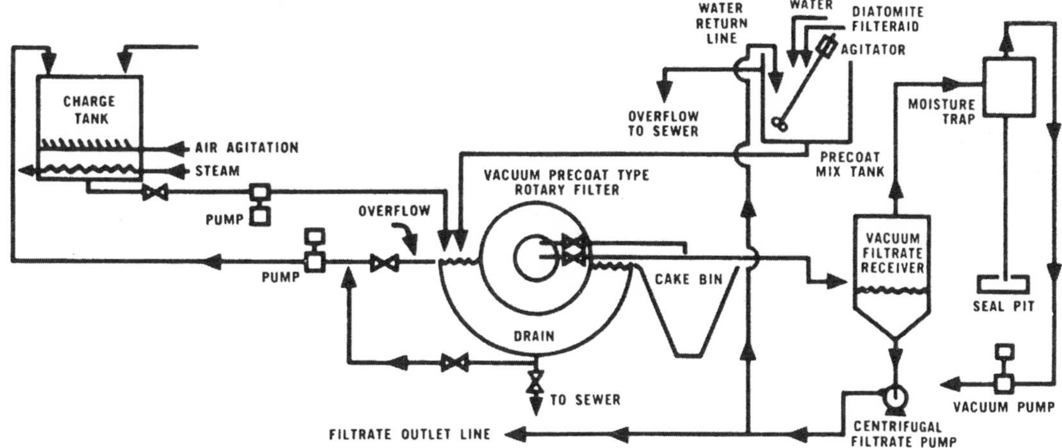

COMPARATIVE FILTERAID CHARACTERISTICS

PRODUCT	COMPOSITION	Number of GRADES AVAILABLE	RELATIVE FLOW RATIO RANGE	PRIME ADVANTAGES AND APPLICATIONS	SOLUBILITY	
					IN ALKALIS (room temp.)	IN ACIDS (room temp.)
DICALITE DIATOMITE FILTERAIDS	Silica	11	1-20	General Filtration for Highest Clarity	Slight in dilute alkalis	Slight in dilute acids
DICALITE PERLITE FILTERAIDS	Glassy Silicate	7	1-17	Outstanding on rotary vacuum. Reduced dosage on both pressure and vacuum filters.	Slight in dilute alkalis	Slight in dilute acids
DICALITE DIABESTOS® FILTERAIDS	Diatomite or Perlite & Asbestos	Numerous	1-20	Precoats coarse screen septums. Prevents bleed-through of solids.	Slight in dilute alkalis	Slight in dilute acids
SOLKA-FLOC® FILTERAIDS	Cellulose	8	7-15	Excellent for precoating coarse screens. Highest purity for absorption of oil from condensate, & removal of iron from caustic.	Slight in dilute or strong alkalis.	Negligible in dilute acids.
DICALITE DIA-FLOC® FILTERAIDS	Diatomite and Solka-Floc	Numerous	1-20	Precoats coarse screen septums. Prevents bleed-through of solids.	Slight in dilute alkalis	Slight in dilute acids

RELATIVE FLOWRATES

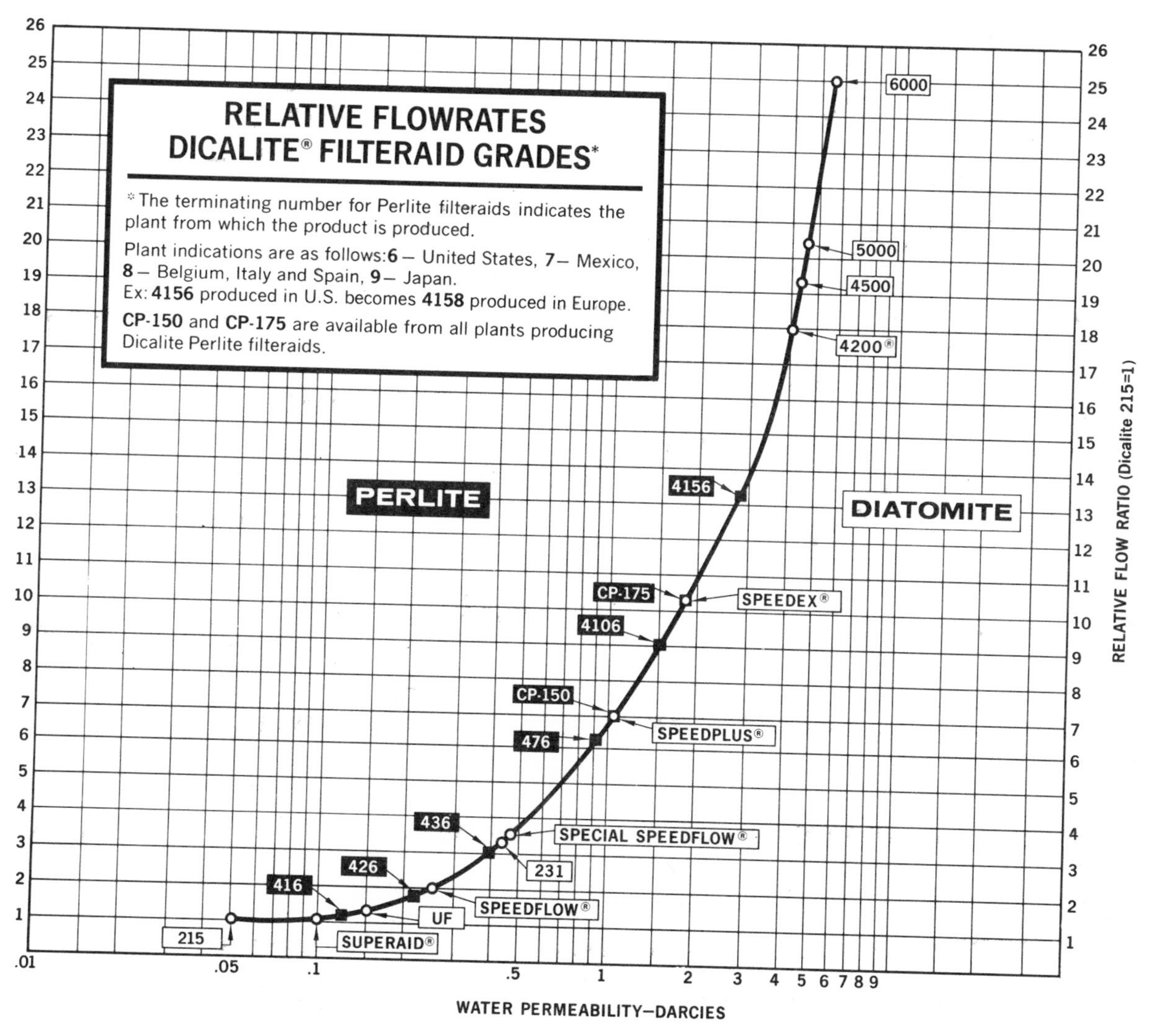

Comparative flowratios of both perlite and diatomite filteraid grades are shown on the above chart.

DIATOMITE PHYSICAL CHARACTERISTICS

Color . . . Calcined, buff to pink; flux-calcined, white

Organic Matter 0.0%

Dry Loose Weight . . . . 9-12 lbs./ft.^{3}

Moisture 0-0.5%

Particle size and size distribution closely controlled according to grade (see next page).

Specific Gravity . . . . . . Calcined, 2.2; Flux-calcined, 2.3

DIATOMITE CHEMICAL CHARACTERISTICS

Dicalite diatomite filteraids are practically pure amorphous silica, containing a small amount of microcrystalline material. They are processed in the temperature range of 1500° to 2000° F, which burns off organic matter. Both calcined and flux-calcined diatomite are non-adsorptive. Dicalite diatomite filteraids have extremely low solubilities in inorganic and organic acids at both low and high temperatures. Solubilities in strong alkalis are relatively low and will vary according to temperature and time of contact. Calcined diatomites indicate a pH range of 6.7 to 7.5 and flux-calcined products show a pH range of 9.0 to 10.5 in a 10% slurry.

Grefco, Inc., Dicalite Division

DICA-SORB

Dica-Sorb is a high-bulk sorbent material that combines inert mineral particles with specially treated fibers. It has the unique ability to repel water and yet absorb from 9 to 15 gallons of petroleum products per bag, depending on viscosity. It floats on water and quickly combines with oil through capillary action to form a cohesive, mat-like mass which can easily be removed from the surface.

Dica-Sorb holds oil tightly and continues to float even when fully saturated. Inert and non-toxic, Dica-Sorb is not hazardous to fish or to other life in treated areas.

Applications

Dica-Sorb can be used wherever oil or oleaginous matter is stored, transported or consumed, as well as for emergency use by fire-fighting and highway safety forces.

Since Dica-Sorb does not absorb water, it is especially useful in marine dock and harbor areas. Dica-Sorb removes floating petroleum waste in ship bilges, and makes an ideal filtering medium to cleanse tanker bilge water.

Dica-Sorb has also been used successfully by municipal and industrial water treatment plants to remove oily waste as a first step in water purification.

How to Use Dica-Sorb

Dica-Sorb weighs as little as four and one half pounds per cubic foot. It is conveniently packaged in poly-lined bags for ease in handling and long storage life. One bag is easily handled by one man. To use, simply open the bag and spread the Dica-Sorb over the oil-contaminated surface.

The oil will be absorbed rapidly and will continue to be absorbed until the material is fully saturated with oil waste. Agitation by stirring, pressure hosing or prop-washing will increase the rate of absorption.

Disposal

When the Dica-Sorb is fully saturated, the mass will coagulate and can be collected against a boom, mesh fence, spillway gate, or other restraining surface. It may be removed from the water by pumper or vacuum truck, seining, shoveling or screening. The absorbed oil will not be released to water, beach area, or to soil in disposal areas.

Dica-Sorb Features

1. Dica-Sorb is non-toxic and harmless to marine life.
2. Dica-Sorb absorbs oil and oil derivatives at the rate of 9 to 15 gallons per bag, depending on viscosity.
3. Dica-Sorb will not absorb water.
4. Dica-Sorb will not release absorbed oil.
5. Dica-Sorb floats even when saturated.
6. Dica-Sorb is easy to collect.
7. Dica-Sorb is available Coast to Coast.

Gulf Degremont, Inc.

Reverse Osmosis Systems

Gulf Degremont's GD-ROTU-1 system is the ideal test unit for obtaining accurate data for the design of full scale water and waste treatment systems using the reverse osmosis process. The unit consists of a three-foot pressure vessel, plunger feed pump with motor and belt drive, pressure relief valve, feed pressure gauge and vessel differential pressure gauge. Requiring only a 230/460 volt, three-phase, 60Hz power supply, the 225-pound unit utilizes a commercially accepted spiral-wound membrane module of production size, which facilitates the dimensioning of the full scale unit. When operated on 77°F. feedwater at 400 psig, the GD-ROTU-1 produces approximately 700 gallons per day of permeate.

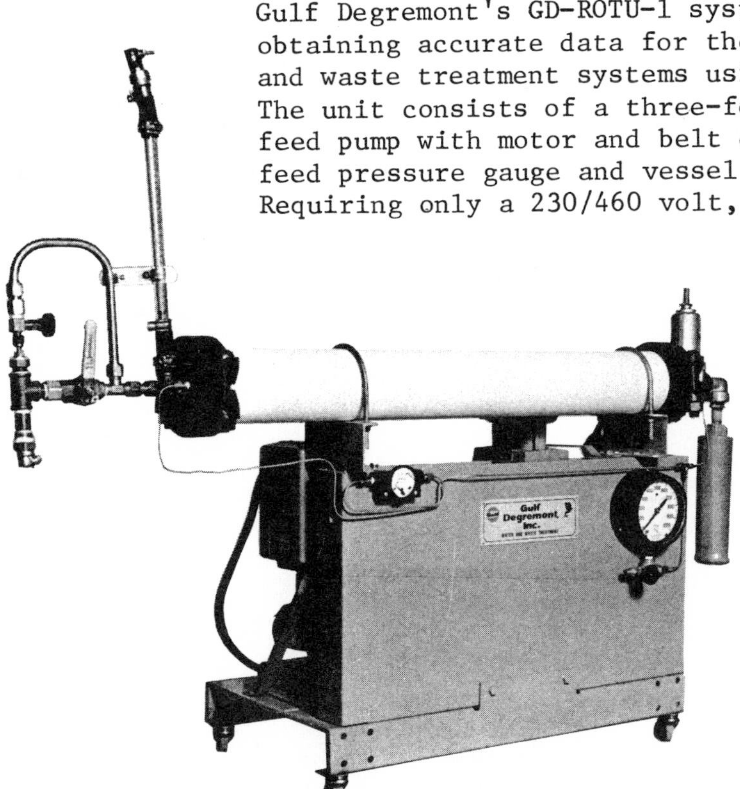

Pilot Unit GD-ROTU-1

Specifications

The GD-ROTU-1 test unit consists of:

1. One 3 ft. long 4" diameter polymer-lined pressure vessel sized to take one ROGA Model 4100 module, complete with end caps secured by Victaulic couplings.
2. Plunger feed pump complete with 2 HP motor and Vee belt drive.
3. Pressure relief valve and flow pulsation dampener.
4. Feed pressure gauge.
5. Vessel differential pressure gauge.
6. Flow indicators on product streams.

RECOMMENDED OPERATIONAL PARAMETERS		
Pressure	(Recommended) (Maximum Feed Pressure)	400 psig 500 psig
Feed Condition	(pH range)	3-6
Temperature	(Range °C)	10-35
Flow Rates	(Feed into unit) (Concentrate out of unit)	5 gpm max. 3 gpm min.

Gulf Degremont, Inc.

BI-FILTER D.C. Hi-Rate, Two-Stage, Deep-Bed Filtration

Gulf Degremont's two compartment, deep-bed, high rate pressure filter system, the Bi-Filter D.C., answers the problem of filtering industrial and sewage plant effluent containing heavy concentrations of suspended solids. The Bi-Filter D.C. combines both roughing (or prefiltration) across a coarse media and polishing across fine sand, thus eliminating reclassification of the separately compartmented media. Optimum flow rates are maintained through each section with a minimum use of backwash water.

Normal performance range for the system involves influent loading up to 250 ppm. Effluent quality can be tailored to almost any process demand.

A special Vibrair underdrain nozzle for use in high rate filtration systems has been developed. During downflow filtration, the nozzle is compressed with the flow of water passing through carefully controlled spaces between spring coils. These spaces are expanded by the reverse flow during backwash and release any trapped dirt.

The heart of the Bi-Filter D.C., the Vibrair Strainer, is a highly efficient underdrain unit for collection of filtered water and distribution of backwash water and scouring air. Constructed of a cast aluminum body and a stainless steel spring, it is installed at high density spacing of 6 to 10 units per four square foot section. Spacing is varied to meet flow conditions.

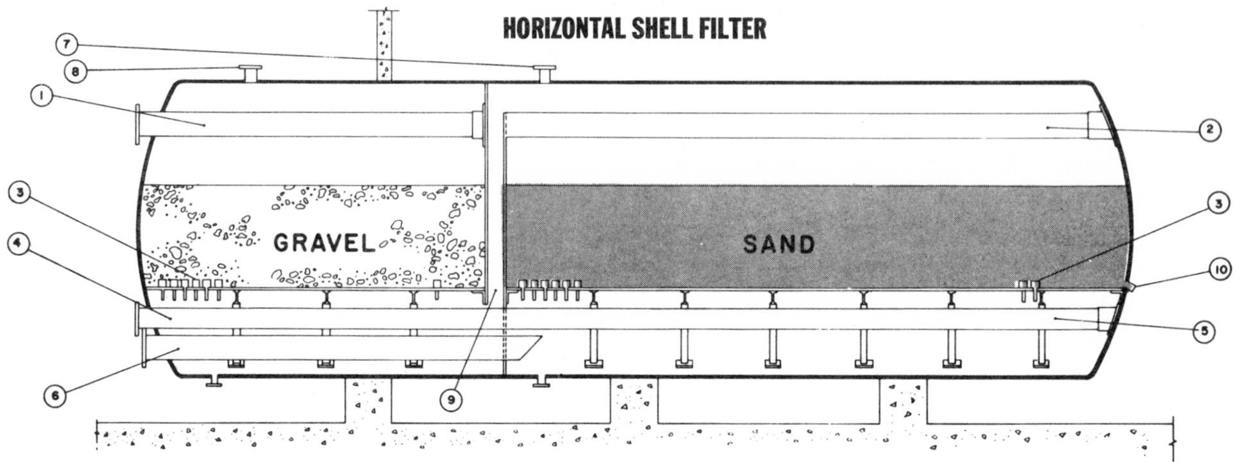

The first bed, coarse media with a diameter of about 1.5 to 4.0 mm, occupies about one-third of the vessel flow area. The fine media, usually about 1.0 mm sand, fills the remaining two-thirds of the vessel. The combination of high flow rates and coarse media drive the impurities deep into the bed, thus lessening the load on the fine media by 25 to 45%. The fine media is used for polishing filtration of the remaining impurities at one-half the rate per square foot of the coarse media.

Gulf Degremont, Inc.

AQUAZUR FILTER UNDERDRAINS

Aquazur underdrains are the result of extensive research and development on sand bed filtration. Construction features of the concrete slab sections, supports and long stem nozzles, insure a high degree of filter efficiency and hydraulic distribution.

Suggested specifications—Gulf Degremont Aquazur filter underdrains have three inch thick double grid reinforced slabs measuring two square feet and are fitted with 25 long stem polypropylene air diffuser nozzles. Spools precast in the slabs allow nozzle installation. Supported by four concrete piers or lateral beams, each slab is anchored by corner tie down bolts and side wall clips.

Degremont long stemmed polystyrene nozzle with sealing ring for reinforced concrete floor.

Section of long-stem nozzle during wash period.

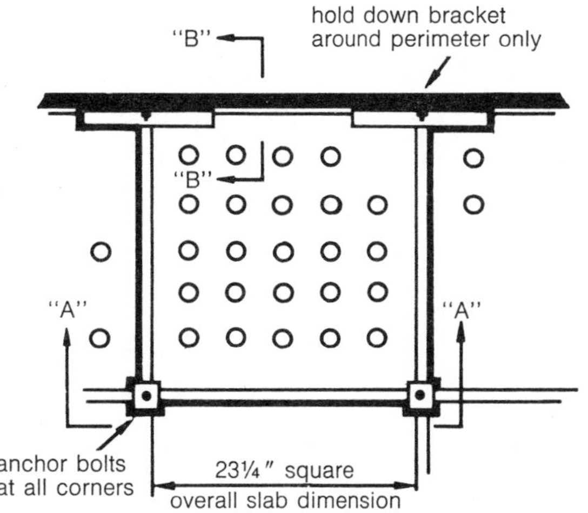

Made of rugged injection molded polyethylene, the Aquazur nozzle eliminates the need for gravel support layers and insures even distribution of air and water during the wash cycle. The air-water wash sequence conserves wash water, improves sand bed cleaning and reduces wash water pollution load.

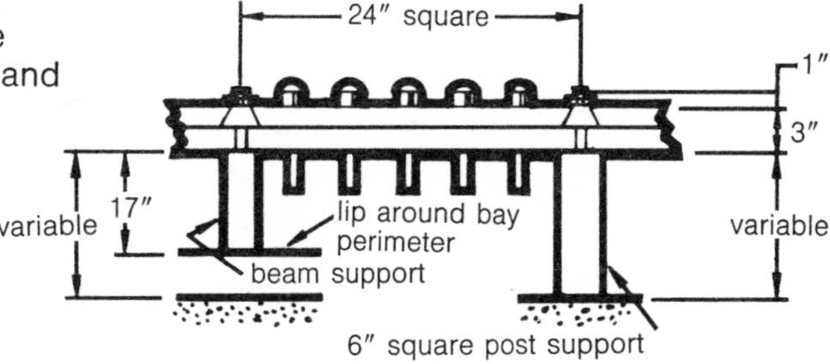

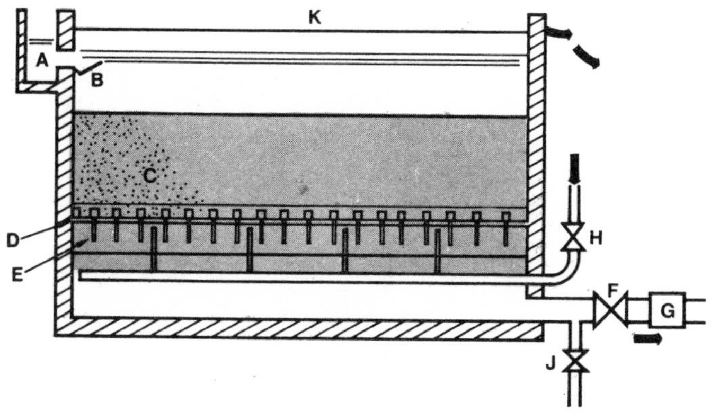

Aquazur Filter Schematic

A. Filter influent channel
B. Filter influent control valve
C. Filter sand bed
D. Underdrain floor slabs
E. Gulf Degremont nozzles
F. Filter effluent valve
G. Filter effluent controller
H. Air scour inlet valve
J. Backwash water inlet
K. Wash trough wier

Gulf Degremont, Inc.

VIBRAIR IV AIR DIFFUSER for Sewage Aeration - Gas Diffusion

Operation: Air under pressure enters the diffuser head; the coil springs expand, allowing air to escape; air bubbles discharge into surrounding liquid. Air bubble generation is enhanced by spring vibration. When the air supply is stopped, the springs contract, thus preventing suspended matter from entering the diffuser and piping.

Gulf Degremont air diffusers can be applied to any of the sewage and biological waste processes where air, enriched-air or oxygen is used in the treatment process. They are widely used in Gulf Degremont Oxycontact and Dia-Pac treatment plants and may be efficiently used in aerated grit chambers, pre-aeration tanks, activated sludge aeration tanks and aerobic digesters when mounted on piping ramps or pipe grid systems.

Vibrair IV Diffusers are made of die cast aluminum body and cover with stainless steel spring to insure long life and a high degree of service efficiency. Spring characteristics allow for the generation of efficient size bubbles and diffusion at varying air flows.

Alternate saddle mounting

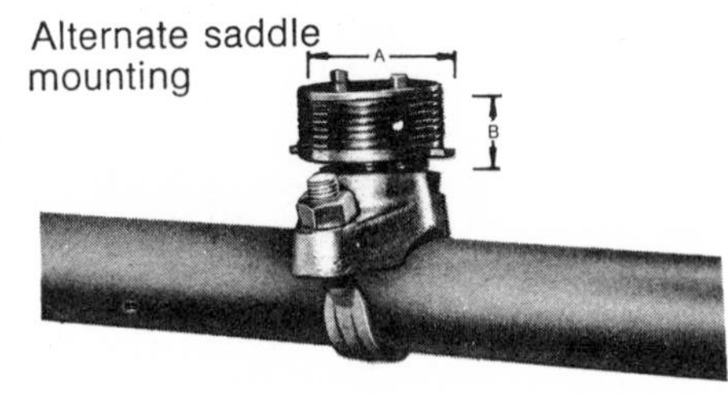

	PIPE CONNECTION	A	B	AIR CAPACITY
Vibrair IV	3/4″ or 1″ N.P.T.	3½″	3″	7 Cfm

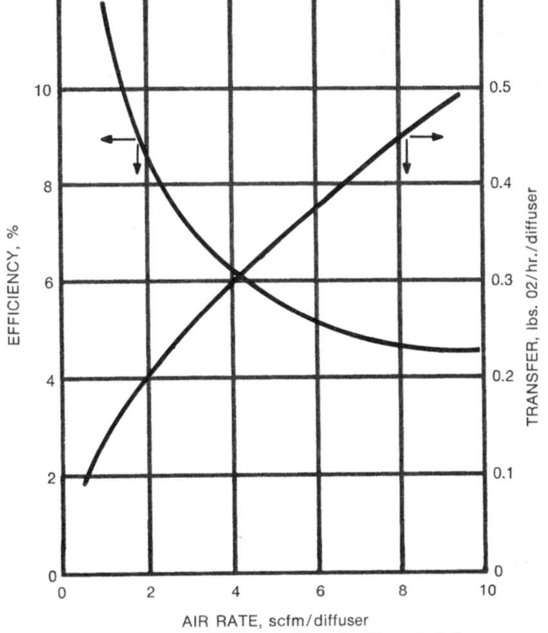

DC-4 AIR DIFFUSER Oxygen Transfer Characteristics.
(Atmospheric air at 7 psig).

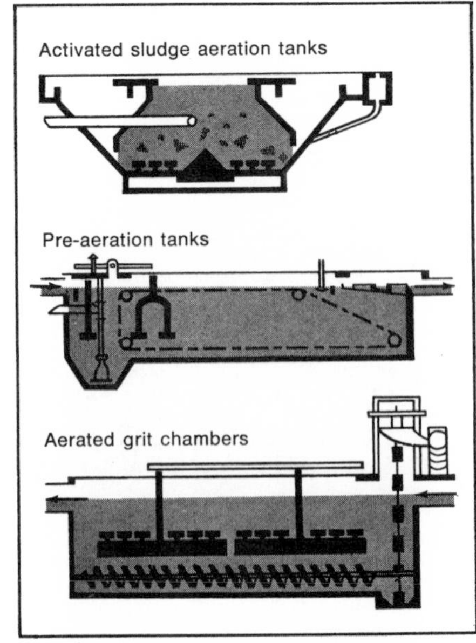

DEGREMONT PULSATOR CLARIFIER

Gulf Degremont's Pulsator is an upflow, rapid rate, sludge blanket type clarifier. Unlike conventional solids contact reactors, the Pulsator does not require any moving mechanical parts under water, resulting in a system with minimal downtime and a consistently higher level of efficiency. The Pulsator clarifier produces a gentle agitation of the floc sludge blanket by pulsating a small amount of water in a central vacuum chamber. Because an absolutely vertical flow of water is obtained, freshly flocculated feed is in contact with the existing sludge blanket to provide nuclei for growth. A Pulsator can be turned down or shut off while still pulsing, in order to retain the sludge blanket, thus providing quick restarting.

The Pulsator clarifier can be contained in any open, flat bottom tank of any configuration and is readily adaptable to existing installations.

Because of its inherent design features, the Pulsator can reduce color and turbidity more effectively than any other process.

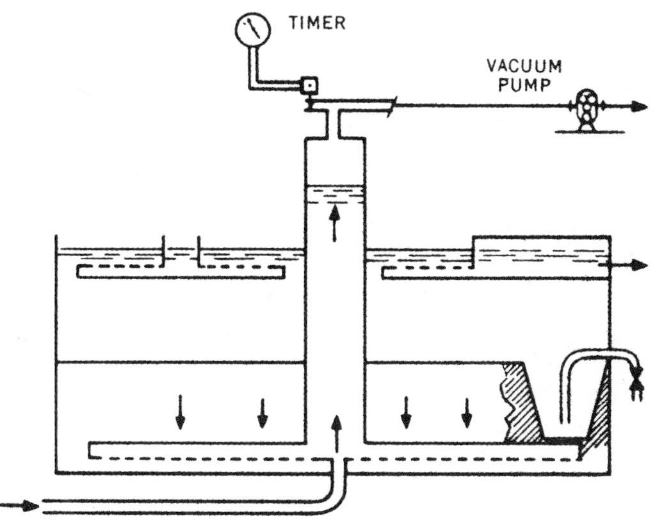

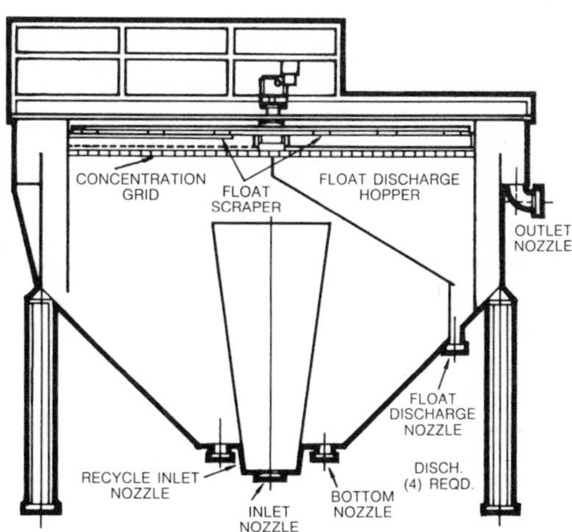

FLOTTAZUR AIR FLOTATION UNITS

Gulf Degremont's Flottazur air flotation units incorporate a new technique of air introduction and recycle water together with a novel method of float removal which increases solids concentrations up to 7%.

REVERSE OSMOSIS

Reverse Osmosis is a continuous, reliable, demineralization and concentration process requiring little operating attention. The heart of the process is the semipermeable GESCO membrane manufactured by Gulf Environmental Systems Company. The membrane has the inherent capability to permit water passage but acts as a highly effective barrier to dissolved solids and suspended solids.

The schematic shown below presents the use of Reverse Osmosis to demineralize or reduce the level of dissolved solids to meet various water requirements of an industrial plant.

Reverse Osmosis can be used for pollution control by reclaiming useable water from plant waste streams or by concentrating process waste streams for recovery of constituents or for other disposal.

A GESCO system can provide simultaneous benefits in the form of waste stream concentration and reclamation of valuable high purity water. Water recoveries in the range of 75 to 95 percent with corresponding waste concentration factors of 4 to 20 are attainable on many streams containing both inorganic and organic solutes. This enables practical recovery of valuable materials or ultimate waste disposal.

Power at 8 cents and chemicals at 2 cents are typical operating costs for each 1,000 gallons of water produced. Membrane life of 3 to 4 years is normal. Operating attention is commonly 15 to 30 minutes each day. Installation is quickly and easily accomplished.

GESCO systems are available in capacities from thousands to millions of gallons per day, or tens to thousands of gallons per minute.

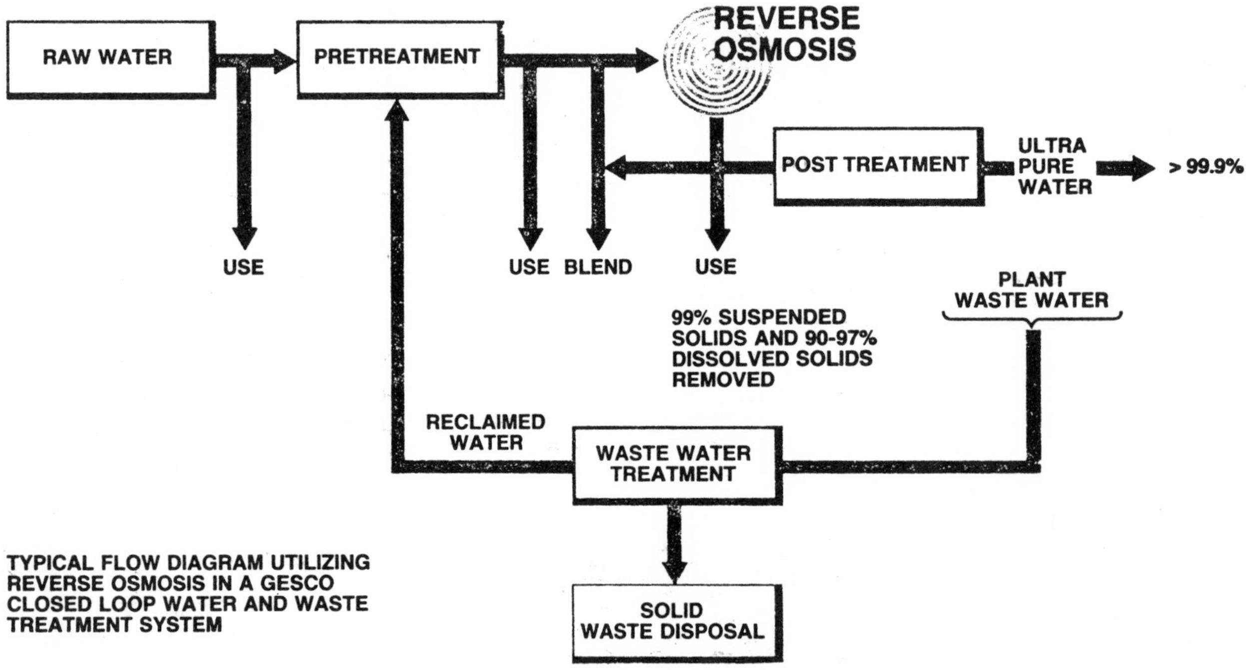

TYPICAL FLOW DIAGRAM UTILIZING REVERSE OSMOSIS IN A GESCO CLOSED LOOP WATER AND WASTE TREATMENT SYSTEM

A **dependable, low cost** method for complete clarification of water or waste water.

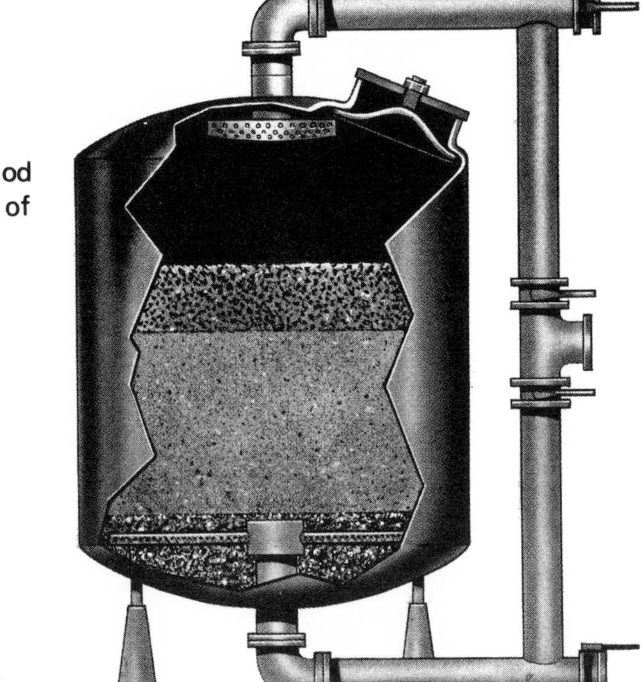

ELECTRO MEDIA is made up of granules which generate permanent electrical charges. These permanent charges have the ability to upset the stability of sub-micron turbidity particles and large colloids causing them to precipitate in the filter bed. The filter bed holds the accumulated solids throughout the depth of the filter bed. These solids are held much more tenatiously by the **ELECTRO MEDIA** filter bed than an equivalent sized non-charged media bed such as a sand or garnet filter bed. During backflush the accumulated solids are quickly scrubbed off the surfaces of the **ELECTRO MEDIA. ELECTRO MEDIA** can be used in conjunction with alum pretreatment when large amounts of color and colloidal sized particles are present. It has been found that much less alum is required using the **ELECTRO MEDIA** filter bed and no residence flocculating time is required after the alum addition.

Performance Data

Very Fast Flow Rate – 20 gpm per square foot.
Easily Clarifies Highly Turbid Water or Waste Without Costly Chemical Additives.
No Breakthrough of Solids – Filter bed blocks itself preventing breakthrough of trapped solids.
Off Stream Time Very Short – Only three minutes backflush time required at 15 gpm backflush rate.
Long Media Life – Only a small amount of media is lost per year due mostly from backflush losses.
Very Low Cost of Operation – Costs are: pumping if required, and a small cost of media addition on a yearly basis.
Simple Operation – Only reverse flow required for media cleaning. This operation is easily automated.

Applications

Surface water supplies (rivers, lakes), boiler feed water, process water, cooling tower color, taste and odor removal, bottling plants, drinking water, plant waste water, sewage effluent, deep well disposal, oil field injection water. **ELECTRO MEDIA** can be added to existing sand filter beds so that flow rates can be greatly increased and clarity improved.

HYDR-O-MATIC self-priming sewage and trash pumps.
6" x 6" handles 4" solids
4" x 4" handles 3" solids
3" x 3" handles 2 1/2" solids

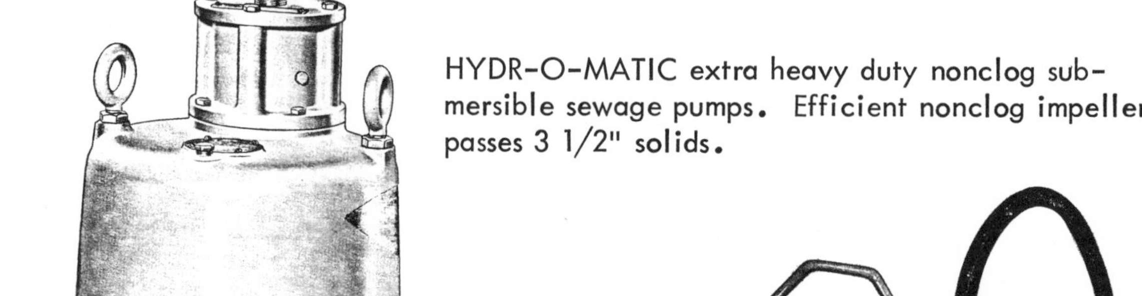

HYDR-O-MATIC extra heavy duty nonclog submersible sewage pumps. Efficient nonclog impeller passes 3 1/2" solids.

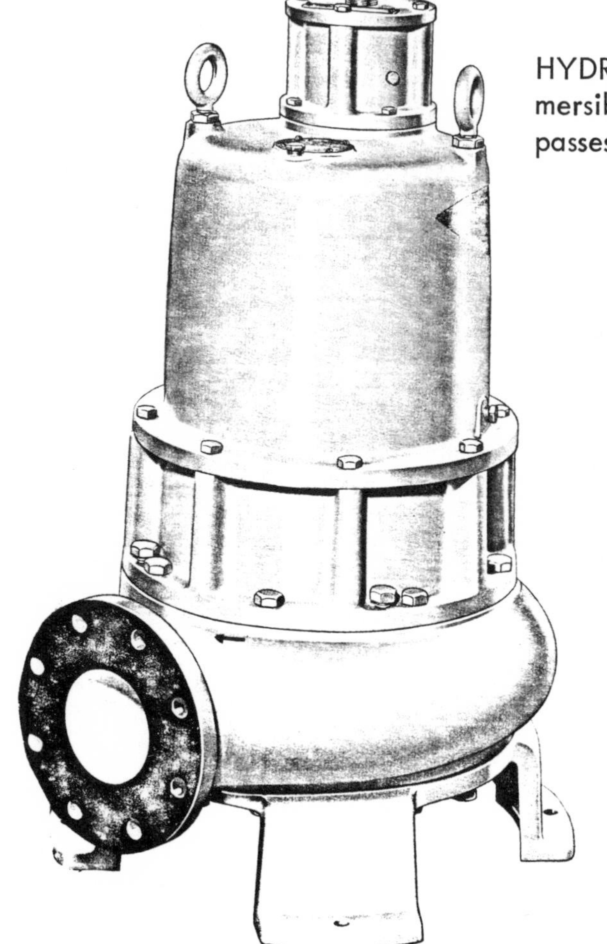

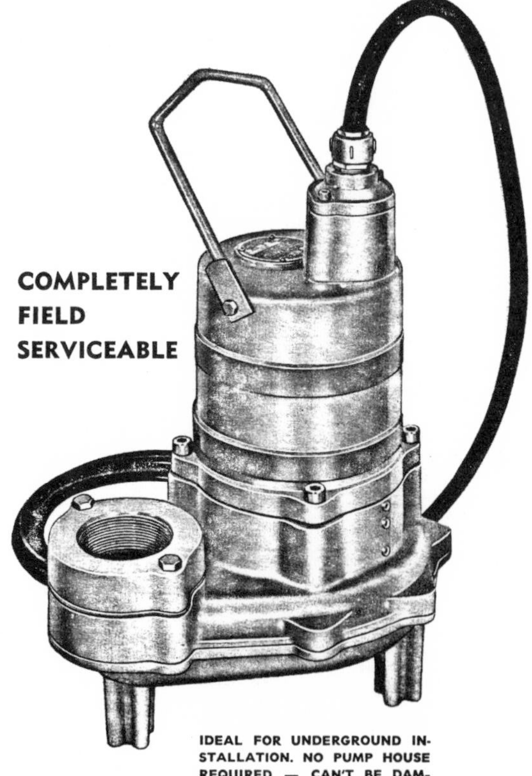

IDEAL FOR UNDERGROUND INSTALLATION. NO PUMP HOUSE REQUIRED — CAN'T BE DAMAGED BY FLOODING.

HIGHWAY UNDERPASSES AND SEWAGE TREATMENT PLANTS.

FURNISHED FOR MANUAL OR AUTOMATIC OPERATION INCLUDING SUMP BASIN, COVERS AND ELECTRICAL CONTROLS.

EFFICIENT NONCLOG IMPELLER DESIGN . . .

HYDR-O-MATIC heavy duty nonclog sewage and industrial submersible pumps for handling sanitary and other drainage.

HYDR-O-MATIC packaged sewage ejector systems. Equipped with self-priming non-clog sewage pumps assuring quick clean-out. Back in service in minutes if clogging occurs.

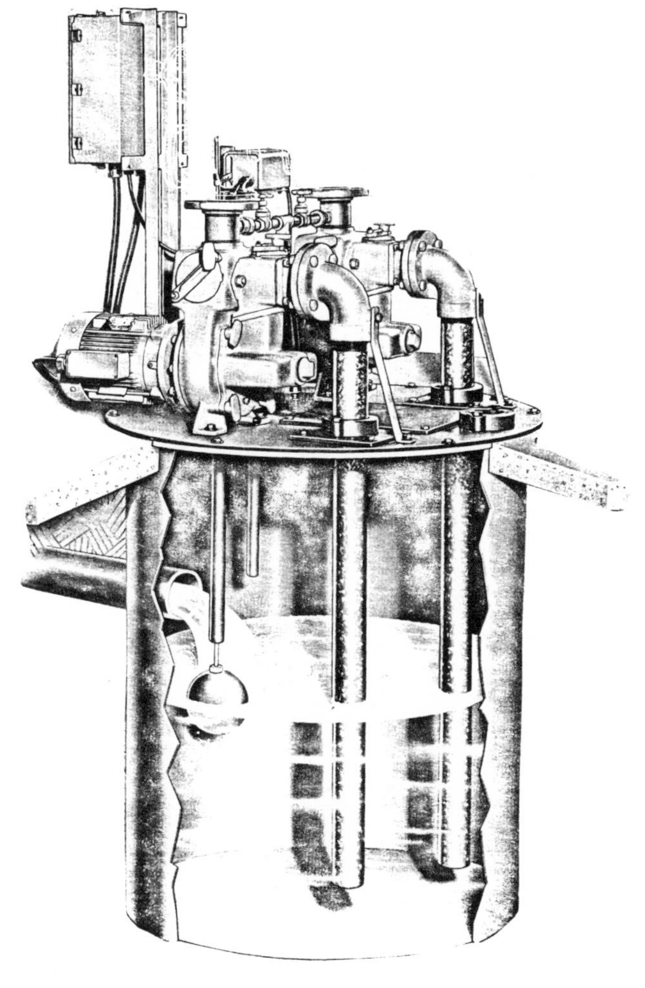

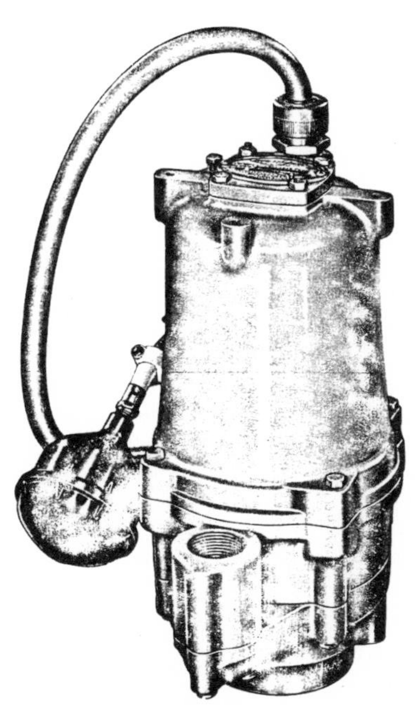

HYDR-O-GRIND submersible grinder pump for handling home and industrial sewage. One combination unit grinds all sewage including rags, wood, plastic, paper and rubber into fine particles and pumps under pressure into sewer main or to treatment plant.

3900 CONTROL replaces rod type float switches, diaphragm switches, and air bubbler systems. For operating sewage pumps to control liquid level in sumps.

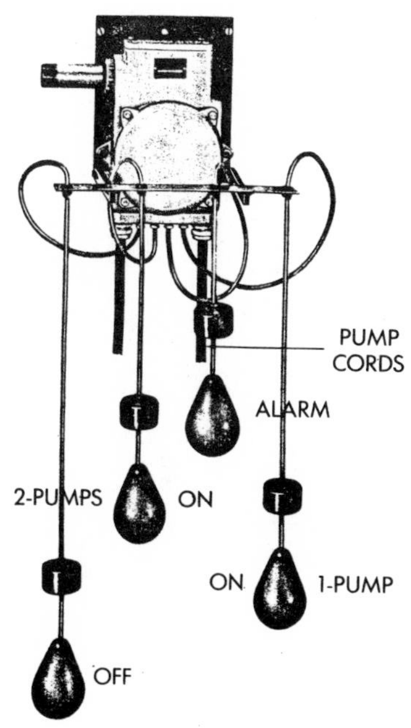

HYDR-O-PRIME self-priming sewage ejector systems. Complete factory built units for use where pumping level is within efficient suction range.

Housing slides back for easy access to pumps and piping.

HYDR-O-RAIL factory built sewage ejector systems. Complete package with outside valve box permitting servicing components without entering sump.

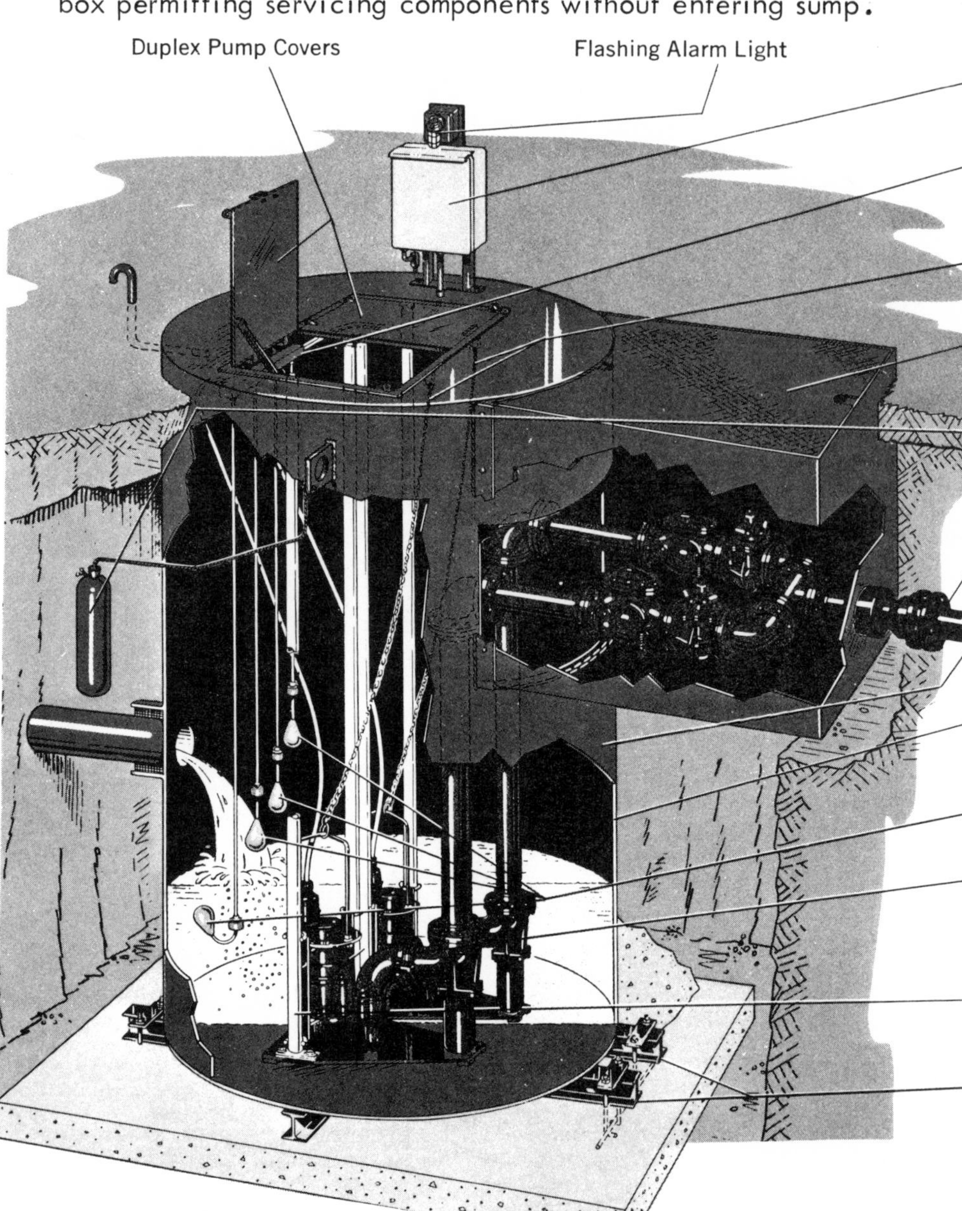

ICI America Inc.

GRANULAR ACTIVATED CARBON FOR WASTEWATER TREATMENT

Adsorption with activated carbon can be an important alternative or supplement to biological treatment. Whether it will be the most economic or the preferred treatment depends on a number of factors. Carbon adsorption is likely to be a preferred method when one or more of the following exist.

- Requirements for consistently high removals of organics and certain other impurities.
- Presence of nonbiodegradable colors or impurities
- Presence of certain toxic pollutants.
- Rapidly changing temperatures and/or pH.
- Limited space for treatment facilities.
- Need for reduced sludge handling.

FLOCCULANTS

ATLASEP flocculants are a new generation of ultra-high molecular weight, organic polyelectrolytes. They have been field demonstrated to be far superior in performance to any flocculants now in commercial use. When used at the correct ionic charge, these extremely long-chain polymers form dense flocs at very low dose levels. ATLASEP flocculants are available over a broad ionic range to provide optimum cost/performance benefits to numerous liquid-solids separation problems.

Product Number	Ionic Character
ATLASEP 1N	Nonionic
ATLASEP 1A1	Weak Anionic Charge
ATLASEP 2A2	Weak-Moderate Anionic Charge
ATLASEP 3A3	Moderate Anionic Charge
ATLASEP 4A4	Strong Anionic Charge
ATLASEP 5A5	Very Strong Anionic Charge
ATLASEP 105C	Moderate Cationic Charge

- The letter designates the ionic character. Higher numbers in the anionic series correspond to higher anionic charge.
- All polymers are supplied in powdered form. They dissolve readily with proper dispersion and gentle agitation.
- Custom-made polymers are available at intermediate charge levels between those shown above if required for specific applications.

CARBON ADSORPTION SYSTEMS

The largest application for the IWT continuous adsorption system is the purification of waste and process liquids by the reduction of organic contaminants.

Organics Readily Removed by Adsorption

1. BOD, COD and Phenols.
2. Saturated oils, alkanes and alkenes.
3. High molecular weight impurities from low molecular weight solvents.
4. Taste and odor in solution.

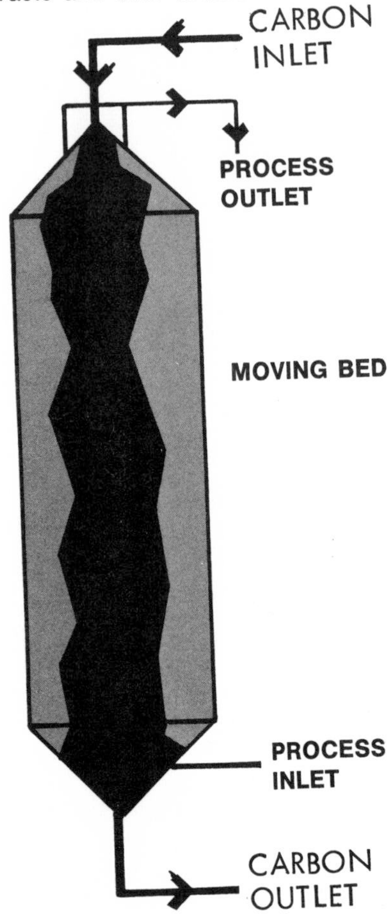

moving bed tower

The IWT moving bed adsorption tower is designed to operate with the process liquid flowing upwardly through the adsorption media (bed). Periodically a portion of the media is sluiced out the bottom of the tower and an equal amount of fresh adsorption media enters the tower through a valve at the top. This method of operation maintains a fresh bed in the tower at all times.

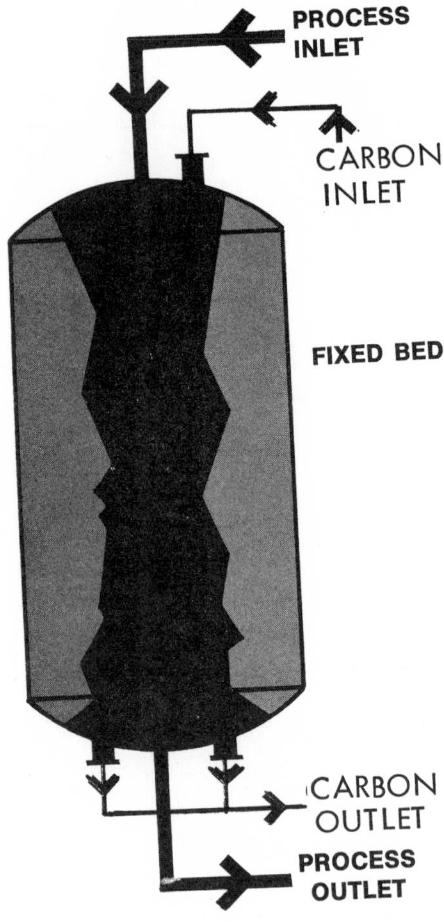

fixed bed column

The IWT fixed bed adsorption column utilizes the down-flow method of operation. As the process liquid enters the top of the column it is dispersed over the entire bed by a hub-radial design distribution system. The liquid is removed from the column at the bottom through a similar system. This type of distribution system insures efficient utilization of the entire carbon bed. When the bed becomes exhausted, it is then sluiced out of the column for rejuvenation or disposal. This type of column is useful when high flow rates of process liquid are involved, and/or low rates of adsorption media consumption are experienced.

Fixed bed columns are operated in series or in parallel depending on customer preference and operating conditions.

REVERSE OSMOSIS SYSTEMS

Applications in Pollution Control:

Reverse Osmosis can be used to concentrate waste streams provided they do not contain oxidizing agents and that they do not deviate from limiting pH ranges. Depending on the suitability of the contaminant and the initial concentration, the amount to which the waste stream can be reduced will vary substantially. Practical limits of concentration are approximately 2 to 5% total dissolved solids.

Reverse Osmosis will effectively remove high molecular weight (above 100) organics to the extent that ultra high purity water can be obtained without elaborate pretreatment equipment.

Membranes are being developed which will selectively reject particular molecules or ions while allowing others to pass. It is anticipated that membranes which have the capability of selectively separating contaminants from valuable materials which are more valuable when isolated will soon be available.

Operation:

To obtain optimum performance, the water temperature should be maintained between 68°F and 85°F (some membranes can be operated at higher temperatures). The temperature can be raised most simply on small systems by blending hot water from a domestic facility. Larger systems may require heaters (1). (see below)

A replaceable cartridge filter (2) removes residual suspended matter.

Acid is fed (3) to prevent scaling as a result of calcium carbonate formation. (Not required with soft or low calcium bicarbonate feed water). The monitor (4) assures that the pH is proper.

The multi-stage pump (5) boosts the pressure to 400 psi plus discharge pressure (maximum discharge pressure of 100 psi). The high pressure forces water through the membrane leaving a brackish water (reject) behind. The regulator (6) controls the pump discharge pressure.

The membranes are contained in modules (7) which are connected in parallel to three headers: feed, reject and product. A monitor (8) measures the conductivity to assure that the effluent purity is proper.

The relief valve (9) protects downstream equipment from over pressure (not required with decarbonator).

When acid feed is used, CO_2 is produced. Approximately 50% diffuses through the membrane. A decarbonator (10) may be required to remove the CO_2. The water falls by gravity to a storage tank (11).

The level switch (12) shuts down the chemical feed (3) and feed pump (5) and decarbonator inlet valve (13) on high level and the transfer pump (14) on low level.

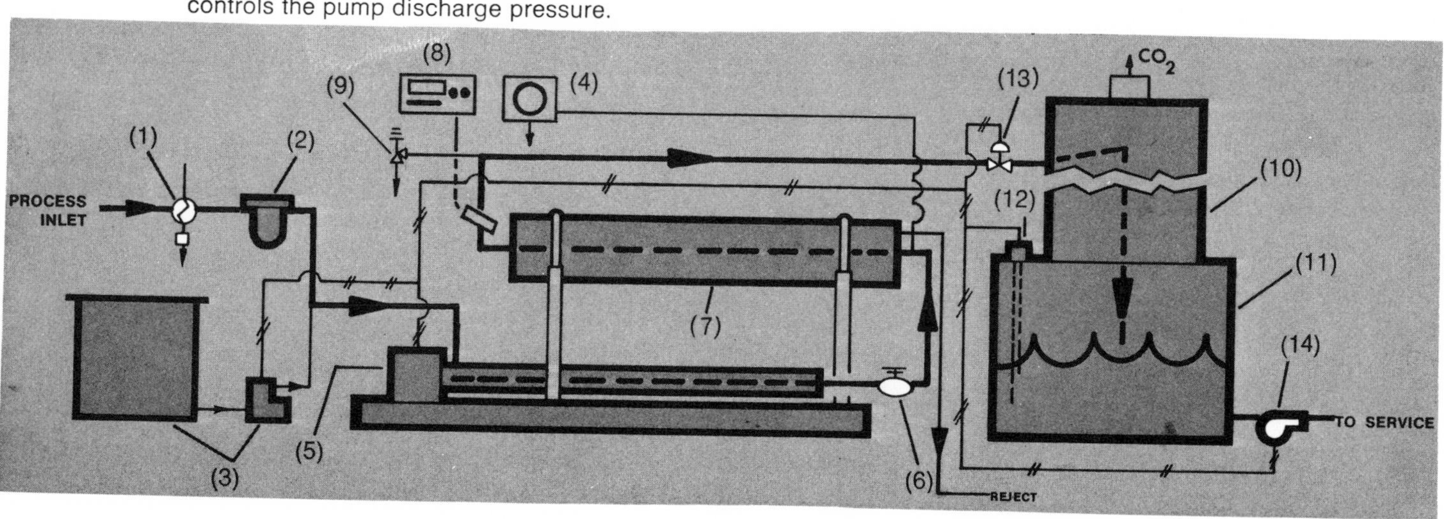

Joy Manufacturing Company, Denver Equipment Division

INDUSTRIAL POLLUTION CONTROL SYSTEMS

Denver uses its 50 years of technology, experience and process know-how in mineral and chemical processing to the solution of certain industrial waste treatment and/or disposal requirements. Many different types of equipment are available designed to handle a wide range of volumes and materials by various methods. Sizes range from very small to very large. The systems and components are of heavy duty construction designed for continuous reliable operation.

Denver process units include:

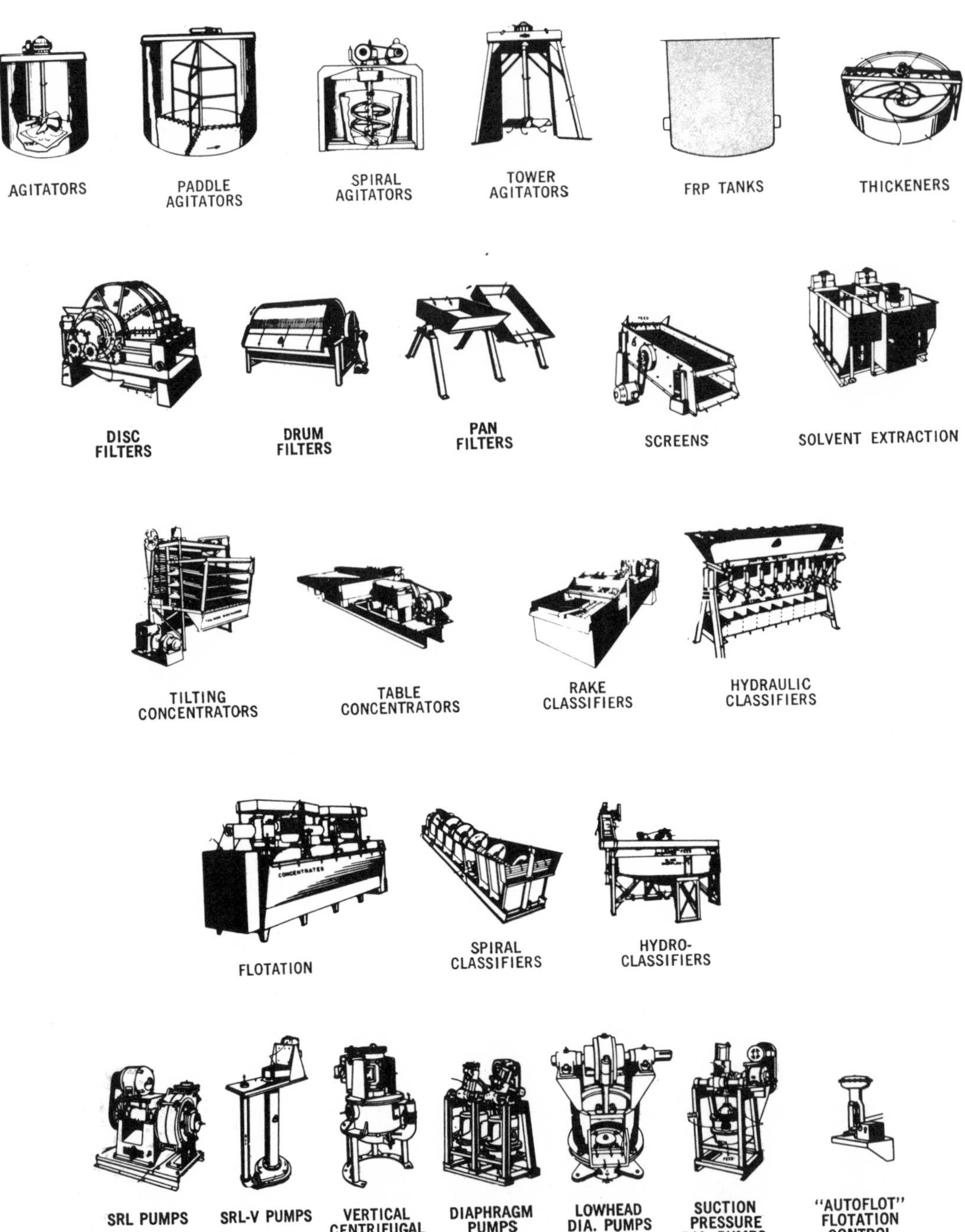

DIFFUSED AERATION SYSTEMS

Keene's Diffused Aeration Systems have four basic design options: fixed, lift-out, single-swing and dual-swing. These four systems, and the modifications available, provide complete flexibility to meet every type of application requirement.

Air Diffusers (AIRCOMB, DIFFUSADOME)

Four basic diffusers are available for use with the diffused aeration systems. All feature control orifices with or without a ball check. Also featured are unique, self-regulating air outlets, of a non-clog design, that permit continuous and even distribution of small air bubbles at varying airflow rates.

Fixed System (MARK "O")

The fixed diffused aeration system is designed for installations where the headers are to remain permanently in the tank. This system consists of an air main connector, control valve, drop pipe, header and air diffusers.

Lift-out (MARK "L")

This system is similar to the fixed system in design and construction with the exception that the entire drop pipe, header and diffuser assembly are provided with a lifting connection, so that it may be readily disconnected and removed from the tank without dewatering.

Single-swing (MARK II)

The single-swing air diffusion unit is designed to allow quick and easy access to the header portion of the air diffuser assembly. The drop pipe is hinged at the center and has a header box with integral lifting device and jack connections. A single jack which can be moved from unit to unit is used for raising the headers.

Dual-swing (MARK IV)

The Mark IV air diffusion unit is the most versatile and adaptable of the Keene line of diffused aeration equipment. This unit is designed with two swing joints which enable the entire unit to be removed from the tank liquid. There are no connections to unhook and the units can be raised or lowered in just a few minutes. A single hoist can be moved from unit to unit.

CIRCULAR CLARIFIERS

Keene Circular Clarifiers are available for primary or secondary settling. This clarifier design utilizes scraper arms driven by a motor/reducer which is mounted on either a full or half bridge structural steel frame. Scraper flights on the arms direct the settled sludge to the center of the tank for removal. Also available is a rapid sludge removal system that directs sludge to suction nozzles on the arms where it is piped to a visual sight well in the center of the tank, or piped directly out of the tank. Primary clarifiers can be provided with a scum baffle just inside the effluent trough. Also available is a surface skimmer that rotates with the scraper arms and directs the scum into a scum collector box. The drive mechanism is protected by an indicating overload device and can be provided with an alarm system.

THICKENERS

Keene thickeners are available in a wide variety of designs and capacities for almost any application. Both full-bridge and half-bridge are available depending on the overall size of the thickening basin.

"AQUARIUS" FLOATING SURFACE AERATORS

Design Features of "Aquarius" Aerator

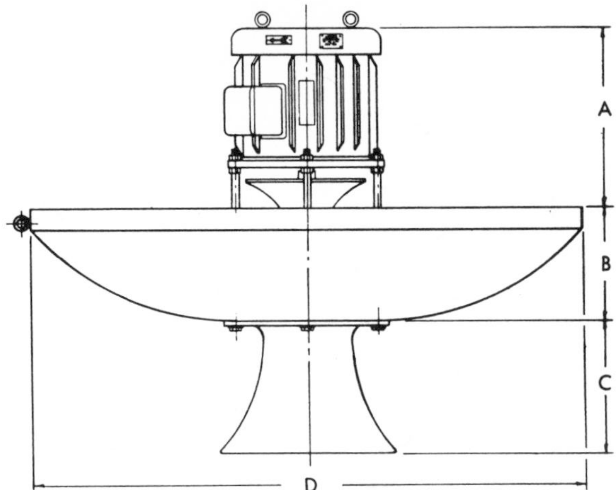

MODEL	APPROX. WT.-LBS.	H.P.	DIMENSIONS			
			A	B	C	D
K-AQ-3	350	3	12-5/8"	8"	12"	4'-2"
K-AQ-5	380	5	12-7/8"	8"	12"	4'-2"
K-AQ-7.5	530	7.5	16-1/2"	10"	15"	5'-0"
K-AQ-10	600	10	17"	10"	15"	5'-0"
K-AQ-15	680	15	17-1/2"	10"	15"	5'-0"
K-AQ-20	820	20	28-1/2"	12"	18"	6'-6"
K-AQ-25	890	25	29"	12"	18"	6'-6"
K-AQ-30	960	30	29-1/2"	12"	18"	6'-6"

The aerator consists of a vertical, low profile, wafer-thin motor; a direct-connected, integral impeller/deflector unit with repelling vanes; a circular, reinforced fiberglass flotation unit filled with closed cell polyurethane foam; intake flare tube; and other accessories as necessary.

The motor is heavy-duty; it consists of a totally enclosed and fan cooled, non-hydroscopic, chemically resistant, round-body motor housing with motor output shaft of 304 stainless steel. The motor is of wafer-thin, low profile type design with aluminum alloy housing for weight reduction to insure maximum overall stability. All joints and mating surfaces are sealed with special sealing compound. It is insulated with Super Class B insulation, is moisture resistant and fungus proof.

There is an integral, rotating, combination pumping impeller and deflector, constructed of urethane rubber structural plastic, cast for tight-fit, direct mounting to the motor shaft. The top plate consists of repelling vanes to discharge liquid away from the motor bearings and shaft seal to maintain a dry operating condition. The entire motor and impeller/deflector assembly are dynamically balanced to insure long life and smooth running operation.

The supporting floating pontoon structure is of high-grade, fiberglass, polyester-reinforced construction. All imbeddable foundation items for anchoring, mooring and attaching components are of stainless steel. The entire flotation unit is filled with closed cell polyurethane foam. The hull of the pontoon is constructed of multiple layers of chopped and woven fiberglass cloth with mat incapsulated in high-strength polyester resin. It is shaped so as to exhibit maximum buoyancy.

A special reinforced, polyester fiberglass uptake flare is accurately fitted and supported by the floating structure to maintain a smooth entrance of liquid unto the impeller, maximizing the efficiency of the unit.

Keene Corporation, Water Pollution Control Division

MECHANICAL SURFACE AERATOR

Materials of Construction — Rotating element completely corrosion resistant, fiberglass construction. All other components below speed reducer of stainless steel or galvanized steel construction. No submerged rotating steel components or weldments.

Rotating element design requiring no rotating baffles to decrease efficiency.

Rotor blades monolithically attached to spandrel so no separately mounted plates can detach during operation.

Rotor of monolithic construction, one piece molded part with no seams submerged in liquid.

Highest quality helical gear speed reducer with minimum service factor in all cases of 2.00.

High exygen transfer efficiency demonstrated and field tested.

Rotating element design with hydraulically efficient curved surface of revolution producing a solid spandrel.

Monolithic blades attached to spandrel and spiral wound for optimizing hydraulic pumping rate.

Stainless steel adjusting rods provided at gear reducer base plate to raise or lower unit thus varying submergence and horsepower.

Streamlined design of rotor prevents hang up of any foreign material such as rags, debris, etc.

Streamlined effect also eliminates ice build up thus allowing for full winter operation.

Rotating element completely non-corrosive and internally steel reinforced with polyurethane foam filler for light weight construction.

Buoyant fiberglass rotor counteracts the downward operating thrust load providing for longer speed reducer bearing life.

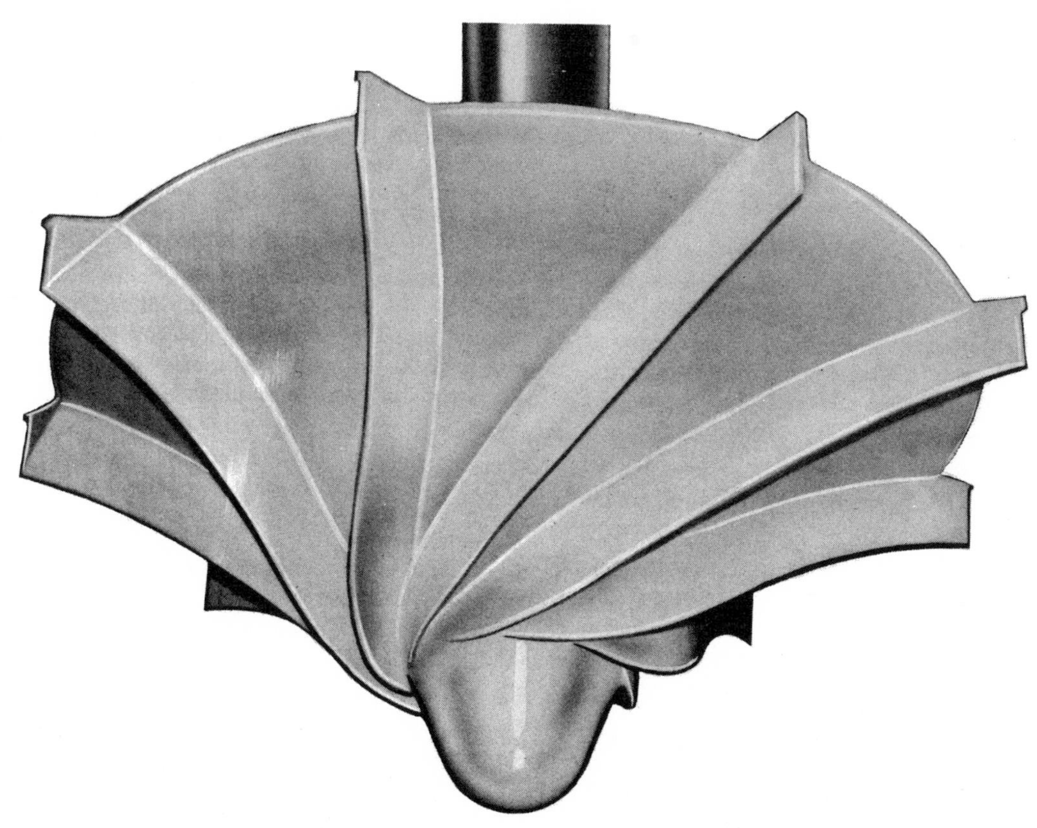

Keene Corporation, Water Pollution Control Division

AMERICAN ROTARY DISTRIBUTORS

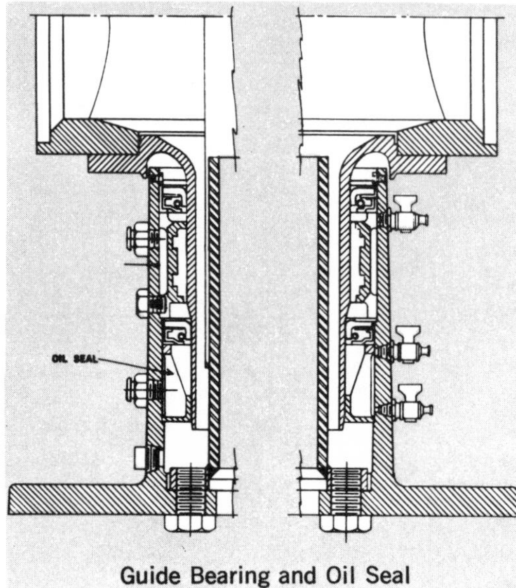

Guide Bearing and Oil Seal

"American" Reaction-Driven and Positive-Drive Rotary Distributors are available for a wide range of requirements for standard and high rate trickling filters, and for sand filters. Design and construction features have contributed to an outstanding record of distribution efficiency and reliability for many years.

EXCLUSIVE OIL SEAL* The oil-immersed lower guide bearing for the rotating member of the center column, is protected from the sewage flow to the distributor by a separate fluid grease seal—a simple, dependable design which eliminates the use of costly mercury or other types of seals. Will not "blow" or fail when operating under surges or high heads. This exclusive feature is furnished on all Reaction and Positive-Drive Distributors. **U.S. Pat. No. 2,379,547*

EXCLUSIVE MULTI-FLO NOZZLE* The first variable-flow combination orifice and discharge nozzle, with self-adjusting spreader, for Reaction-Driven Rotary Distributors. Exceptionally low head losses over a wide range of flows. Eliminates overflow controls in most cases, and provides excellent distribution. Adds greatly to operating efficiency. Has proved its worth in many installations.

**U.S. Pat. No. 3,066,871*

Comparison of the distribution pattern of Multi-Flo Nozzle on the left and Fixed Orifice Nozzle on the right.

Table No. 1	REACTION-DRIVEN ROTARY DISTRIBUTORS				
SIZE OF CENTER COLUMN	MAX. FILTER BED DIA.	MAX. FLOW - GPM - CTR. COL. at 4 ft/sec.	at 6 ft/sec.	MAX. ARM PIPE SIZE	APPROX. NET AREA-SQ.IN.
6" RL	36'	360	550	5"	28
6" RI	60'	360	550	5"	28
8" RI	56'	550	800	6"	50
8" RH	90'	550	900	6"	50
10" RI	56'	800	1200	8"	78
10" RH	156'	800	1500	8"	78
15" RH	166'	2000	3300	10"	177
20" RH	206'	3700	6000	14"	315

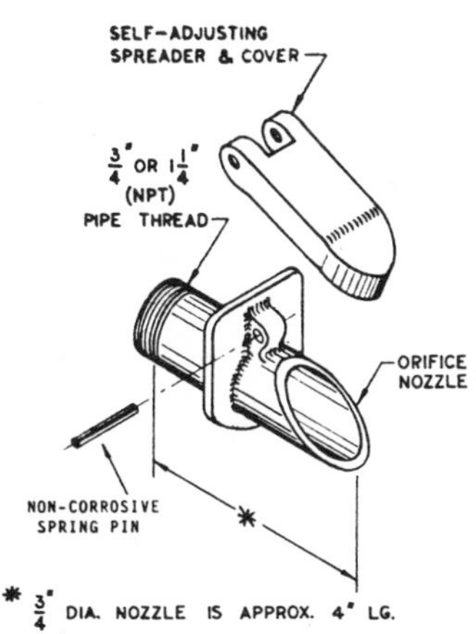

* 3/4" DIA. NOZZLE IS APPROX. 4" LG.
1 1/4" DIA. NOZZLE IS APPROX. 6" LG.

VARIABLE-FLOW COMBINATION ORIFICE AND DISCHARGE NOZZLE

for Reaction-Driven Rotary Distributors

The discharge of sewage through the Multi-Flo Nozzle automatically positions self-adjusting spreader. This action maintains exceptionally broad-area pattern of distribution throughout wide variation of flow. Minimizes clogging and cleaning. Spray shield protects pipe arms.

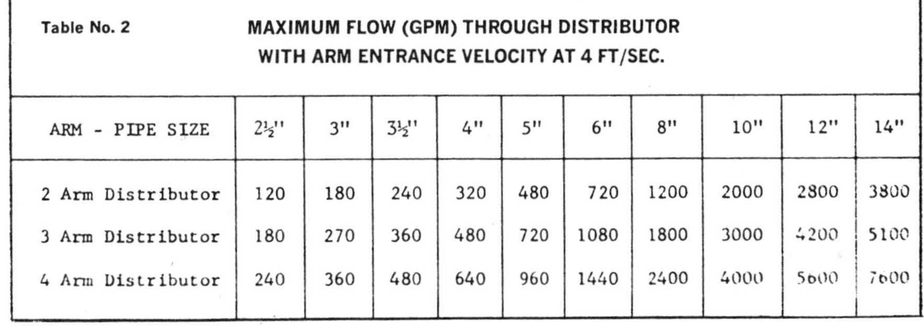

Table No. 2 — MAXIMUM FLOW (GPM) THROUGH DISTRIBUTOR WITH ARM ENTRANCE VELOCITY AT 4 FT/SEC.

ARM - PIPE SIZE	2½"	3"	3½"	4"	5"	6"	8"	10"	12"	14"
2 Arm Distributor	120	180	240	320	480	720	1200	2000	2800	3800
3 Arm Distributor	180	270	360	480	720	1080	1800	3000	4200	5100
4 Arm Distributor	240	360	480	640	960	1440	2400	4000	5600	7600

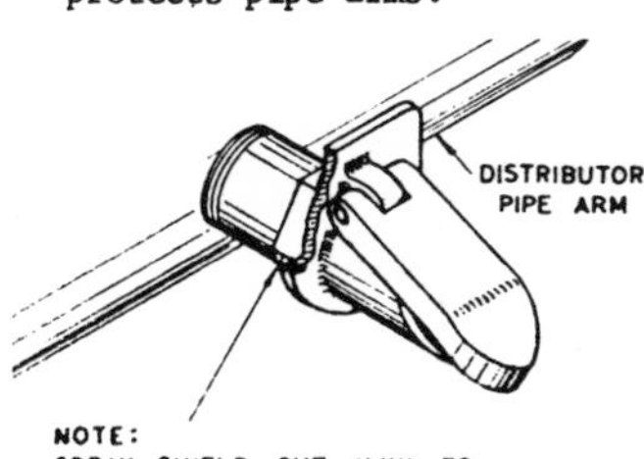

NOTE:
SPRAY SHIELD CUT AWAY TO SHOW HEX SHAPE FOR WRENCH

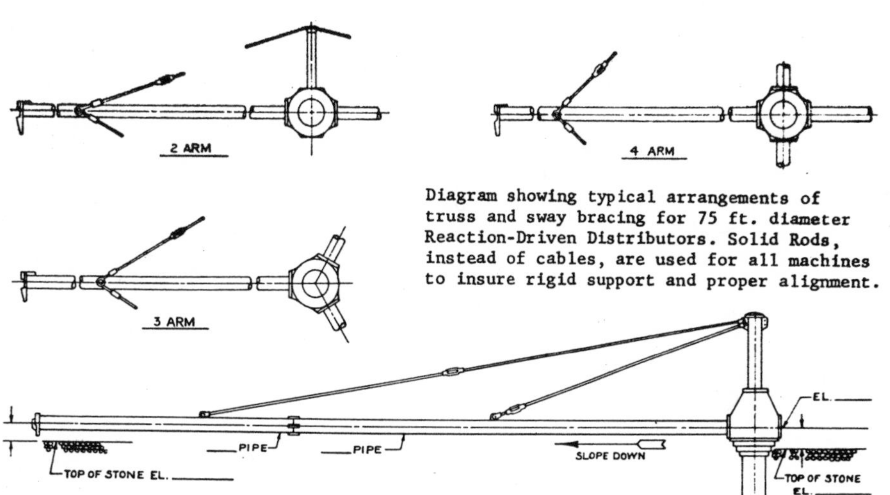

Diagram showing typical arrangements of truss and sway bracing for 75 ft. diameter Reaction-Driven Distributors. Solid Rods, instead of cables, are used for all machines to insure rigid support and proper alignment.

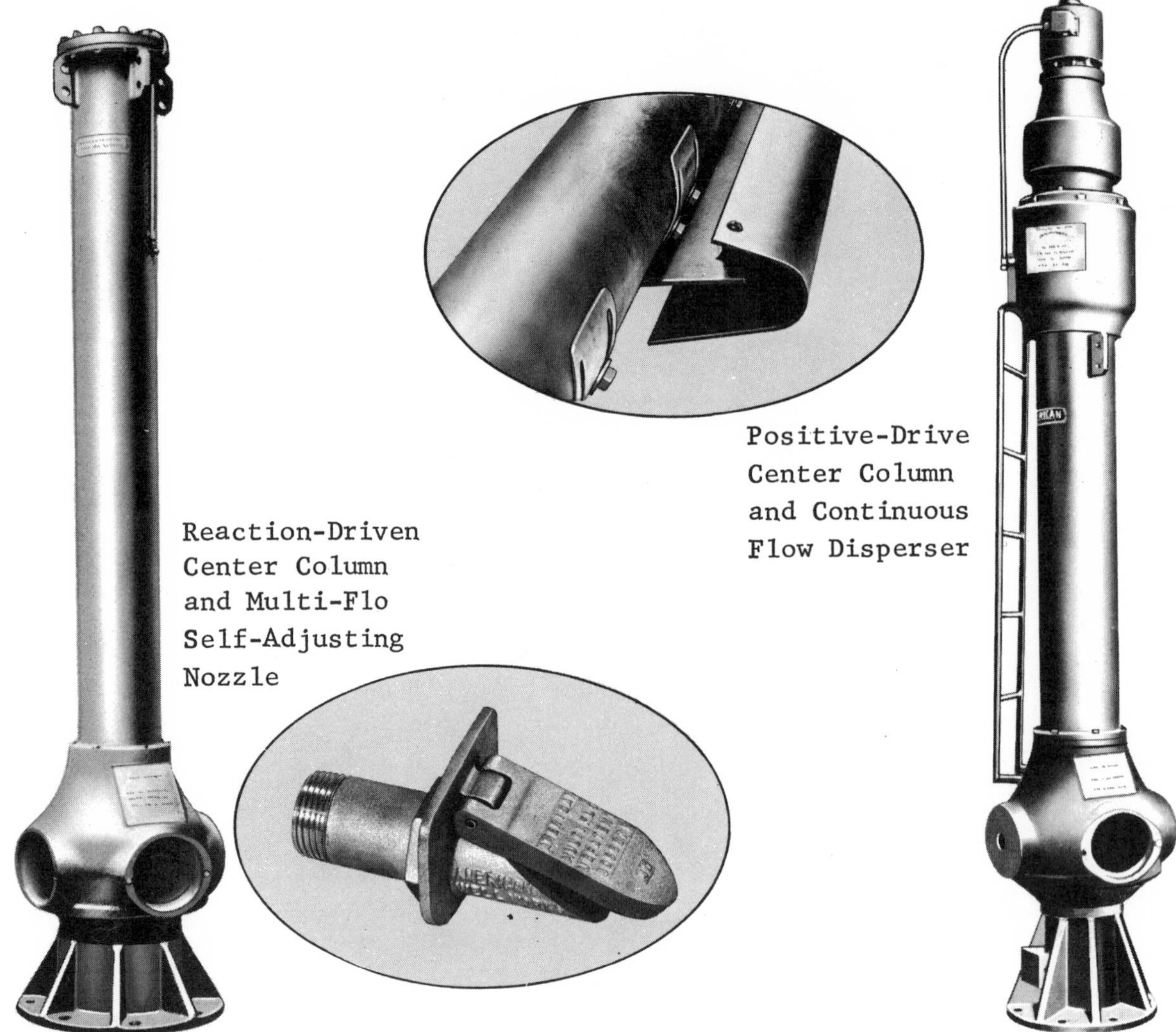

Reaction-Driven Center Column and Multi-Flo Self-Adjusting Nozzle

Positive-Drive Center Column and Continuous Flow Disperser

REACTION-DRIVEN ROTARY DISTRIBUTORS

Used when the hydraulic gradient provides the required net head to maintain the reaction force of the liquid jetting from the orifices for proper rotation.

These machines are furnished with two, three, or four arms. However, the three-arm machine with Multi-Flo Nozzles can handle a flow variation in the magnitude of 4:1 or greater; and it has the advantages of minimum wind effect, relatively low starting inertia, perfect balance, and higher frequency dosing of the filter bed.

Low-head-loss, single or alternating Dosing Siphons may be used to provide proper dosage at low flows.

Fixed orifice nozzles are used on some dosing siphon applications, and for sand filters. In the latter case, pipe arms are furnished with continuous spreader plates.

POSITIVE-DRIVE ROTARY DISTRIBUTORS

Used when the hydraulic gradient cannot provide sufficient net head to operate a Reaction-Driven machine throughout the required range of flow.

Weather-proof motorized drive, operating continuously at low power, maintains ideal dosing conditions for the filter. Two-arm machines satisfy most requirements, as the adjustable orifices and continuous flow disperser along each pipe arm accomodate low heads and wide range of flows.

Minimum basic head requirements for the motor-driven machine are: Foundation Pier and Distributor Port Losses (page 4); Starting and Entrance Losses to Distributor Arms (page 5); Sufficient additional head to fill the pipe arms at maximum flow.

Successfully used to solve many problems. Consult the factory for recommendations.

TRICKLING FILTER PLANTS

The trickling filter's function is the oxidation of organic matter and the biological coagulation of finely divided solids as a secondary step in treating settled sewage. This is expressed as BOD removal. The BOD reduction is accomplished by the microscopic organisms living on the slime clinging to the filter media. The amount of BOD removed is proportional to the number of organisms present which in turn are proportional to the volume of rock media.

Maximum benefit of a trickling filter is obtainable by preceding the filter by primary sedimentation to remove settleable solids and separable greases, followed by secondary sedimentation to remove humus sludge from the filter effluent.

A properly designed and constructed filter provides many years of trouble-free service under adverse conditions. Therefore, only the best equipment and materials should be used in constructing trickling filters. The components of the filter must not only withstand severe outdoor service but also any deterioration by the sewage or its decomposition products or by industrial wastes being treated in conjunction with domestic sewage.

TYPES OF FILTERS

Standard Filters: Standard or "low rate" filters are generally considered to be filters loaded at or below 600 lbs. of BOD per acre ft. (4 MGAD at 15 lbs. BOD per 1000 cu. ft.). Although many standard rate filters are carrying heavier loadings, they are not designed for greater loadings initially than that noted above.

High Rate Filters: These are filters that are loaded at or above 3000 lbs. of BOD per acre ft. (10 MGAD at 30 lbs. BOD per 1000 cu. ft.).

Roughing Filters: The high rate filter when used as a roughing filter may be loaded as high as 15,000 lbs. BOD per acre ft. and can remove many pounds of BOD even with an efficiency as low as 30%. Such filters are effective in relieving an existing overloaded treatment plant or as the primary filter in a two-stage filter installation.

Standard vs. High Rate: The choice between standard or high rate filters is dependant on whether the outlet stream requirements warrant discharge of a well nitrified effluent or whether an effluent equally low in BOD, but not having the added stability provided by nitrate nitrogen, will satisfy stream requirements. In localities where streams possess reasonable dilution and carrying capacity, there is little reason to deny use of a high rate type filter, thereby depending upon the receiving stream to complete the nitrification cycle. It must be remembered that the relative cost of removing BOD increases out of proportion to the amount removed as the filter loading drops.

Recirculation: This may be applied either to high rate or standard rate filters. By utilizing recirculation, whereby the applied sewage is reduced in strength, extremely strong wastes can be treated satisfactorily and economically. A 1:1 recirculation ratio is usually satisfactory for domestic sewage containing small amounts of industrial wastes, although the receiving stream requirements must govern. This ratio is desirable since the rate of flow through the plant of mixed recirculation and plant inflow never exceeds the designed maximum rate of flow for the plant without recirculation, thereby eliminating the necessity of increasing the volume of the primary and secondary settling tanks.

Keene Corporation, Water Pollution Control Division

AERATED GRIT SYSTEMS

Keene aerated grit systems are basically designed to separate the grit from the putrescible matter and remove the clean, inoffensive grit from the raw sewage flow.
The design criteria, based on theory and practice, establish the correct flow-depth-length-width ratios, also the necessary volume of air to maintain an optimum velocity for dependable grit washing. The air is dispersed through non-clog Keene Aircomb Diffusers, selected and spaced to maintain velocity requirements. Ideally the washed, settled grit is delivered, by screw conveyor, to a Keene bucket elevator for ultimate surface disposal. Alternately a clam shell bucket or air lift pump may be used to remove the settled grit.

KEENE AIRCOMB DIFFUSER

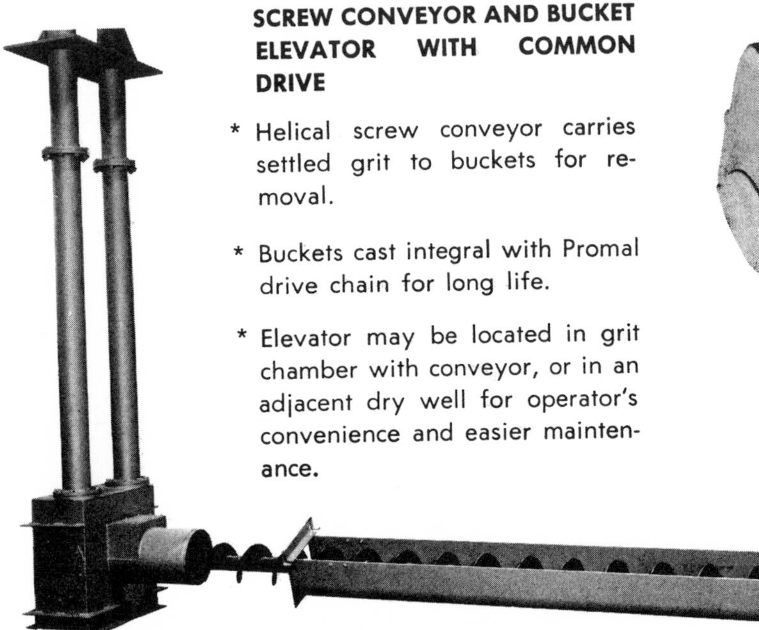

SCREW CONVEYOR AND BUCKET ELEVATOR WITH COMMON DRIVE

* Helical screw conveyor carries settled grit to buckets for removal.
* Buckets cast integral with Promal drive chain for long life.
* Elevator may be located in grit chamber with conveyor, or in an adjacent dry well for operator's convenience and easier maintenance.

* Serration on both sides of the Aircomb provide simple, multiple, variable-flow orifices for controlled aeration.
* Peaked dome design prevents accumulation of solids.
* Velocities are self-cleaning.
* Variable diffusion, capacity as high as 5:1 if necessary.
* Fully proved in the field.
* Take apart by simply removing top cap. Put together just as easy.

STATIC AERATOR

Kenics Corporation has applied its Static Mixer principle to aerated water treatment systems called Static Aerators. These units, constructed of polyethylene, are 12" in diameter and 5' in length. The units are individually anchored at the bottom of the basin and are supplied by an air distribution system which feeds air into the bottom of each unit. A significant feature of the air distribution system is the use of a large air orifice. Hole diameters are 3/8 inch and not subject to fouling.

Installation of Static Aerator Systems is simple and straightforward. The air supply (blowers) are installed on land in a convenient location and the compressed air is piped to the aerators via inexpensive pipe; e.g. steel headers and polyethylene aerator distribution lines. The anchor attachment for the aerators is a simple clamp attachment that holds the aerator and distribution pipe in proper position.

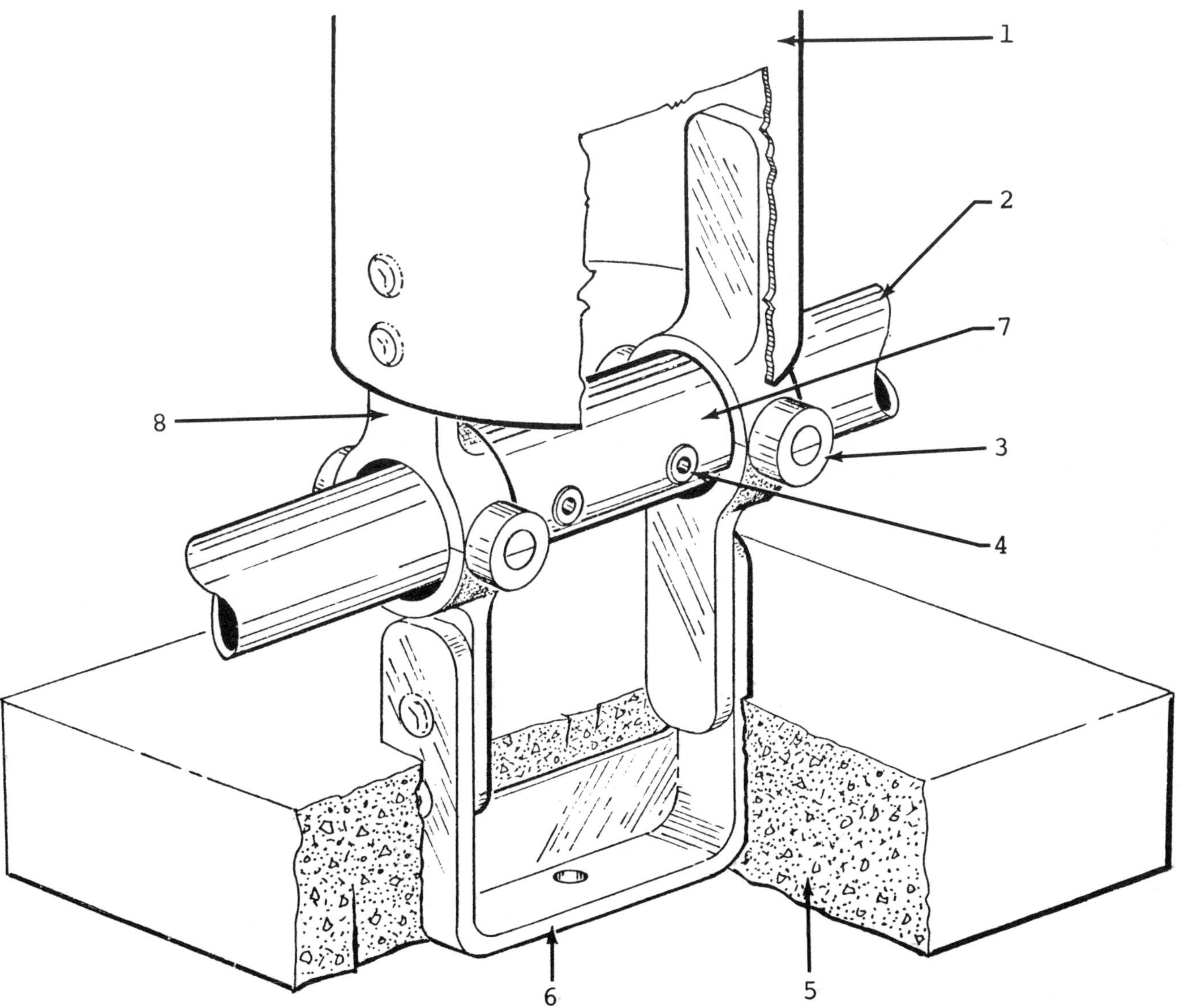

1- Aerator
2- Polyethylene Pipe
3- Locking Sleeve
4- Rivet
5- Concrete Block
6- Anchor Bracket
7- Pipe Clamp
8- Aerator Mounting Yoke

PRINCIPLE OF OPERATION

The significant oxygen transfer in Static Aerator units is due to the actual contacting of the air-water mixture. Air rising up through the aerator causes the system to act on the "air-lift pump" principle. Thus, an air-water mixture is caused to rise concurrently through the Static Aerator unit. During this sequence, the oxygen rich air and low oxygen water are vigorously contacted to give the outlet liquid a high dissolved oxygen content. The profile of the circulation in the basin is ideal, regardless of basin depth; unlike surface aerators which act on the surface water and thus have a limited depth in which they can effectively perform. The Static Aerator unit takes the bottom basin water and aerates the water while pumping it toward the surface. The high D.O. water exiting the Static Aerator unit continues to be pumped to the surface. The circulation loop is completed by high D.O. surface water flowing down and dispersing in the areas between the Static Aerator units and the low D.O. water flowing into the aerators.

The ability of the Static Aerator to operate at virtually any practical depth minimizes the problem of lagoon aeration. Lagoon dimensions are normally limited by factors external to the aeration process, such as water table or the economics of constructing deep lagoons.

Excellent circulation is provided in all parts of the lagoon since the aerator is an effective air lift pump. Oxygen is evenly distributed resulting in more complete biological action on all waste.

The Static Aerator System can be easily adjusted to correspond to the level of BOD at any given time by simply adjusting the air supply. This does not reduce the uniformity of oxygen in the liquid; it lowers operating costs and elevates efficiency while maintaining effective aeration and mixing.

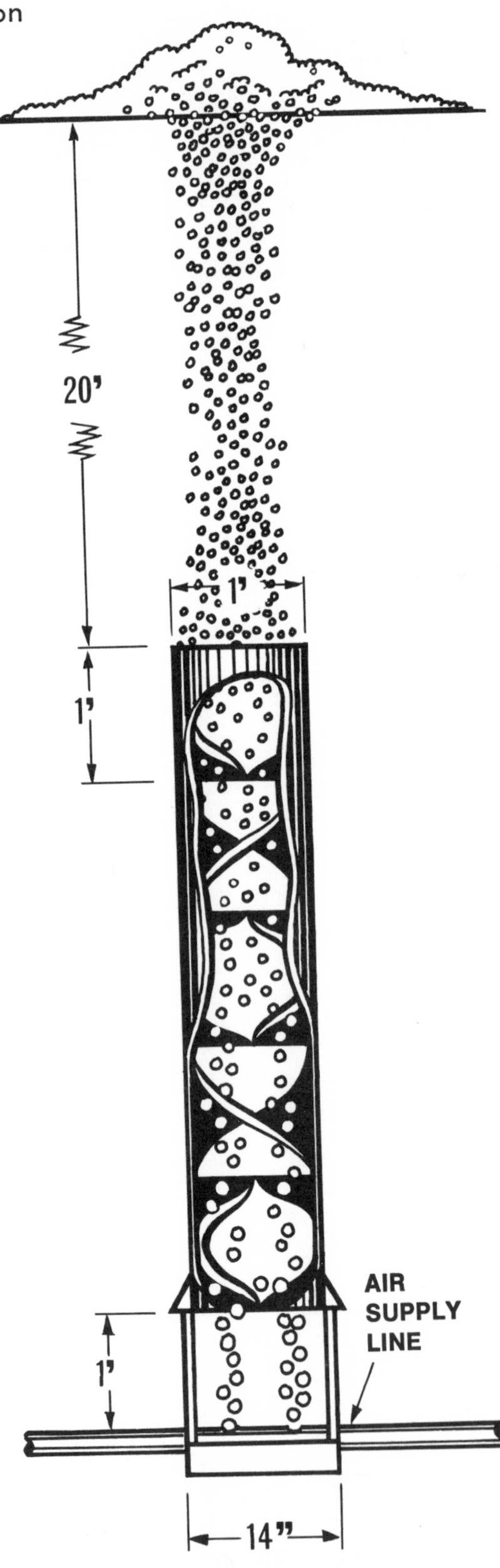

Sub-surface Static Aerator unit uses five Static Mixer elements to intimately contact liquid with rising air for high efficiency aeration with no moving parts.

Koch Engineering Company, Inc.

FLEXIRINGS FOR TRICKLING FILTERS

FLEXIRINGS are "random dumped" into the trickling filter shell, the non-symmetrical configuration of the resulting bed reduces channeling and insures that a maximum percentage of the available packing area is utilized for growth and oxygen transfer.

The random position of the individual rings insures a tortuosity of liquid and air that prolongs residence time and therefore increases B.O.D. removal.

Installation and Bed Depth

Since FLEXIRINGS are self-supporting, the simplest and most economical tower construction may be employed. There is no need for expensive retaining wall construction. It is even possible to utilize an inexpensive, thin fiberglass shell. Configuration may be circular, square or rectangular to suit conditions.

With regard to packing installation, there is no field-welding of sheets required, no assembly and stacking of sections, no sawing to fit circular configurations in the case of round towers. After construction of the trickling filter tower, it is only necessary to rapidly dump the FLEXIRINGS at random into the tower to the desired depth. Since there is no danger of breakage, no special precautions have to be taken.

Due to the selection of polypropylene and the rigid FLEXIRING design, FLEXIRING packing can be installed in beds of 25 feet or deeper without the requirement (when using plastic type sheets) of installing cross-members at regular intervals. Only a simple grating set on concrete piers is required to support the media at the base of the tower.

Filter Depths: The previously noted disadvantageous limitation of the rock pile as to depth (usually 6 ft.) and consequent expansion in diameter requiring considerable area is completely reversed in the case of the new media. With the introduction of plastic media it has been possible to go to much greater depths taking advantage of the fact that B.O.D. removals are a function of filter depth. The design of higher towers rather than increases in diameter represents usage of a lesser volume of media per unit of B.O.D. removed. Bed depths usually exceed 15 ft. with 20-25 ft. commonly standardized on; with certain wastes higher bed depths may be indicated.

With increased heights, installation care must be exercised in the case of certain sheet-type packing due to the lack of self-supporting structural strength when made of such weaker materials; internal reinforcing members may accordingly be required in such cases where the higher bed depths are desired. There have also been cases of structural failure of sheet-type packing media due to either chemical or thermal effects. This problem is completely eliminated with FLEXIRING media.

TABLE 1—KOCH FLEXIRINGS PHYSICAL DATA—COMPARISON CHART

PRODUCT	SURFACE AREA Sq. Ft./Cu. Ft. OF PACKING	VOIDS %	WEIGHT LBS. Per Cu. Ft.	MATERIAL	MAXIMUM RECOMMEND TEMPERATURE—°F
FLEXIRING	28	97	4	Polypropylene (Other materials available)	265°
Competitive	27	94	6.3	Saran	120°

TABLE 1 shows the basic physical characteristics of the 3½ inch FLEXIRING in comparison with corrugated-sheet type packing media. FLEXIRINGS rate superior in Surface Area (28 sq. ft. per cubic ft. of packing), in percent Voids (97%), in Weight per cubic ft. (4 lbs. per cu. ft.) and in Temperature suitability (to 265°).

HARDINGE FLOCCULATION UNITS

The object of coagulation or flocculation is to compact the colloidal turbidity or color particles of a liquid into heavier particles of floc which will readily settle. Suitable chemicals are added which, with slow stirring, produce a floc which will entrap the fine materials and allow them to settle. The choice of coagulants is governed by the characteristics of the material to be treated. The amount required varies with the turbidity, organic material, mineral constituents, and temperature.

The weatherproof gear drive mechanism of the Hardinge Flocculation Unit is supported from the top of the flocculation tank. A vertical torque tube is suspended from the drive mechanism and supports the arms which carry the paddles. Paddles are individually mounted and free swinging. There are no submerged bearings, sprockets or chains.

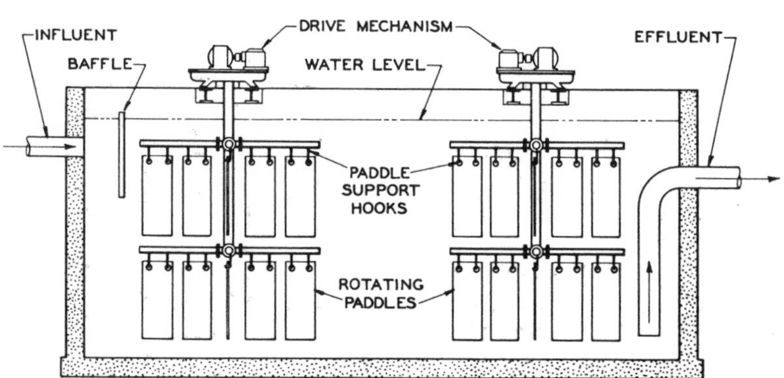

Hardinge Flocculation Units impart a gentle stirring action which promotes rapid flocculation. Speed is selected for controlled impingement and maximum agglomeration. The number of paddles used can be varied to suit unusual operating conditions.

This equipment is usually followed by a clarifier.

Koppers Company, Inc., Metal Products Division

AUTOMATIC BACKWASH FILTER

When light and fine suspended particles in a liquid cannot be economically removed by sedimentation processes, sand filters are used. Removal of the last traces of suspended solids is a final treatment step in processing municipal and industrial water supplies, chemical processing, and industrial wastes.

Hardinge automatic backwash filters supply a continuous flow of filtered water with no shutdown time for cleaning. Operational supervision is negligible. Power consumption is extremely low. Filtration is through an 11" bed of sand. Cleaning of the sand is accomplished, one compartment at a time, by twin pumps mounted on the traveling backwash carriage.

ABW Filters are used for municipal and industrial water supplies, removing carbonate contaminants from brines, final treatment of sewage effluents.

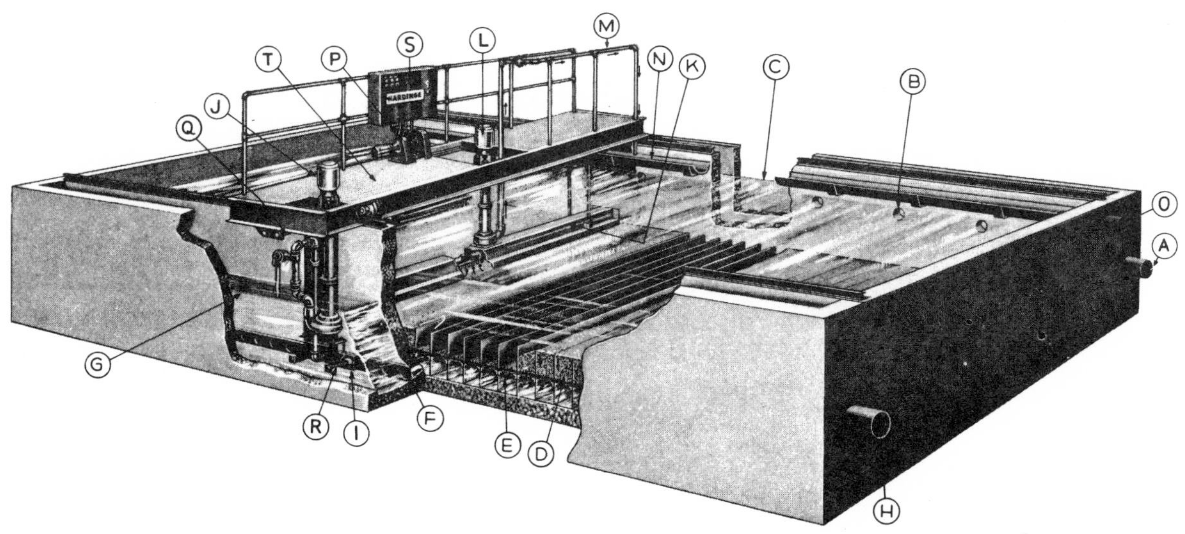

A. Influent line.
B. Influent ports.
C. Influent channel.
D. Compartmented filter bed.
E. Sectionalized under-drain.
F. Effluent and backwash ports.
G. Effluent channel.
H. Effluent discharge line.
I. Backwash valve.
J. Backwash pump assembly.
K. Washwater hood.
L. Washwater pump assembly.
M. Washwater discharge pipe.
N. Washwater trough.
O. Washwater discharge.
P. Mechanism drive motor.
Q. Backwash support retaining springs.
R. Pressure control springs.
S. Control instrumentation.
T. Traveling backwash mechanism.

HARDINGE CLARIFIERS

The effect of gravitational sedimentation is utilized in the process of clarification. The design of the clarifier tanks is such that the velocity flow through them is reduced to a point which allows the heavier solids to settle. Mechanisms are supplied which will continuously collect, concentrate, and remove the settled solids without manual attention. Where there is floating material to be removed, skimming devices are included.

The installation can be made in either circular or rectangular tanks. Tank construction can be of concrete, steel, wood or tile. Any of the submerged parts of the collecting mechanism can be furnished, if desired, of materials which will resist the corrosive characteristics of the material to be handled.

RECTANGULAR CLARIFIER

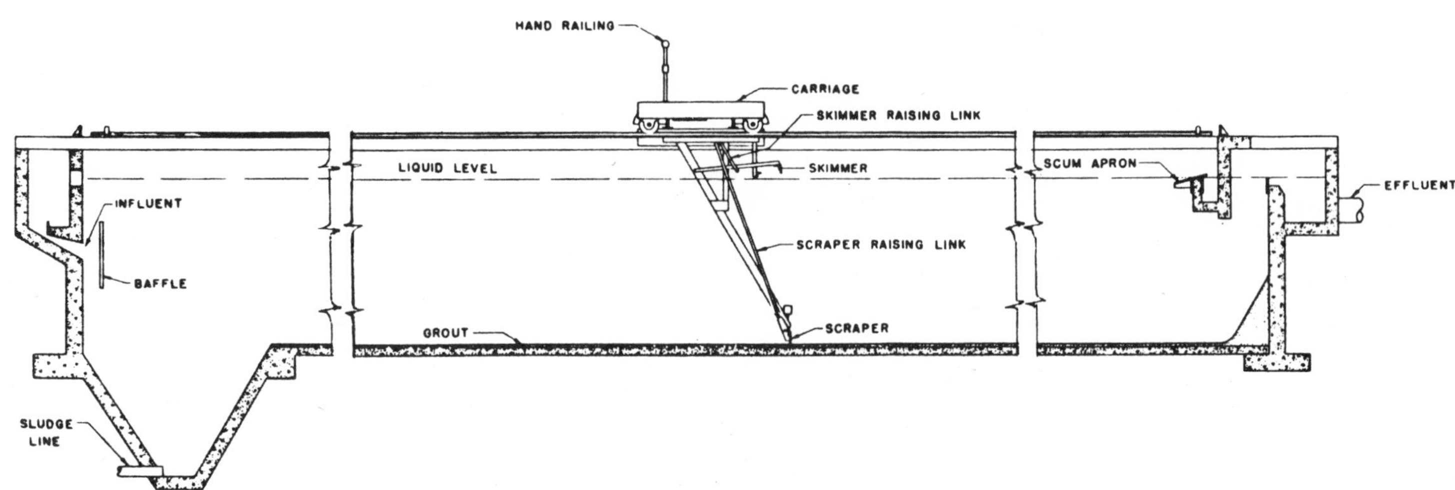

The Hardinge Rectangular Clarifier consists of a carriage which spans the width of a rectangular tank and supports the sludge scraper and skimmer. As the automatically controlled mechanism travels back and forth over the tank, all settled solids are moved into a sludge hopper at the influent end, and all floating material is delivered into a transverse trough near the effluent end. For extra wide tanks, a separately driven cross screw sludge conveyor is included.

This equipment can be fitted into either new or existing tanks. Sizes may range up to 80 feet wide and 400 feet long. By using the Hardinge Duplex Clarifier, two adjacent tanks can be operated with one mechanism.

COMPLETE TREATING SYSTEM FLOW SHEETS

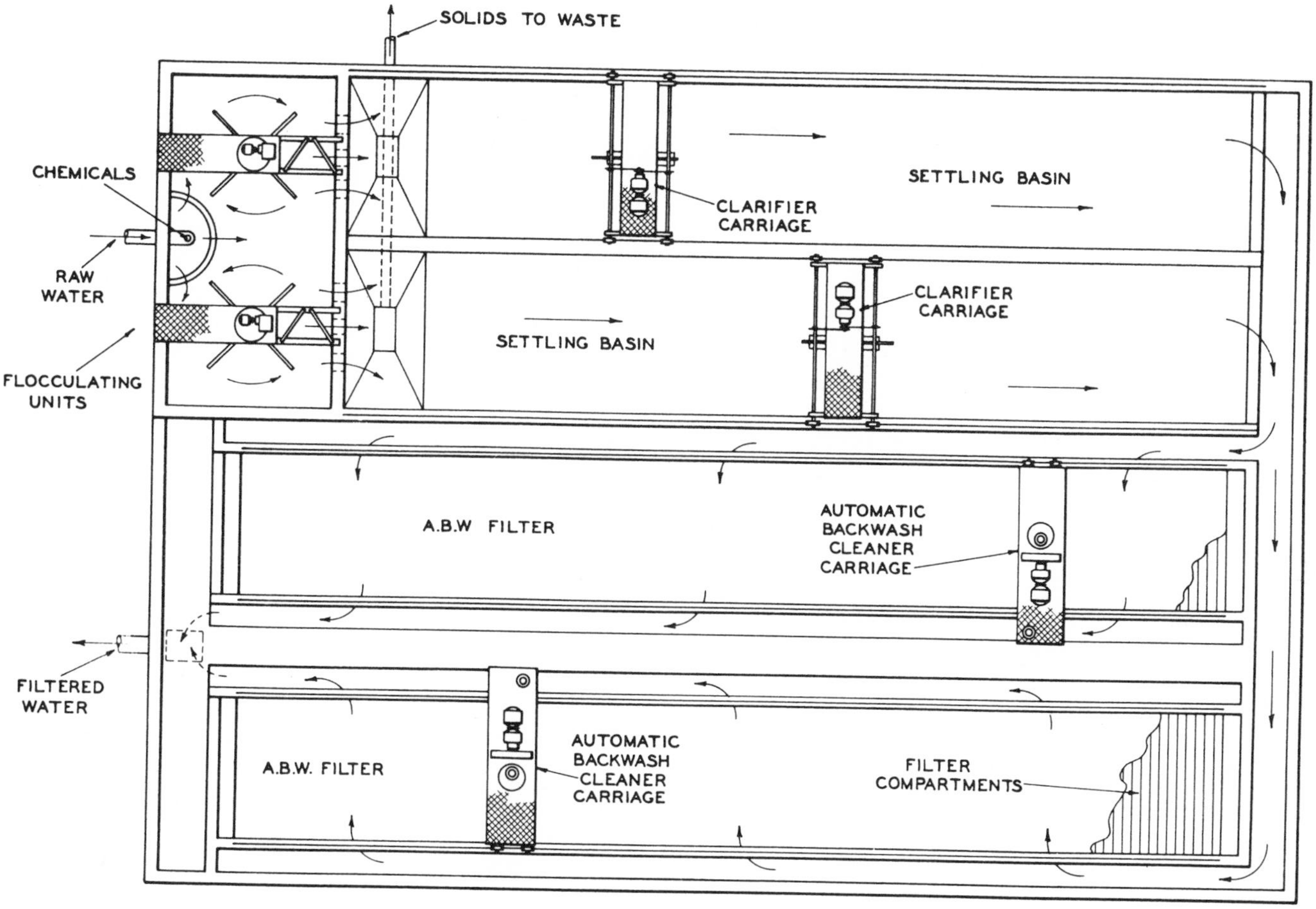

SOLIDS TO WASTE
CHEMICALS
RAW WATER
FLOCCULATING UNITS
SETTLING BASIN
CLARIFIER CARRIAGE
SETTLING BASIN
CLARIFIER CARRIAGE
A.B.W FILTER
AUTOMATIC BACKWASH CLEANER CARRIAGE
FILTERED WATER
A.B.W. FILTER
AUTOMATIC BACKWASH CLEANER CARRIAGE
FILTER COMPARTMENTS

FLOCCULATION UNITS
RECTANGULAR CLARIFIER
INFLUENT CHANNEL
A.B.W. FILTER
EFFLUENT CHANNEL
EFFLUENT CHANNEL
A.B.W. FILTER
INFLUENT CHANNEL
INFLUENT CHANNEL
WASTE CHANNEL
FLOCCULATION UNITS
RECTANGULAR CLARIFIER
A.B.W. FILTER
EFFLUENT CHANNEL
EFFLUENT CHANNEL
A.B.W. FILTER
INFLUENT CHANNEL

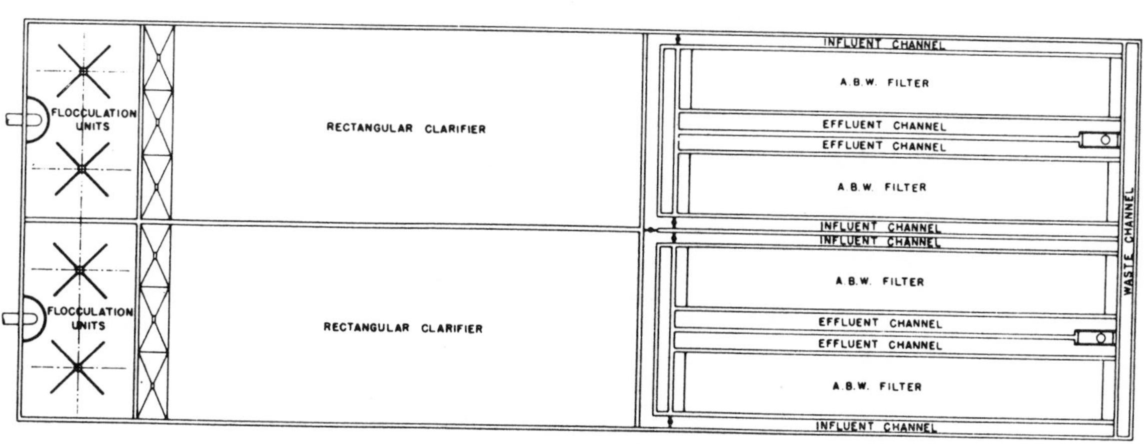

PLAN FLOW DIAGRAM FOR MULTIPLE EQUIPMENT

SPECIFICATIONS FOR HARDINGE CFNR CLARIFIERS

BEAM SUPPORTED TYPE—UP TO 60 FT. DIAMETER

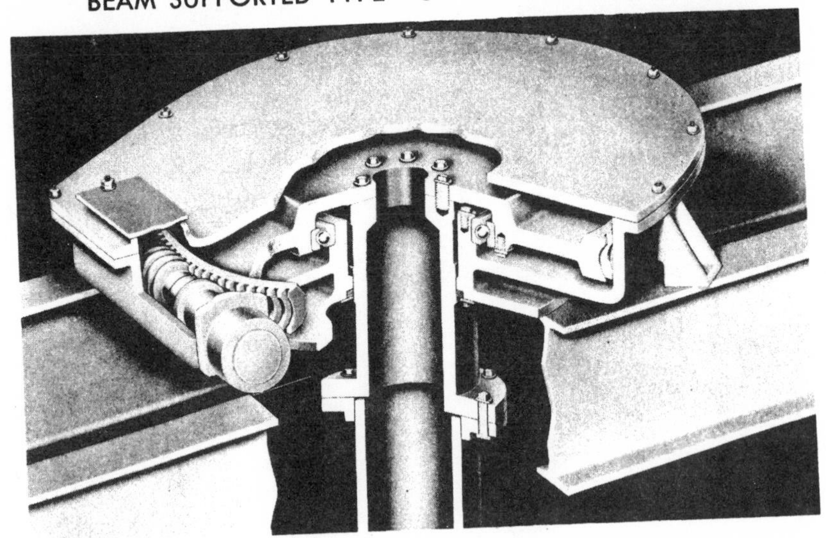

GENERAL

The clarifier shall be a steel beam supported rotating type. The sludge collecting scrapers shall be suspended from and rotated by a centrally located drive mechanism. The clarifier shall include the steel beam assembly which spans the tank, a motor drive unit, drive gearing and housing, overload protection, sludge scrapers, influent well, and, when required, a skimmer and scum hopper. It shall be as manufactured by Hardinge Company.

DRAWINGS

The general arrangement of the equipment shall be according to the plan and specifications on the drawings.

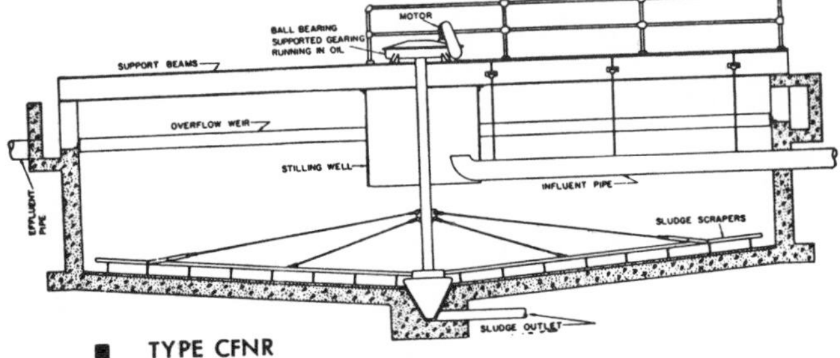

■ TYPE CFNR

SLUDGE REMOVAL MECHANISM

The sludge removal mechanism shall consist of structural arms attached to the bottom of a vertical torque tube with vertical and horizontal reinforcing between the arms and torque tube. The sludge scrapers shall attach to the underside of the arms, and shall have adjustable levelling strips to insure close running clearance with the tank bottom.

DRIVE UNIT

The drive unit shall be carried on steel beams. The gearing shall consist of a Type GA Meehanite cut tooth worm gear, a worm of hardened forged steel, Meehanite or bronze, worm shaft, sprockets, roller chain, and gear motor. The gear and worm shall be enclosed in a weatherproof housing of cast Meehanite and steel, and shall operate in an oil bath. The housing shall include an inspection door conveniently located over the worm.

The motor shall be a geared head, weatherproof motor.

MECHANISM SUPPORT
The revolving mechanism shall be supported by and rotate on a replaceable ring type ball thrust bearing designed to resist all vertical and lateral stresses developed by the sludge scrapers.

WALKWAY AND RAILING
The access walkway from side of tank to mechanism shall be steel grating. Railings will protect both sides of walkway and shall be of standard pipe railing.

INFLUENT STILLING WELL
The central stilling well indicated on the plans shall be. . . . feet in diameter.

OVERLOAD PROTECTION
A shear pin shall be included on the gearing to prevent damage to the mechanism. An ammeter shall be provided to indicate loading of the sludge scrapers, and a magnetic starting switch shall be furnished for under voltage and overload protection of the motor.

As an alternate, if a more elaborate and expensive overload protection device is wanted, an alarm bell may be hooked into the circuit to warn the operator of excessive power load, or an automatic raising device may be installed to lift the scrapers in heavy duty applications.

SKIMMER AND SCUM HOPPER (When required)
A skimmer and suitable scum hopper shall be included as part of the equipment. The skimmer shall rotate with the vertical shaft. The scum hopper shall be of steel construction and cantilevered from the side of the tank, and include a partially submerged adjustable apron on the approach side.

LUBRICATION
All bearing surfaces shall be provided with suitable means for lubrication.

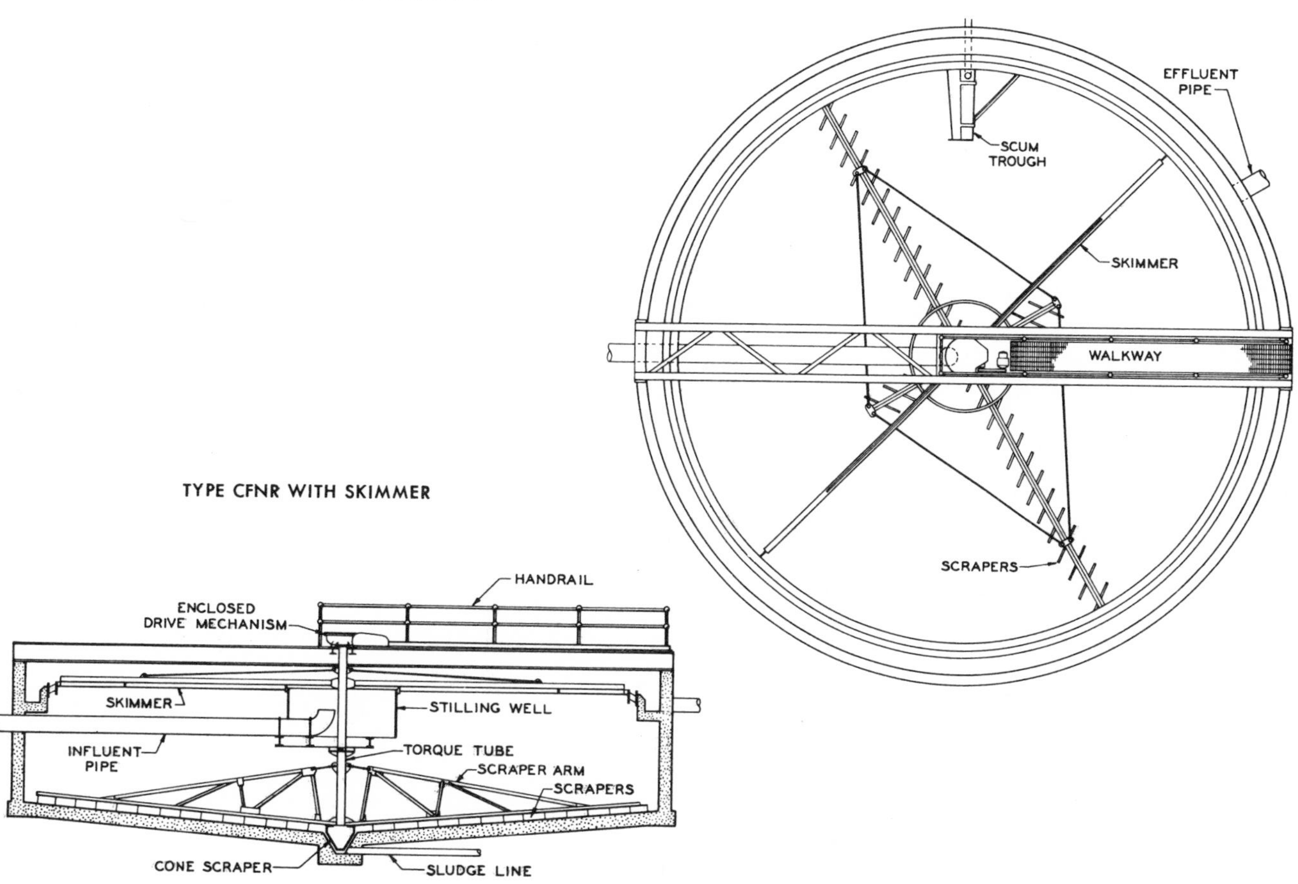

TYPE CFNR WITH SKIMMER

SPECIFICATIONS FOR HARDINGE CFCNR CLARIFIERS

CENTER COLUMN TYPE—60 FT. TO 200 FT. DIAMETER

GENERAL
The Center Column Clarifier shall consist of a rotating sludge collecting mechanism which operates in a circular tank to collect and move all settled solids to a central outlet.

The weight of the mechanism shall be carried by the center column. The torque tube which encircles the center column shall attach to the underside of the drive gear and will support the rake arms.

The clarifier shall include the steel beam assembly which extends from the tank wall to the center pier, a motor drive unit, drive gearing and housing, overload protection, sludge scrapers, influent stilling well, and, when required, a skimmer and scum hopper.

DRAWINGS
The general arrangement of the equipment shall be according to the plan and specifications as shown on the drawings.

SLUDGE REMOVAL MECHANISM
The bottom sludge removal mechanism shall consist of structural steel arms braced horizontally and vertically, and rigidly attached to the torque tube. The sludge scrapers shall attach to the underside of these arms. Each scraper shall be equipped with an adjustable steel levelling strip to insure close running clearance with the tank bottom.

DRIVE UNIT
The drive unit shall consist of a . . . . . . H.P. geared motor, suitable for . . . . . . phase, ... cycle, ... volts current.

The gearing shall be arranged to rotate the mechanism at approximately . . . . . . R.P.M.

GEARING AND MECHANISM SUPPORT
The rotating mechanism shall be supported on a ring type ball bearing. The gearing shall be enclosed in a weatherproof housing and shall be oil lubricated.

OVERLOAD PROTECTION
The overload protection shall consist of a shear pin on the drive shaft, an overload relay starting switch for the motor, and an ammeter to indicate the load on the scrapers.

BEAMS AND WALKWAY
The beams extending from the tank wall to the center pier shall carry a steel grating walkway and standard pipe handrailing on each side of the walkway and around the centrally located drive mechanism.

INFLUENT STILLING WELL
To insure uniform distribution of the incoming feed, the central stilling well shall be . . . . . . feet in diameter.

CENTER STEEL COLUMN
A center steel column shall be furnished to support the vertical load of the mechanism and to absorb the torque of the drive. This column shall also serve as a riser pipe for the incoming feed.

SKIMMER AND SCUM HOPPER (When required)
The skimmer shall be supported by a structural frame attached to one of the scraper support arms. The scum hopper shall be of steel construction and cantilevered from the side of the tank.

LUBRICATION
All bearing surfaces shall be provided with suitable means for lubrication.

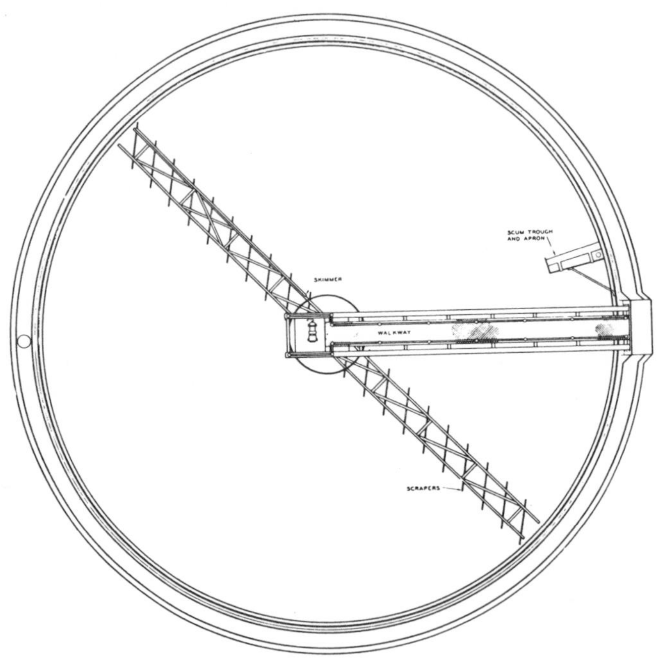

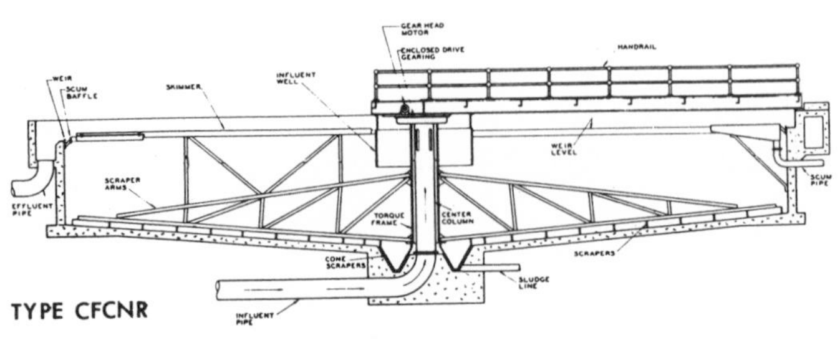

TYPE CFCNR

ROTOR AERATION IN THE OXIDATION DITCH

The Lakeside Oxidation Ditch process is a modified form of the activated sludge process and may be classified in the extended aeration group. This type of plant produces BOD reductions of 90% or better at low first costs.

The principal components of the Oxidation Ditch are the aeration ditch, cage rotor, final settling tank, return sludge pump or air lift, and excess sludge handling facilities. See the Figure.

The raw sewage passes directly through a bar screen to the ditch. The ditch forms the aeration basin and here the sewage is mixed with previously formed active organisms. The cage rotor is the aeration device which entrains the necessary oxygen into the liquid and keeps the contents mixed and moving. The mixed liquor in the ditch flows to the clarifier for separation. Quiescent conditions in the clarifier afford separation of the solids, formed in the ditch, from the liquid. The clarified liquid passes over the effluent weir and is discharged directly to the receiving stream.

The settled sludge is removed from the bottom of the clarifier by an air lift or pump and is returned to the ditch. All sludge formed in the process is returned to the ditch. Scum which floats to the surface of the clarifier race is removed and also returned to the Oxidation Ditch for further treatment.

Since the Oxidation Ditch is operated as a closed system, it will be necessary to remove some sludge from the process periodically. Excess sludge may be dried directly on sludge drying beds or stored in a holding tank for later disposal.

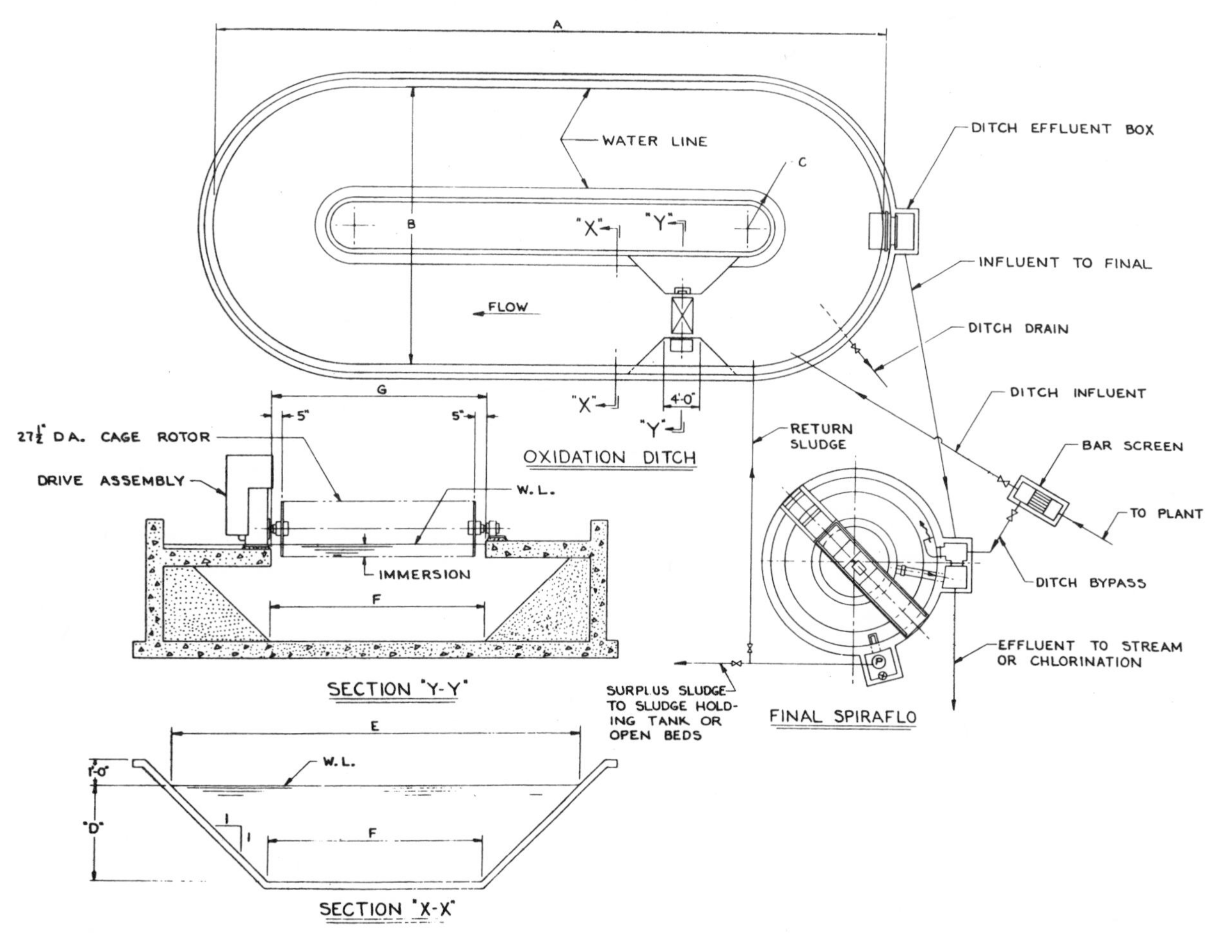

Marolf, Inc.

MARJECT PACKAGED WASTEWATER LIFT STATION

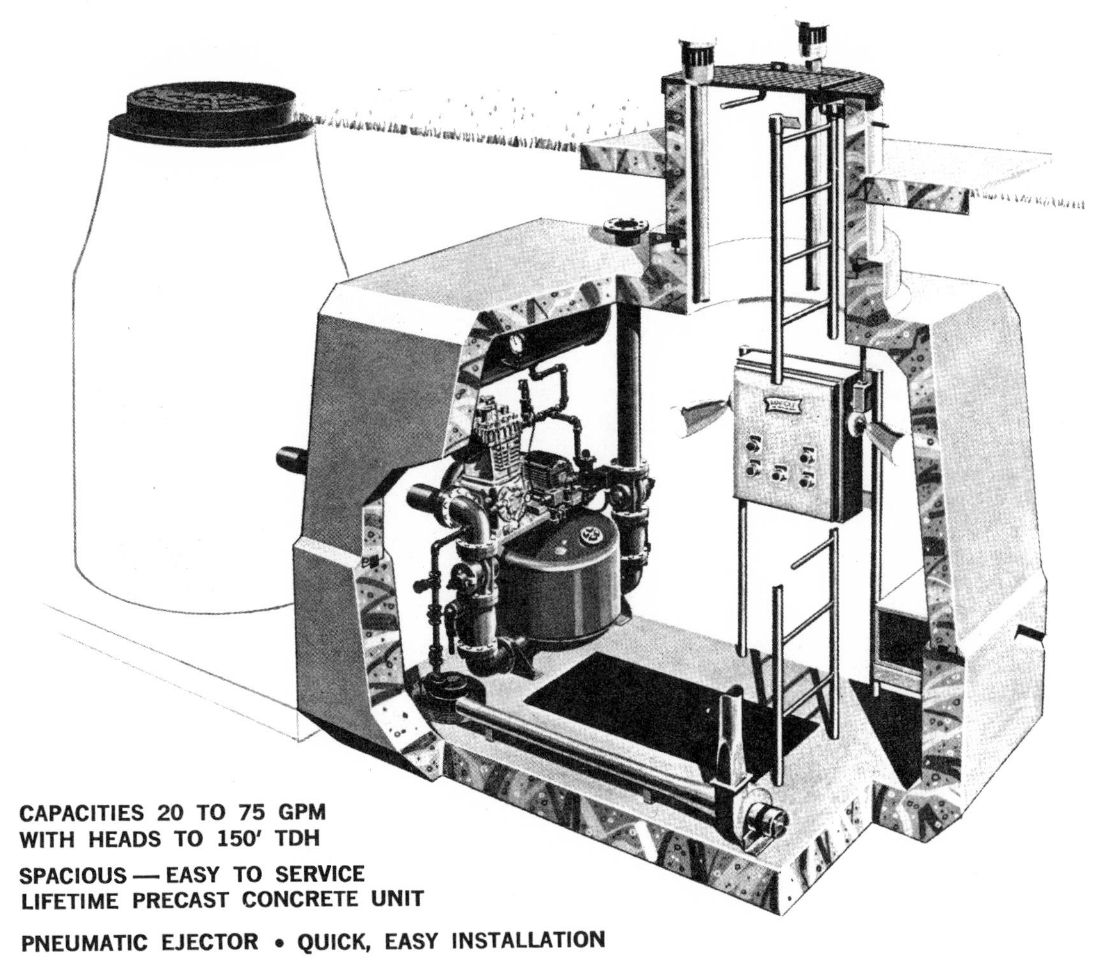

CAPACITIES 20 TO 75 GPM WITH HEADS TO 150' TDH

SPACIOUS — EASY TO SERVICE LIFETIME PRECAST CONCRETE UNIT

PNEUMATIC EJECTOR • QUICK, EASY INSTALLATION

The Marject lift station is designed for handling small flows with solids up to 3 1/2". It is odor-free because the pneumatic ejectors are hermetically sealed. Quick cycling prevents accumulation of solids. "Automatic takeover" controls can be provided. Marject lift station is made of 5000 psi precast concrete, the most durable material for such application where moisture prevails. Thick bottom sections of concrete encasement add ballast strength and serve as rigid platform for pre-mounted equipment. Floor has built-in slope for draining.

AUTOMATIC OPERATION

The electro-controlled receiver fills by gravity through an inlet check valve. When the incoming liquid reaches the electrode, the compressor forces air into the receiver, which then ejects the liquid through the discharge check valve. Once the liquid in the receiver reaches a predetermined level (timed pump-down) the compressed air supply stops and the residue air is vented to atmosphere. The unit is then ready for the next cycle.

HEAVY-DUTY EJECTORS

The air compressors are single-stage, multi-cylinder type with oil pressure lubrication. They have oversize, tapered roller bearings; counterweighted crankshafts dynamically balanced for vibrationless operation; and more efficient cooling through vertical cooling fins. Quiet operation results from cushioned steel valves. The receivers are epoxy-lined for smooth, liquid flow and long life. Either simplex or duplex receivers available with dual compressors. Air storage tank provided where large air requirements needed.

Marolf, Inc.

MARMERSIBLE PACKAGED WASTEWATER LIFT STATION

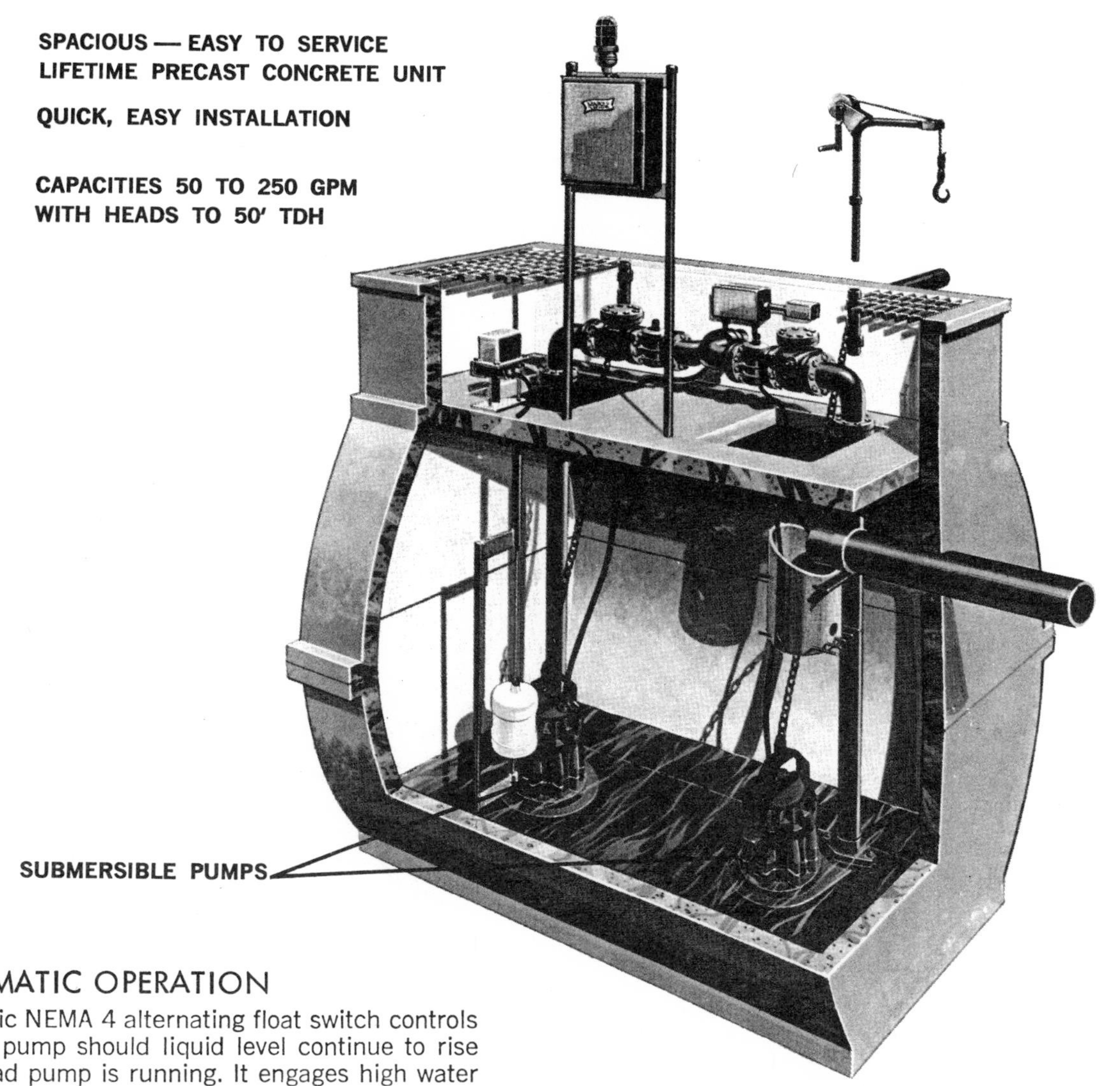

AUTOMATIC OPERATION

Automatic NEMA 4 alternating float switch controls standby pump should liquid level continue to rise while lead pump is running. It engages high water alarm light should abnormal level occur in the wet well. Non-corrosive float is resistant to fatty acids, hydrogen sulphide, chlorine water and other compounds common to wastewater.

Neoprene gasketed NEMA 4 control panel has: separate U/L listed circuit breakers for each control, alarm outlet, or other auxiliary circuit; magnetic starter with thermal overload protection in each leg; H-O-A selector switch with each motor, and 100-watt weatherproof high water level alarm light with red lens and guard.

HEAVY-DUTY PUMPS

Heavy-duty submersible pumps available in 3" or 4" discharge sizes with a range from ¾ to 5 hp. Solids can pass through without aid of screens or shredders. Pump impellers fully enclosed and dynamically balanced with extra smooth passageway. Flood-proof motors are completely sealed against water, moisture, dirt and sewage gas. Pumps don't need to be fully submerged to dissipate heat. Pumps will operate with low water levels.

The Marmersible submersible lift station is of an advanced rectangular design for easy access to components for servicing. Round models available for smaller units. Self-contained wet well with a dry valve chamber above.
Shipped as a totally complete unit or with on-site assembly by Marolf personnel. All components and the structure of the unit covered by "unified" responsibility. Marmersible chamber is made of 5000 psi precast concrete, a durable material for below ground structures. No base slab required for installation.

Marolf, Inc.

MARIZONTAL PACKAGED WASTEWATER LIFT STATIONS

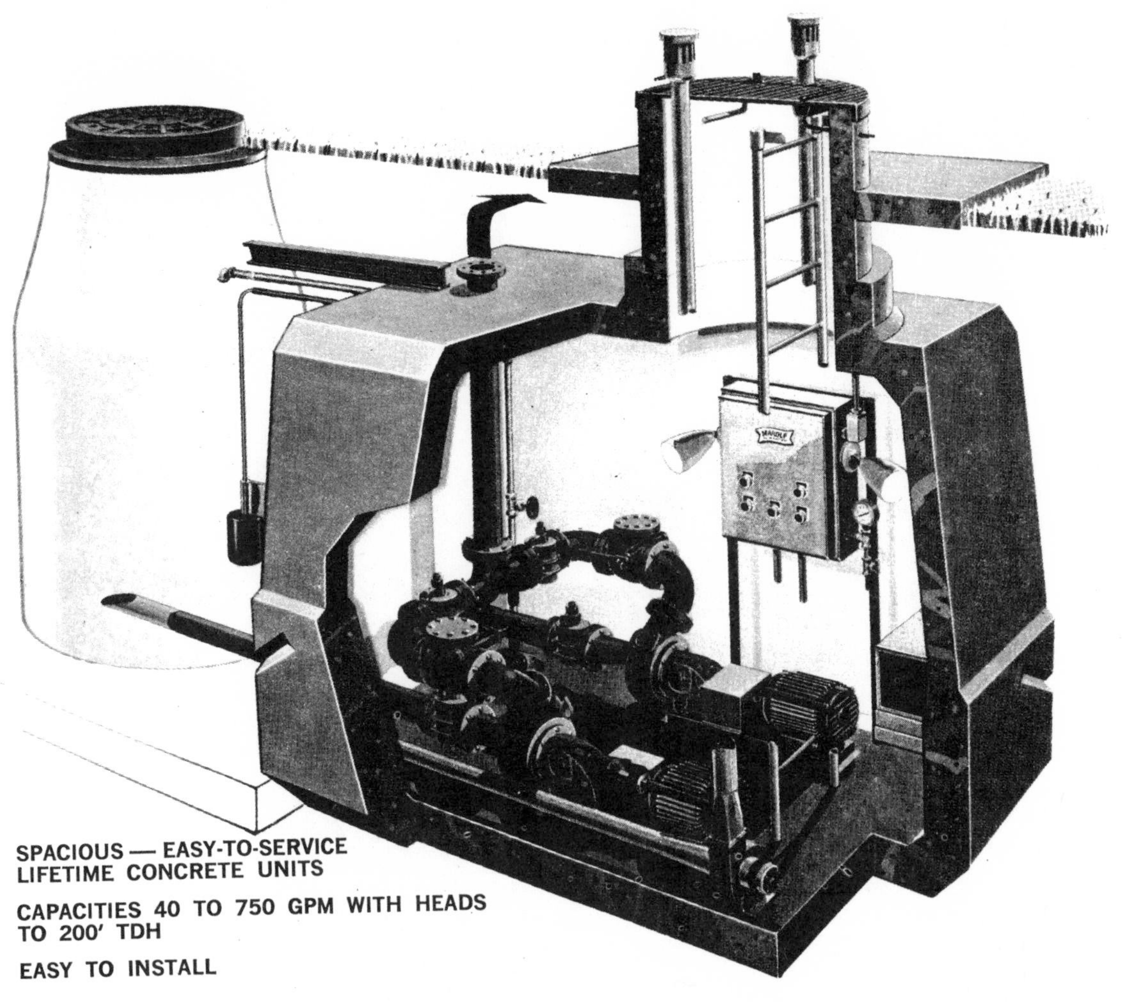

ADVANCED PACKAGED DESIGN

The Marizontal dry-pit lift station is of an advanced rectangular design which comprises a complete self-contained unit. It is shipped by truck to the site. With the aid of the proper lifting sling, it can easily be placed in excavation. No base slab is required under the station. Even the pre-cut wet well piping is included in the packaged system. Large chamber size simplifies incorporation of variable speed equipment.

Pre-cast, 5000 psi concrete is the most durable material known for underground use especially where moisture and variations in chemical composition of the soil are present. Concrete has been known to last for ages beyond the life span of man. The thicker bottom section of the encasement adds ballast and strength, and serves as a rigid platform upon which equipment is mounted. This also eliminates the need for base slabs.

The station can be installed either just below the surface or up to 30 feet below the ground level. Initial start-up and operation are done by our factory service department. We also train your operators whenever necessary to insure utmost satisfaction.

DRYNESS INSURED

To insure dryness, a sump is cast in the base. A dehumidifier is also provided. Vital steel components used in the inside are hot-dipped galvanized for longest possible life. All components are standard name brands made to rigid specifications for quality and performance. When parts are needed you do not find yourself in "left field" with "discontinued" makes.

BUILT TO LAST

All pre-cast concrete 36″ entrance tubes are painted, waterproofed, match-marked for easy installation, rubber-jointed, drilled and grouted. This makes the entrance tube integral with the pre-cast station walls. These stations are built to last — not rust out in a few years. The selection of concrete is sound engineering practice. No field welding or external anodes are required.

Marolf, Inc.

MARFAK PACKAGED WASTEWATER LIFT STATION

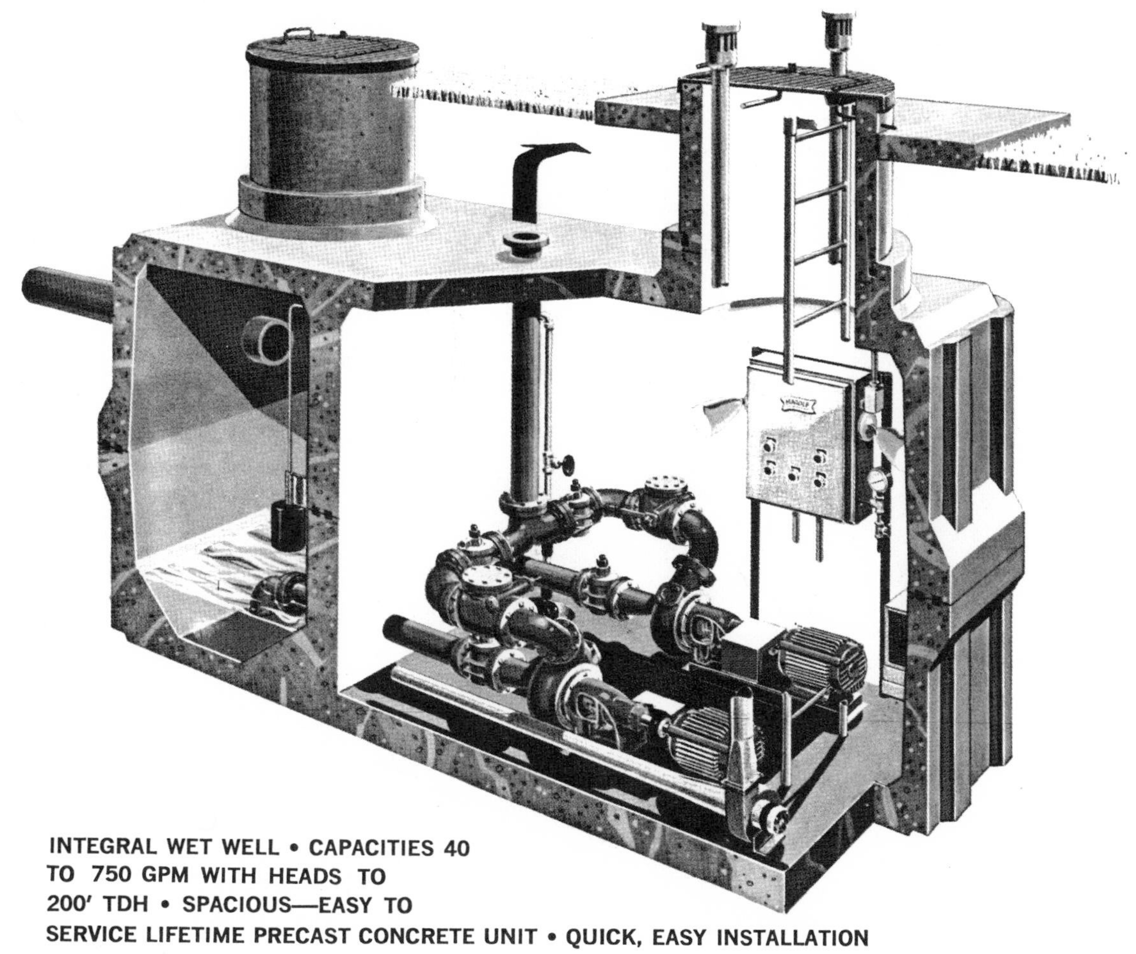

INTEGRAL WET WELL • CAPACITIES 40 TO 750 GPM WITH HEADS TO 200' TDH • SPACIOUS—EASY TO SERVICE LIFETIME PRECAST CONCRETE UNIT • QUICK, EASY INSTALLATION

The combination of lift station and integral wet well in the Marpak unit provides compactness and economy. Advanced rectangular design allows easy access to components for servicing in dry pump chamber, which is separate from the wet well. Requires no base slab. Entire lift station package comes complete with piping, control panel, sump discharge, precast structure and control piping. No field welding or external anodes required. Has straightline suction piping. Marpak is made of 5000 psi precast concrete--the most durable material for such application where moisture prevails. Thick bottom sections of concrete encasement add ballast strength and serve as rigid platform for premounted equipment.

AUTOMATIC OPERATION

The MARPROBE compressorless, adjustable liquid level control provides simplified, dependable, automatic operation. A pressure sensing probe relays wastewater pressure information to the control panel for system operation. Sump pump, blower, dehumidifier and interior lights operate automatically.

Neoprene gasketed NEMA 12 control panel has: separate U/L listed circuit breakers for each control, alarm outlet, or other auxiliary circuit; magnetic starter with thermal overload protection in each leg; H-O-A switch with each motor; and 100-watt weatherproof high water level alarm light with red lens and guard.

HEAVY-DUTY PUMPS

Two non-clogging centrifugal pumps with 2-bladed impeller utilized for 3" size solids handling. A pull-out back permits servicing of rotating assembly without disturbing connections. These are the benefits you need in solids handling pumps. They are made of high-grade, ground and polished steel with renewable shaft sleeve to insure long packing box life and shaft protection.

Extra-large, heavy-duty bearings outside pump casing — safe from dirt and grit — designed for continuous operation.

Marolf, Inc.

MARPRIME PACKAGED WASTEWATER LIFT STATION

SELF-PRIMING

CAPACITIES 40 TO 1000 GPM WITH HEADS TO 80′ TDH

SPACIOUS — EASY TO SERVICE LIFETIME PRECAST CONCRETE UNIT

QUICK, EASY INSTALLATION

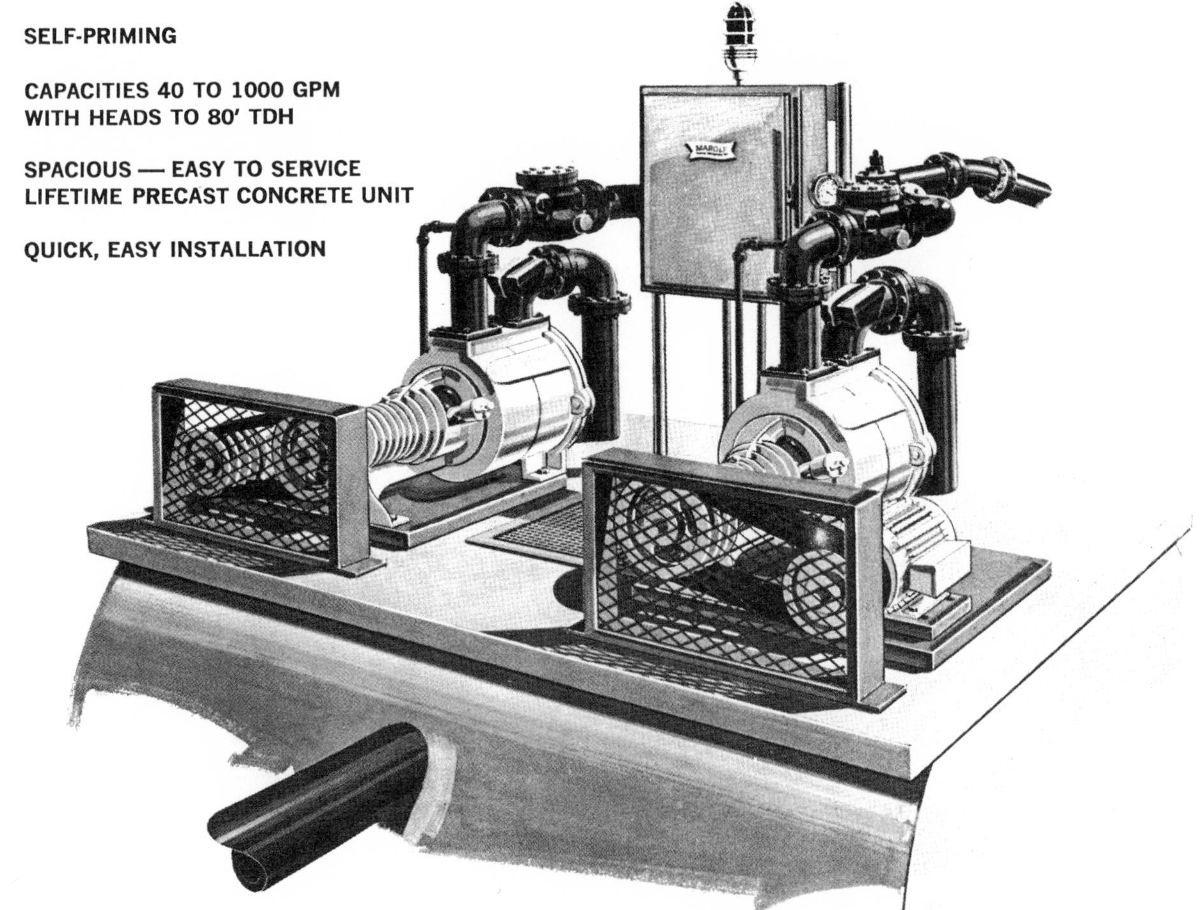

The Marprime lift station is designed for handling solids up to 3" for lift to 12'. Pumps are both self-priming and non-clogging. Shipped as a complete package with piping, control panel, sump discharge, precast structure and control conduit. All components are covered by the "unified" responsibility offered by Marolf. No field welding or external anodes are required.

Marprime lift stations are made of 5000 psi precast concrete, the most durable material for such application where chemicals and moisture prevail.

AUTOMATIC OPERATION

A non-corrosive PVC float governs mechanical alternator for sending signal to a float switch. Switch starts standby pump should the liquid level in the wet well continue to rise while the lead pump is running. Should abnormal liquid level occur in the wet well, a highwater alarm is engaged. NEMA 4 control panel has: U/L circuit breakers for each control, alarm outlet or auxiliary circuit; magnetic starter with thermal overload protection in each phase; and a Hand-Off Auto selector switch for each motor.

SELF-PRIMING PUMPS

Non-clogging, self-priming centrifugal pumps will handle solids up to 3″. They have a dynamic suction lift of 12′ (over 12′ consult the factory). Pumps are protected against leakage by a double mechanical seal with built-in suction check valve. Easy inspection is possible with removable hatch-type end cover, replaceable wearplate, and a two-vane, semi-open impeller with a renewable shaft sleeve made of high carbon steel. The pumps are mounted horizontally on top of the lift station and are multiple V-belt or direct-coupling connected. Heat-dissipating fins keep pump bearings cool. Gate valves are flanged AWWA, IBBM, NRS double-disk type. Plug valves can be provided when desired. The check valves are flanged AWWA, IBBM type.

Marolf, Inc.

STRESS-KEY PACKAGED SEWAGE TREATMENT PLANTS

For restaurants, motels, campgrounds, entertainment facilities . . . subdivisions, mobile home parks, shopping centers . . . industrial plants, laundries, dairies . . . schools, hospitals, nursing homes . . . modern decentralized municipal systems.

"Stress-Key" design permits immediate installation of our heavily reinforced, highly dense, pre-cast concrete wall sections on your concrete base slab. A combination Cap-Beam or Walkway pre-cast beam set on the top of the panels locks the entire wall system.

Plant Advantages

- Odor-free, quiet
- Saves 50% installation time
- Saves 20% to 50% in power operating cost
- Expandable — reduces initial investment
- Dependable delivery — installed and running within 4 to 8 weeks of official approval, depending upon size . . . start-up and operation instruction provided
- Below-ground or handsome above-ground installation
- Meets pollution-control codes
- Efficient — 94%+ BOD removal
- Sizes range from 15,000 to 1 million or over gallons per day
- Factory-trained preventive maintenance service available
- Certificate of Performance by the National Sanitation Foundation — complete data upon request
- Warranty — 1-year guarantee of all mechanical equipment
- Automated for minimum attention

Pre-Cast Concrete Advantages

- Cost 15% to 20% less than steel or cast-in-place concrete
- Lifetime durability — no rust
- Strong — 5000 psi
- Rigid — modular designs
- Pre-shrinking assures a crack-free structure
- Rectilinear — takes less space than round tanks
- Expandable — design construction simplifies additions
- Usable for storing most liquids or dry materials

The wall sections are pre-cast with a deep, continuous groove on both edges. A steel Stress-Key wedge, shaped like a barbell, is inserted in each hole, locking together adjoining sections. Grout is poured into the grooves, completing the structure.

The cap-beam or walkway knits the entire structure into a solid mass. A portion of the walkway beam is hollowed to receive concrete along its entire length. This forms a bond not found in any similar construction.

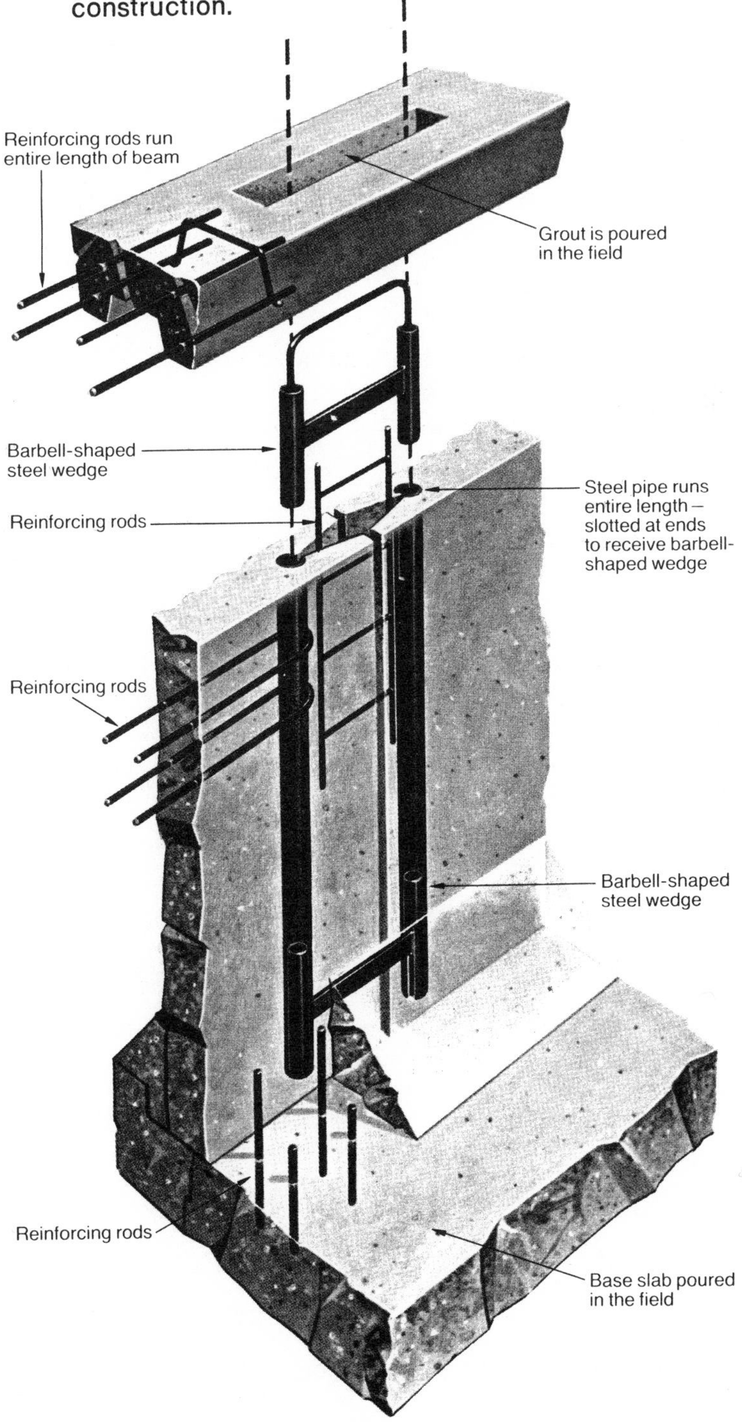

IPC ADVANCED WASTEWATER TREATMENT SYSTEM

The Independent Physical Chemical System developed by Met-Pro is a new, reliable advanced wastewater treatment system that produces a consistently high quality effluent from the day the plant goes into operation. It can be easily installed in the field requiring only a minimum amount of space. It can be adjusted easily to meet changing load conditions.

The IPC System does all the work of primary, secondary and tertiary treatment in one integrated, independent system. The troublesome upsets and operating controls associated with biological systems are thus eliminated.

WHERE IT SERVES

Small Communities - where it will provide an effluent low in B.O.D., suspended solids, phosphates and bacteria as required by local, state and federal regulatory agencies.

Developers - permitting them to have immediate sewage treatment so that construction can proceed without costly delays.

Trailer Camps - where it makes the difference between being in operation or shut down because of inadequate wastewater treatment.

Contractors - in areas not served by an existing sewage treatment plant the benefits are two-fold: it solves the wastewater treatment problem and produces a high quality effluent with reuse possibilities.

How It Works

This advanced wastewater treatment system incorporates coagulant addition to raw wastewater, flash mixing, flocçulation, clarification, granular carbon contacting, filtration and disinfection to produce an effluent of very low B.O.D., suspended solids, phosphates and bacteria content. Nitrogen removal can be included if reduction of nitrogenous oxygen demand is necessary.

Met-Pro's System is "advanced" because it employs the most recent sophisticated technology developed in the field of physical-chemical treatment. It combines this technology with field proven modules that have been designed for compatibility. The result is a compact system that provides, consistently, a very high quality effluent.

Plant Capacity (Gals./Day)	Space Req'd (Sq. Ft.)	Number of Skids	Total Power Requirements (Horsepower)	Shipping Weight (Lbs.)
5,000	208	1	8	11,000
10,000	208	1	9	12,000
15,000	208	1	10	14,000
25,000	264	1	14	19,000
35,000	304	1	15	20,000
50,000	367	1	19	23,000
75,000	480	2	25	27,000
100,000	590	4	28	37,000
150,000	816	4	36	47,000
300,000	1632	8	62	84,000

Met–Pro Systems Division

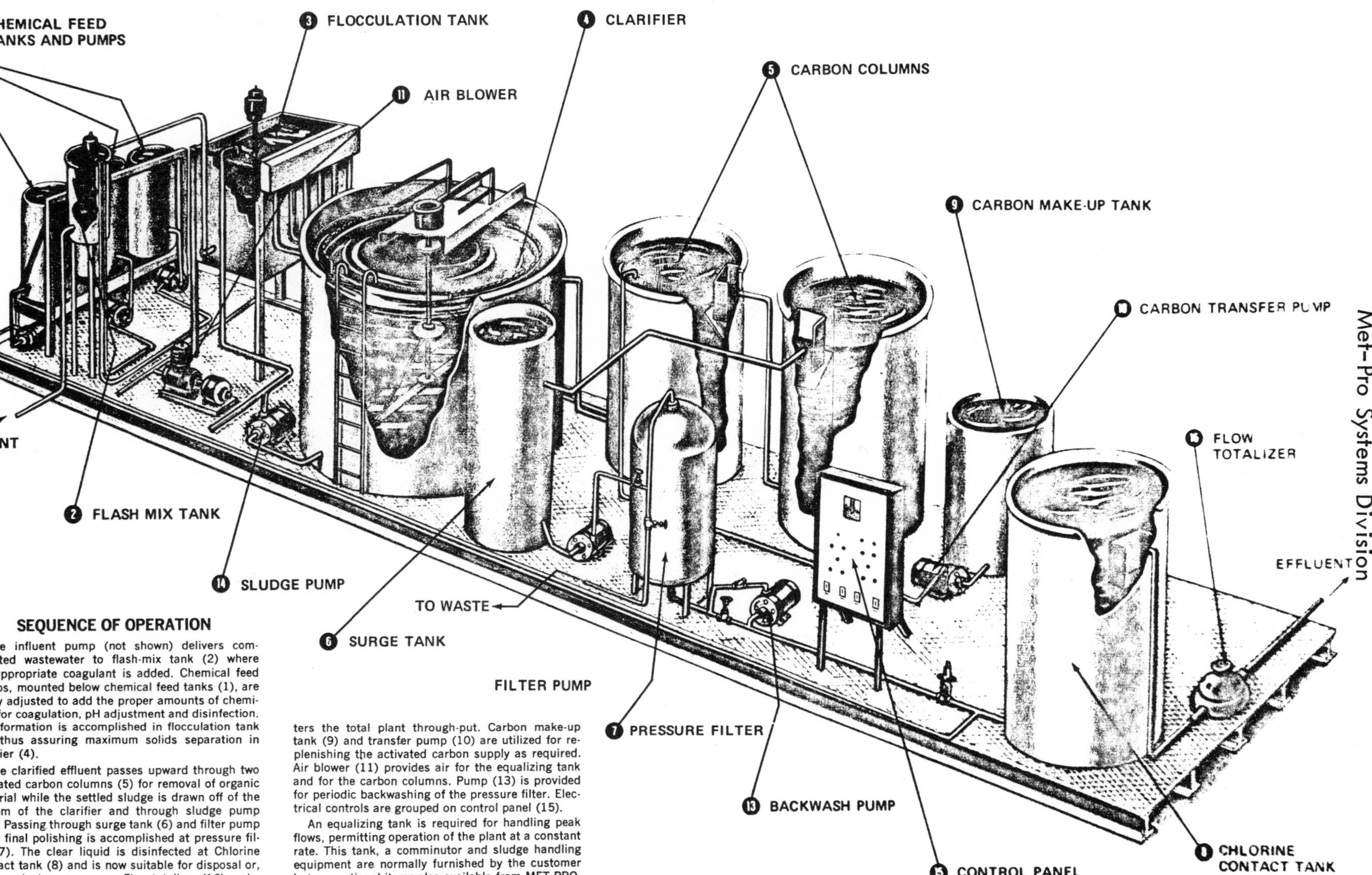

SEQUENCE OF OPERATION

The influent pump (not shown) delivers comminuted wastewater to flash-mix tank (2) where the appropriate coagulant is added. Chemical feed pumps, mounted below chemical feed tanks (1), are easily adjusted to add the proper amounts of chemicals for coagulation, pH adjustment and disinfection. Floc formation is accomplished in flocculation tank (3), thus assuring maximum solids separation in clarifier (4).

The clarified effluent passes upward through two activated carbon columns (5) for removal of organic material while the settled sludge is drawn off of the bottom of the clarifier and through sludge pump (14). Passing through surge tank (6) and filter pump (12), final polishing is accomplished at pressure filter (7). The clear liquid is disinfected at Chlorine Contact tank (8) and is now suitable for disposal or, in many instances, re-use. Flow totalizer (16) registers the total plant through-put. Carbon make-up tank (9) and transfer pump (10) are utilized for replenishing the activated carbon supply as required. Air blower (11) provides air for the equalizing tank and for the carbon columns. Pump (13) is provided for periodic backwashing of the pressure filter. Electrical controls are grouped on control panel (15).

An equalizing tank is required for handling peak flows, permitting operation of the plant at a constant rate. This tank, a comminutor and sludge handling equipment are normally furnished by the customer but are optional items also available from MET-PRO.

SERIES 14000

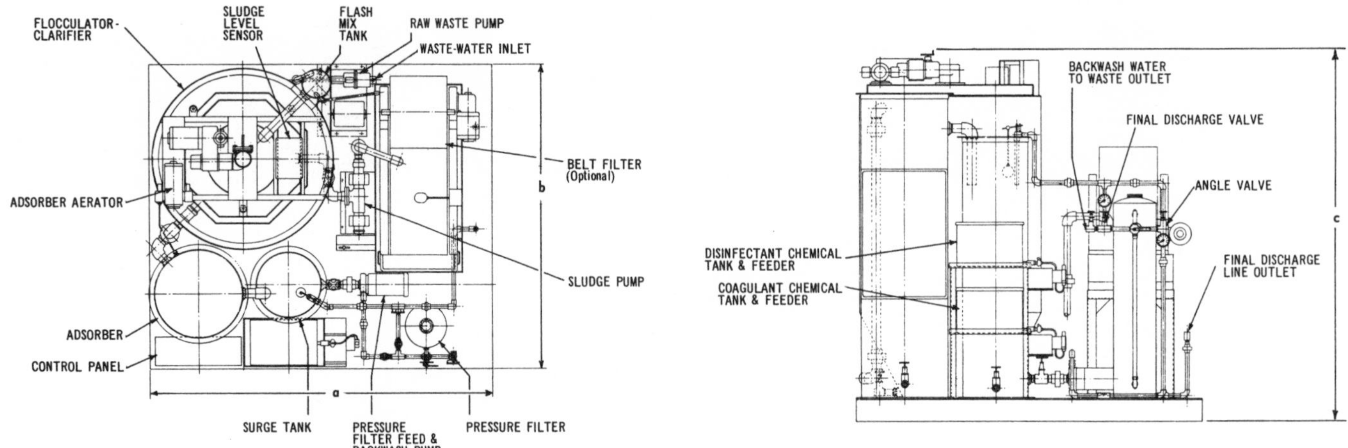

ENGINEERING DATA

Model No.	Capacity GPD	Length a	Width b	Height c	Shipping Wt. (lbs.)	Total Horsepower	Pipe Connections	
							Inlet	Outlet
14007	7,000	8'–0"	7'–0"	8'–8"	5,700	2.75	3/4"	3/4"
14015	15,000	13'–0"	8'–0"	8'–8"	10,000	2.75	3/4"	1"
14025	25,000	20'–0"	8'–0"	8'–8"	15,000	3.50	1½"	1"

POWER REQUIREMENT: 220 V, 3 Phase, 60 Hertz standard; others available.

DESCRIPTION OF OPERATION

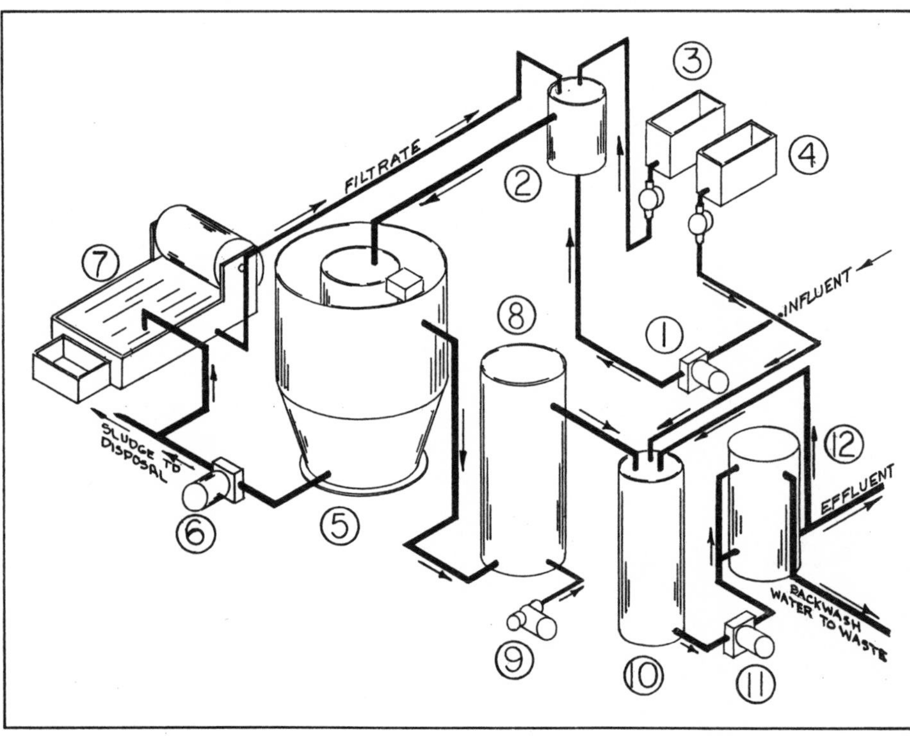

Comminuted sewage is pumped from an equalizing tank (not shown) by raw waste pump (1) to flash mix tank (2). Coagulant feeder (3) delivers a proportioned amount of chemical solution to the flash mix tank where intimate contacting is accomplished by means of high speed agitator. From the flash mixer, the wastewater flows by gravity into the flocculating section of clarifier (5) where gentle agitation promotes floc formation. As sedimentation takes place in the clarifier the settleable solids are collected and are pumped (6) to disposal. A disposable media filter (7) is offered as an option for sludge concentrating prior to ultimate disposal.

The clarifier effluent flows upflow through granular carbon adsorber (8) for removal of dissolved organic materials and into surge tank (10). Adsorber aerator (9) proportions air into the bottom of the adsorber, maintaining an aerobic fluidized carbon bed. Filter pump (11) delivers the waste through pressure filter (12) for final polishing. Disinfectant feeder (4) pumps a controlled amount of chlorine solution into the surge tank for disinfection.

LIGHTNIN MIXERS AND AERATORS

Surface Aerators: The simplicity and maximum efficiency of surface aeration makes it the most commonly used mechanical aeration system. The large diameter, low speed, high efficiency Lightnin turbines combine maximum oxygen transfer with effective mixing and solids suspension. Lightnin impellers provide superior performance over wide ranges of variations in liquid levels, and biological solids concentrations. Surface aerators are usually mounted on a simple fixed platform. Lightnin will recommend a platform design for specific requirements.

Submerged aerator

Submerged aerators transfer oxygen and mix with turbines dispersing air near the basin bottom. The design of the system gives high oxygen transfer and solids suspension while eliminating noise and spray problems. High uptake rate wastes can be handled by the unique design which allows high power to be applied to high gas rates in small volumes of waste. Lightnin submerged aerators offer their greatest advantage where oxygen demand varies over a wide range. Oxygen transfer capabilities from zero to 100% of design capacity can be obtained simply by varying the gas rate to the submerged turbine. The constant speed of the turbine itself provides mixing and solids suspension at any air rate. Submerged turbine aeration basins can be completely covered and vented to prevent air pollution.

Combined Aerators: Combined aerators use both surface and submerged turbines on a common shaft. In processes having predictable variation, to economize on operating costs, oxygen transfer can be proportioned between the surface aerator and submerged aerator. During peak operating periods, both impellers are used to provide maximum oxygen transfer.

Series DAT Aerators: The high pumping capacity Lightnin DAT Aerator is specifically intended for waste treatment systems designed to treat high strength wastes in tanks over 20 feet deep. The Lightnin drive with a down flow, high efficiency impeller operating in a draft tube, disperses air from a sparge ring located below the impeller.

MicroFLOC TECHNOLOGY

Tube-Type Clarification: It has long been recognized that a settling basin should be as shallow as possible and that detention times of only a few minutes could be used in very shallow basins. Neptune MicroFLOC has found small diameter tubes to provide optimum hydraulic conditions for sedimentation.

By inclining the tubes 60 degrees in the direction of flow, continuous gravity drainage of the settleable materials from the tubes can be achieved. The incoming solids settle to the tube bottom and then exit from the tubes by sliding downward. A flow pattern is established in which the solids settling to the tube bottom are trapped in a downward flowing stream of concentrated solids. This countercurrent flow of solids aids in agglomerating particles into larger and heavier particles which settle against the velocity of the upward flowing water. This self-cleaning system of clarification allows increased capacities in less space.

Mixed-Media Filtration: When water to be filtered contains substantial amounts of solids, a conventional filter will blind at the surface in a very short time, even at low filtration rates. The reasons are apparent from an examination of a cross-section of such a filter.

An ideal filter would be the inverse of a sand filter. It would have the coarsest material at the top and the finest on the bottom. The grain and pore sizes would be uniformly graded from coarse to fine, from top to bottom.

Neptune MicroFLOC has developed a unique method of constructing a filter so that it closely approaches this ideal configuration. In this mixed-media filter, three or more materials of different specific gravities are allowed to intermix and no attempt is made to maintain separate or discrete layers of the different materials. The result is a filter in which coarse, lighter materials occupy most of the upper portion of the bed, with the intermediate materials in the center and the smaller, heavier particles at the bottom.

Mixed-media filters permit operation at much higher flow rates than conventional filters. Hydraulics often are designed for 8 gpm per square foot. Other features include longer filter runs, reduced backwash water requirements and greater filtration efficiency for consistently high quality finished water.

Coagulant Control Center: Neptune MicroFLOC has developed the Coagulant Control Center with a pilot filter system to provide a continuous and fully reliable method of determining the adequacy of the coagulant dosage.

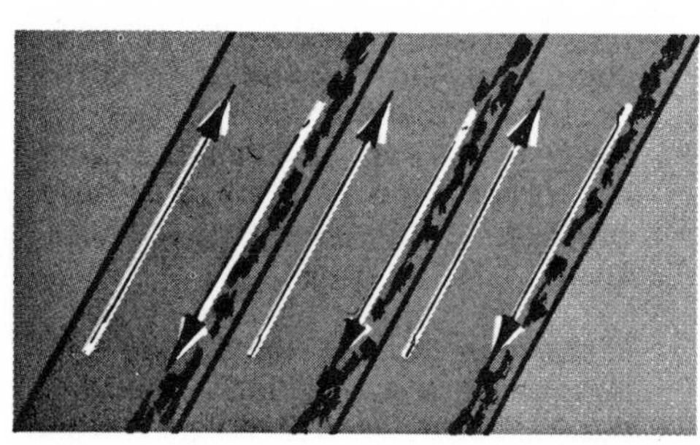

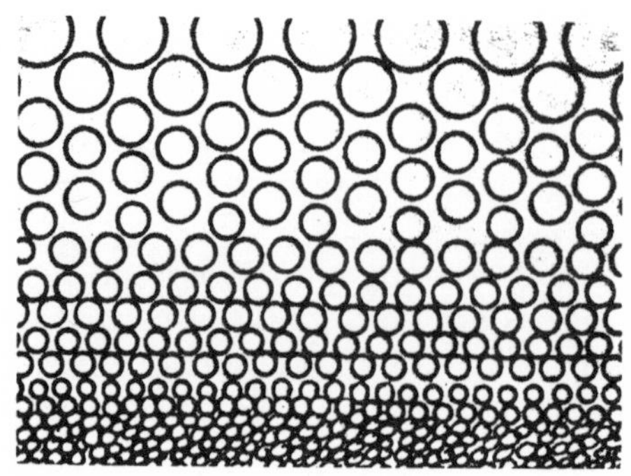

WATER BOY

Water Boy is a compact, complete factory built water treatment plant. The complete water treatment process -- coagulation, flocculation, clarification and filtration -- has been integrated into one small steel package, designed to permit either permanent or portable installation. Water Boy requires a minimum of operation skill and attention, yet provides optimum trouble free service.

Here's how Water Boy works. The required chemicals are introduced as raw water enters the plant. The chemically dosed water then enters the flash mix chamber, which can also serve as a fluidized calcite column in low alkalinity areas. Next, the treated water flows over a calibrated weir to a mechanical flocculator, and on to the unique MicroFLOC tube clarifier. Clarified water enters the mixed media filter chamber and passes through the filter bed to the underdrain system, where it's pumped to the combination clearwell and backwash tank.

Filter backwash is automatic, initiated on headloss. Positive sludge removal from the tube clarifier is integrated in the backwash, and final backwash water refills the settling chamber, cutting plant downtime and wasted water. All pumps, motors and controls are mounted integrally. Only load center electrical connections and influent, effluent, backwash and waste pipelines are needed to place the unit in service.

60 GPM unit above provides water for an oil company campsite in Alaska. Ten, 20, and 100 GPM units are also available.

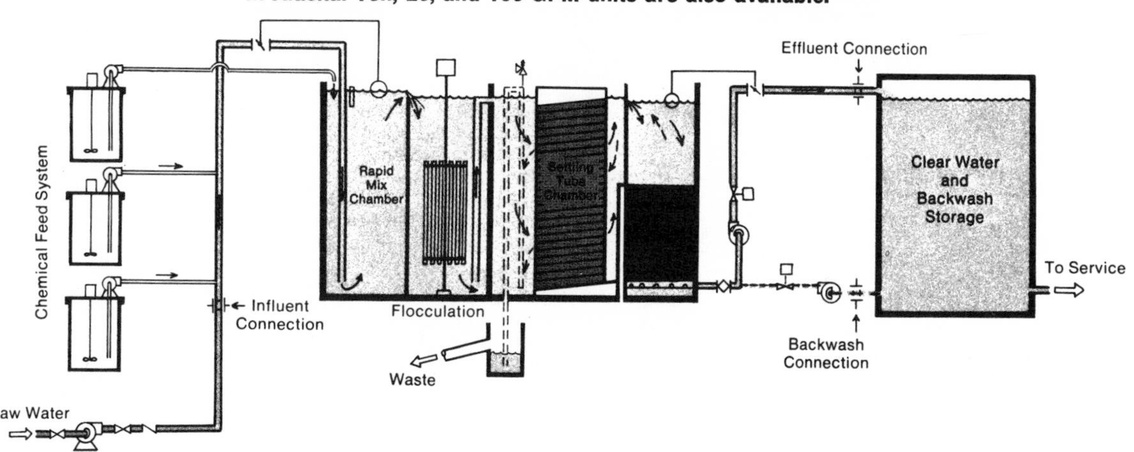

Water Treatment Systems

Neptune MicroFLOC has introduced the Aquarius concept of water treatment utilizing advanced techniques for clarification and filtration. This concept applies to the expansion of existing plants and to the design of new plants from 10 gpm to many mgd in size. The concept features: (1) Control of Coagulation (2) Efficient tube settling (3) Mixed-media filtration.

This concept can be applied to increase the capacity of existing plants without increasing space requirements, and at low capital expenditure.

First, by placing MicroFLOC tube settling modules in an existing clarifier, its capacity can be more than doubled. Depending upon the raw water conditions and the capacity desired, a portion or all of the surface area of the clarifier or rectangular flow basin might be covered with the tube modules.

Second, plant filtration capacity can be expanded through the installation of MicroFLOC mixed media in the plant filters. Filtration rates of 5 to 8 gpm/sq. ft. are standard and higher quality water is produced. Chemical costs can also be reduced and improved plant performance obtained, through the use of continuous coagulation control.

The Aquarius concept applies to new plant design. It couples the tube clarifier to the mixed media filter for most efficient production of high quality water at low capital cost. This concept is readily adaptable to any type of new plant construction, either in concrete or prefabricated steel. In concrete, the tube clarifier can be designed as a single unit, or several units in parallel to handle any given rate of flow or raw water condition. The mixed media filters can be built singly or in multiples to accomodate the effluent from the clarifiers. Automatic control of coagulation can also be provided.

Mixed-Media Beds Approach the 'Ideal' in Filter Design

The mixed media concept of water filtration offers a virtually unlimited variety of combinations of materials to suit specific uses. Actually, several hundred formulae are available from Neptune MicroFLOC, designed to control such factors as turbidity loading, operating rate, capacity for storing impurities, filtration efficiency, surge resistance and efficiency in the use of chemicals for secondary flocculation.

For example, in potable water plants, there is a need for great stability and high filter efficiency. If activated carbon is to be applied directly to the bed, then a fine bottom medium becomes essential to prevent carbon breakthrough. If the plant is to operate at variable rates, surge resistance is important. If the valves are to be manually operated, length of run is a factor. If there are algae or fibers in the raw water, the particle size at the top of the bed may have to be increased to provide storage capacity without unduly shortening the length of the run. For many industrial applications, simple high rate operation and minimum capital cost may be most important.

The MicroFLOC principle of grading the size of the particles from coarse to fine in the direction of flow is shown in an idealized drawing (Figure 1). The coarser particles are trapped in the larger voids at the top of the bed, while the finer particles penetrate to the bottom of the bed where they are held in the more finely divided media. Thus there is storage of waste material throughout the depth of the bed.

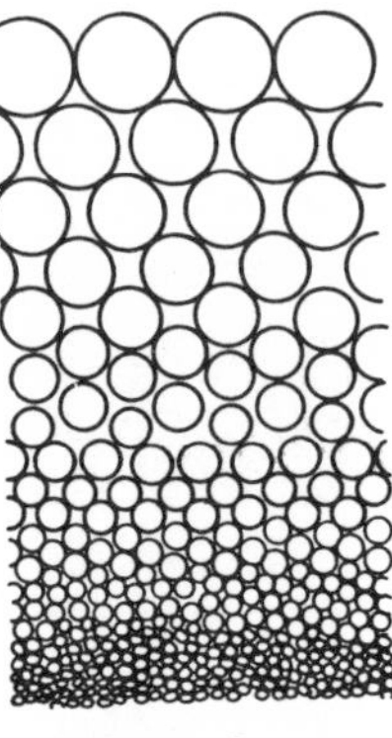

Fig. 1

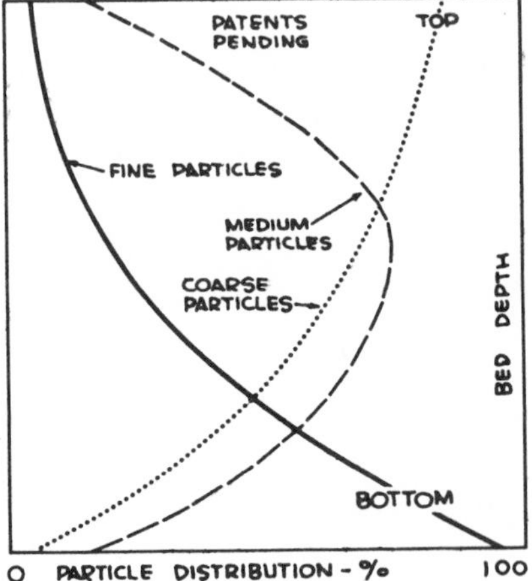

Fig. 2

The other diagram (Figure 2) shows how particles of differing sizes and specific gravities are actually mixed throughout the bed. Note that controlled mixing of materials is achieved, rather than stratification by type of material. At all points in the bed there is some of each component, but the percentage of each changes with bed depth. Thus most of the coarse particles are at the top, most of the fine particles at the bottom, with medium-sized particles predominating at the center. There is steadily increasing efficiency of filtration in the direction of flow.

Neptune MicroFLOC Inc.

MODULAR AQUARIUS - Tailored Water Treatment
in Steel or Concrete - 150 GPM to 5 MGD

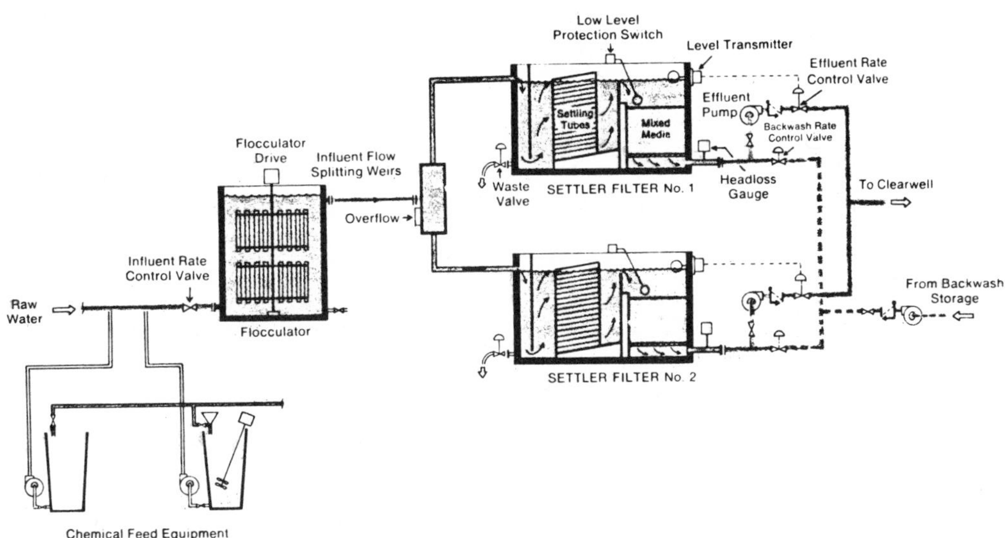

A typical flow schematic for a modular Aquarius plant, either in concrete or steel, is illustrated. The raw water is pumped or flows by gravity to the treatment system. Influent flow is set at a fixed, but adjustable, rate with the flow started or stopped by clearwell level. Influent rate can be set by either a manual valve, by a flow limiting device, or by a limit stop on the influent valve. Coagulating and sterilizing chemicals are added upstream of the influent flow control system. Typically alum, a polymer and chlorine, together with a pH adjustment chemical, are used. The chemically treated water then passes to a mechanical flocculator designed to rapidly form a highly settleable floc.

Flow from the flocculator is divided equally to each settler/filter unit by submerged weirs. From the flow splitting weir, water enters sedimentation chambers equipped with Neptune MicroFLOC's unique settling tubes. These essentially horizontal settling tubes provide a multiplicity of shallow depth settling basins approximately 1 inch in size, operating in parallel. Depending upon the application, either 24 inch or 39 inch long tubes are used. The overflow rate based on the surface area of the settling tubes is less than 300 gpd per square foot. The settling tubes are made of ABS or PVC plastic, meeting standards for municipal systems.

Advantages:

1. Modular Aquarius provides complete treatment -- coagulation, flocculation, settling and filtration. Components are performance-matched.
2. Compact design utilizes only one-third to one-half the space of conventional plants. Units can be arranged to suit available space.
3. Installation is simplified with utilization of factory-assembled steel tankage or simple concrete structures.
4. Operation is simple, yet automatic, with backwash initiated automatically by filter headloss or pushbutton.
5. A positive system of sludge removal is provided, which is coordinated with filter backwash.
6. Capital investment is low with quality of performance high. Expansion is simplified with modular design.

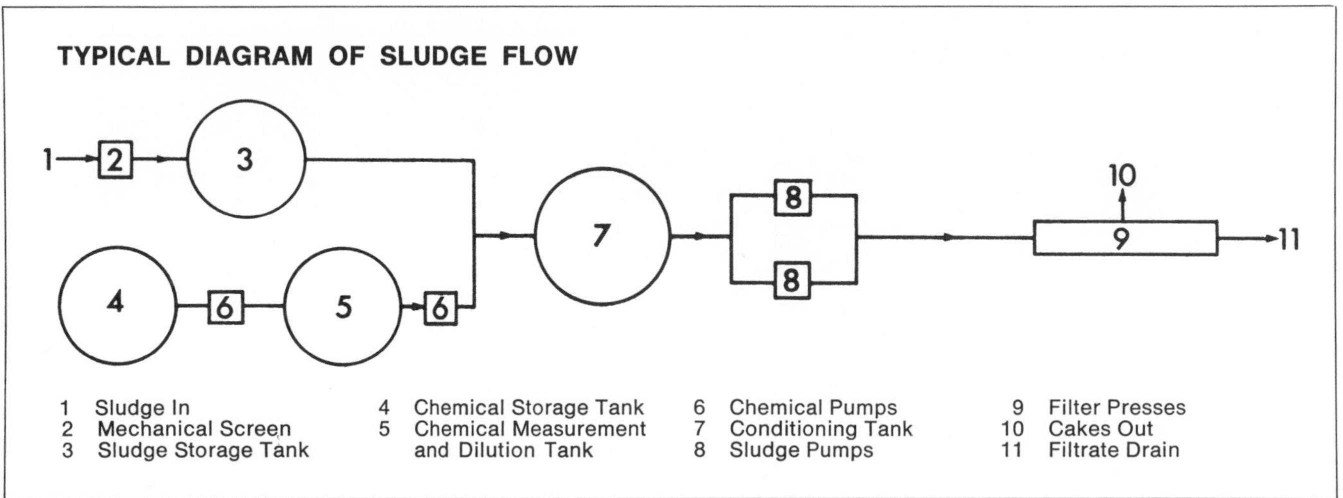

FILTER PRESSES

Nichols/E & J manufacture a comprehensive range of Filter Presses for effluent and sewage treatment. Presses are available which produce 12, 20, 24, 30, 36 or 48 inch square cakes. These cakes are normally 1 inch or 1 1/4 inches thick but this can be varied to suit individual requirements.

The 12, 20, and 24 inch presses are normally supplied with up to 60 chambers and the 30, 36 and 48 inch presses with up to 80 chambers. The number of chambers can be arranged as desired and presses are available up to 100 chambers.

Presses are supplied with single feed input at the fixed end, but special applications requiring feed from both ends can be catered for on larger presses.

FRAME STRUCTURE

The Filter Presses can be supplied with either side bar or overhead suspension for the plates and can be either floor standing or completely supported from overhead steelwork. The plates on the overslung type are supported by hanger brackets from an overhead beam. These brackets incorporate ball and roller bearings for easy movement of the plates. Individual plates can be removed from the frame without dismantling the overhead beam or side bars.

FILTER PRESS PLATES

The Filter Press plates or trays are cast in the Nichols/E & J foundry from high quality Grade 17 cast iron and are of rigid and robust construction. After casting the plates are fully machined on both sides to very close tolerances. Each plate can be provided with drain cocks to isolate it to prevent contamination of the filtrate in the event of a burst filter cloth, or with an internal porting arrangement to provide a single drain point for all the plates at the end of the press.

Steel-reinforced, rubber molded Filter Press plates are now available. Weighing less than half of the equivalent cast iron they are virtually unbreakable, having rubber staybosses to reduce flexing if they are subjected to unequal pressures. These plates offer several other advantages over cast iron plates including far better filtration flow due to the cylindrical pip surface; prolonged filter cloth life; increased cubic capacity; improved cake release and freedom from corrosion and contamination.

Fully Automatic 80-chamber 1¼″ cake
133 cu. ft. capacity 4′ Filter Press

Nichols/E & J have now produced the answer to the demand for a fast, reliable, single-lever Filter Press plate-opening mechanism. By a simple shuttle device fitted to either side, an 80-chamber press can be fully opened and closed in about 9 minutes. This device is actuated by a 1 h.p. motor which drives a crank imparting a reciprocating motion. A simple cam catch and release mechanism, moving along each side of the press picks up each plate in turn and moves it forward, allowing the filter cake to drop out.

CAKE SIZE Sq. Ins.	NO. OF CHAMBERS	FILTER AREA Sq. Ft.	VOLUME Cu. Ft. 1″Cake	1¼″Cake
12	1	2	0·083	0·104
	10	20	0·830	1·04
	15	30	1·245	1·56
	20	40	1·66	2·08
20	1	5·6	0·23	0·29
	10	56	2·3	2·9
	20	112	4·6	5·8
	30	168	6·9	8·7
24	1	8	0·335	0·419
	20	160	6·7	8·375
	30	240	10·05	12·56
	40	320	13·4	16·75
30	1	12·5	0·52	0·65
	20	250	10·4	13
	30	375	15·6	20·5
	40	500	20·8	26
36	1	18	0·75	0·937
	20	360	15	18·75
	40	720	30	37·5
	60	1080	45	56·25
	80	1440	60	75
48	1	32	1·35	1·67
	20	640	27	33·40
	40	1280	54	66·80
	60	1920	81	100·20
	80	2560	108	133·60

Osmonics, Inc.

REVERSE OSMOSIS SYSTEMS

OSMO® –30043 for Pure Water, Plating Rinse Re-use,† Sugar Concentration, etc.

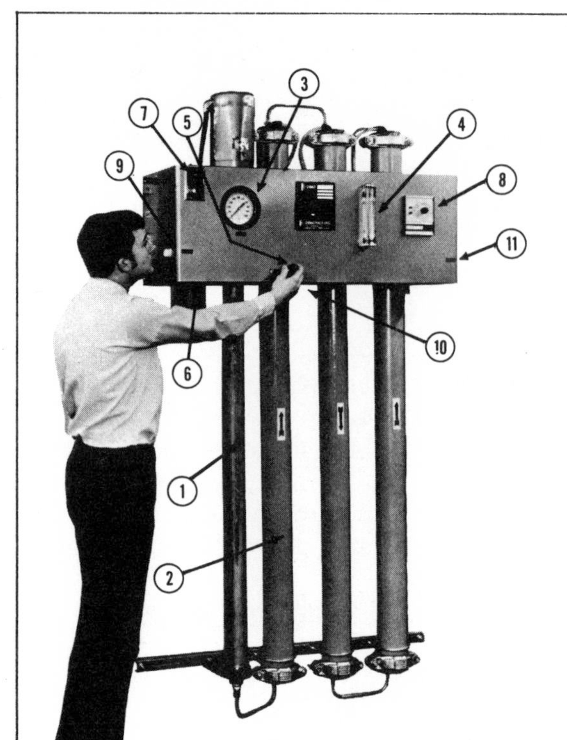

1. STAGED CENTRIFUGAL PUMP
2. PRESSURE VESSEL
3. PRESSURE GAUGE
4. FLOW METERS (Optional)
5. PRESET CONTROL VALVE
6. PREFILTER
7. STARTER
8. CONDUCTIVITY MONITOR (Optional)
9. FEED INLET
10. CONCENTRATE OR BLOWBY
11. PERMEATE OR PURE WATER

		OSMO® MODULE TYPE				
	Pressure Psi	504-97	654-97	334-97	334-89	334-0
Permeate Rate* *Gal./hr.*	200	13-16	17-21	10-16	14-22	55-77
	400	24-30	30-37	18-28	25-38	NA
	600	29-36	37-46	23-33	30-43	NA
Economic Life *Months*	200	36	36	36	36	6-36
	400	24-36	24-36	24-36	24-36	NA
	600	18-24	18-24	18-24	18-24	NA
NaCl Rejection %		94-98	94-98	94-98	85-91	0-10
Organic Cut-off *Molecular Wt.*		200	200	200	400	1000

*Assumes 1000 ppm T.D.S. @ 68°F

SOME APPLICATIONS FOR REVERSE OSMOSIS ARE:

A. WATER PURIFICATION

1. Salt separations, including chloride, sodium, calcium, fluoride, sulphates, phosphates, nitrates, carbonates, aluminum, gold, silver, nickel, copper, etc.
2. Organic separations
3. Sediment removal
4. Water reuse
5. Pretreatment for de-ionizers
6. Bacteria, spores, virus, pyrogen and algae removal
7. Boiler feed
8. Humidifier feed
9. Car wash rinses

B. WASTE WATER TREATMENT

1. Separation of plating salts
2. Reclamation of rinse waters and complete water reuse.
3. Reclamation of valuable metals from plating and photography
4. Upgrading waste water to meet State and Federal standards
5. Separation of electro-deposition paint
6. Lowering BOD and COD in waste water
7. Removing detergents and pesticides
8. Removing radioactive elements
9. Removing phosphates and nitrates.

TYPICAL MEMBRANE REJECTIONS

SALTS

CATIONS Name	Symbol	Percent Rejection	Percent Passage (Average)	Maximum Concentration Percent
Sodium	Na^{+}	94-96	5	3-4
Calcium	Ca^{+2}	96-98	3	*
Magnesium	Mg^{+2}	96-98	3	*
Potassium	K^{+1}	94-96	5	3-4
Iron	Fe^{+2}	98-99	2	*
Manganese	Mn^{+2}	98-99	2	*
Aluminum	Al^{+3}	99+	1	5-10
Ammonium	NH_4^{+1}	88-95	8	3-4
Copper	Cu^{+2}	98-99	1	8-10
Nickel	Ni^{+2}	98-99	1	10-12
Zinc	Zn^{+2}	98-99	1	10-12
Strontium	Sr^{+2}	96-99	3	—
Hardness	Ca and Mg	96-98	3	*
Cadmium	Cd^{+2}	96-98	3	8-10
Silver	Ag^{+1}	94-96	5	*
Mercury	Hg^{+2}	96-98	3	—
ANIONS				
Chloride	Cl^{-1}	94-95	5	3-4
Bicarbonate	HCO_3^{-1}	95-96	4	5-8
Sulfate	SO_4^{-2}	99+	1	8-12
Nitrate	NO_3^{-1}	93-96	6	3-4
Fluoride	F^{-1}	94-96	5	3-4
Silicate	SiO_2^{-2}	95-97	4	—
Phosphate	PO_4^{-3}	99+	1	10-14
Bromide	Br^{-1}	94-96	5	3-4
Borate	$B_4O_7^{-2}$	35-70**	—	—
Chromate	CrO_4^{-2}	90-98	6	8-12
Cyanide	CN^{-1}	90-95**	—	4-12
Sulfite	SO_3^{-2}	98-99	1	8-12
Thiosulfate	$S_2O_3^{-2}$	99+	1	10-14
Ferrocyanide	$Fe(CN)_6^{-3}$	99+	1	8-14

*Must watch for precipitation, other ion controls maximum concentration.
**Extremely dependent on pH; tends to be an exception to the rule.

The following are typical rejections of salts and organics using the OSMO®-334-97 membrane module. As can be seen di-valent ions tend to reject better than mono-valent ions. If mono-valent ions are combined with di-valent ions the rejection will be controlled by the di-valent ion.

For estimating purposes, take an average of the feed and the concentrate and use this average concentration to figure the expected purity of the permeate.

Salts complexed with organics of large molecular weights will tend to act like the organics they are complexed with.

ORGANICS:

	Molecular Weight	Percent Rejection	Maximum Concentration Percent
Sucrose sugar	342	100	25
Lactose sugar	360	100	25
Protein	10,000 Up	100	10-20
Glucose	198	99.9	25
Phenol	94	***	—
Acetic acid	60	***	—
Formaldehyde	30	***	—
Dyes	400 to 900	100	—
Biochemical Oxygen Demand	(BOD)	90-99	—
Chemical Oxygen Demand	(COD)	80-95	—
Urea	60	40-60	Reacts similar to a salt
Bacteria & virus	50,000-500,000	100	—
Pyrogen	1000-5000	100	—

***Permeate is enriched in material due to preferential passage through the membrane.

GASES, DISSOLVED:

Carbon dioxide	CO_2	30-50%
Oxygen	O_2	Enriched in permeate
Chlorine	Cl_2	30-70%

THE MODULE is a spiral wound or scroll configuration which contains a high percentage of membrane area per volume and is very easy to replace. The membrane with a polyester backing is wrapped around the PVC center tube and sealed with epoxy around the edges. Polypropylene turbulence promotors act as spacers between the membrane. The spiral is superior in all applications where the feed solution can be filtered to 100 mesh or 140 microns.

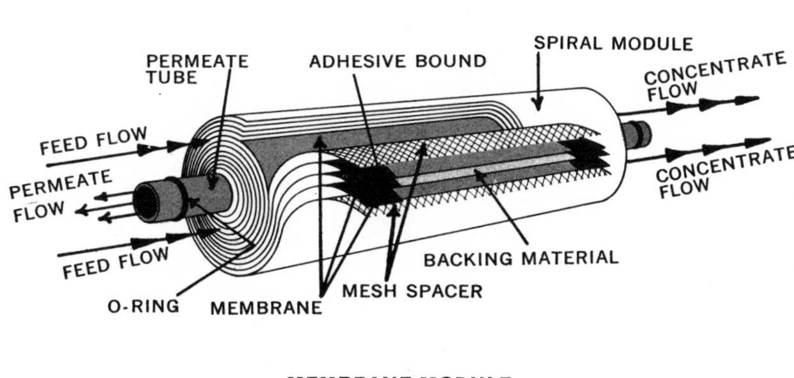

MEMBRANE MODULE

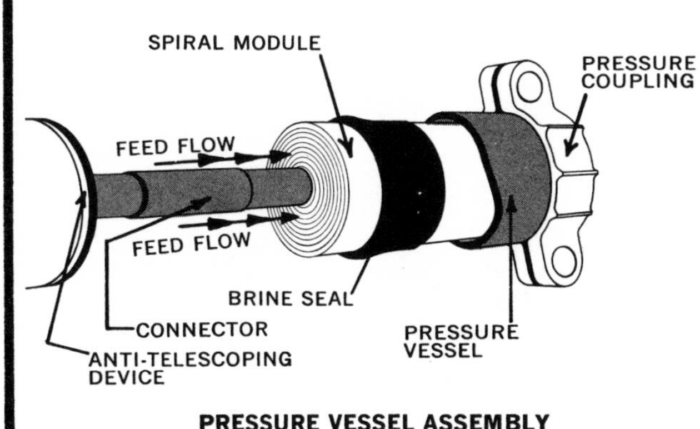

PRESSURE VESSEL ASSEMBLY

Ozone Research and Equipment Corporation

OREC 03D-AR SERIES PROCESS OZONATORS

Internal View 03D-AR Series

APPLICATIONS

- TERTIARY WASTE TREATMENT
- WATER PURIFICATION
- POLLUTION CONTROL
- ODOR CONTROL
- INDUSTRIAL PROCESSES
- AIR PURIFICATION
- CONTAINER STERILIZATION
- FOOD PRESERVATION

The OREC stainless steel tubular ozone generators, with dielectrics, employ the corona discharge (silent arc) principle of ozone generation. Each generator is individual and independent; simple to service and economical to replace, without costly down time.

Operating instruments and controls are front mounted on the instrument panel. Panel mounted air flowrator, flow valve, pressure regulator, pressure gages, wattmeter and variable voltage transformer control allow selection of a wide range of ozone concentrations and outputs. A momentary depress switch adjusts the remote, motor driven variable voltage transformer.* Integral ozone measurement instrumentation allows chemical analysis of the ozone stream by the classical KI method. All components in contact with the ozone stream are ozone resistant. Electrical interlocks interrupt power when the case or instrument panel is opened. The instrument panel is hinged for easy access. The chassis is grounded to the power supply and all electrical components are grounded to the frame.

03D-AR SERIES SPECIFICATIONS

MODEL	POUNDS / DAY * OZONE	GRAMS / HOUR OZONE	POUNDS / HOUR OZONE	AIR FLOW SCFM	COOLING WATER GPH**	WATER PIPING	OZONE PIPING	DIMENSIONS W' x H' x D'	NET WEIGHT (Approx.)
03D5-AR	5	95	.21	4.6	62	½	½	6 x 5 x 4.5	1170
03D10-AR	10	189	.42	9.2	125	¾	¾	6 x 5 x 4.5	1410
03D15-AR	15	284	.63	13.9	187	¾	¾	6 x 5 x 4.5	1650
03D20-AR	20	378	.83	18.5	250	¾	¾	6 x 5 x 4.5	1890
03D25-AR	25	473	1.04	23.1	312	¾	1.0	6 x 5 x 4.5	2130
03D30-AR	30	568	1.25	27.7	375	1.0	1.0	6 x 6 x 8.5	2375
03D40-AR	40	757	1.67	37.0	500	1.0	1.0	6 x 6 x 8.5	3680
03D50-AR	50	946	2.08	46.3	625	1.0	1.5	6 x 6 x 8.5	4160
03D60-AR	60	1135	2.5	55.5	750	1.0	1.5	6 x 6 x 8.5	4650
03D70-AR	70	1324	2.92	64.8	875	1.0	1.5	6 x 6 x 8.5	5130
03D80-AR	80	1513	3.34	74.0	1000	1.0	1.5	6 x 6 x 8.5	5610
03D90-AR	90	1703	3.75	83.3	1125	1.5	1.5	6 x 6 x 8.5	6090
03D100-AR	100	1892	4.17	92.5	1250	1.5	1.5	6 x 6 x 8.5	6575

* At 1% wt Ozone Concentration in Air. ** At 70F° Water Temperature.

Power requirement is a function of operating variables. Total power required including all accessory equipment is 11 ± 1.5 KWH per pound ozone produced. The Ozonator per se is a capacitance load and has a leading power factor of approximately .6. Standard power source is 230 or 440 volt 50/60 Hz. Reduce output 16% at 50 Hz.

Models 03D5 and 03D10 also available in 120 volt 50/60 Hz. Maximum operating pressure, standard models, 30 PSIG.

Peabody Barnes

SUBMERSIBLE SEWAGE PUMPS

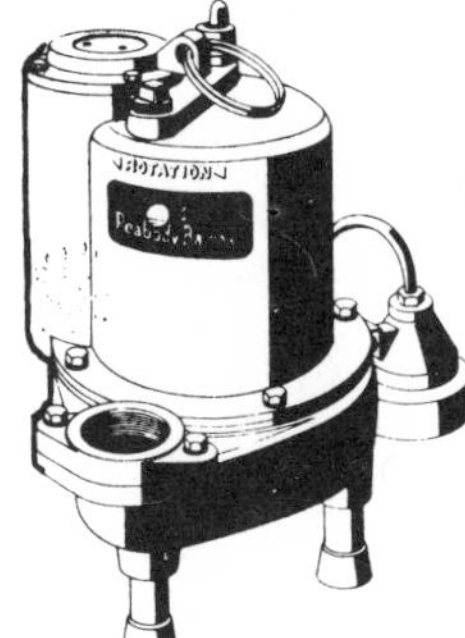

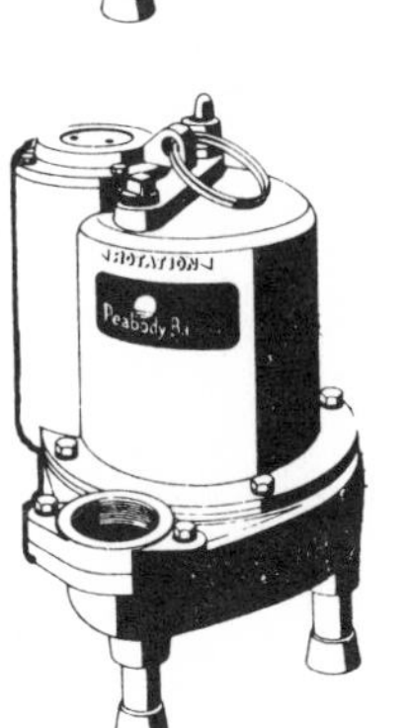

Model No.	Size	Dia. Solids	Motor Data HP	Volts	Phase	Wt. Lbs.
SE-51	2"	2"	½	115	1	80
SE-53	3"	2"	½	115	1	80
SE-52	2"	2"	½	230	1	80
SE-54	3"	2"	½	230	1	80
SE-55	3"	2"	½	230	3	80
SE-56	3"	2"	½	460	3	80
SE-73	3"	2"	¾	230	3	80
SE-74	3"	2"	¾	460	3	80
SE-75	3"	2"	¾	230	1	82
SE-101	3"	2"	1	230	1	85
SE-103	3"	2"	1	230	3	85
SE-104	3"	2"	1	460	3	85

3450 RPM

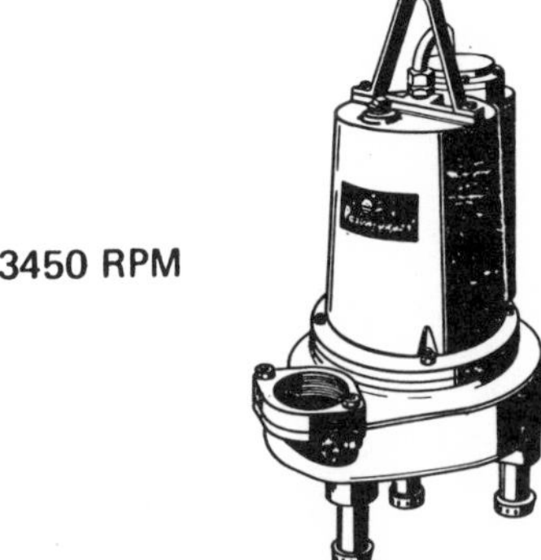

Model No.	Size	Dia. Solids	Motor Data HP	Volts	Phase	Wt. Lbs.
SEH101	2"	1¾"	1	230	1	85
SEH105	2"	1¾"	1	230	3	85
SEH106	2"	1¾"	1	460	3	85
SEH151	2"	1¾"	1½	230	1	175
SEH155	2"	1¾"	1½	230	3	175
SEH156	2"	1¾"	1½	460	3	175
SEH201	2"	1¾"	2	230	1	175
SEH205	2"	1¾"	2	230	3	175
SEH206	2"	1¾"	2	460	3	175

MODEL NO.	SIZE	SOLIDS	MOTOR DATA HP	RPM	VOLTS	PHASE	LBS.
SEH302	4"	3"	3	1750	230	3	280
SEH304	4"	3"	3	1750	460	3	280
SEH502	4"	3"	5	1750	230	3	310
SEH504	4"	3"	5	1750	460	3	310
SEH752	4"	3"	7½	1750	230	3	340
SEH754	4"	3"	7½	1750	460	3	340
SEH1002	4"	3"	10	1750	230	3	400
SEH1004	4"	3"	10	1750	460	3	400
SE151	3"	2½"	1½	1725	230	1	175
SE153	3"	2½"	1½	1725	230	3	175
SE154	3"	2½"	1½	1725	460	3	175
SE201	3"	2½"	2	1725	230	1	175
SE203	3"	2½"	2	1725	230	3	175
SE204	3"	2½"	2	1725	460	3	175
SE302	4"	3½"	3	1140	230	3	280
SE304	4"	3½"	3	1140	460	3	280
SE502	4"	3½"	5	1140	230	3	310
SE504	4"	3½"	5	1140	460	3	310
SE752	4"	3½"	7½	1140	230	3	340
SE754	4"	3½"	7½	1140	460	3	340

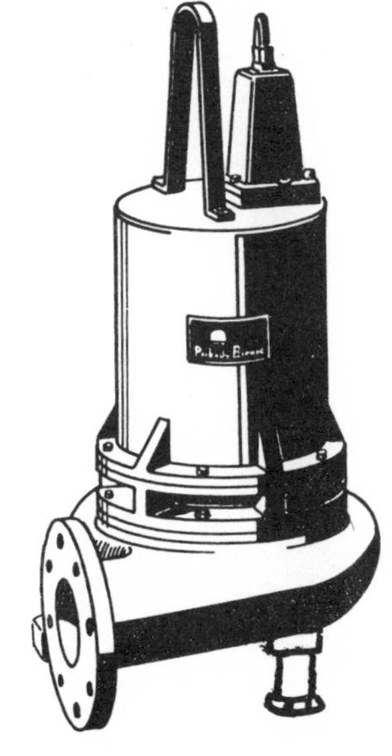

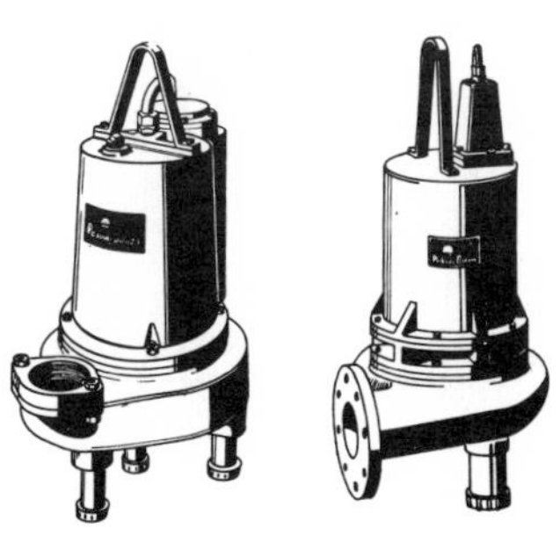

SUBMERSIBLE SEWAGE EJECTORS

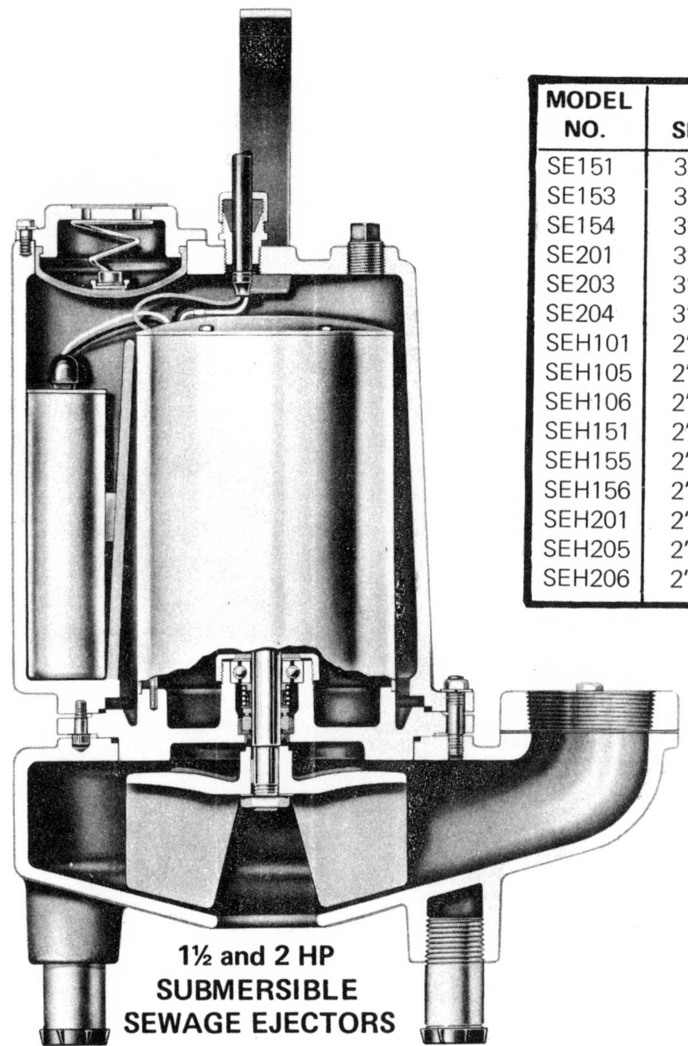

1½ and 2 HP SUBMERSIBLE SEWAGE EJECTORS

MODEL NO.	SIZE	SOLIDS	MOTOR DATA			
			HP	RPM	VOLTS	PHASE
SE151	3"	2½"	1½	1725	230	1
SE153	3"	2½"	1½	1725	230	3
SE154	3"	2½"	1½	1725	460	3
SE201	3"	2½"	2	1725	230	1
SE203	3"	2½"	2	1725	230	3
SE204	3"	2½"	2	1725	460	3
SEH101	2"	1¾"	1	3450	230	1
SEH105	2"	1¾"	1	3450	230	3
SEH106	2"	1¾"	1	3450	460	3
SEH151	2"	1¾"	1½	3450	230	1
SEH155	2"	1¾"	1½	3450	230	3
SEH156	2"	1¾"	1½	3450	460	3
SEH201	2"	1¾"	2	3450	230	1
SEH205	2"	1¾"	2	3450	230	3
SEH206	2"	1¾"	2	3450	460	3

SEWAGE PUMP CONTROLS

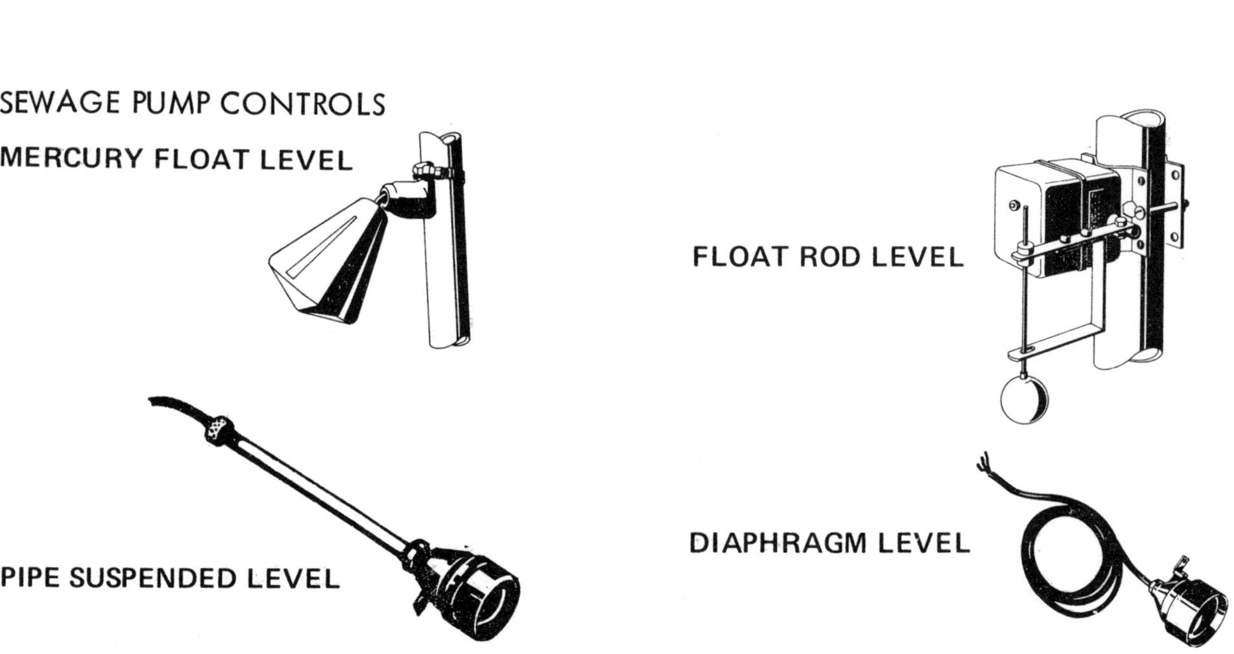

MERCURY FLOAT LEVEL

FLOAT ROD LEVEL

PIPE SUSPENDED LEVEL

DIAPHRAGM LEVEL

Peabody Barnes

SELF-PRIMING CENTRFUGAL SEWAGE PUMPS

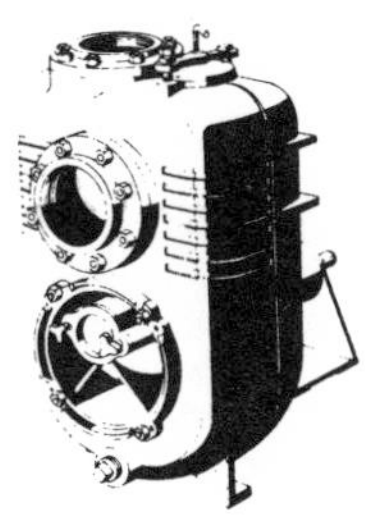

MODEL NO.	SIZE	DIA. SOLIDS	APPROX. SHP. WT.
3TCU	3"	1 ¾"	325 lbs.
3SCU-1	3"	2 ½"	330 lbs.
4SCU	4"	3"	525 lbs.
6SCU	6"	3"	960 lbs.

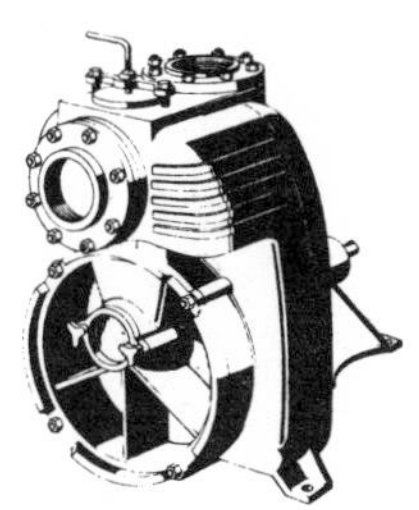

UNIVERSAL DRIVE - ADAPTS TO ANY POWER SOURCE

RUGGED CAST IRON CONSTRUCTION

PATENTED, PRESSURE-EQUALIZING DOUBLE OIL-FILLED SEAL

1. Prevents liquid and grit from entering seal chamber.
2. Oil cools and lubricates seal faces.
3. Maintenance-free, not damaged even if pump runs dry for extended periods. You get seal life not possible in any other pump.

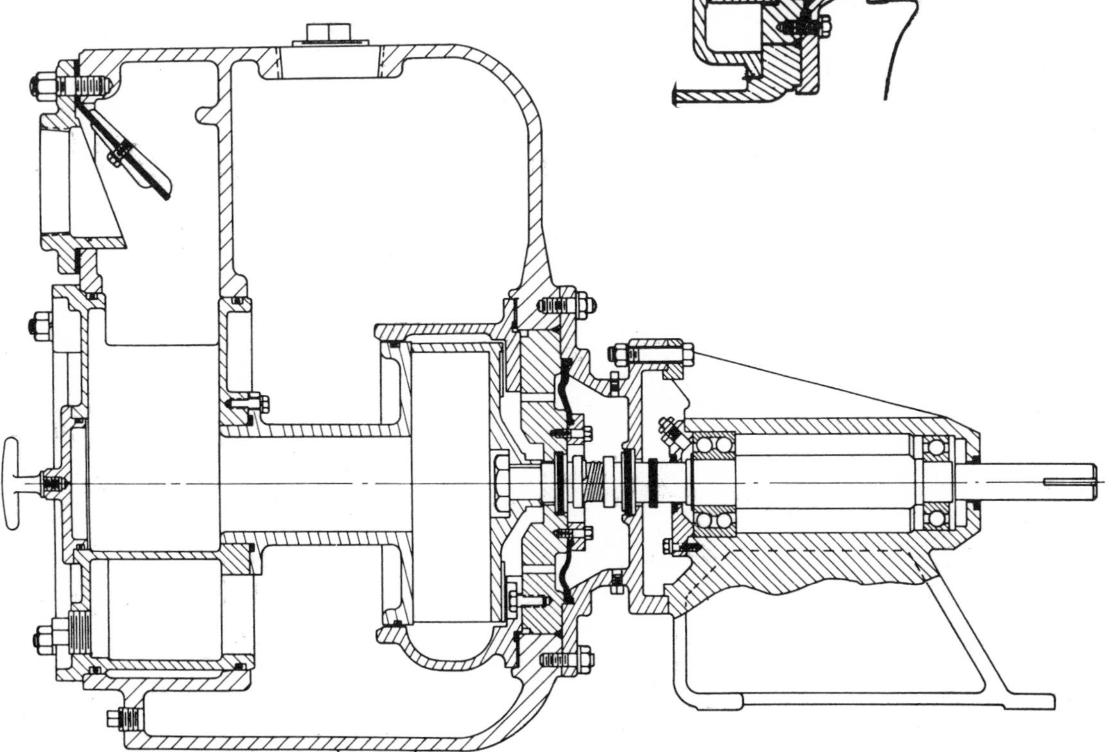

SECTION DRAWING FOR MODEL 3SCU SEWAGE PUMP

PREFABRICATED WASTE TREATMENT PLANT

5000 P.S.I. concrete construction has a life expectancy of one hundred years—many times that of steel chambers. Concrete eliminates the possibility of electrolysis which corrodes and on occasion has perforated steel units in as little as two years.

Compact, completely engineered units eliminate the need for a costly, separate job-built wet well. Close coupled chambers mean smaller excavation and eliminates the possibility of backfill damage to suction line connections.

Completely equipped units further simplify the installation. Only the inlet, the discharge and the electrical lines need connecting in the field. Location for inlets are optional.

Dual pump construction ensures full standby capacity for design loads and automatic double capacity for unforeseen peak conditions. Pumps start alternately to provide equal use and wear. Should the lead pump fail, an alarm is sounded and the second pump automatically takes over. Provisions for remote alarms are included as standard equipment. Pumps and equipment can be removed or serviced without disturbing attached piping.

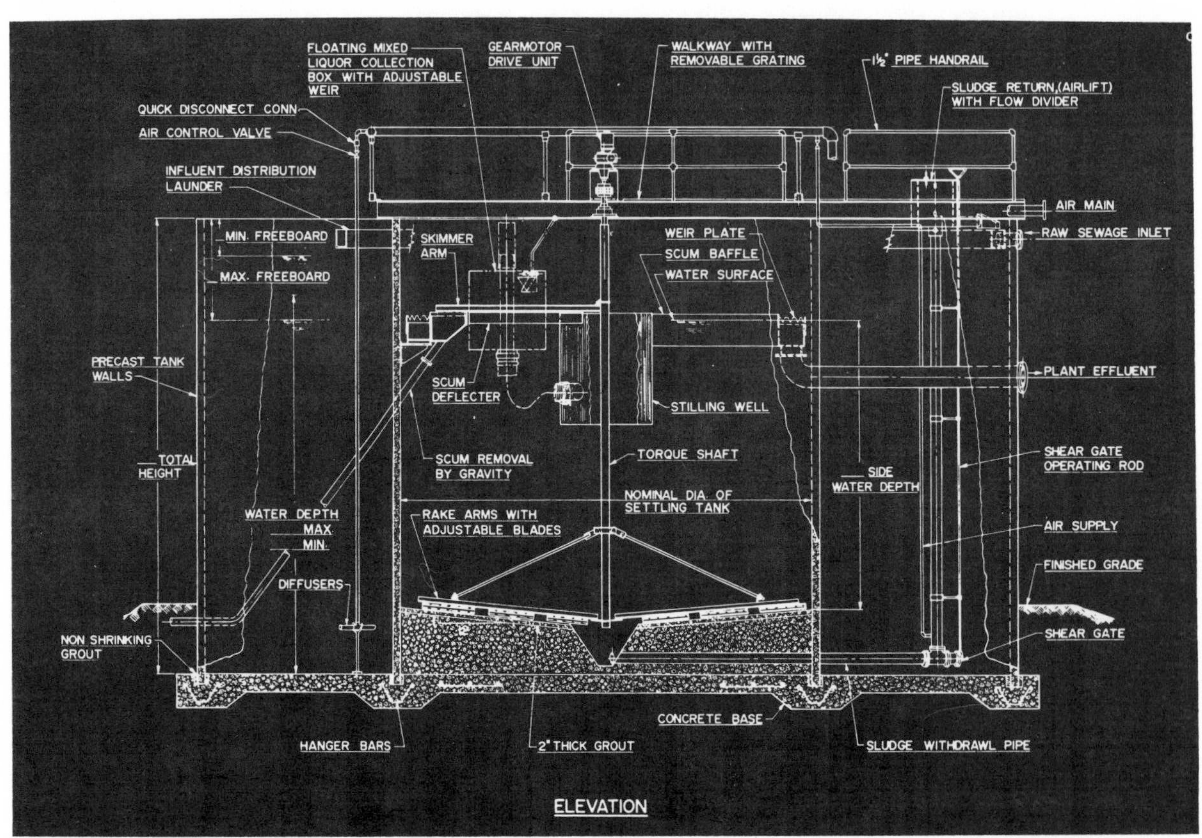

Submersible concrete stations and duplex sewage pumping stations are also available.

The V-notch Variable Orifice consists of a precisely grooved plug sliding in a fitted ring. This gives an exact orifice size for high accuracy. Repeating plug position duplicates orifice size. 3-inch plug travel, compared with a fraction of an inch for needle valves, gives wide operating range. It also gives accuracy and ease of adjustment in setting the feed rate, whether manual or automatic. The V-notch won't stick: It's made of corrosion-resistant, self-lubricating plastic. The relatively large, V-shaped opening is less prone to clogging than the long, narrow opening around comparable needle or disk valves.

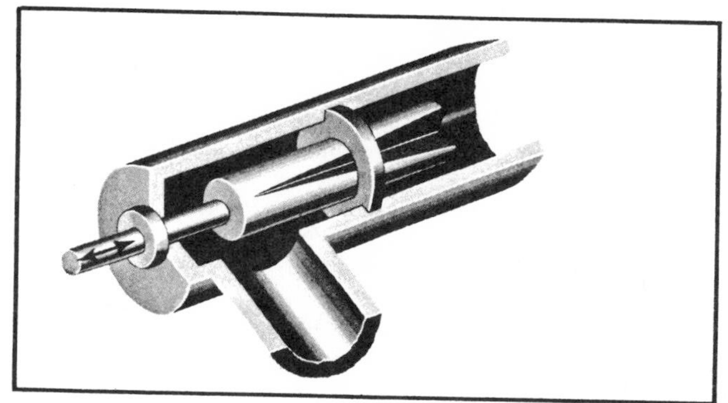

INJECTORS

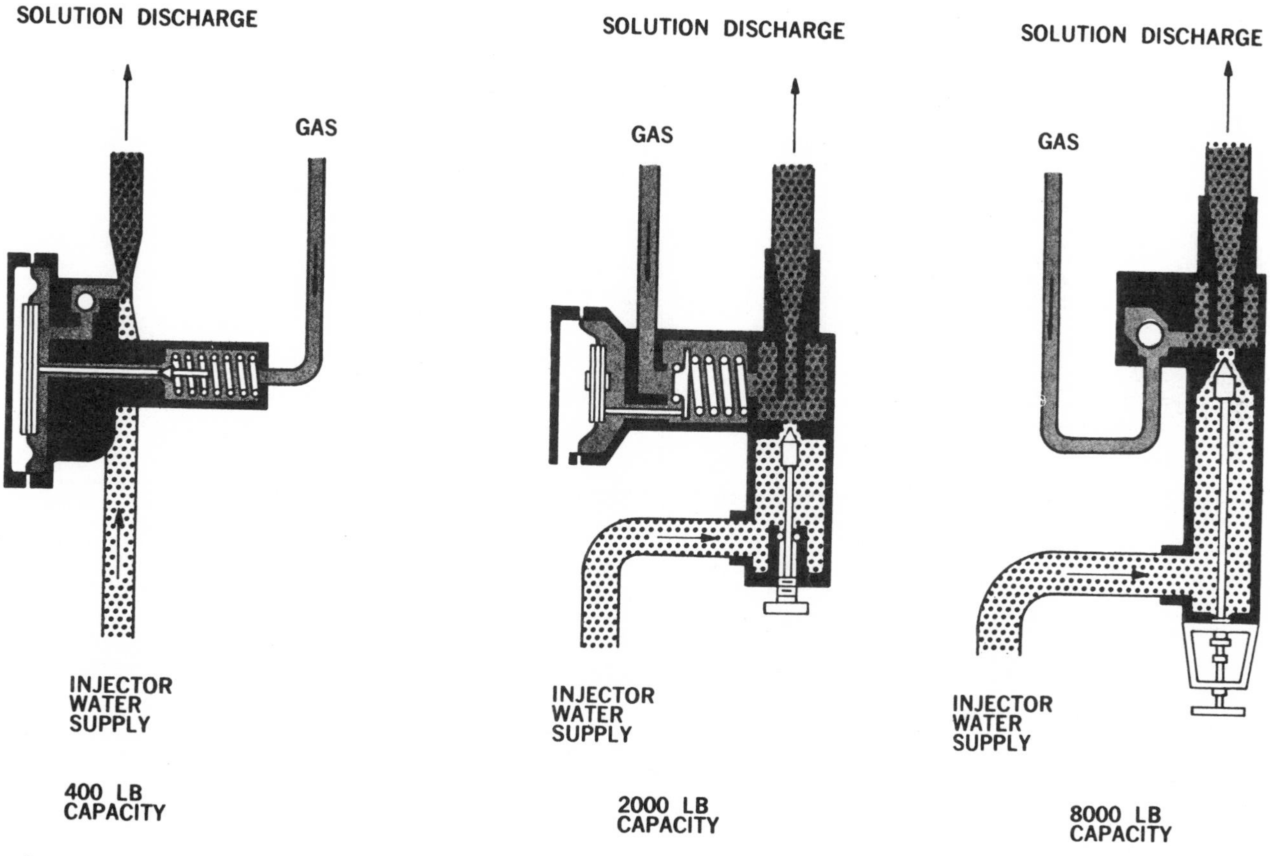

SOLUTION DISCHARGE
GAS
INJECTOR WATER SUPPLY
400 LB CAPACITY

SOLUTION DISCHARGE
GAS
INJECTOR WATER SUPPLY
2000 LB CAPACITY

SOLUTION DISCHARGE
GAS
INJECTOR WATER SUPPLY
8000 LB CAPACITY

1798

OPERATION OF SERIES V-800 CHLORINATION MODULES

The three Series V-800 Chlorination Modules operate by a vacuum developed at an aspirator-type injector. Gas enters through a spring-diaphragm pressure regulating valve which regulates the gas to a fixed reference vacuum and which opens only under normal operating vacuum. This valve also regulates the vacuum ahead of the V-notch orifice and shuts off the gas if an interruption of the injector-water supply should destroy operating vacuum. Gas next passes through a rotameter to the V-notch orifice. Here feed rate is changed, manually or automatically, by changing orifice area or by varying the differential across the orifice. After the orifice, gas passes through a vacuum regulating valve which maintains a constant vacuum across the orifice.

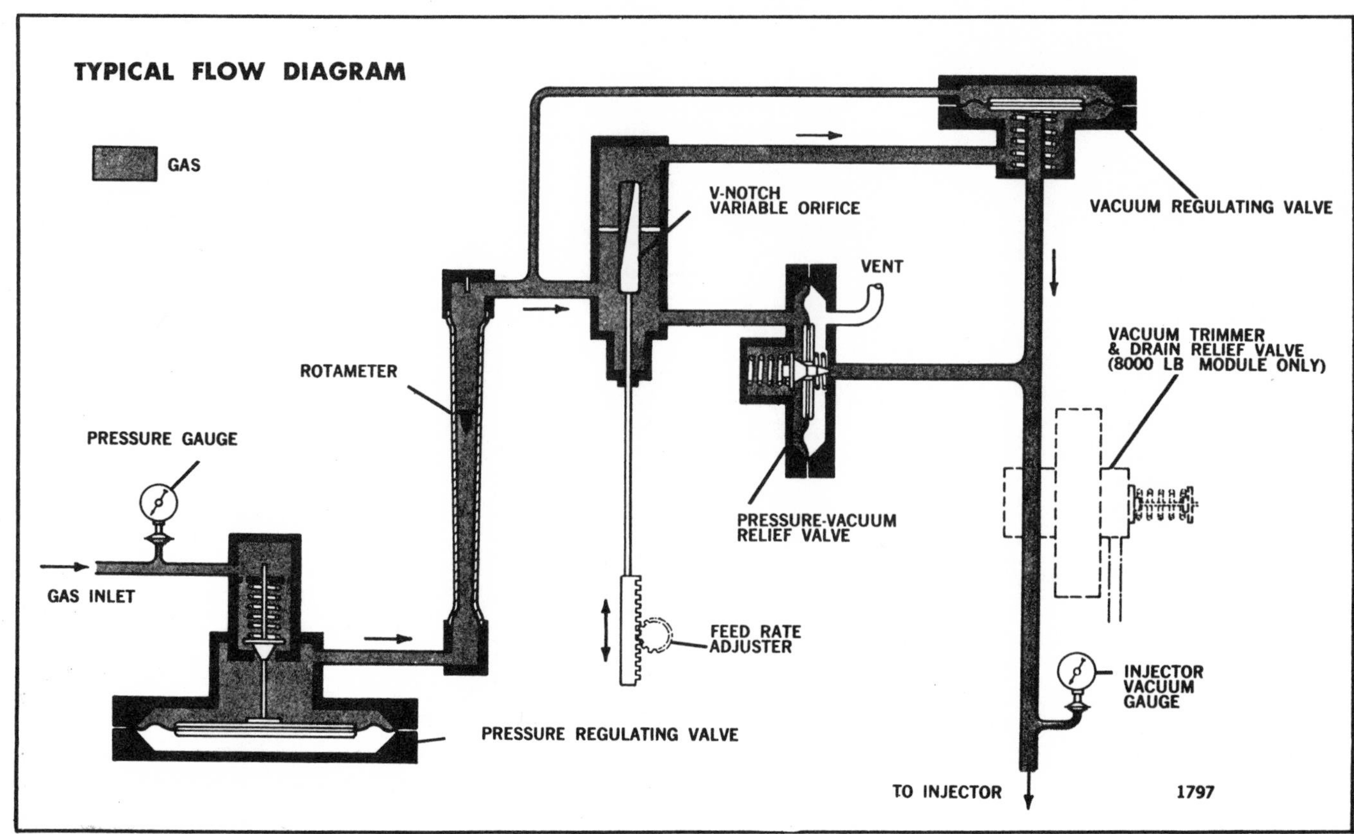

TECHNICAL DATA

	400 lb. Module	2000 lb. Module	8000 lb. Module
overall dimensions	All modules are the same height, 5′8¼″, and the same depth, 15″. Widths are:		
	20½″	27½″	34½″
shipping weights	120 lb	220 lb	300 lb

TECHNICAL DATA — SERIES V-800 MODULE

	400-lb MODULE	2000-lb MODULE	8000-lb MODULE
accuracy	4% of indicated flow rate		
capacities	(rotameter sizes in lb per 24 hours with 20 to 1 feed range)		
	3, 10, 20, 30, 50, 75, 100, 150, 200, 250, 300, 400	50, 75, 100, 150, 250, 500, 1000, 1500, 2000	1000, 2000, 3000, 4000, 5000, 6000, 8000
injector size	1″ fixed throat	up to 400 lb: 1″ fixed throat; over 400 lb: 2″ adjustable plug	3 or 4″ adjustable plug
injector operating water supply	25-300 psi	25-300 psi	25-150 psi
pressure at point of application	100 psi	100 psi	50 psi
	Note: Injector operating water supply must be reasonably clean. Required pressure and flow depend on injector size, chlornator capacity, and pressure at point of application. A booster pump before or a solution pump after the injector will allow application against pressures higher than those shown above.		
vent connection	½″	¾″	1″
inlet connection	¾″	¾″	1″
electrical requirements	110-120 volt, ac or dc, for heater in chlorine inlet line		
	30 watts	30 watts	75 watts
	Other requirements for control circuits in semi-automatic and automatic control, and for such accessories as alarms, evaporators, external chlorine pressure reducing valves, and booster or solution pumps.		

HOUSED CHLORINATION SYSTEM — SERIES 85-800

The Wallace & Tiernan Housed Chlorination System is a complete gas metering system in a weatherproof chlorinator house. It is assembled and shipped as a unit with internal piping and wiring completed. Installation requires only moving the house to its location and connecting to an electric power supply, water supply, and the point of chlorine application.

The building is fiber glass with aluminum trim. It has a finished steel base and a gabled roof with lifting rings. The double walls are of sandwich-type construction. An exhaust fan, a vent, and an internal light are included. Two 1000-watt space heaters, with a thermostat and low-temperature alarm switch, provide controlled heat for the interior. Components are wired to a central terminal box with a coded strip for making external connections.

The chlorination system consists of a V-notch Chlorinator with separate chlorine pressure and vacuum regulating valves. Chlorinator capacity can be 3 to 400 lb of chlorine per 24 hours. Operating range is 20 to 1. The chlorinator can respond to manual adjustment or to automatic control by an electric or a variable vacuum flow-proportional signal. Alarm contacts are provided to warn when the chlorinator-operating vacuum gets too high or too low.

A booster pump can be provided. It is available in several capacities, can be self-priming or non-self-priming, and is bronze fitted. It has a magnetic starter. A 1½-inch strainer and a sample valve are provided. The system includes a Wallace & Tiernan Two-cylinder Scale. It reads out in net pounds the amount of chlorine remaining. Readout is continuous, direct, and on a separate dial for each cylinder.

DISSOLVED AIR FLOTATION SYSTEMS

Dissolved air flotation is a very practical technique for separating suspended particles and colloidal materials from liquids in general and from industrial waste streams in particular. By attaching air bubbles to the suspended particles until their combined net specific gravity becomes less than that of water, the particles are made buoyant. These buoyant particles then rise quickly to the surface, forming a "float" which can be removed by skimming. This separated solid material is available for reuse or disposal. Clarified effluent is generally acceptable for safe disposal, return to process, etc.

To produce the microscopic air bubbles required for effective flotation, the liquid is pressurized, and air is introduced until the liquid approaches saturation. For example, at 50 pounds pressure about five times more air can be dissolved in water than at atmospheric pressure. Subsequently, when pressure is released, the liquid becomes supersaturated and the volume of air in excess of the atmospheric saturation concentration comes out of solution in the form of microscopic bubbles. These bubbles may range in size from thirty to one hundred twenty microns, depending on factors such as the type of release, pressure, presence of chemicals, etc.

Most of the bubbles become attached to the solid particles by surface energies while others become entrapped by the solids or by the hydrous oxide flocs.

Particles of colloidal nature are normally too small to permit the formation of a satisfactory air-particle bond. They must first be coagulated by a chemical such as an aluminum or iron compound and then absorbed by the hydrous metal oxide floc generated by these compounds. Frequently, a coagulant aid is used to agglomerate the hydrous oxide flocs, to enhance particle size and promote the rate of flotation.

PERFORMANCE COMPARISON OF FAVAIR MARK II DESIGN WITH CONVENTIONAL FLOTATION

Favair's most outstanding feature — the ability to operate at a high loading rate — is clearly illustrated by this graph. The curves were obtained by first operating a standard flotation unit in the conventional manner and then converting it to the patented Favair design. All tests were conducted on a paper mill waste containing 27% ash, primarily clay and TiO_2. To obtain an acceptable effluent (in this case 30 ppm residual S.S.), the conventional unit could not be operated above 2.3 gpm per square foot. The conversion to the latest Favair design permitted an increase in the loading rate by a factor of 2.35 or 5.4 gpm per square foot. Another important advantage — the ability to cope with reasonable hydraulic overloads and still provide a degree of clarification — is illustrated by the more acute angle of the Mark II curve.

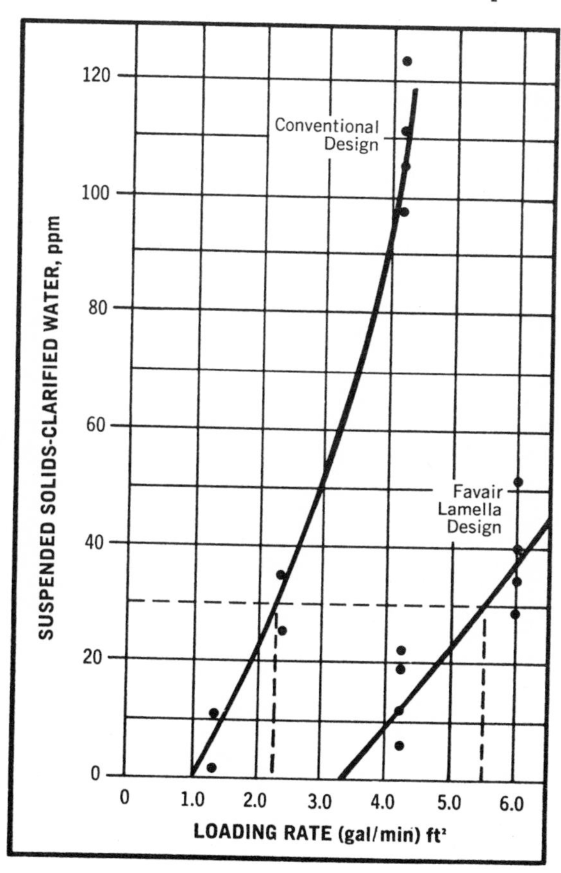

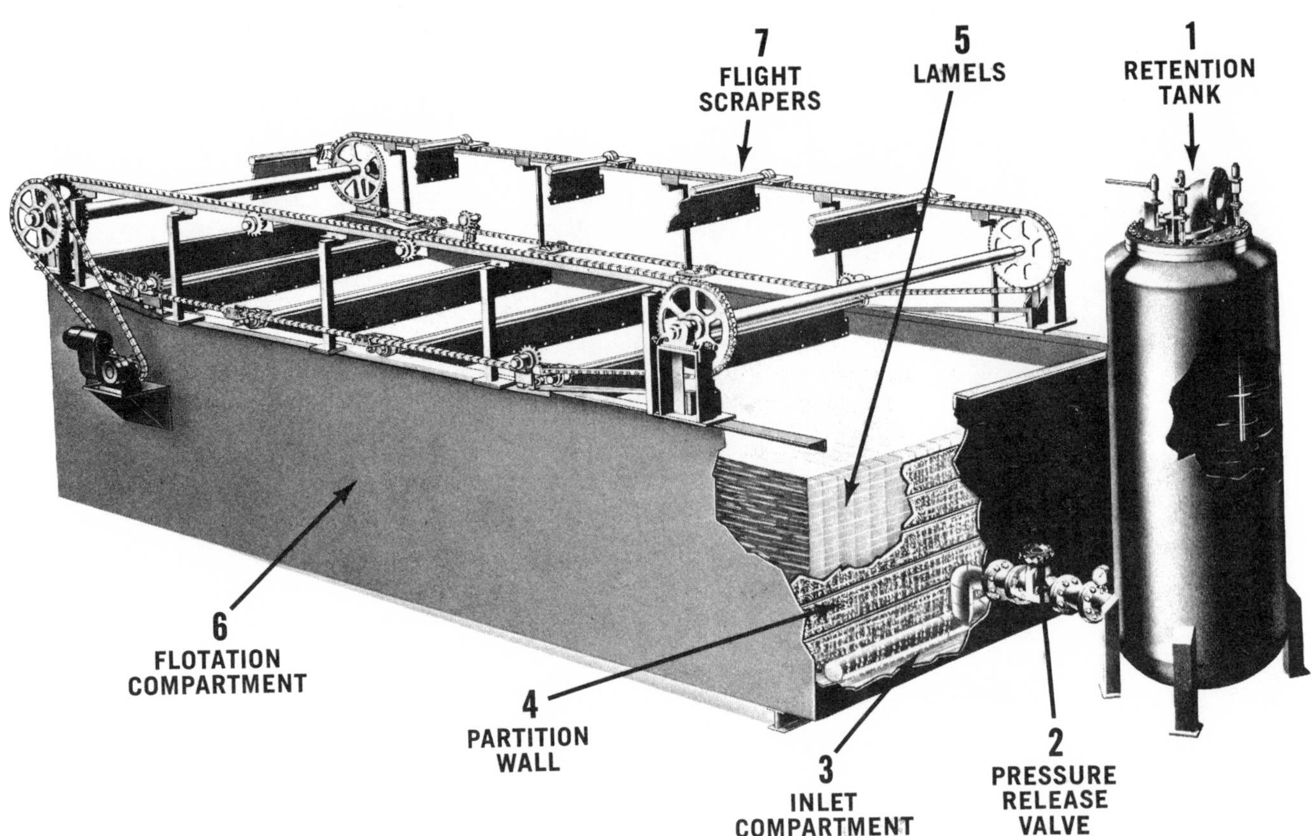

HOW THE FAVAIR® MARK II IMPROVES THE FLOTATION PROCESS

The flotation section of the Favair Mark II system is unique. It contains a number of vertical plates which are mounted longitudinally to the direction of liquid flow. Technically each plate is called a lamel. Correct spacing of these lamels is imperative for satisfactory operation. This design is based on Reynold's theory that by decreasing the hydraulic radius (cross section) of a channel it is possible to increase the velocity of the liquid without exceeding the limit (Reynold's number) between laminar and turbulent flow. Therefore, by dividing the flotation compartment into a number of narrow channels, it is possible to increase the total flow without producing disturbing turbulence. Thus, we are approaching the conditions which heretofore have been attainable only in the laboratory. This design is covered by U.S., as well as by foreign patents.

To complement these advances the Favair Mark II has also been equipped with a new and exclusive "Ferris Wheel" "no flop" skimmer. This smoothly working system eliminates the possibility of flow disturbances which can be encountered with conventional skimmer systems.

HOW THE FAVAIR® MARK II FLOTATION SYSTEMS OPERATES

The liquid to be treated is pumped at approximately 50 psi into the Favair retention tank (1). As the liquid enters the tank, air is introduced and is dissolved to a concentration approaching saturation. Chemicals may also be added in the retention tank, if needed to promote floc formation.

From the retention tank, the air-saturated waste liquid is directed through pressure release valves (2) into the inlet compartment (3) of the flotation tank. As the liquid mixture is depressurized, the dissolved air comes out of solution and forms microscopic bubbles which attach themselves to the suspended and colloidal contaminants. This liquid mixture flows through and over an inlet partition wall (4) and then on between the lamels (5) in the main flotation compartment. The solids, with air bubbles attached, rise to the surface where they are pushed by the flight scrapers of the skimmer system (7) up a collection ramp and into the solids recovery compartment for reuse or disposal. The effluent, a clear subnatant, free of suspended solids leaves the flotation compartment passing through ducts to the treated liquid compartment where it flows over a weir and thence to reuse or waste.

The Permutit Precipitator

The Permutit Precipitator offers a more efficient means for removing impurities from a liquid by precipitation, adsorption, settling and upward filtration. It requires less space, less chemicals and less time than any previous design of reaction and settling tank.

The savings in ground space requirements are approximately 50%. The savings in certain chemicals and adsorbents vary from 10% to over 40%. The savings in time of treatment vary from 50% to 75%.

Permutit Precipitator units are available in sizes having daily capacities varying from 1000 gallons to any reasonable unit capacity. Precipitator batteries having daily capacities of up to 120,000,000 gallons are in use in various industries and municipalities.

The chief applications are in the following fields of water treatment:

- Water softening
- Removal of turbidity, color, taste and odor
- Reduction of alkalinity
- Removal of silica
- Removal of iron and manganese

Principles of Operation

When precipitates are formed in water as the result of chemical reactions, they normally settle slowly toward the bottom of the container. In the Permutit Precipitator, the precipitates are not permitted to settle to the bottom but, by a combination of mechanical agitation and hydraulic flow, are kept suspended within controlled limits. The velocity of flow of water through this equipment is so controlled that the precipitates are kept in suspension up to a level where the water passing upward through them leaves them sharply behind, and flows from the top of the equipment as a clear liquid.

This operation is based upon the well-known phenomenon that a particle is supported by an upward moving stream of water if the velocity is high enough so that the friction of the water on the particle exceeds the pull of gravity. Thus, if the velocity is first high and then gradually decreased, a point is reached at which the particle becomes too heavy to be supported by the frictional effect of the passing water and remains behind, thus becoming separated from the water which previously supported it.

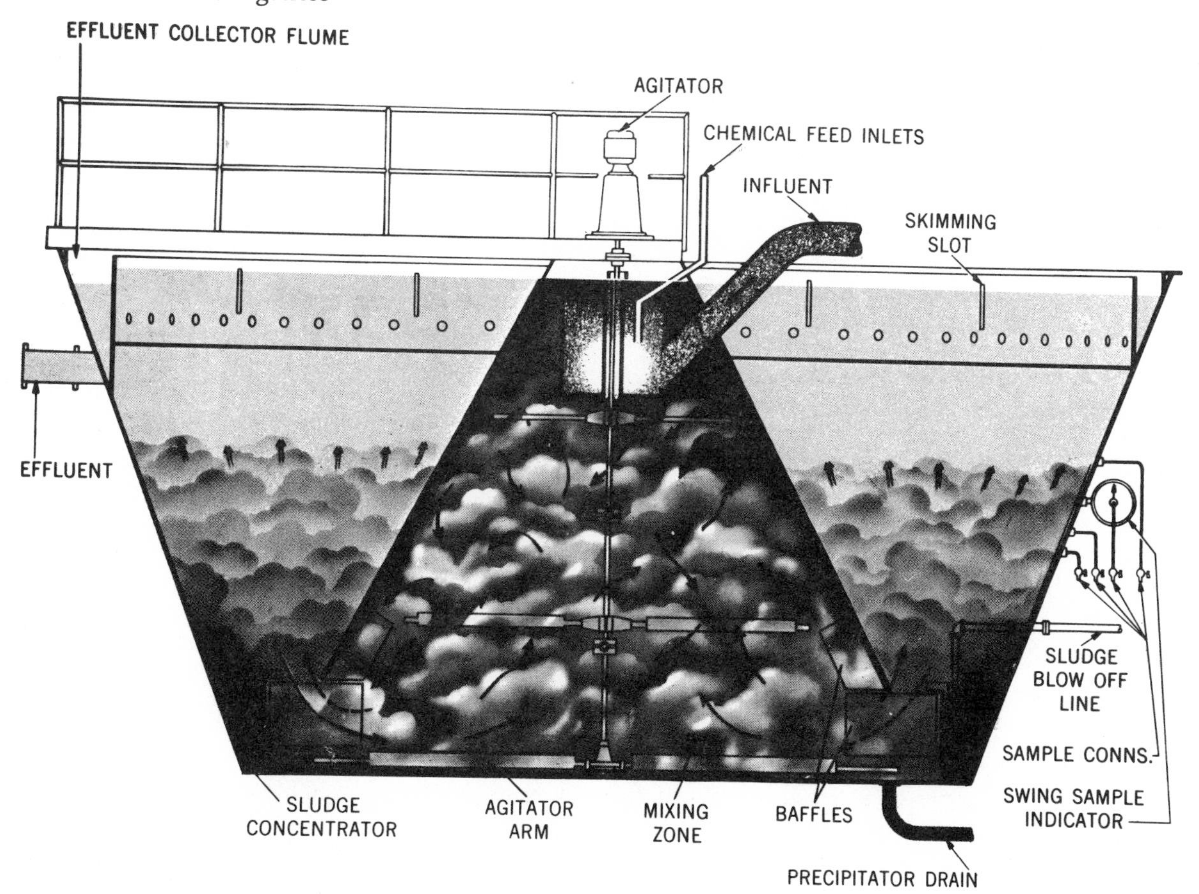

Section of Precipitator shows thorough mixing of chemicals with influent and previously formed sludge

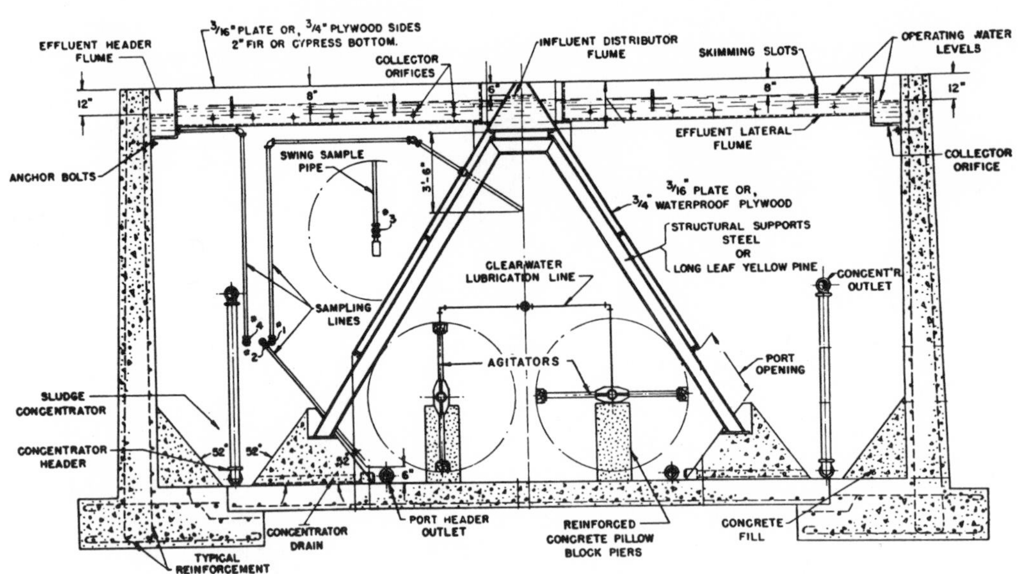

BASIC DESIGN OF PERMUTIT HORIZONTAL PRECIPITATOR *showing agitators and the sludge filter consisting of a wedge-shaped section with flat, inclined sides*

Special Applications of the Permutit Precipitator

TURBIDITY REMOVAL

The Precipitator principle of a self-contained suspension filter is particularly efficient in removing turbidity caused by particles of such small diameter that they are difficult to remove by processes of straight filtration. The raw water and the coagulant (alum or iron salts, etc., as the water may require, plus pH correctants to cause optimum coagulation) are fed simultaneously at the top of the mixing chamber. The ensuing mixing causes flocculation to take place with the formation of a large, tough floc and in the upward flowing filter zone of the Precipitator, the maximum adsorptive and straining effect of the floc in removing turbidity is obtained. Final polishing of the water may then be accomplished with filters.

REMOVAL OF COLOR

Color is removed from water by the use of coagulants applied in the pH range found most favorable, in addition to which certain waters may require clay, chlorine or activated carbon. The Precipitator is particularly applicable for color removal because the intimate contact that is effected between the water and the adsorbents in the mixing and filter zones insures the maximum utilization of their adsorbent properties.

SILICA REMOVAL

One of the methods for the removal of silica from water has been the use of metallic hydroxide sludges as silica adsorbents. Precipitator installations, due to the efficient use of the adsorbent powers of the sludge, have proved to be of great value for silica removal.

TASTE AND ODOR REMOVAL

Activated carbon is widely used for the removal of tastes and odors from water supplies. The principle involved is one of adsorption of the odoriferous or taste producing substances by the activated carbon. When activated carbon is employed in the older types of coagulation and settling tanks or basins, it settles out in the sludge before its adsorbent powers have been fully utilized.

Due to the prolonged and intimate contact with agitated and suspended sludge that is afforded by the Permutit Precipitator, the adsorbent properties of the activated carbon are utilized to the fullest extent. In practice, this has been found to reduce the dosage of activated carbon by 40% or more, thus effecting notable economies in the use of this adsorbent.

The Permutit Company, Division of Sybron Corporation

Vertical Precipitators For Softening or Coagulation

Where vertical units are desired two basic designs are used depending upon the flow rate and the process involved. In either design a surface rate of 1.35 gpm per square foot of surface area is used for softening, and 1.0 gpm per square foot of surface area is used for coagulation.

Standard flow rate selections are listed in the following tabulations.

VERTICAL PRECIPITATORS

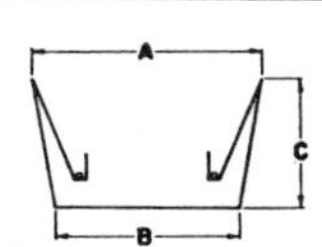

Basic Design Number I

gpm to be treated*	Coagulation A	Coagulation B	Coagulation C	Softening A	Softening B	Softening C
50	10'-0"	4'-6"	10'-0"	8'-0"	5'-0"	13'-0"
75	12'-0"	5'-9"	10'-0"	9'-6"	6'-6"	13'-0"
100	13'-6"	7'-0"	10'-0"	11'-0"	7'-6"	13'-0"
150	16'-9"	8'-6"	10'-0"	13'-0"	10'-0"	13'-0"
200	18'-9"	10'-6"	10'-0"	15'-0"	11'-6"	13'-0"
250	21'-3"	11'-6"	10'-0"	16'-6"	13'-0"	13'-0"
300	22'-3"	13'-6"	10'-6"	18'-6"	14'-0"	13'-0"
350	24'-0"	14'-6"	11'-0"	21'-0"	14'-0"	13'-0"
400				23'-0"	14'-0"	13'-0"
500				25'-0"	16'-0"	13'-0"

*Output from Precipitator, provided waste water not in excess of 10% of flow rate.

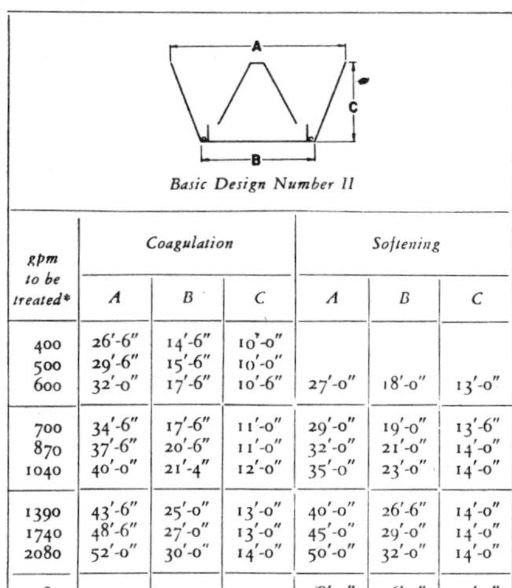

Basic Design Number II

gpm to be treated*	Coagulation A	Coagulation B	Coagulation C	Softening A	Softening B	Softening C
400	26'-6"	14'-6"	10'-0"			
500	29'-6"	15'-6"	10'-0"			
600	32'-0"	17'-6"	10'-6"	27'-0"	18'-0"	13'-0"
700	34'-6"	17'-6"	11'-0"	29'-0"	19'-0"	13'-6"
870	37'-6"	20'-6"	11'-0"	32'-0"	21'-0"	14'-0"
1040	40'-0"	21'-4"	12'-0"	35'-0"	23'-0"	14'-0"
1390	43'-6"	25'-0"	13'-0"	40'-0"	26'-6"	14'-0"
1740	48'-6"	27'-0"	13'-0"	45'-0"	29'-0"	14'-0"
2080	52'-0"	30'-0"	14'-0"	50'-0"	32'-0"	14'-0"
2780				58'-0"	36'-0"	14'-0"
3470				65'-0"	43'-0"	14'-0"

*Output from Precipitator, provided waste water not in excess of 10% of flow rate. Larger units also available.

HORIZONTAL PRECIPITATORS

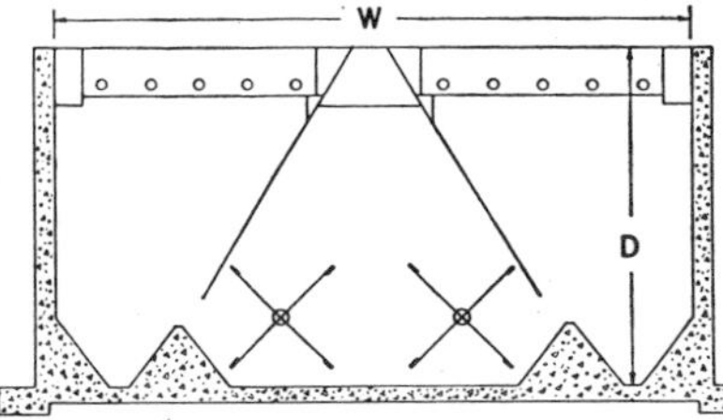

Fig. 7 DIMENSION DRAWING

Section No.	Depth "D"	Width "W"	gpm per foot of length
12	12'0"	18'6"	24.7
13	13'0"	27'0"	38.3
15	15'0"	33'6"	46.5
17	17'0"	48'0"	67.0

EXAMPLE

1500 gpm to be treated in No. 12 Precipitator; the gpm per foot of length in the table is 24.7, therefore 1500 ÷ 24.7 = 60'9", the required length of the Precipitator.

Other possibilities:

For No. 13 — 1500 ÷ 38.3 = 39'3" long

For No. 15 — 1500 ÷ 46.5 = 32'3" long

For No. 17 — 1500 ÷ 67.0 = 22'6" long

These computations show that a Precipitator for treating 1500 gpm may be obtained in lengths ranging from 60'9" to 22'6". By consulting the tabulations at the right, the most economical section can be obtained. In this case it is section No. 13.

RECOMMENDED SIZES

The following tables provide dimensions for single and double unit Permutit Precipitators, for the more common sizes of plants using the most economical sections.

gpm to be treated	SINGLE UNITS Section No.	Depth	Width	Length
300	12	12'0"	18'6"	12'3"
350	12	12'0"	18'6"	14'3"
400	12	12'0"	18'6"	16'3"
500	12	12'0"	18'6"	20'3"
600	13	13'0"	27'0"	15'9"
700	13	13'0"	27'0"	18'3"
870	13	13'0"	27'0"	22'9"
1040	13	13'0"	27'0"	27'0"
1390	13	13'0"	27'0"	36'3"
1740	13	13'0"	27'0"	45'6"
2080	13	13'0"	27'0"	54'3"
2780	15	15'0"	33'6"	60'0"
3470	15	15'0"	33'6"	74'6"
5210	17	17'0"	48'0"	78'0"
6940	17	17'0"	48'0"	104'0"

gpm to be treated	DOUBLE UNITS Section No.	Depth	Width	Length
600	12	12'0"	18'6"	12'3"
700	12	12'0"	18'6"	14'3"
870	12	12'0"	18'6"	18'3"
1040	12	12'0"	18'6"	20'3"
1390	13	13'0"	27'0"	18'3"
1740	13	13'0"	27'0"	22'9"
2080	13	13'0"	27'0"	27'0"
2780	13	13'0"	27'0"	36'3"
3470	13	13'0"	27'0"	45'6"
5210	15	15'0"	33'6"	57'0"
6940	15	15'0"	33'6"	74'6"

PERMUFLOC

The High-Rate Permufloc System operates at flow rates from 4 to 6 gallons per-minute, per-square-foot, instead of the 2 to 3 gallons per-minute, per-square-foot rate of conventional gravity filters. To do this, the process uses special dual-media filter beds, polyelectrolyte coagulant aids and a continuous monitoring system.

The Dual Filter Bed consists of a bottom layer of fine filter material and an upper layer of coarser, carbonaceous filter media. This is just the reverse of a conventional filter bed which is graded from fine sand on top to coarse material on the bottom. The Permufloc filter bed permits suspended material to penetrate deeply into the coarse layer and, therefore, provides greater storage capacity and much longer filter runs than a conventional filter bed.

Controlled Flocculation is an integral and vital part of the Permufloc process. Polyelectrolyte type coagulant aids are used to produce proper flocculation. Introduced directly ahead of the filter bed these polyelectrolytes cause a flocculation-filtration phenomena to occur within the filter bed itself, in addition to the floc normally formed in the sedimentation basin. At the same time, the polyelectrolytes "toughen" the flocculated material so that it is retained within the filter media at high flow rates.

Total System Control: Permufloc dual media bed achieves the optimum combination of storage capacity with reasonable head loss and length of filter run. However, like any high-rate filtration system it does require protection from flow surges to produce these optimum results.

In particular, effluent controllers must be free from hunting or oscillation to prevent premature turbidity breakthrough. For this reason butterfly valve type controllers are not recommended. They are not considered sufficiently stable to prevent surging because of the dynamic unbalance of forces created by the flowing water and the butterfly valve vane itself.

Smooth Stable Flow is assured with a Permufloc System because the filter controls are designed to match the process control. This is accomplished by utilizing the patented Simplex Type "P" Controller which modulates the water flow by constricting the annular area between the pipe I.D. and a nose cone. Thus, the flow is controlled equally at all points and the controller is dynamically, as well as statically, balanced at all rates of flow. This controller is recommended for use with the Permufloc high rate fIltration process, since it affords maximum protection to the bed throughout the filter run. It adjusts smoothly to any desired changes in flow rate and maintains the set rate with complete absence of flow pulsations as the head loss gradually increases over the filter run.

CHEVRON TUBE SETTLER

"Chevron" is the name The Permutit Company has given its new proprietary tube settling device for high efficiency, continuous water clarification.

Its design encompasses all of the technical and practical factors required to produce the optimum possible results in both existing clarifiers and in clarifiers specifically designed for use with Permutit Chevron Tube Settlers. The installation of Chevron settlers in existing plants can often double capacity at a fraction of the cost of adding a new clarifier. In original installations, specifically designed to maximize the Chevron settler's capability, plant size and costs are very materially reduced. Plant operators will welcome the degree of simplification in clarifier operation that Chevron tube settlers also provide.

The figure shows the unique cross section shape from which the Chevron Tube Settler derives its name as well as its efficiency.

OPTIMUM TUBE SETTLER SHAPE AND SIZE

Since the highest velocites can be obtained by decreasing the equivalent diameter, the best shape for a tube of a given area is the shape that has the largest perimeter. The Chevron shape provides the highest perimeter of common shapes as shown in the following table:

Shape	Area, $Inches^2$	Perimeter, Inches	Equivalent Diameter, Inches
Circle	4	7	2.3
Hexagon	4	7.6	2.1
Diamond	4	8	2.0
Square	4	8	2.0
Chevron (2")	4	10	1.6
Chevron (1")	1.4	6	0.94

The Chevron has the largest perimeter and therefore, the lowest equivalent diameter of the tubes having an area of 4 square inches. Compared to a square of the same area, for example, the perimeter of the Chevron is 25% larger than the perimeter of the square and the Chevron would therefore give significantly better performance. In order to take full advantage of the benefits associated with a low equivalent diameter, the area of the Permutit Chevron has been reduced to 1.4 square inches. A comprehensive analysis of all technical, manufacturing and application considerations indicates this to be the overall optimum cross section.

The Permutit Company, Division of Sybron Corporation

AUTOMATIC VALVELESS GRAVITY FILTER

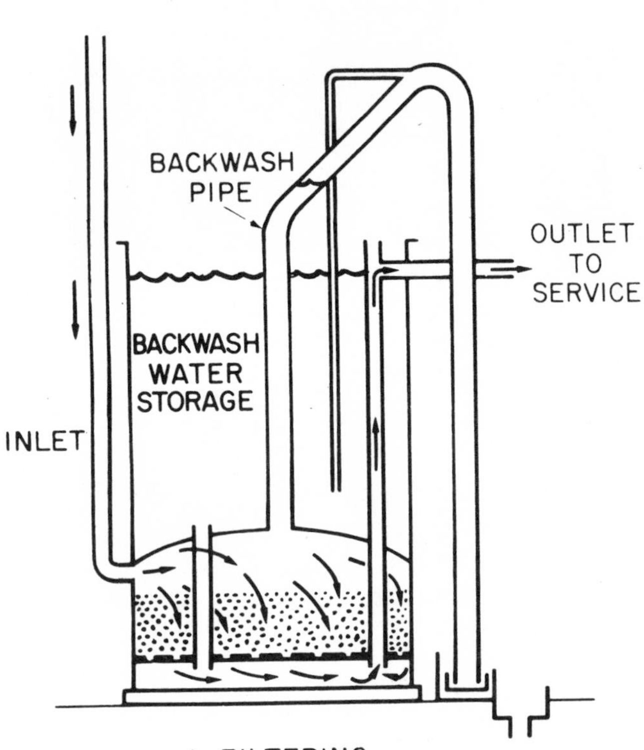

1. FILTERING

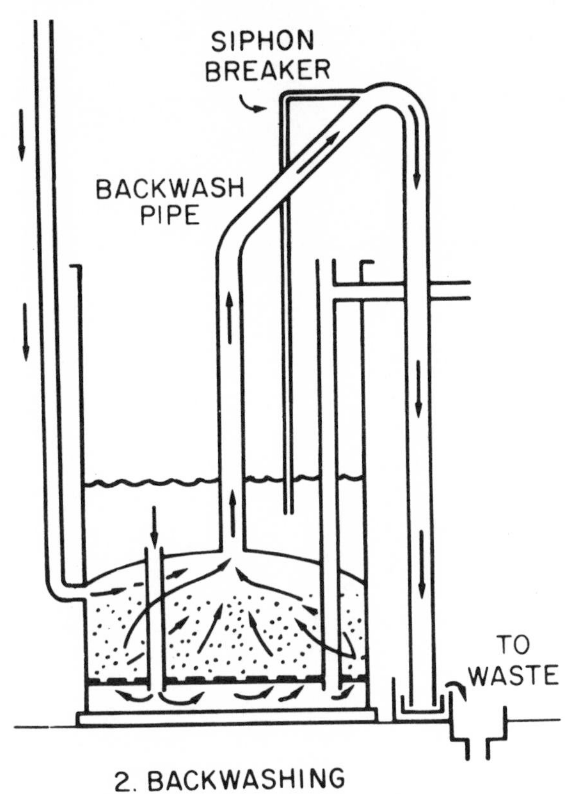

2. BACKWASHING

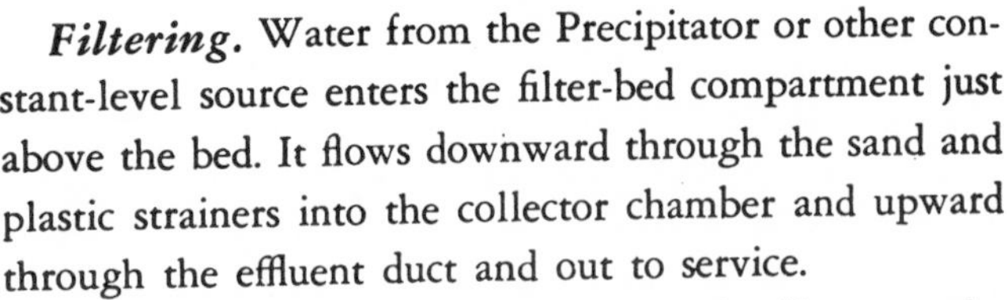

Filtering. Water from the Precipitator or other constant-level source enters the filter-bed compartment just above the bed. It flows downward through the sand and plastic strainers into the collector chamber and upward through the effluent duct and out to service.

As the filter bed collects dirt during the filter run, the head loss increases, and the water level slowly rises in the inlet pipe and in the backwash pipe. Just before the water passes over into the downward section of the backwash pipe, a self-actuated primer system, not shown, evacuates air from the backwash pipe. This pulls water rapidly over into the pipe so that a large volume of water flows down the backwash pipe and starts the siphon action that backwashes the filter.

Backwashing. The backwash pipe carries approximately 6 times as much water as the inlet pipe. This reduces the pressure immediately above the sand bed and draws the water from the backwash storage space down through the ducts and upwards through the strainers expanding the sand bed and cleaning it and then discharging to waste. This backwash action continues until the level of the water in the backwash storage space drops below the end of the siphon breaker. This admits air to the top of the backwash pipe and backwashing stops.

The inlet water then resumes its downward gravity flow automatically rinsing the bed. The rinse water then flows up into the backwash storage space where it is stored for the next backwash. When the level reaches the outlet level, it goes no higher, and all filtered water flows to service.

OPERATING CONDITIONS

Flow Rate: Standard design: ½ to 3 gpm per sq. ft.
Special design: up to 6 gpm per sq. ft.

Length of service runs: 1 to 2 days

Head loss: Predetermined loss usually 4-5 ft.

Wash Rate: 18 gpm per sq. ft. at start diminishing to 12 gpm per sq. ft. Average is 15 gpm per sq. ft.

FLOW AND WASH RATES

Diam. Filter*	*Flow Rate at 3 gpm per sq. ft.*	*Flow rate at 2 gpm per sq. ft.*	*Wash Rate at 21 gpm per sq. ft.*
4 ft.	38 gpm	25 gpm	264 gpm
5 ft.	59 gpm	39 gpm	431 gpm
6 ft.	85 gpm	57 gpm	622 gpm
7 ft.	116 gpm	77 gpm	852 gpm
8 ft.	151 gpm	100 gpm	1100 gpm
9 ft.	191 gpm	128 gpm	1407 gpm
10 ft.	236 gpm	157 gpm	1722 gpm
11 ft.	285 gpm	190 gpm	2095 gpm
12 ft.	340 gpm	226 gpm	2487 gpm
13 ft.	398 gpm	265 gpm	2928 gpm
14 ft.	460 gpm	307 gpm	3397 gpm
15 ft.	530 gpm	354 gpm	3894 gpm
16 ft.	603 gpm	402 gpm	4219 gpm
18 ft.	765 gpm	509 gpm	5326 gpm
20 ft.	942 gpm	628 gpm	6586 gpm
22 ft.	1140 gpm	760 gpm	8000 gpm
25 ft.	1473 gpm	980 gpm	10309 gpm

*For all diameters: Straight Height is 14 ft. 10¼ in.; Overall Height is 20 ft. 10¼ in. Units 10 ft. diameter and under are shipped with tank assembled; large tank sizes are shipped unassembled for field assembly.

LOSS-OF-HEAD PRINCIPLE

Operation of filters on a time-cycle basis, the conventional method, does not allow for variations in influent quality such as those caused by variations in the quality of the raw water itself or in the effectiveness of the pre-treatment. Operation by inspection also introduces errors since the appearance of the bed does not necessarily indicate its general condition. Loss of head, however, indicates the true condition of the bed and is generally accepted as the most accurate control factor.

The head loss at which a Permutit Valveless Filter initiates backwashing is determined by the height of the inverted U turn at the top of the backwash pipe. Since the level of water in this pipe is proportional to the head loss across the filter bed, the higher this turn, the greater the loss of head. Optimum height of the turn is generally 4 to 5 feet above the top level of the backwash storage water.

ADVANTAGES

Manpower: None required since there are no valves to operate, gauges to watch, pumps to start and stop, or decisions to be made.

Rotary Wash: Not required – the unit operates only to a 4′ or 5′ headloss, and experience with the valveless filters has shown that under these conditions the "schmutzdecke" does not become sufficiently packed to require a surface wash to break the surface mat and keep the bed clean.

Valves: None required. The entire operation takes place hydraulically, and change of flow direction is accomplished by atmospheric pressure.

Rate of Flow Controllers: No mechanical, air-operated or hydraulically-operated device is required as the unit filters whatever volume is delivered to it.

Rate of Flow Indicators: None are required since the plant inlet flow is divided evenly among the units. The rate to each unit can easily be calculated from the plant inlet meter by dividing by the number of filters in operation.

Loss of Head Indicator: None are required since the unit is its own loss-of-head indicator. The elevation of the water in the backwash pipe is the loss of head. Since the unit washes whenever the head loss reaches the proper point, loss of head information is not really needed.

Backwash Rate Controller: No mechanical, air-operated, or hydraulically-operated device is required since a backwash regulator is designed into the unit. Adjustment is made only at the time of installation.

Backwash Rate Indicator: None is required since the backwash regulator insures proper rate.

Pumps: None are required for filter operation. Since the backwash water is always stored in the unit at the proper elevation, no wash pump is needed. Since rotary wash is not required, a high pressure pump for the purpose is eliminated.

Operating tables: None are required since there is nothing to operate.

Electricity, instrument air or pressure water: None. All operations are controlled by the unit itself.

Maintenance: None required. Since there are no moving parts, there is no wear hence no maintenance other than normal painting procedures.

If the requirement is simply to introduce air to oxidize iron or manganese, a sniffler valve or direct pressure aeration may be preferred to avoid repumping. However, care must be taken to avoid supersaturation when the pressure is reduced. Supersaturation causes "milky water" and filter troubles from air binding or upset beds.

A unique Permutit development, the bypass pressure saturator, provides aeration without the danger of supersaturation by introducing air to only a predetermined portion of the total flow.

To reduce CO_2 or taste and odor or to accomplish any of these at the same time that you oxidize iron or manganese, a Permutit natural draft type aerator is recommended. Permutit natural draft aerators entrain air from the atmosphere. They are of the open type in which water cascades over trays of graded coke or a series of slat trays. Construction may be of wood, rubber-lined steel, stainless steel, aluminum, or Permuglas plastic. Splash aprons reduce splashing and permit free circulation of air. Formation of iron or manganese film on coke or slat surfaces hastens oxidation reactions. Some reduction of CO_2 also takes place, raising pH and further helping oxidation.

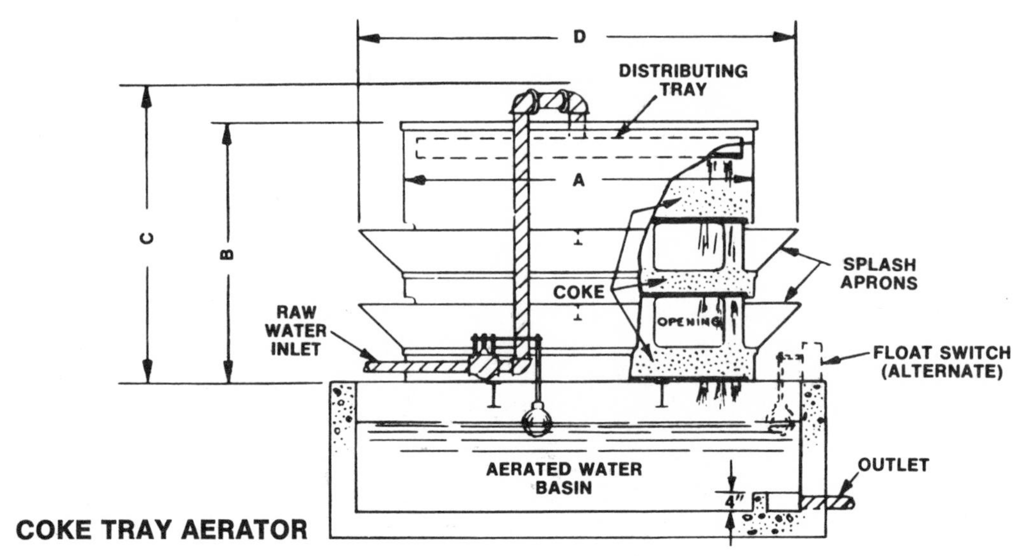

COKE TRAY AERATOR

SQUARE STEEL

Aerator No.	Capacities Max. G.P.M.	Operating Wt. Lbs.	Area In Sq. Ft.	Size Pipe Inlet	Dimensions			
					A	B	C	D
1	180	3500	9	3″	3′-0″	6′-2″	6′-4″	5′-4″
2	320	5200	16	4″	4′-0″	6′-2″	6′-6″	6′-4″
3	500	7500	25	5″	5′-0″	6′-2″	6′-6″	7′-4″
4	720	10500	36	6″	6′-0″	6′-2″	6′-7″	8′-4″
5	980	14000	49	8″	7′-0″	6′-2″	6′-10″	9′-4″
6	1280	17000	64	8″	8′-0″	6′-2″	6′-10″	10′-4″
7	1620	22000	81	10″	9′-0″	6′-2″	7′-1″	11′-4″
8	2000	26000	100	10″	10′-0″	6′-2″	7′-1″	12′-4″
9	2420	30000	121	12″	11′-0″	6′-2″	7′-7″	13′-4″

CYLINDRICAL STEEL

Aerator No.	Capacities Max. G.P.M.	Operating Wt. Lbs.	Area In Sq. Ft.	Size Pipe Inlet	Dimensions			
					A	B	C	D
1	142	2900	7.1	3″	3′-0″	6′-10″	7′-0″	4′-10″
2	252	4500	12.6	4″	4′-0″	6′-10″	7′-2″	5′-10″
3	392	6400	19.6	5″	5′-0″	6′-10″	7′-2″	6′-10″
4	566	8500	28.3	6″	6′-0″	6′-10″	7′-3″	7′-10″
5	770	11000	38.5	6″	7′-0″	6′-10″	7′-3″	8′-10″
6	1006	13900	50.3	8″	8′-0″	6′-10″	7′-6″	9′-10″
7	1272	17100	63.6	8″	9′-0″	6′-10″	7′-6″	10′-10″
8	1570	20500	78.5	10″	10′-0″	6′-10″	7′-9″	11′-10″
9	1900	24500	95.0	10″	11′-0″	6′-10″	7′-9″	12′-10″
10	2260	28600	113.0	12″	12′-0″	6′-10″	8′-3″	13′-10″

In a Permutit degasifier, water flows down through Raschig rings or tray packing while a large volume of air is forced or induced to flow counter current to the water. This air current is maintained by a blower (forced draft type) or fan (induced draft type). The freshest air contacts the effluent water and gives lowest final gas content. Proper packing design breaks water into thin films, promotes rapid and efficient gas removal.

Permutit degasifiers are totally enclosed to avoid splashing and spray losses. Cylindrical units are usually more economical, but square or rectangular shapes are available and should be used when iron is present. Sizes and capacities are practically unlimited, depending on requirements, and there is a wide choice of construction materials. The outer tank may be of wood, rubber-lined steel, stainless steel, Permuglas plastic, or aluminum. Permutit degasifiers can be used for water with CO_2 as high as 3500 ppm.

Although degasifiers provide very complete aeration, their higher cost is unwarranted in many cases where simple aeration achieves the objective.

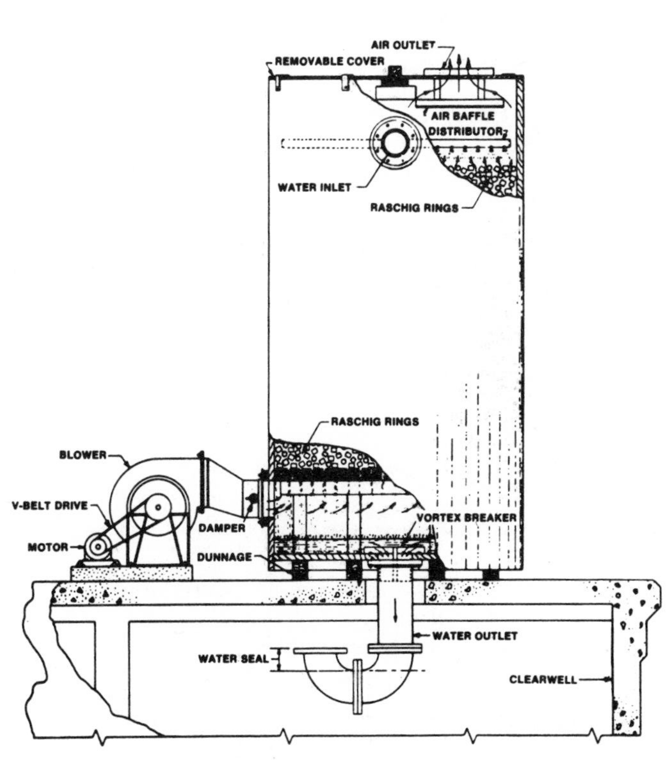

PERMUTIT DEGASIFIER with RASCHIG RINGS

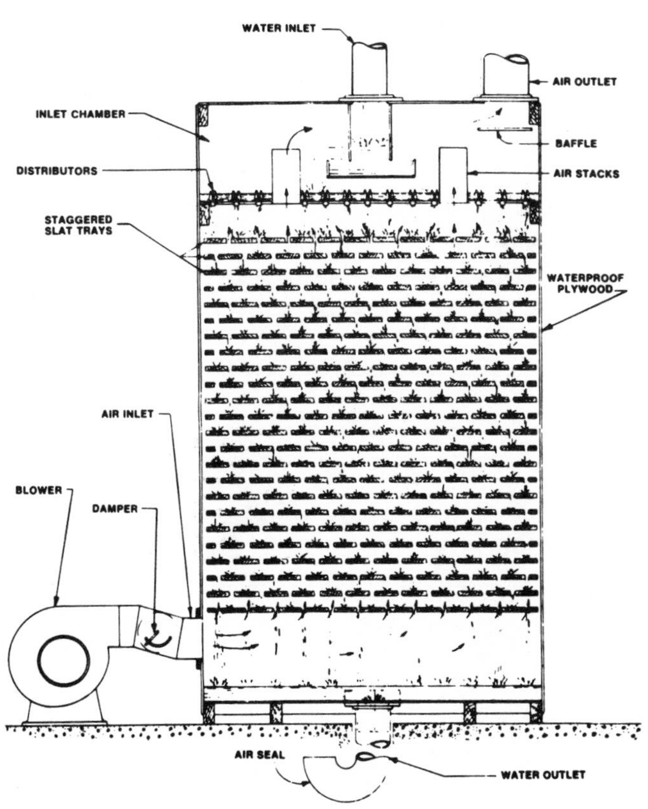

SECTION THROUGH PERMUTIT DEGASIFIER RECTANGULAR DESIGN
(ARROWS INDICATE AIR FLOW)

The Permutit Company, Division of Sybron Corporation

PermuRO REVERSE OSMOSIS PROCESS SYSTEMS

PermuRO Reverse Osmosis is one of the most significant innovations yet achieved in the technology of fluid treatment. Essentially an ionic sieve, the Reverse Osmosis process rejects dissolved organic materials and minerals as well as minute particulate contaminates to provide virtually pure, sterile water.

APPLICATIONS FOR PermuRO IN WASTE TREATMENT:

- Process Water Recovery
- Aqueous Waste Concentration
- Water Purification for Disposal
- Product Separation and Recovery
- Concentration of Waste Regenerants
- Tertiary Treatment

PERMASEP® Reverse Osmosis Barriers are fine hollow fibers, (85 micron) bound into a cylindrical bundle containing many square feet of surface area. This fiber barrier assembly is installed in a pressure shell fitted with a feed water inlet, concentrate outlet and purified water outlet. Feed water under pressure circulates around the fibers. Pure water passes through the walls of the fibers and flows through the fibers' bore. Contaminants remain on the outside of the hollow fibers, increasing in concentration. This concentrate is discharged through the concentrate outlet.

PERMASEP hollow fiber gives:

- Highest production per unit of space occupied.
- Assured performance and long life.
- Broadest pH range of any RO membrane: 4 to 10.
- Usable product discharge pressure.
- Inert to microbial attack.

PERMASEP is a registered trademark of the E. I. DuPont De Nemours Company (Inc.) for its Permeators.

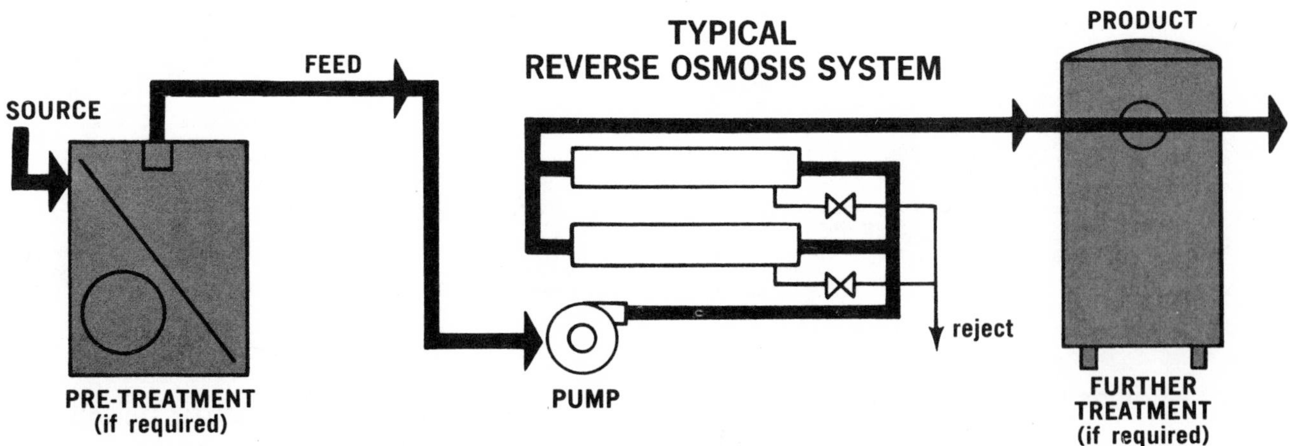

Polymetrics, Inc.

SPIRAL REVERSE OSMOSIS SYSTEM

The operation of a Polymetrics Spiral Reverse Osmosis System is quite simple. Supply water first passes through a roughing filter (25 micron) after which a trace of acid is added. This lowers pH and serves to prevent calcium carbonate precipitation, as well as substantially extending the life expectancy of the membranes. When concentrations of calcium and sulfate are exceptionally high in the supply water, an additional agent may also be added to inhibit precipitation.

The water next passes through a multistage centrifugal pump. This adaptation of a deep well pump has no reciprocating parts, insuring silent, vibration-free operation. Pumps with horsepower from 1/2 to 100 are utilized in standard Polymetrics systems. Parallel (standby) pumps are available.

Water then passes into a housing which contains spirally wrapped modules. A portion of the water permeates the membrane and is collected for use. The remaining water, containing the rejected impurities (or "concentrate"), enters a second or third housing where the process is repeated. Again, a portion goes to use while the concentrate goes to drain with the rejected impurities. (A portion of the concentrate may be recirculated to provide required minimum flow.)

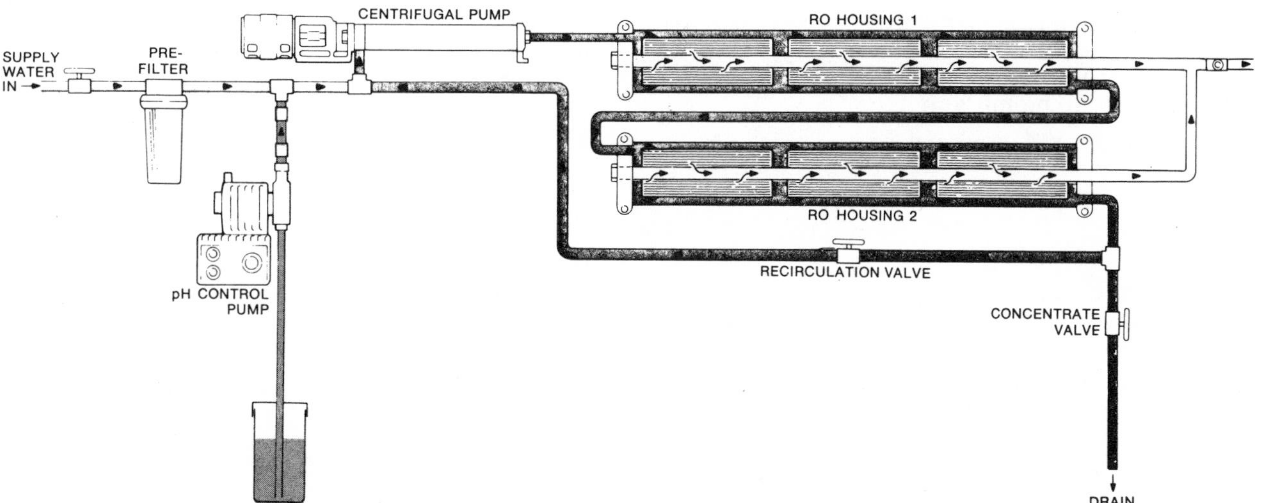

The basic building block, the spiral module, consists primarily of a sandwich of two sheets of membrane separated by a porous support or backing material. The backing material supports the membrane under operating pressures and provides a path for collecting the purified water. The sandwich is sealed around three sides while the fourth is attached to a perforated plastic (PVC) tube. The sealed membrane package is rolled around the PVC tube in the form of a spiral.

The membrane material is generally a derivative of cellulose acetate. Polymetrics also offers a complete line of reverse osmosis systems utilizing the new DuPont Permasep B-9 permeators which offer additional operational and cost advantages.

Raypak, Inc.

ROpak TURBOFLO REVERSE OSMOSIS SYSTEMS

Applications: (a) Supply water purification, (b) food products recovery in food processing, (c) sewage water treatment and (d) industrial wastewater treatment.

Raypak's engineers offer a low cost method of trying ROpak reverse osmosis under actual conditions in your plant before making a major investment. This is accomplished with the ROpak Single Core Pilot Plant which provides flux rates from 1000 to 3000 GPD, based on 90% NaCl rejection. This compact unit is also available with 0%, 50%, 80%, 95% and 98.5% NaCl rejection membranes, plus selective molecular separation membranes. The nominal 1000 GPD unit is entirely self-contained, and compact enough to load into a station wagon or pickup. Standard instrumentation consists of high pressure feed gauge, concentrate back pressure gauge and concentrate flow meter. Optional instrumentation available includes permeate, feed and concentrate, conductivity meters, permeate flow meter, strip recorders, feed thermometer and feed pH meter and controller.

Economic feasibility

- Lowest energy per gallon of permeate
- Longer membrane life
- Highest flux rate per sq. ft. of membrane, to 34 GPD
- Less prefiltration required
- Eliminates unscheduled production stoppages
- Lower overall operating costs
- Broadest range of applications in RO industry
- Widest range of membrane composition

Technically superior

- Handles greater pH range (3.0 to 8.75, depending on solution)
- Best hydraulic performance
- Operating pressures to 1500 PSIG
- Membrane subject only to compressive forces
- No size limitation due to modular concept
- Wide range of membrane selection to 98.5% or more NaCl rejection.

Lowest Operating and Maintenance Costs

- Turbulence inhibits membrane fouling
- Membrane leaks easily detected and repaired
- Reusable membrane core
- Simple membrane replacement
- Quality components and hardware

ECONOMIC ANALYSIS

Plant Model	Nominal Capacity (Based on Membrane) 1000 U.S. Gal./Day @ % Rej.				Depreciation 10 yr.	Power $.01/KWH	Water Maintenance Misc. Costs	Membrane Replacement 3 yr. Life	Total	Operating Costs Per 1000 Gal. per Membrane Type			
	80% NaCl	90% NaCl	95% NaCl	98.5% NaCl	$/day	$/day	$/day	$/day	$/day	80% NaCl	90% NaCl	95% NaCl	98.5% NaCl
3005	7	6	5	3.5	3.39	.96	2.00	1.68	8.03	1.15	1.34	1.61	2.29
3006	29	25	21	15	7.20	3.84	4.25	7.19	22.48	.78	.90	1.07	1.50
3007	59	50	42	30	13.75	5.76	7.10	14.38	40.99	.69	.82	.98	1.37
3008	126	108	90	63	21.09	10.80	10.70	31.00	73.59	.58	.68	.82	1.17
3011	252	216	180	126	39.96	21.12	15.00	54.99	131.07	.52	.61	.73	1.04
3013	504	432	360	252	72.31	41.24	20.00	109.95	243.50	.48	.56	.68	.97
3015	840	720	600	420	115.26	74.44	30.00	167.58	387.28	.45	.54	.65	.92
3017	1,680	1,440	1,200	840	210.81	144.40	50.00	342.47	747.68	.44	.52	.62	.89
3019	3,360	2,880	2,400	1,680	400.24	276.00	80.00	657.53	1,413.77	.42	.49	.60	.84

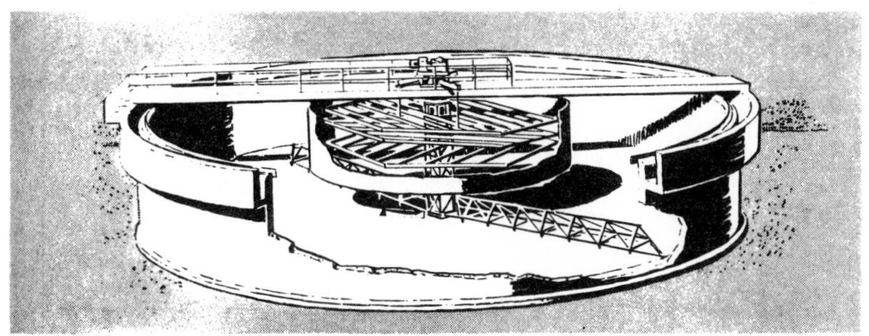

nu-treat flocculator-clarifier ▲

REX Nu-Treat is a combination flocculator-clarifier that assures most efficient mixing, optimum floc formation combined with optimum sedimentation and sludge removal. A unique nutating flocculating motion imparts a thorough, gentle, uniform mixing action that assures complete flocculation yet no harmful currents are set up in the sedimentation zone.

Nu-Treat can be used in both water and waste treatment in round, square or rectangular tanks. Simple rugged design cuts both maintenance and operating costs.

VERTI-FLO clarifier

The Rex Verti-Flo is an outstanding improvement in rectangular settling tank design offering many advantages for separation of solids from liquids. Verti-Flo provides the proper settling conditions assuring maximum clarification per unit of tank volume. Inlet, vertical and weir velocities are all extremely low and carefully controlled, thus assuring far more suitable settling conditions than can exist in conventional rectangular tanks.

Verti-Flo transforms the conventional horizontal-flow tank into a vertical-flow tank by a unique cellular system of collecting troughs, weirs and partitioning baffles whereby a single large horizontal settling zone is divided into a series of small, vertical-flow cells. With this design, maximum use of settling area and settling volume are available, short circuiting and flow disturbing eddies are minimized. Smaller tanks can be used with consequent savings in construction and equipment costs. Existing rectangular tanks, converted to Verti-Flo, will have at least double the original capacity and provide a far clearer effluent.

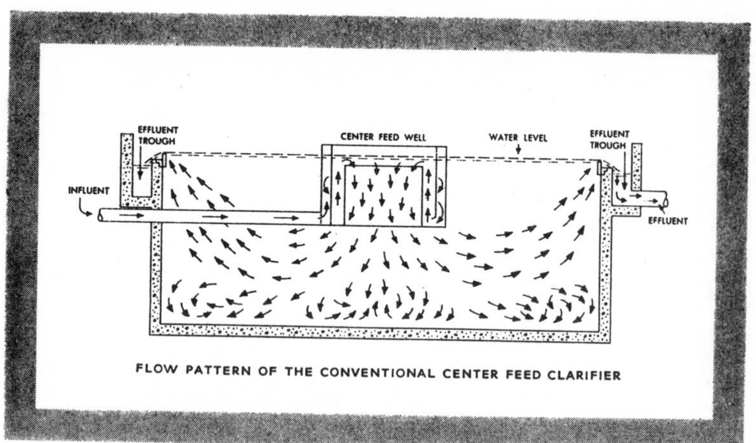

FLOW PATTERN OF THE CONVENTIONAL CENTER FEED CLARIFIER

center feed clarifier

In the conventional center feed type tank, flow is brought into the tank from the side or the tank bottom into a center feed well.

Flow is distributed from the center well laterally through the tank to an effluent trough entirely around the outer tank periphery.

The center feed tank is generally used where solids are relatively heavy and when skimming of the liquid surface is required.

Settled solids can be removed from the tank by circular scraper mechanism or REX Tow-Bro Sludge Remover.

RIM-FLO clarifier

An exclusive REX development for circular tanks, the Rim-Flo Clarifier assures greatly increased capacity and improved settling conditions.

The peripheral feed, converging flow principle of Rim-Flo, provides optimum hydraulic stability that assures most effective settling conditions. Flow is distributed around the outer periphery of the tank in a circular feed channel. Flow is then introduced uniformly into the tank through metering orifices in the feed channel and immediately diffused between the peripheral skirt baffle and tank wall before entering the settling zone. Entrance velocities are extremely low and the flow enters the clarification zone under ideal flow conditions. Solids promptly drop out of suspension onto the tank floor for rapid removal. In side-by-side tests with conventional circular center-feed tanks, it has provided up to double capacity. Many industrial waste treatment operations have been able to effect substantial costs savings with Rim-Flo because its high capacity and performance permit the use of smaller basins to achieve the desired results. Installed in existing circular tanks, Rim-Flo has provided sufficient additional capacity to eliminate overload conditions. REX Rim-Flo is ideally suited for use with REX Tow-Bro Sludge Removers.

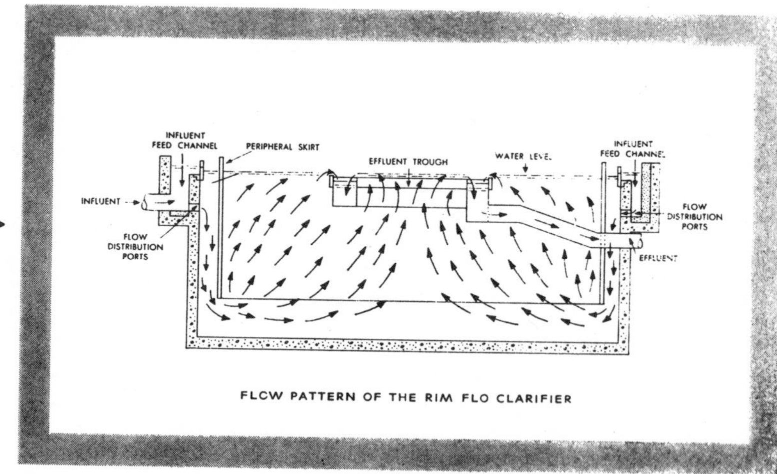

FLCW PATTERN OF THE RIM FLO CLARIFIER

FLOCTROL

REX Floctrol provides the ultimate in thorough flocculation . . . assures maximum floc build up with optimum use of tank volume and chemicals thereby effecting a considerable saving in construction and chemical costs. Floctrol is a unique combination of scientifically designed and proportioned paddle wheels, rotating baffles, and fixed partition walls with center ports so arranged that thorough mixing is assured. Short-circuiting through the chamber is held to the minimum.

A Floctrol Basin is separated into at least three, and sometimes four, mixing zones depending on basin length.

Each mixing zone is separated by a solid baffle wall having a suitable port concentric with the mixer shaft. Port area and velocities are selected to cause a slight head loss between mixing zones thus controlling flow with a minimum of short circuit. Rotating baffles mounted on the mixer shaft on each side of the port diffuse the flow throughout the entire cross section of the chamber and direct the liquid flow into the paddle wheel to assure uniform contact of liquid with paddle wheel blades.

Sectional illustration showing paddle blades, circular rotating baffles, and fixed partition walls with center ports to form one mixing zone. Tapered mixing assuring maximum and thorough mixing with a minimum of horsepower is accomplished by varying the paddle area as the flow progresses through the basin. ▶

slo-mixers ▲

REX Slo-Mix Flocculators are particularly suited for applications where grease and scum are present in the flow. Rows of paddle wheels are arranged axial to the flow with partitioning baffles between stages for optimum mixing. The baffle construction of Slo-Mix allows passage of floatable materials to subsequent units for removal. Progressive decrease of mixing is accomplished by a reduction in the number of paddle blades per arm per mixing zone.

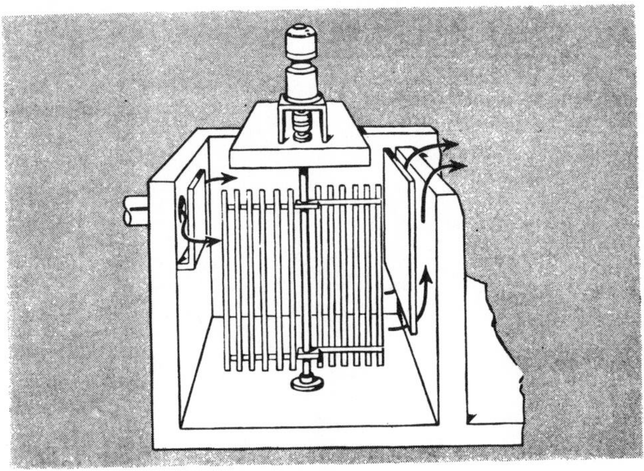

▲

vertical-paddle flocculators

REX Vertical-Paddle Flocculators are shaft-mounted on a vertical motor and gear reducer to operate in a square compartment. This arrangement is used primarily to meet conditions not lending themselves to conventional design practice.

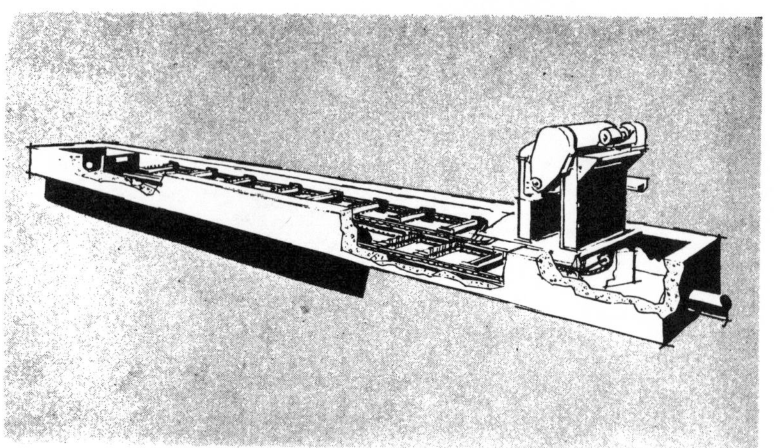

chain and bucket type collectors

This type of collecting mechanism is recommended for use in plants where deep channels are required and where moderate to unusually large amounts of solids must be removed and where solids can not be scraped up an incline. Formed steel buckets, mounted on two strands of REX Z-Metal Chain, remove the heavy solids from the channel floor and elevate them out of the flow. These solids may be discharged into storage hoppers, onto a conveyor system, or directly into trucks. Standard and heavy-duty styles of bucket collectors are available.

chain and scraper type collectors

This rugged style of collector is available in three service classifications—light, medium and heavy duty. Amount and type of solids determine selection.

Steel channel scrapers mounted on REX Z-Metal Chain scrape the heavy solids along the channel floor and either up an inclined deck for removal or into a sump. Where the solids are scraped up an incline, they may be discharged into conveyors, trucks or storage facilities. In some heavy-duty installations, solids are scraped into a sump and removed through pumping or may be elevated out of the channel by chain and bucket elevators or collectors. Dewatering of solids can be accomplished with REX Washers or De-waterers.

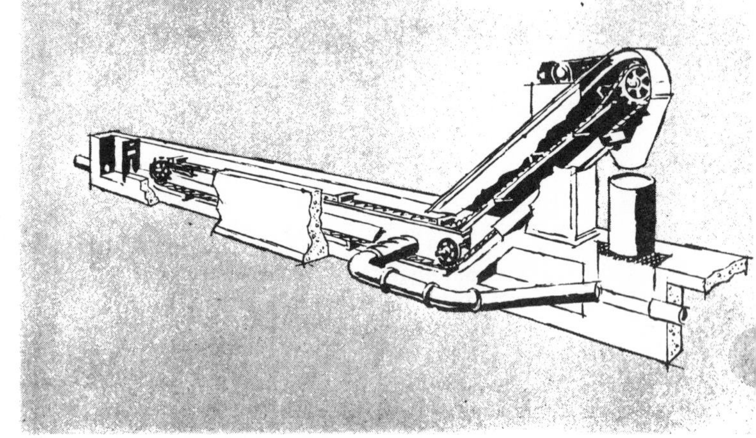

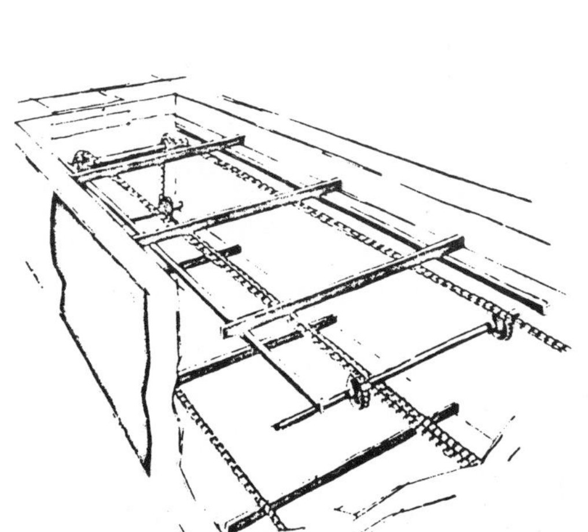

conveyor sludge collectors

REX Conveyor Sludge Collectors are used in rectangular settling basins to remove both settled and floated solids. They are ruggedly built for long, economical service.

In primary type tanks, REX Conveyor Sludge Collectors provide a positive, efficient and flexible means of sludge and scum removal. Scraper flights mounted on 2 strands of REX Z-Metal Chains scrape the settled sludge from the tank floor to sludge collection hoppers where the sludge is withdrawn from the tank. On the return run, the flights skim the surface to remove floatable solids and scum and deliver this material to an auxiliary skimmer for subsequent removal. Where skimming is not required the return run of the flights is submerged. REX Conveyor Sludge Collectors are available in a wide variety of types and sizes to suit any plant requirement.

screw conveyor collectors

For smaller settling and clarification tanks, either concrete or steel, REX Screw Conveyor Collectors provide an economical means of sludge collection. In addition where large volumes of sludge are not anticipated, screw conveyors are used as cross collectors with conveyor sludge collectors.

▲

circular scraper solids collectors

REX Circular Scraper Type Solids Collectors are designed for use in conventional circular settling tanks or in Rim-Flo Clarifiers. They provide both efficient and economical solids and scum collection and removal. The rugged, balanced design of the equipment assures long service life with minimum operating and maintenance costs and high capacity solids removal from the tank bottom. Where floatable solids are present, a skimmer arm is provided to remove this scum from the surface and deliver it to a scum hopper.

REX circular scraper mechanisms are available in many styles and sizes . . . half bridge . . . full bridge . . . siphon feed . . . side feed, etc. A wide range of torque designs allows adaptability to removing solids of varying characteristics.

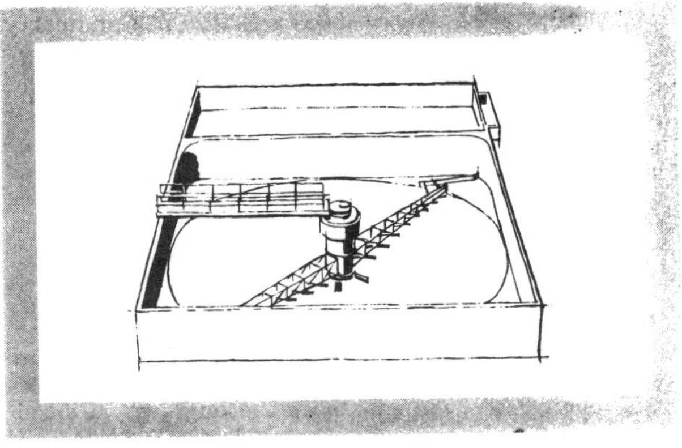

square tank ▲ scraper-type collectors

REX Scraper-Type Solids Collectors for square tanks are similar to circular scraper sludge collectors except that they have pivoted arms at the ends of the revolving scraper arms to assure complete removal of settled solids from tank corners. Drive unit is located on a center pier. Settle solids are scraped to a sludge hopper for removal.

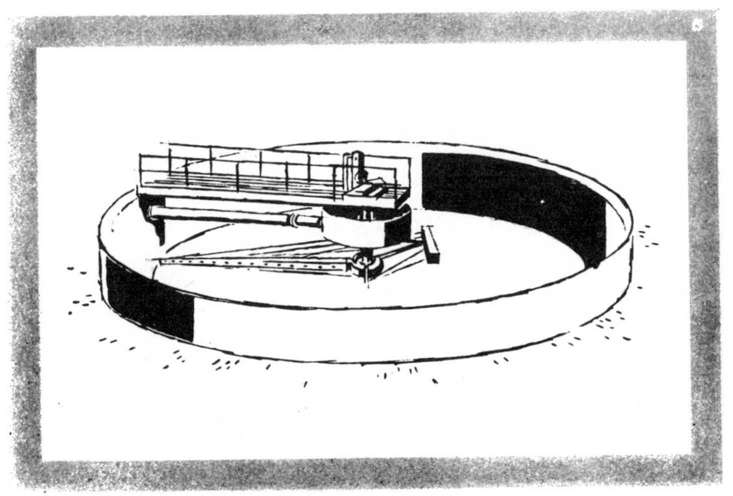

UNITUBE TOW-BRO®

(solids removers)

A unique development of REX Engineers, Unitube Tow-Bro provides a positive means of solids removal from circular or square settling tank through a gentle suction action instead of a scraping action. Basically the unit consists of a revolving header arm equipped with scientifically spaced and sized orifices. As the arm revolves, settled sludge is gently and uniformly removed from the tank bottom.

◄ The header arm of the Unitube Tow-Bro is of an exclusive, rectangular, peaked-roof design. The arm decreases in cross section from the center of the tank to the outer tip. Orifices in the tube increase progressively in size from the smallest at the widest arm cross section to the largest at the outer end of the arm. The balance between orifice size and taper of the arm assures uniform flow velocities in the tube and complete solids removal from the tank bottom in one revolution. Because there is a minimum of agitation, Tow-Bro is particularly effective for handling light delicate solids. In addition, with Tow-Bro maximum solids concentrations and less contamination of effluent by rising solids is assured.

Tow-Bro is available for many sizes and types of settling tanks . . . circular, square, rectangular, or dumbbell shaped.

FLOAT-TREAT SYSTEM

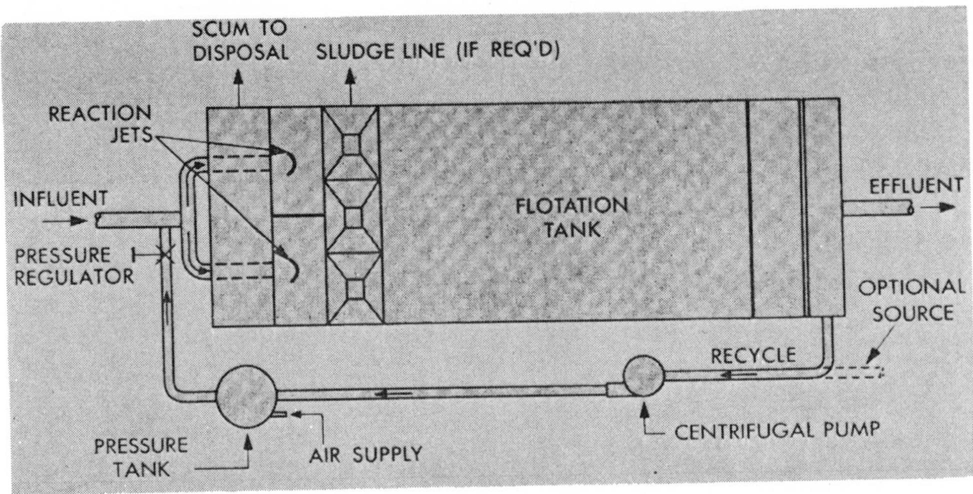

THE REX FLOAT-TREAT SYSTEM provides effective removal of suspended solids from liquids by dissolved air flotation. Flotation of the solids is accomplished by the introduction of millions of microscopic air bubbles into the process water. As these bubbles rise, they attach themselves to the particles in suspension and carry them to the surface for removal.

An important principle of the Rex Float-Treat System is the introduction of air into the recirculated effluent, and blending the recirculated flow with the raw flow. The recycle liquid is pumped at relatively high pressures to a suitable pressure-retention tank where air is introduced. While flowing through the pressure tank, the maximum possible quantity of air is dissolved in the recycle stream.

The air-saturated stream is then blended with the raw waste and fed into the Float-Treat tank. As the stream enters the tank, it impinges on circular, concave baffles . . . reaction jets . . . which assure thorough, rapid blending and dispersion throughout the cross section of the tank.

As the pressurized flow is restored to atmospheric pressure, millions of microscopic bubbles are formed. These air bubbles attach themselves to suspended particles in the flow and lift the particles to the surface for effective removal.

REX Float-Treat has proven effective in conjunction with chemically treating industrial wastes. Generally, chemicals are introduced into a flocculating tank to allow floc formation before the dissolved air bubbles come in contact with the solid.

Chemical flocs that would normally settle or rise very slowly in gravity settling or flotation tanks are caused to rise rapidly through the effect of the buoyant air bubbles. This increases tank capacity and usually results in a substantial saving in chemicals; and a higher degree of treatment.

FLOAT-TREAT SEPARATOR

The Rex Float-Treat Separator removes solids from liquids at low cost, and in the shortest possible time. Suspended solids that would normally settle or rise slowly in gravity settling or flotation tanks rise rapidly in the Float-Treat Separator. Float-Treat also provides for the removal of heavy, inorganic, non-floatable solids by settling. This classification of floatable and settleable solids is an added advantage of the Rex Float-Treat.

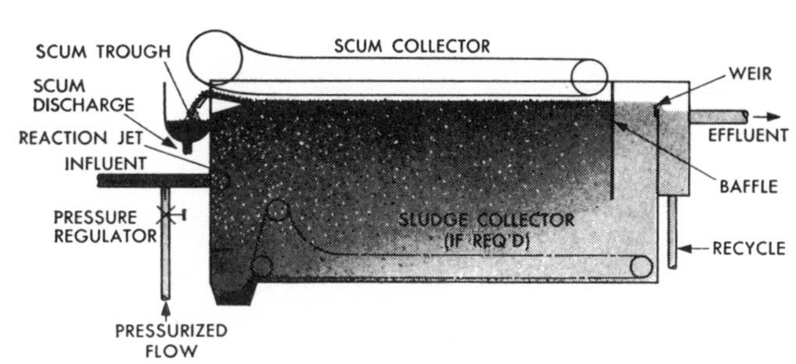

FLOAT-TREAT THICKENER

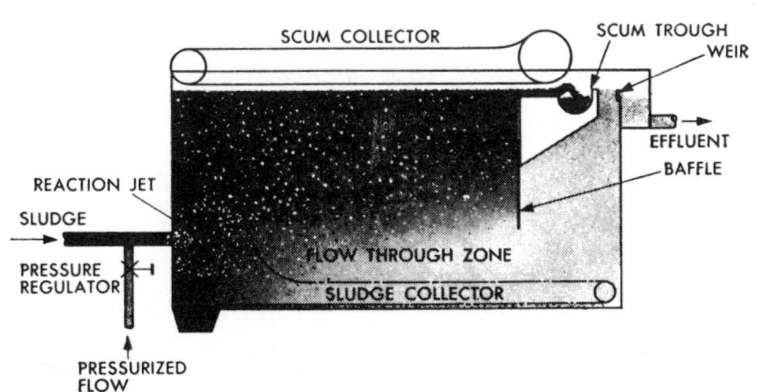

The Float-Treat Thickener provides unusually high solids concentrations when thickening activated sludge or other biological sludges. The compacting effect of the continuous supply of air bubbles in this system results in a sludge density much greater than that normally obtained by gravity thickeners. Solids concentrations of 4% and more with activated sludge are common.

Either mixed liquor, return sludge, or mixtures of raw and biological sludges can be classified and thickened in the Rex Float-Treat Thickener.

Unique Drawstring Principle Assures Positive Cloth Alignment and End Seal

This patented principle eliminates any necessity for exactly tailoring the cloth width, and compensates for any fabric shrinkage or elongation. Cloth is tracked by a tensioning cord laced into each cloth end loop. The tension on the cord keeps them running in grooved guide rolls which free-wheel on both return and tension rolls. Since cloth is wide enough to lap over the edges of the drum, and the guide rolls feed it onto the drum in correct position, it remains in track without any need for electrical alignment controls, tracking grooves or training bars.

The drawstring principle creates a positive seal against end leakage of solids or loss of vacuum. It also assures rapid, simple removal of the filter cloth for changing. Cloth lap joint is made by means of a stainless steel clipper lacing.

Continuous Dewatering with Positive Discharge: (Unit illustrated on page 187) As the drum travels through the slurry tank (1) the specially designed rotary vacuum valve (2) distributes vacuum to each drainage section of the filter drum (3). As the drum rotates, solids adhere to the cloth forming the desired cake. Dewatering starts when the drum emerges from the tank.

Liquid passes through the cloth, supporting drainage grids (4), drain lines, and the rotary vacuum valve to the vacuum receiver. As the cloth carrying the dewatered cake leaves the drum, it passes over the discharge roll (5) where cake is released from the cloth and discharged either into a receiving hopper or a belt conveyor. Complete cake removal is facilitated by the oscillating cloth movement resulting from the double reversed raised helix built into the discharge roll (5). The discharge roll, which is power driven, rotates at a slightly higher speed than the drum insuring a positive lift-off from the drum, a taut filter cloth, and positive tracking.

Cloth then travels to the return roll (6) and on to the tension roll (7). During this portion of its travel, the cloth is continuously (or intermittently) front and back washed by high velocity spray nozzles (8 and 9). After passage over the tension roll (7), cloth is reapplied to the drum to repeat the filtration cycle. Cloth wash water is collected in the trough (10) to prevent any possibility of diluting the slurry or filtrate. Separately driven oscillating type agitator (11) provides uniform thorough mixing in the slurry tank.

Cloth binding is eliminated with VACU-TREAT since the cloth is completely removed from the drum during each cycle and any solids adhering to the cloth are washed away by the sprays. Since binding is no problem, a much finer mesh cloth can be used to assure maximum solids retention and a filtrate of maximum clarity.

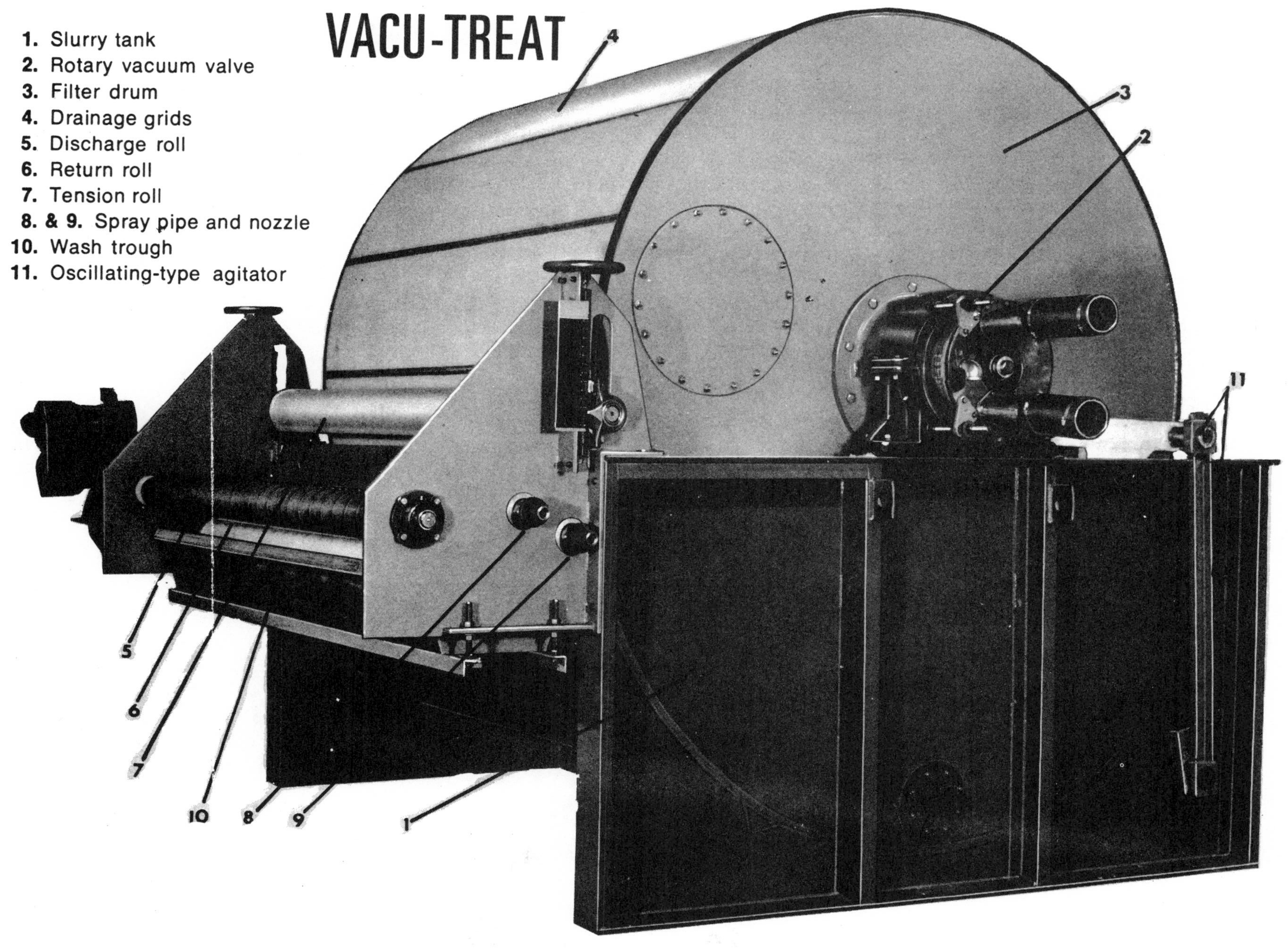
VACU-TREAT
1. Slurry tank
2. Rotary vacuum valve
3. Filter drum
4. Drainage grids
5. Discharge roll
6. Return roll
7. Tension roll
8. & 9. Spray pipe and nozzle
10. Wash trough
11. Oscillating-type agitator
4
3
2
11
5
6
7
10
8
9
1

Because of the improved cake removal method, it is possible to discharge much thinner cakes than with conventional drum filters, thus decreasing cycle time and permitting higher design and operating capacities per unit area. The result is a dryer cake that is easier and less costly to dispose of, more economical to incinerate, and requiring less conditioning chemicals in most applications.

Rex VACU-TREAT can be installed as a single unit for smaller plants or in multiples for the largest of waste treatment systems. A wide range of filtering areas are available with maximum areas of up to 900 square feet. Drum diameters range from 3 to 12 feet with drum face width up to 24 feet.

4. Drainage grids
5. Discharge roll
6. Return roll
7. Tension roll
8. & 9. Spray pipe and nozzle

Drum assembly illustrating arrangement of grid sections and leading and lagging (optional) drain lines.

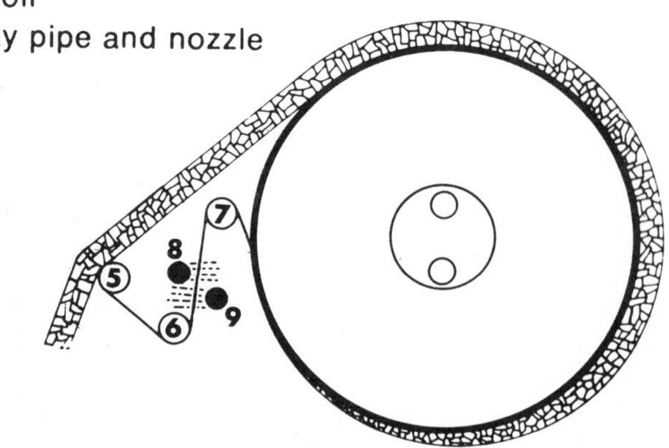

SKIMMERS

REX Skimming Equipment is available in six basic types and each type is manufactured in several styles to assure the widest choice to meet any operating condition.

REX Roto-Skim: Simplest and most economical of REX Skimming Equipment, the Roto-Skim is widely used in municipal and industrial waste treatment installations. It consists of a slotted steel pipe installed with the edge slightly above the liquid surface, so that when the edge is rotated below the surface, floatable solids and some liquid flow into the pipe openings and are flushed out of the tank for disposal. The Roto-Skim is available in four basic types: lever, rack and pinion, worm gear, and motor operated.

Scraper Type Power Skimmers: Among the most commonly used power skimmers are REX Chain and Scraper types. These are available in three basic styles: (1) Auxiliary Power Skimmer; (2) Rigid Wood Flight Full Length Surface Skimmer; (3) Pivoted Steel Flight Full Length Surface Skimmer.

The auxilary power skimmer is used with a conventional 4-shaft sludge collector and conveys the skimmings from liquid level, up a dewatering beach and discharges into a trough, either adequately sloped or equipped with a cross screw conveyor. Rigid and pivoted flight full length skimmers are used where separate skimming and sludge scraping are required.

REX Spiral Skimmers: This unit is primarily used for light frothy scums. The power operated, revolving spiral-shaped blade, mounted ahead of a scum trough, skims the floating material from the tank surface and deposits it into the trough, with flushing or spray water as required.

Circular Skimmers: Circular skimmers are used where small volumes of skimmings are anticipated. Common applications are primary clarifiers in sewage plants or industrial waste chemical coagulation plants where only small amounts of oils must be skimmed.

Roll Skimmers: This skimmer finds wide application in the refining, petro-chemical, automotive, steel mill, machine shop and similar industries for free mineral oil removal from liquid surfaces. The skimming unit revolves down into an oil film on the water surface and preferentially picks up an oil film on the roll surface. The adhering oil film is doctored off into a trough for further removal. These units are highly efficient for oil removal but are not recommended for animal greases or sewage scums.

Belt Skimmers: These units are similar in application and operation to the roll skimmers except that a belt is used instead of a roll. An important advantage of the belt skimmer is that it combines oil skimming and lifting of the skimmed oil to a higher point in one operation.

The belt and roll skimmer will remove oil containing less than 5% water, while conventional decanting type skimmers remove oil containing as much as 80% water.

ROTO-SKIM

REX Roto-Skim Pipes are available in diameters from eight to twenty inches depending upon capacity requirements. The diameter is governed by the variation in liquid level, basin width, and travel length of the skimmings in the pipe.

The pipe has a 60-degree slot forming two weirs. The skimmings with water, flow over the weir edge after the pipe is tipped.

Roto-Skim revolves in and is supported at each end by a rolled steel collar which is welded to an adjustable steel mounting plate. A marine plywood filler for the open end support provides a watertight connection to the tank wall without need for grouting.

RACK AND PINION ROTO-SKIM

This skimmer is widely used where the floating material is a fluid and excess quantities of flushing water are not required. The A.P.I. oil-water separator is an example. The close control of pipe rotation permits more accurate decanting of the oil off the liquid surface, reducing the amount of water removed. A result of 20% oil — 80% water mixture by volume is common.

The Rack and Pinion unit is also used where a basin freeboard exceeds three feet or where more accurate control of skimming is required.

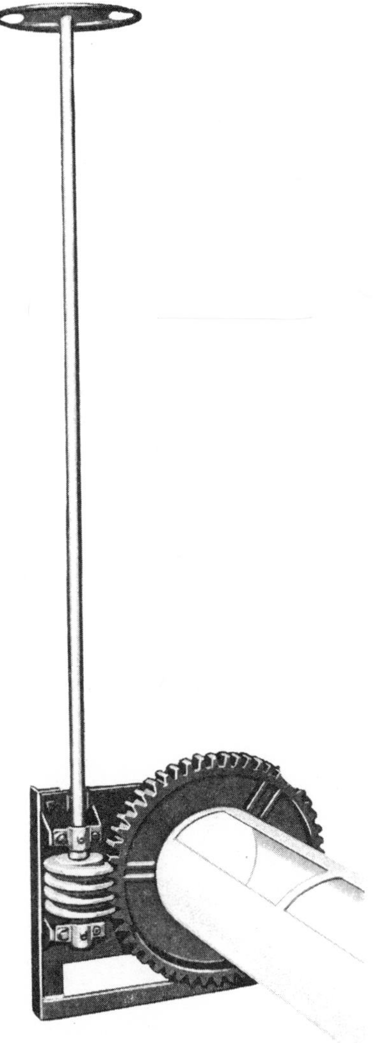

WORM-GEAR OPERATED ROTO-SKIM

Worm gear operated pipes are used where precision skimming is required and where skimmings are fluid such as oil on the surface of an A.P.I. separator. Where oil surface elevation remains constant worm gear pipes can allow continuous removal. Operating mechanism construction is a cast iron cut tooth worm with a steel cut tooth double thread worm wheel operating on a bronze bushing to allow free and easy operation.

TELESCOPIC SCUM PIPE

Where troublesome scums form in small collection boxes or inlet channels, positive removal is accomplished with a simple telescopic decanting funnel. The funnel is lowered just below the liquid level to remove the floating solids and adequate water for flushing purposes. The unit is manually operated with a 12-inch diameter non-rising hand wheel. The height of the funnel is rapidly adjustable — requiring only one turn per 3/4 inch of vertical travel. A seamless brass slip tube and neoprene gasket provide smooth, effortless operation.

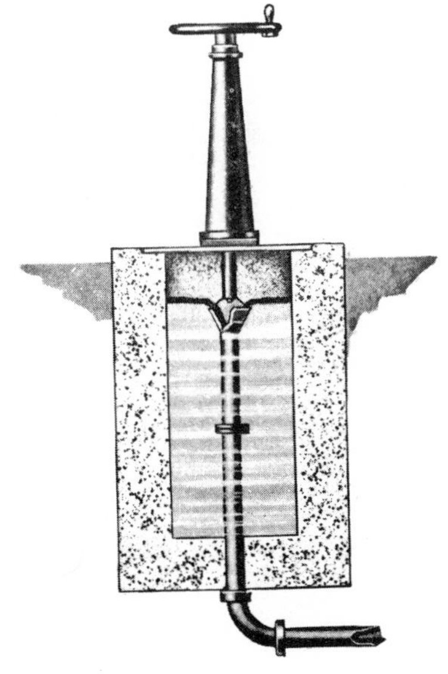

MOTOR OPERATED ROTO-SKIM

Where automatic operation of skimmer pipes is required, two Motor-Operated designs are available. In one design, illustrated above, Roto-Skim offers an economical selection. The scum pipe is rotated with a variable speed drive unit connected to an intermediate linkage system. The degree of pipe rotation is simply adjusted in the linkage and can be preset to any desired level. An optional motor operated Roto-Skim consists of a power source mounted on a stand which turns the pipe through a worm gear arrangement.

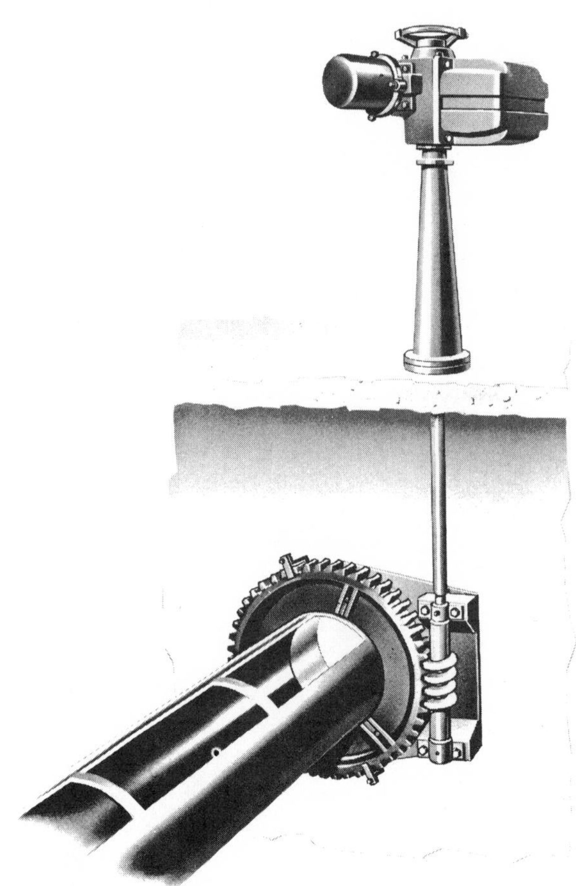

In most clarifiers a skimmer is required to move floatable solids to the skim pipe or into a scum trough. REX offers a wide range of mechanical power skimmers to efficiently perform this function.

Power skimming equipment include pivoted or rigid type flight scrapers that skim the entire tank surface; and/or collecting auxiliary power skimmers that discharge skimmings into a skimming trough. Power skimmers are used where large volumes of heavy skimmings are anticipated.

RIGID FLIGHT TYPE

Rex Chainbelt Inc.

The rigid flight type skimmer consists of wood flights mounted on two strands of REX Super Z-metal Chain. This skimmer design includes two types. A four shaft collector may be used where skimming is accomplished on the return run toward the effluent end of the basin, as shown below in the photo of the partially dewatered gravity oil water separators in a major eastern refinery. Where separate collection of settled solids is provided, a rigid flight surface skimmer is used. The REX Gravity Grease Separator for packing plants is a typical example.

For both four shaft and separate skimmer designs flights are of durable redwood, drilled and notched, and cut to exact length at the factory to assure accurate, fast field installation. Bearings and sprockets are all assembled on the shaftings to also assure fast field installation. Chain attachments extend the full depth of the flight for longest life of both attachment and flight.

SPRING BRASS SQUEEGEES

Spring brass squeegees can be provided for two flights on a conveyor sludge collector to provide positive cleaning of side walls and assure full width skimming on return run of conveyor mechanism.

PIVOTED FLIGHT TYPE

The full length skimming provided by this efficient type is particularly suitable for float-treat sludge thickening or separator applications, installations where only floatable solids are to be removed, or where separate removal mechanisms are required for floatable and settleable solids in gravity separators.

The end-hung chain attachment permits the scraper blades to pivot as they ride up the scum beach or trough, thus maintaining full contact with the skimming surface, and effecting maximum dewatering, automatically. In addition, the pivoted design permits effective cleaning of the flight immediately after discharge of the skimmings into the scum trough, and "soft" re-entry into the liquid at the effluent end of the basin. The chain is completely out of the water at all times — particularly important for some industrial applications involving corrosive skimmings. A variable speed reducer programmed on time clock controls is recommended for most applications.

This close-up view clearly indicates the effective action of the end-hung pivot attachment link.

POWER-OPERATED SPIRAL SKIMMERS

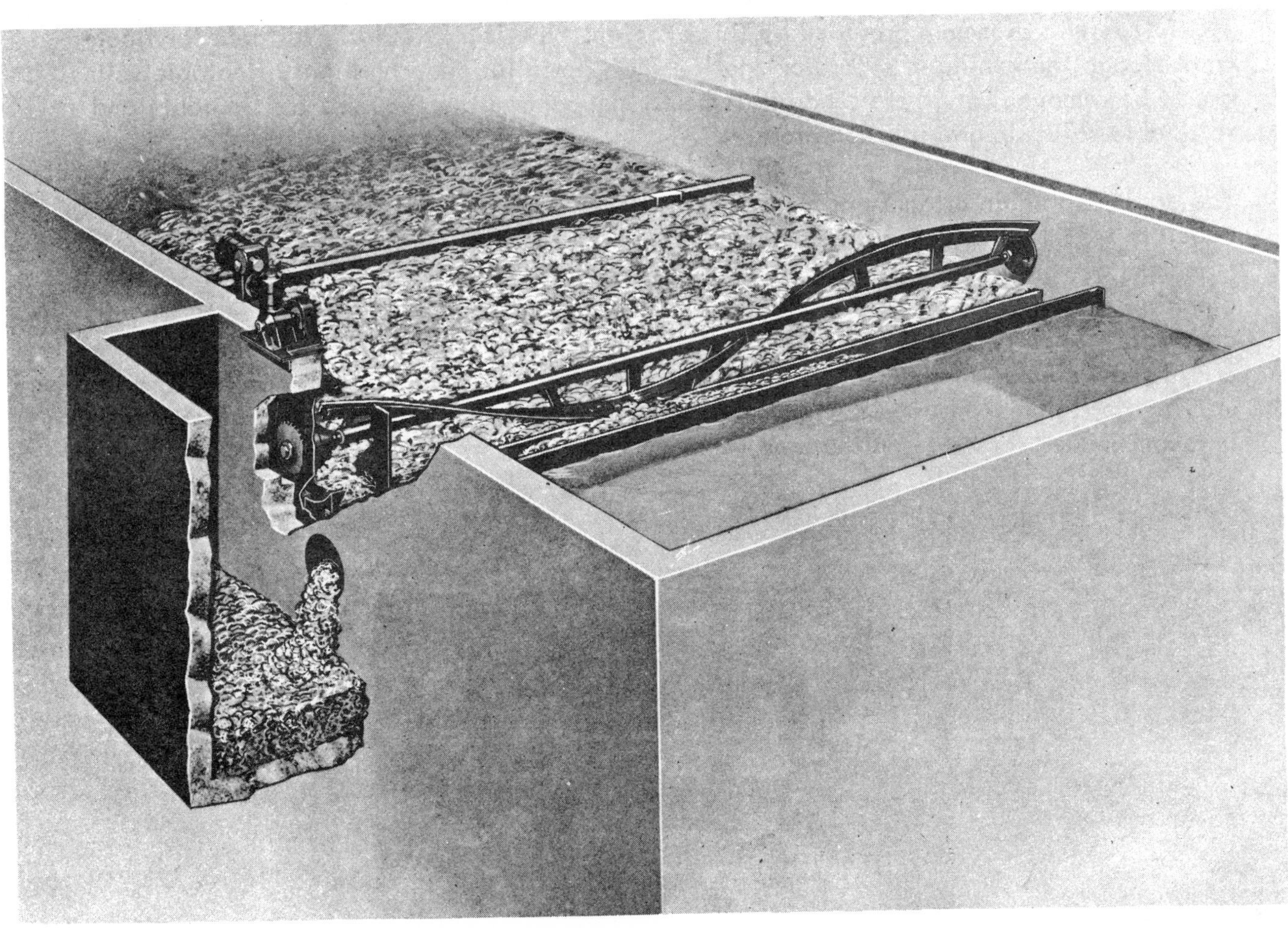

REX Power Operated Spiral Skimmers are primarily used in specialized applications, particularly those involving light, frothy scums that would not float into a scum pipe of their own accord. Potato starch waste or detergent type foams are typical examples.

Rubber tipped spiral blades revolve to lift the scum from the water surface and discharge it into a scum trough located immediately behind the blade. A complete 360° spiral is provided to more effectively remove material from the liquid surface. A curved metal or concrete beach is also recommended to assure positive pickup. A screw conveyor in the trough (or an additional flush water source) is required to remove the scum from the trough.

The spiral skimmer can be equipped for completely automatic operation.

CIRCULAR POWER SKIMMERS

The circular skimmer consists of an arm spanning the radius of the tank. The tip of the arm has a special pivoting blade approximately three feet long, offset from the main arm. A "pocket" is formed to positively trip the skimmings as the revolving arm approaches the skimmer beach. The beach is stepped to properly guide the blade up out of the water, for correct discharge of the skimmings. The blade rides over a cleaning lip immediately after discharge to eliminate skimmings carry-over.

ROLL SKIMMERS

These highly effective units find wide application in steel mills; refineries, chemical, automotive or metal working plants for economical skimming of oils from water. The total volume of skimmed oil is substantially reduced, and normally contains less than 5% water. This permits direct pumping of skimmed oil to recovery units without further decanting. Pump size and oil collection pits can be much smaller assuring maximum cost savings. Oil removal is automatic and eliminates costly manpower normally used in common decanting type skimming operations.

The roll skimmer consists of a pipe that is rotated into the oil film on the water surface. As the roll revolves, oil is selectively picked up on the roll surface. This adhering oil film is removed by a doctor blade, and deposited in a trough for further disposal.

Roll skimmer capacities vary, depending upon viscosity of the oil, thickness of oil film on the water surface, and length of roll skimmer.

These skimmers are available in a number of basic styles, both fixed, single roll and pontoon mounted, single or double roll. All models can be equipped for time-clock operation.

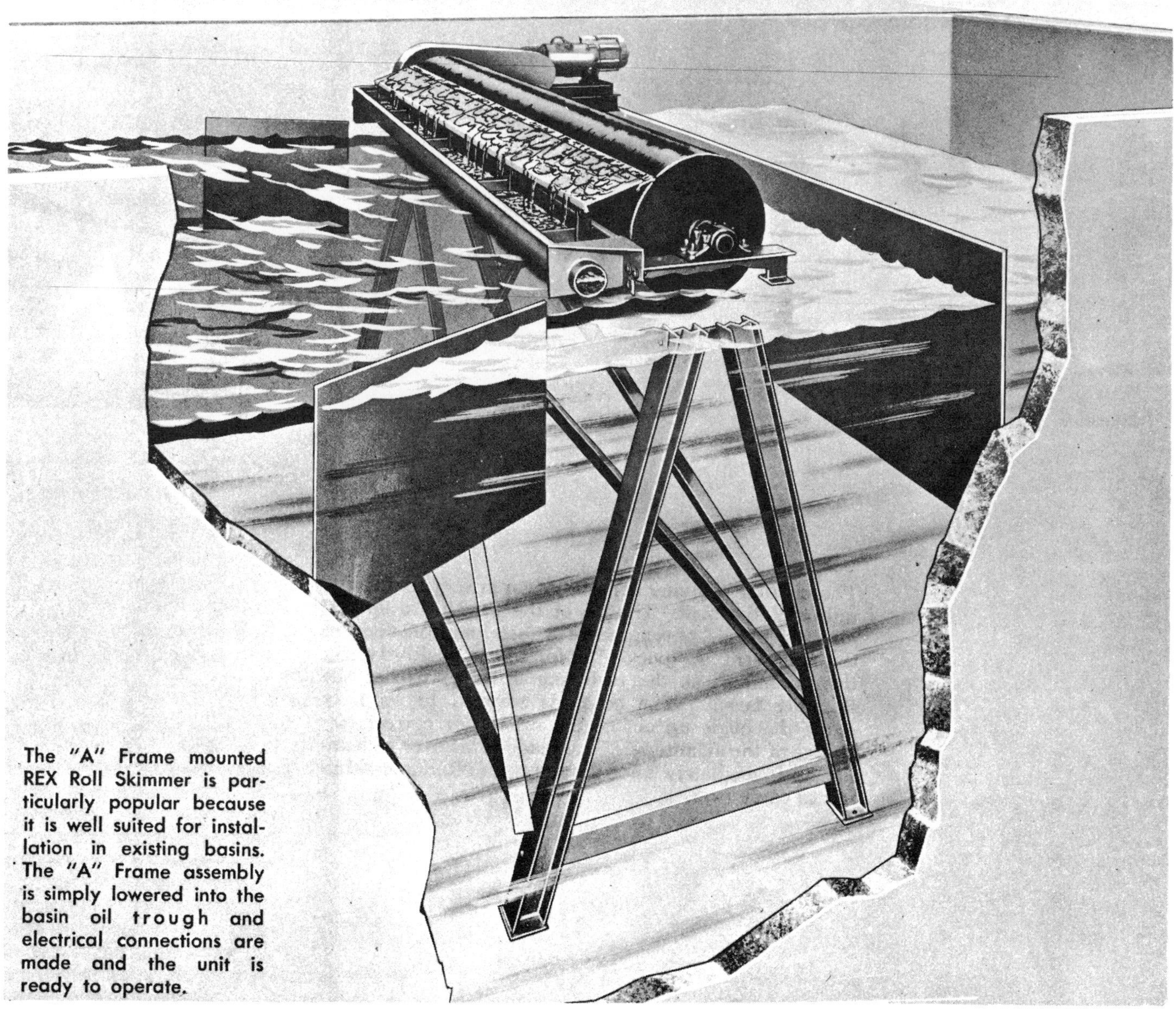

The "A" Frame mounted REX Roll Skimmer is particularly popular because it is well suited for installation in existing basins. The "A" Frame assembly is simply lowered into the basin oil trough and electrical connections are made and the unit is ready to operate.

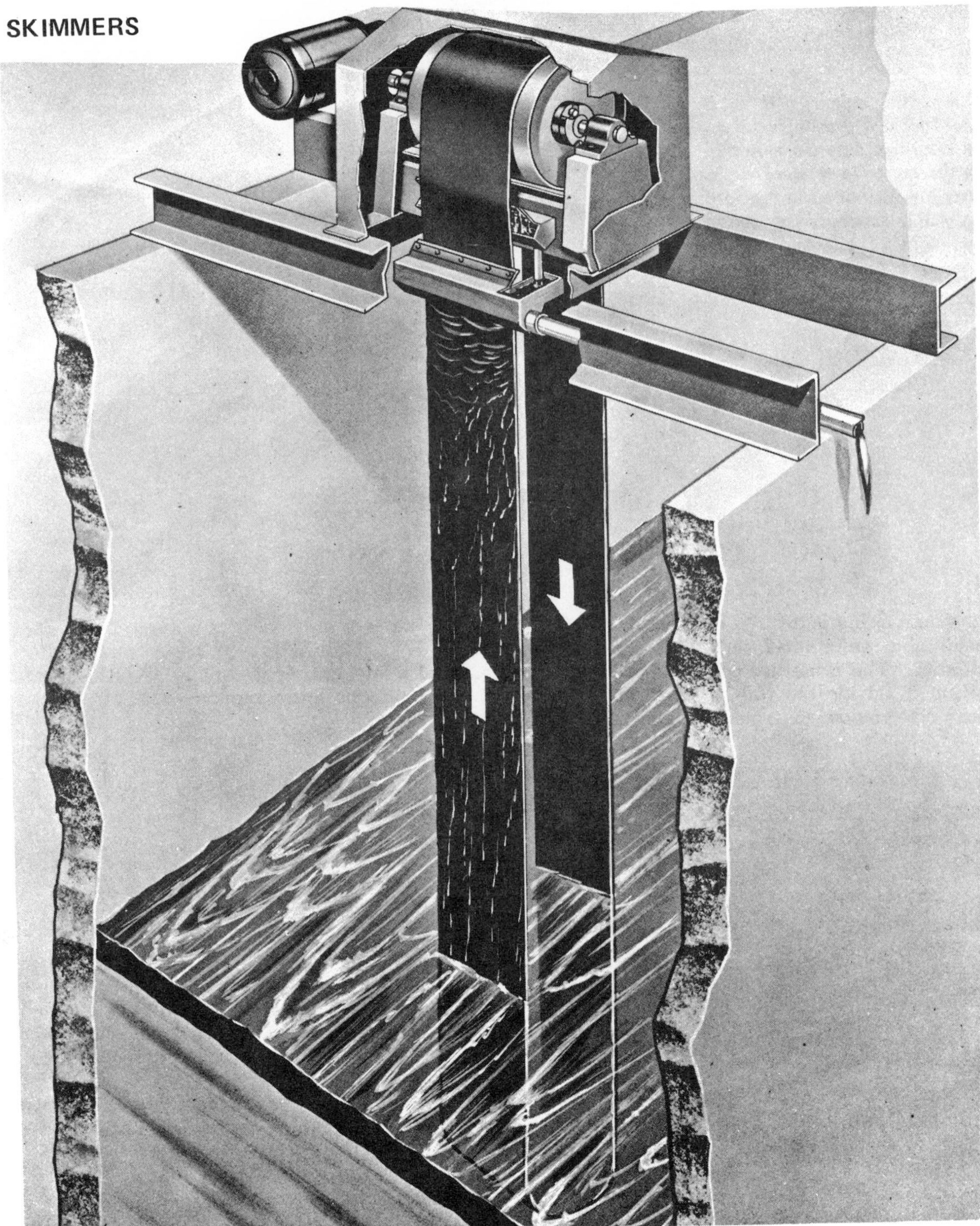

Important operating advantages of the REX Belt Skimmer are:

(1) It will skim the oil and elevate it to a higher elevation in one step.

(2) Permits wide flexibility for water level variations.

(3) Eliminates need for expensive concrete oil storage sumps and pumping equipment.

The belt skimmer consists of an endless Hycar belt that revolves in a channel. As the belt revolves, the oil film is picked up and elevated to primary and secondary doctor blades which remove the oil from both sides of the belt. The oil then flows into a trough and container.

The belt skimmer is designed as a standard unit, floor mounted directly on an operating platform, or cantilevered from the top of the basin wall. An intermediate housing section is included to develop the exact discharge height for a specific application. Total belt heights of 15 to 20 feet are common.

REX Belt Skimmers provide an effective means of skimming free oils from mill scale pits in steel mills and oily waste sumps in refineries, petro-chemical plants, machine shops, and railroad and aircraft maintenance shops. Like the roll skimmers, they restrict the amount of water skimmed off with the oil to the minimum, approximately 5%.

REX REACTORS

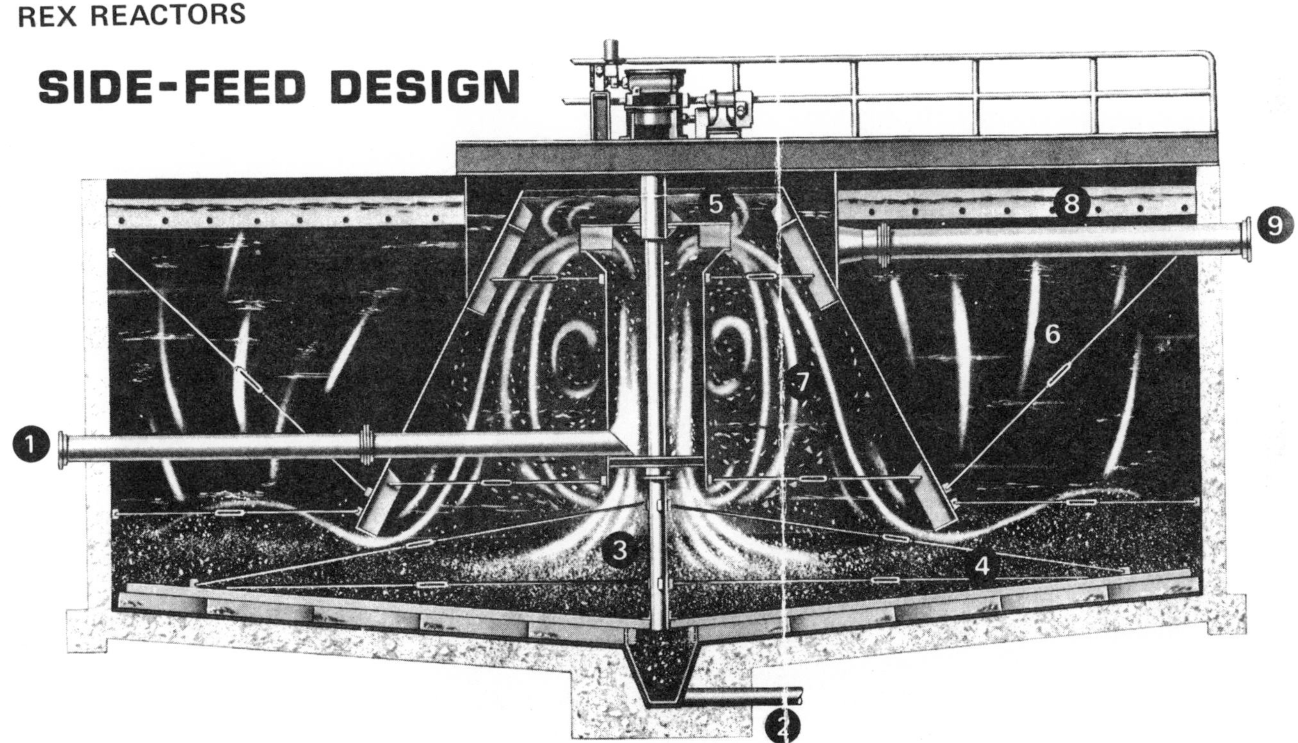

1 INFLUENT PIPE

2 SLUDGE DRAWOFF

3 SOLIDS RECIRCULATION

4 CIRCULAR SCRAPER SLUDGE COLLECTOR

5 TURBINE AGITATOR

6 CLARIFICATION ZONE

7 FLOCCULATION ZONE

8 RADIAL LAUNDERS

9 EFFLUENT LINE

WALL MOUNTED ROLL SKIMMERS

The wall mounted skimmer is most appropriate for new basin construction where bearings may be mounted directly on the concrete or steel walls. The roll may be positioned in the fore bay of an A.P.I. separator to scoop off high volumes of oil, and at the effluent end to automatically remove the oil.

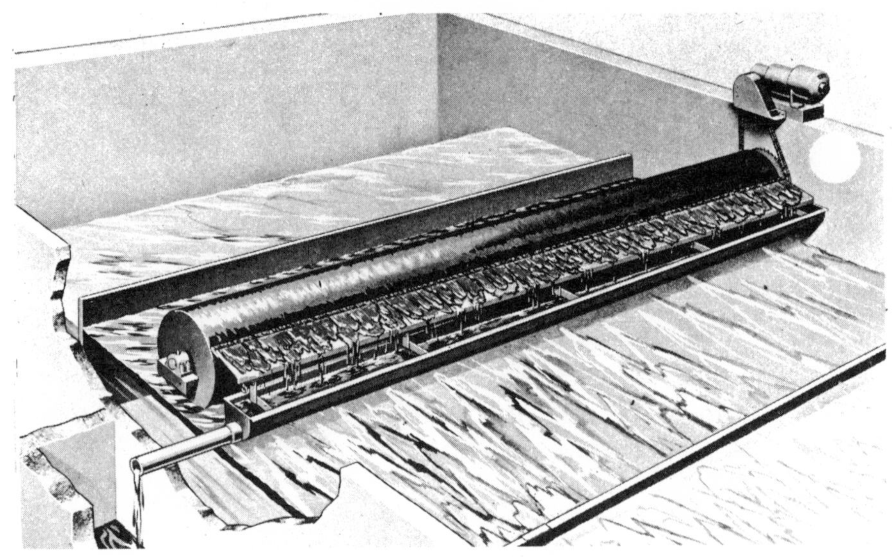

PONTOON MOUNTED ROLL SKIMMERS

Where skimming is desired over many points over a large surface area, such as a lagoon, or where considerable water level variations can be expected, the REX Roll Skimmer can be mounted on pontoons. The units are complete with oil storage sump pump and float control. Pontoons are positioned astride the roll to offer maximum stability. Pontoons are compartmentalized, amply sized and provide generous displacement.

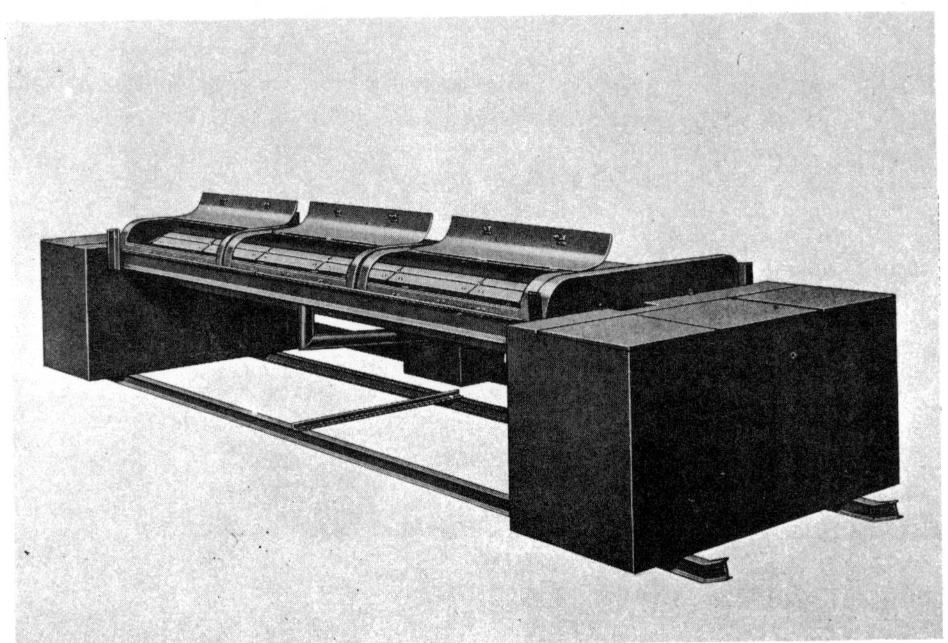

The Pontoon Mounted Single Roll Skimmer is used where the oil normally moves toward the unit from one direction. These units are available with rolls varying in length from 4 to 18 feet. Protective hood over the roll is an optional feature.

MILL SCALE COLLECTOR

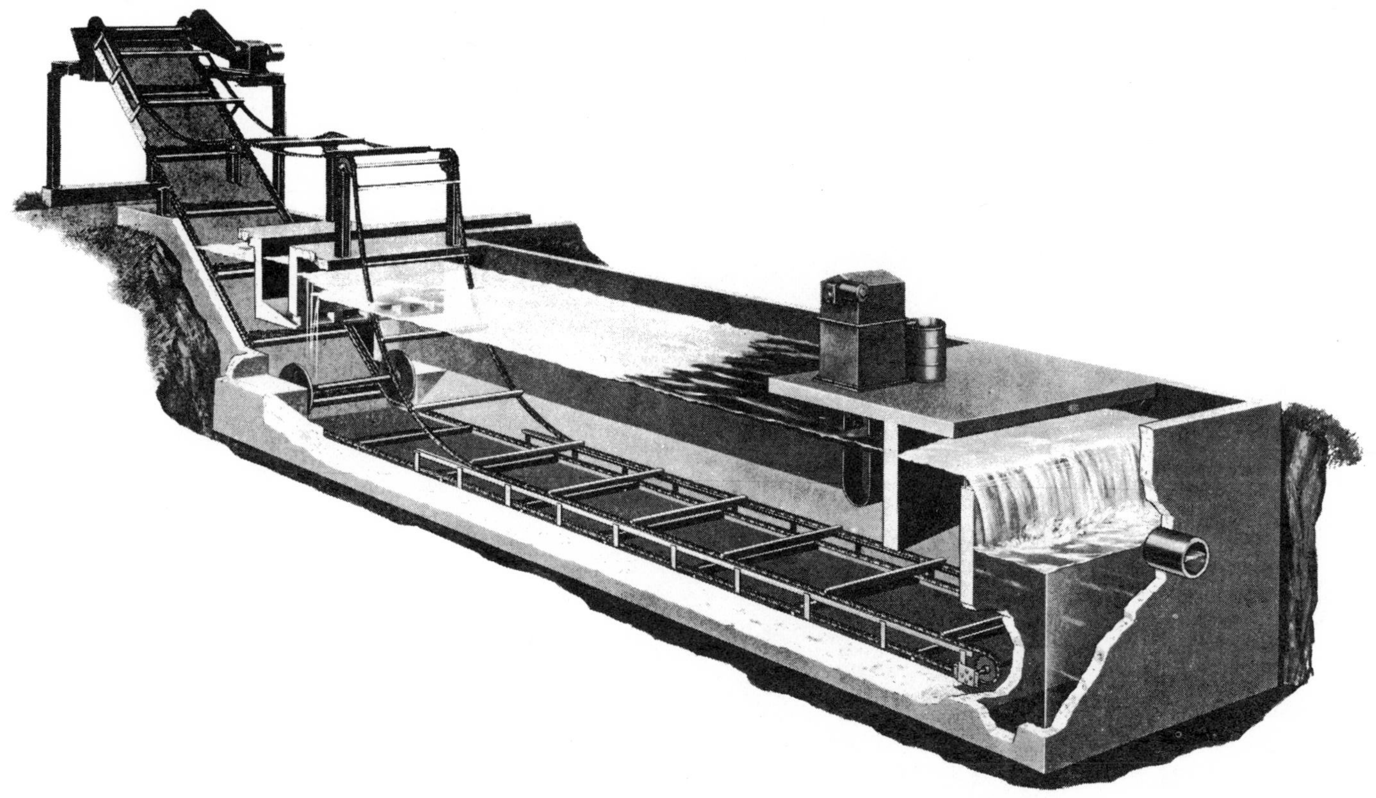

Rex Mill Scale Collectors offer steel mills a threefold advantage. They increase scale recovery, permit reuse of cooling water and abate pollution. Chain scraper collectors scrape settled mill scale from tank floor and scrape them up an incline for discharge into dump cars or a belt conveyor. Scale is automatically dewatered as it is dragged up the incline. Heavy duty chain, rugged flight design, high surface hardness on all working parts and a generous reducer size and motor horsepower assure balanced design. A belt skimmer provides automatic removal of floating oil accumulations.

SIPHON-FEED DESIGN

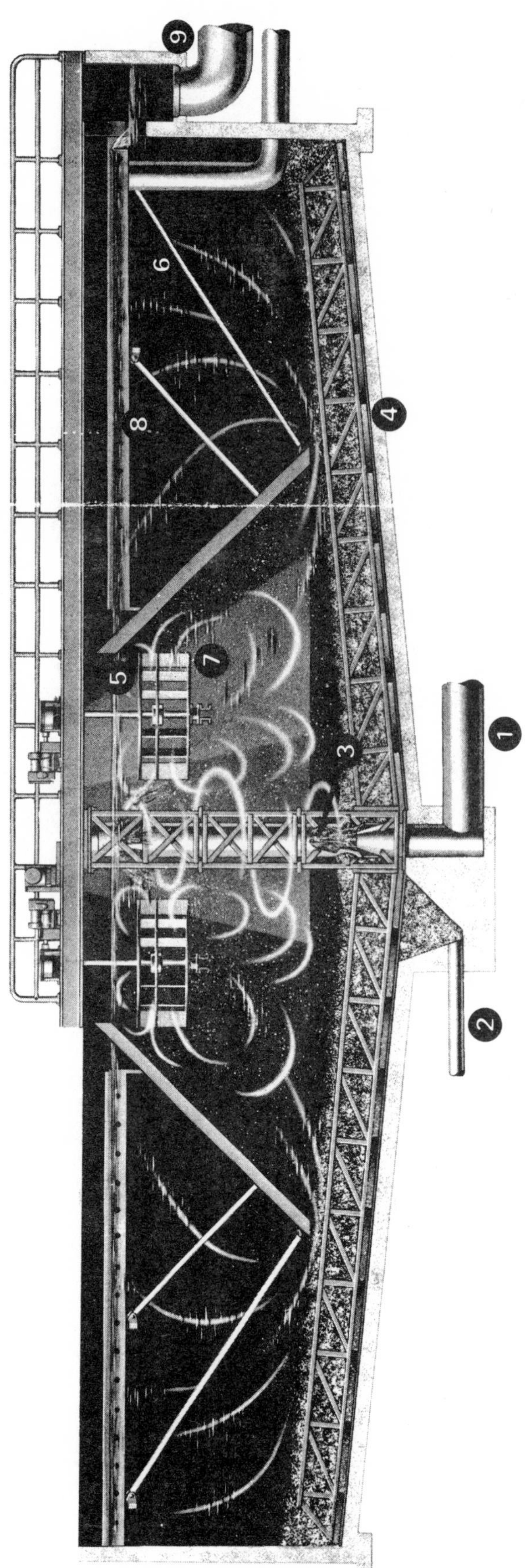

1 Influent Pipe
2 Sludge Drawoff
3 Solids Recirculation
4 Circular Scraper Sludge Collector
5 Turbine Agitator
6 Clarification Zone
7 Flocculation Zone
8 Radial Launders
9 Effluent Line

Rex Reactors provide a unique design for effective recirculation and mixing of previously formed sludge solids, or floc, with the raw water.

In the siphon-feed design a venturi-like constriction is placed in the center pier influent pipe and orifices are located immediately above the constriction. The kinetic energy created by the increased flow velocity of incoming raw flow causes the previously formed settled solids to be recirculated through the orifices. Thus the volume of flow recirculated is always in proportion to the volume of incoming raw water to assure completely "controlled recirculation." Size of the influent-pipe is based on a velocity gradient to provide initial mixing in the pipe.

Primary mixing is a high-energy rapid mix by two turbine agitators to assure rapid dispersion of chemicals. The turbines also accomplish recirculation by drawing up some of the lighter floc particles. The amount of mixing or agitation is controlled by varying the agitator speed.

The gentle rolling action imparted by the agitators combined with the recirculating floc provides optimum coagulation and agglomeration of precipitates in the secondary zone. The flocculating action is gentle so there is no shearing — no floc damage. As floc particles grow, they become too heavy to recirculate, and drop to the tank bottom for removal by the collector mechanism.

As the flow reaches the bottom periphery of the conical hood, velocities are at the minimum. This entrance velocity is the most critical design point in the chamber to avoid short circuiting and reduced efficiencies. As the flow passes upward, it slows even more due to the outward taper of the hood which creates a constantly enlarging flow area. This results in dynamic separation of the floc particles from the treated water as the subsidence velocity of the particle becomes greater than the upflow velocity.

Effluent launders are arranged to assure the upward flow to a series of submerged orifices in the launders. Orifices operate under a head loss to assure an even distribution of flow throughout system of launders. V-notch weirs may be furnished in the launders for effluent take-off.

Where a sludge blanket is maintained the blanket acts as a filter to remove impurities and light precipitates by agglomerating them with the precipitate suspended in the sludge blanket. The blanket can be controlled by the rate of sludge withdrawal to eliminate the chance of hydraulic upsets caused by flow variations.

In the side-feed arrangement, recirculation is accomplished by the pumping action of a single agitator through a central recirculation drum.

Raw water and chemicals enter the recirculation drum and immediately contact the heavy pre-formed floc being recirculated from the tank bottom by the agitator. The mixture is then drawn up through the recirculation drum — the size of which is based on a velocity gradient to provide initial mixing. Optimum mixing and rapid dispersion are accomplished as the mixture passes through the turbine agitator. The hydraulic flow induced by the pumping action of the agitator provides a gentle flocculation motion resulting in optimum floc growth. Subsequent flow is similar to that in the siphon feed arrangement.

SATELLITE TREATMENT SYSTEMS

REX SATELLITE TREATMENT PLANTS offer an economical solution to the problem of treating combined sewage or storm water overflows . . . provide an effective alternate to the separation of sewers and allow flexibility in both treatment and design.

This new concept in effective storm water handling is based on a high-rate treatment principle of handling large volumes of wastewaters encountered in heavy rainfall periods. To assure this required fast treatment, Rex Satellite Plants employ a unique combination of screening, flocculation, and dissolved-air flotation.

Wastewaters are initially screened with mechanically cleaned bar screens having 1" clear openings to remove large particles from the flow. In the Satellite Plant the first operation consists of fine screening with a revolving drum screen to remove finer particulate matter. From the screen chamber, the waste water flows by gravity to a Float-Treat System consisting of a Turbine Flocculator and Float-Treat Separator. In the Float-Treat System, chemicals are used to aid in producing a floc which is subsequently floated to the surface of the separator. Chemicals reduce the total plant area required while providing better treatment on a consistent basis. Floated solids are removed with a skimmer mechanism while heavier solids, such as grit, settle in the separator and are removed from the tank bottom with a conveyor sludge collector.

The effluent from the Satellite Treatment Plant may be discharged directly to a water course after disinfection. The screenings and both settled and floated solids may be directed to a conventional treatment plant.

The efficiency of the high-rate treatment provided by Rex Satellite Plants is evidenced by results from field operations with suspended solids removals running from 75% to 90% and COD and BOD removals running as high as 65%.

These high removal rates have been achieved on a consistent basis under first flush and extended flow conditions. Using the combined advantages of different treatment methods, the Rex Satellite Treatment Plants provide this effective treatment in a minimum area.

The treatment system is also designed to minimize operating costs. Headloss through the entire plant is held to a minimum. Less volume of waste is produced by the plant, thus reducing storage volume and disposal costs. The equipment operates only during storm overflows and is designed to start operation on a dry tank basis. Therefore, if desired, the tanks can be drained following a storm flush.

The ability of the Rex system to achieve a high level of treatment in a small area permits location of plants at overflow discharge points. Automatic operation allows remote location without operator attendance. Standard package arrangements are available up to capacities of 20MGD. For larger overflows, modular units are used. The individual treatment systems are automatically brought into service as the flow increases and taken off the line when the flow decreases. Chemical conditioning is used to achieve a consistent, high level of treatment over the wide range of influent conditions.

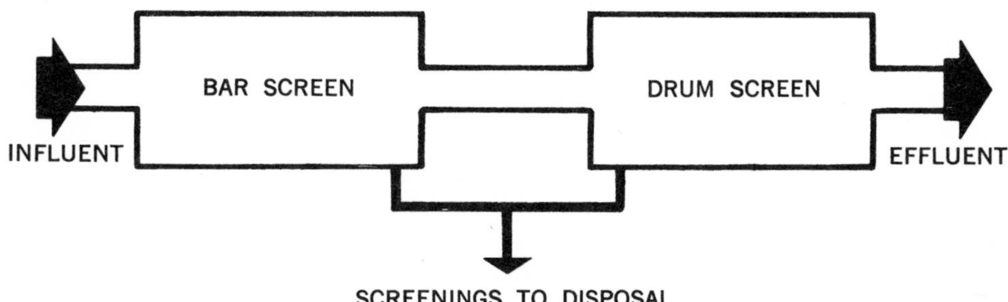

Where treatment equivalent to conventional primary treatment is acceptable, the Rex Satellite system is available with a single step drum following the removal of coarse solids with a conventional bar screen.

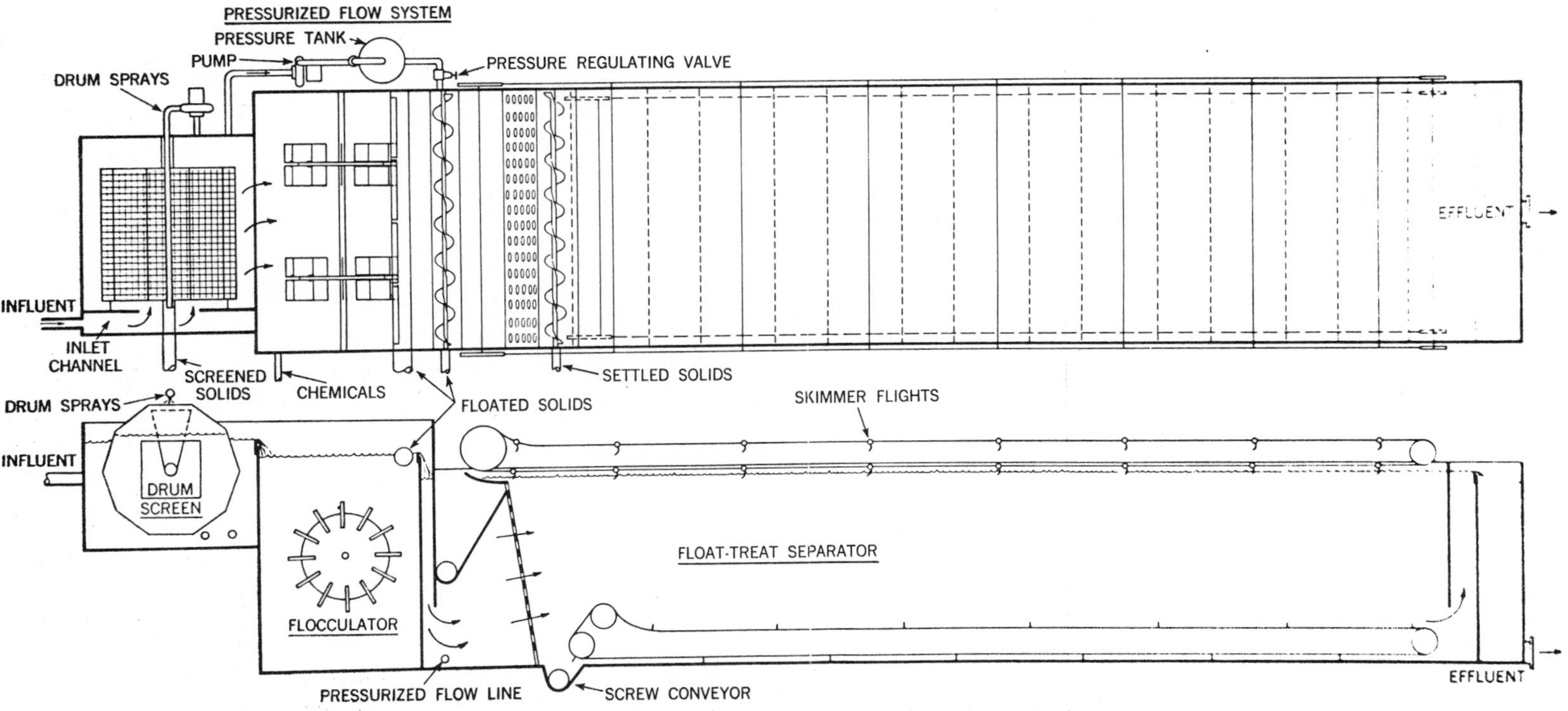

Flow entering the Rex Satellite Plant goes into the screen inlet channel. The flow enters the drum and after screening, goes to the flocculation chamber. Solids collected by the screen are flushed and removed by the drum sprays. Chemicals are added and mixed in the flocculation chamber before the flow enters the Float-Treat Separator.

A portion of the flow leaving the drum screen is pumped to the pressure tank where air is dissolved into the water. The pressurized flow passes through the pressure reduction valve and is then blended with the effluent from the flocculation chamber. The tiny bubbles formed from the pressurized flow attach themselves to the particles in the waste water carrying them to the surface. Skimmer flights remove the floated solids and a bottom collector removes heavier particles that will not float. The effluent from the plant is then disinfected and discharged.

Rex Chainbelt Inc.

REX DRUM SCREENS

After initial coarse screening, the raw waste enters the open end of the Rex Drum Screen. The water passes thru the screen media and into the screened water chamber around the drum. As the flow passes thru the screen, a controlled mat of solids is allowed to accumulate — capturing finer particles and increasing screen efficiency.

As the drum rotates, solids are carried up out of the flow and flushed with a regulated overhead spray from the screen into a screenings trough. Approximately ½ of 1% of the total flow in the form of pumped screened water is required for cleaning and flushing, thus keeping volume of waste solids to a minimum.

The drum is designed to provide long-life with a minimum of maintenance. Shaft bearings, requiring no lubrication, provide years of trouble-free service under demanding conditions. Special lifting blades on the drum interior assure complete removal of all solids. Standard wire mesh screen is used for the screening surface in panels that are bolted to the drum.

A unique double-lipped wiping seal effectively isolates the influent from the screened water. The seal is located to avoid direct flow impingement to reduce gritty abrasion. The flexible design makes installation easier.

REX TURBINE FLOCCULATORS

Rex Turbine Flocculators of the horizontal or vertical shaft design assures optimum distribution of mixing energy throughout the entire flocculation chamber. Design of turbines provide maximum floc buildup for subsequent removal. Thorough mixing action assures complete utilization of chemicals — keeping operating costs to a minimum. Mixing rates are carefully controlled to minimize power requirements over a wide range of operating conditions.

TYPICAL
FLOAT-TREAT SEPARATOR

REX FLOAT-TREAT SEPARATORS

Optimum removal of suspended solids over a wide range of flows in a minimum amount of space is provided by these efficient units. Screened water is pumped to a pressure tank where it is saturated with air. The air saturated stream is introduced thru a pressure reduction valve into the Float-Treat Tank and blended with the effluent from the flocculation chamber. As the stream enters the tank, millions of air bubbles are released. These bubbles attach themselves to suspended particulate matter and chemically formed floc and float to the surface.

The rapid rise of the bubbles provides maximum capacity for tank size reducing the area required for treatment. Inlet structure and tank design are controlled to provide the optimum in hydraulic characteristics. The high concentration achieved in the floated blanket keeps the volume of sludge to be disposed to approximately 1% of the total flow. This can easily be handled in the city's conventional treatment system.

The Rex Float-Treat principle has been proven in hundreds of installations.

Richards of Rockford, Inc.

RICH-AIRATOR MK II -- MECHANICAL SURFACE AERATOR
Floating and Stationary Mount

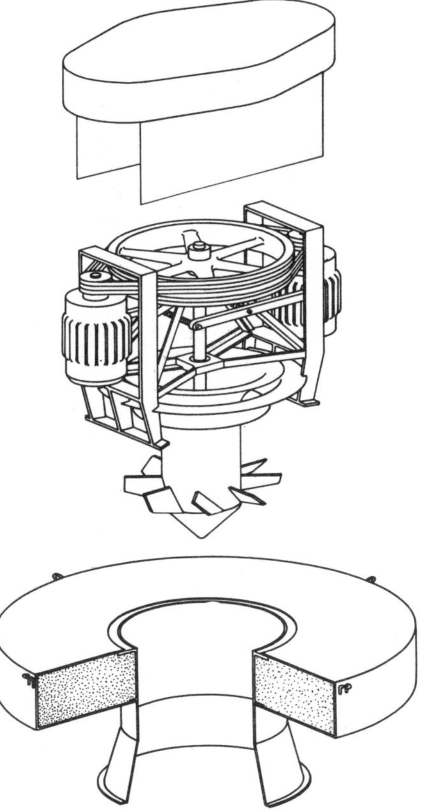

DESIGN

- Aerator: Spool type, axial flow ducted turbine.
- Size: 20 to 150 HP.
- Motors: TEFC totally enclosed fan cooled Class B, 60 Hertz, 230/440 Volts, rated for severe duty. Service Factor of 1.15.
- Turbine Unit: Shaft of 17-4 Stainless Steel. High ramp blades for maximum efficiency.
- Float: All Stainless Steel filled with polyurethane foam.
- Speed Reducer: Service Factor 3.0! No sensitivity to shock or side loading.
- Gears: None.
- Couplings: None.
- Design Life: Years of trouble free service from a truly simple machine.
- Maintenance: Purge motor and turbine shaft bearings with waterproof lithium base grease annually and inspect.

PERFORMANCE

- Oxygen Transfer: Standard Condition Rates tops in industry. Field Transfer Rates based on data derived from waste analysis.
- Induced Flow: In excess of 1,600,000 Gal./HP/Hr. Superior Mixing.
- Guarantee: Full one year mechanical warrantee on materials and workmanship.

FIXED MOUNT RICH-AIRator

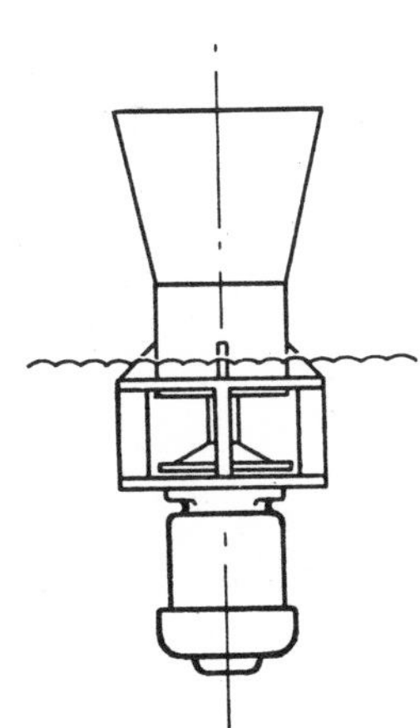

The floating RICH-AIRator comprises three basic elements, namely an electric motor, an axial flow pump, and a supporting float. Each of these elements has been combined into a rugged unit of exceptional strength, with a durability to meet the severe demands for continuous all-weather applications.
Manufactured in a range of sizes from 1 to 75 horsepower.

The compact RICH-AIRator modules accomodate easily to a broad range of waste treatment arrangements and capacities. Systems can be easily designed to meet the most rigid requirements of basin configurations and sizes.

The RICHARDS pumping system generates a three-dimensional circulation pattern, causing not only a continual turnover from bottom to surface, but also a circular movement of the wastewater in a horizontal plane. This mixing pattern assures complete dispersion of the dissolved oxygen throughout the tank or lagoon.

HYDRO-SWEEP SLUDGE COLLECTORS
for primary and secondary clarifiers

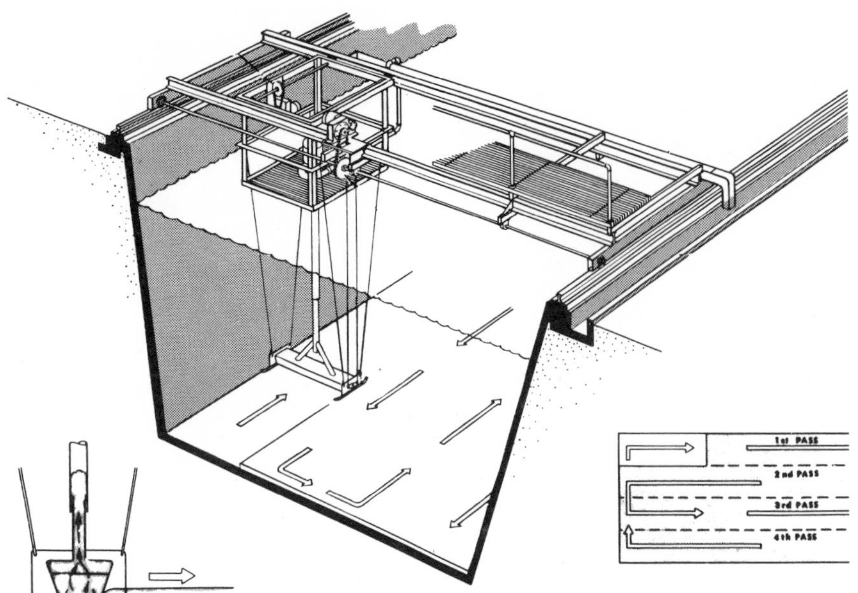

MODEL RH-S RECTANGULAR CLARIFIER INSTALLATION
For installation in new tanks or existing basins. Travel-truss construction enables sweep head to traverse laterally. An advantage of the rectangular system is the capability for low cost expansion at a later date by basin length extension. Surface skimmers are optional. Pat. #3416176

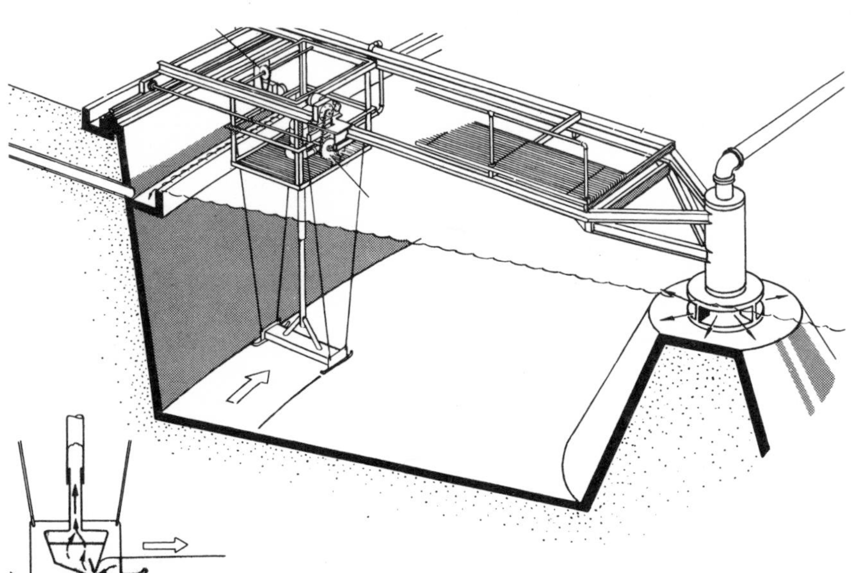

MODEL CH-S CIRCULAR CLARIFIER INSTALLATION
For installation in new or existing circular tanks. Flow is introduced through a central pier, which also supports the rotating bridge at the center. The motor and solids handling unit is truss-supported. The sweep head revolves in one continuous revolution from central pier to outside wall. Surface skimmers are optional. Pat. #3416176

HYDRO-SWEEP® SLUDGE RECOVERY SYSTEMS

Hydro-Sweep® is a universal system that can be installed in most basins or tanks.

Other clarifier mechanisms are generally designed to operate within specific settling tanks—as a unit.

Pick-Up head goes to the sludge instead of plowing the sludge to a discharge point.

Less stir up and riling of solids.

High biological concentrations are picked up virtually unchanged from their settled state for recirculation to the aeration basin.

Sludge capture velocities are under control at all times.

Hydro-Sweep® systems are designed so as not to have to operate continuously.

The entire system is automatic requiring no special operator.

Richards of Rockford, Inc.

RICH-PACK - A Modular Waste Treatment System Package

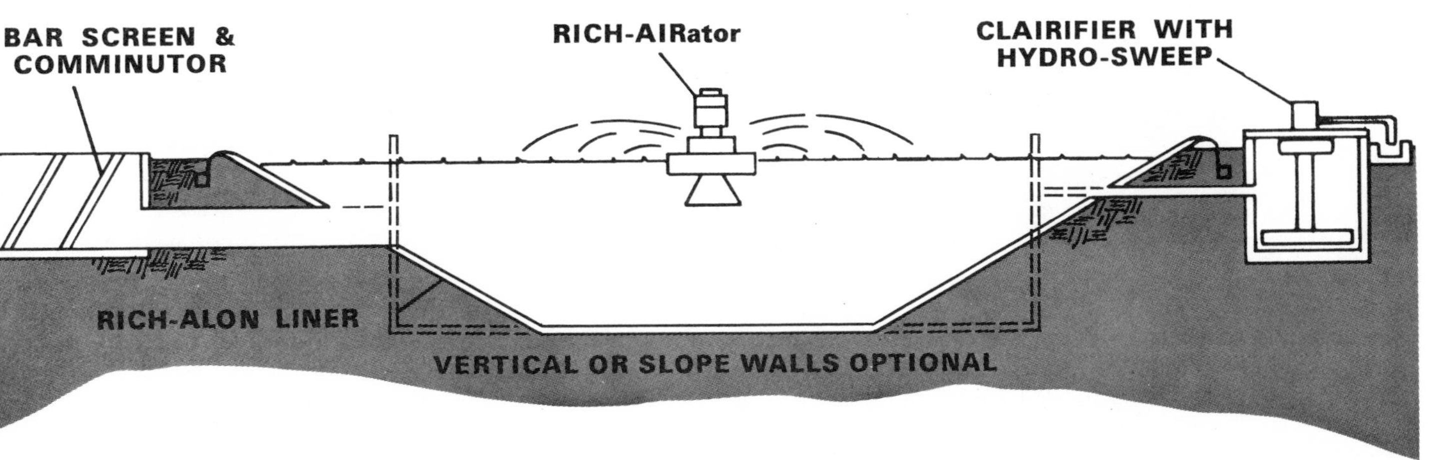

A MODULAR PACKAGE PROVIDES THIS BASIC EQUIPMENT:

RICH-ALON -- Rugged impermeable, prefabricated Aeration Basin Liner....
Available in a wide range of sizes
Replaces expensive reinforced concrete
Prevents ground percolation and basin erosion

RICH-AIRator -- A Floating Mechanical Aerator....
Sufficient horsepower to assure ample bio-oxygen supply, complete circulation, and mix

HYDRO-SWEEP -- Clarifier Sludge Collector....
Rectangular or Circular
Completely automatic
Recirculation control of aerobic sludge, where required

CONTROL PANEL -- NEMA 4 Weatherproof or NEMA 12 Dustproof....
Automated, prewired with 110V control circuits and transformer
Time clock operates plant with minimum of supervision and servicing

OPTIONAL EQUIPMENT --
Surface Skimmer for Clarifier Mechanism
Chlorine Contact Tank
Weir Box/V-Notch Weir Plate
Aerobic Digester
Laboratory Test Kit

Richards of Rockford, Inc.

KOOL-FLOW
For Thermal Pollution Control

Fully assembled KOOL-FLOW unit ready for final testing

The KOOL-FLOW's high thermal dissipation ability is achieved by a unique combination of a high-efficiency pump and exposure of the water to the surrounding atmosphere. The annular nozzle forms extremely unstable, high-velocity sheets of water that disintegrate into droplets. The degree of atomization is minimized and the over-all system efficiency is achieved by the high rate of droplet production from the pump.
Tests conducted in wind tunnel evaluations indicate minimum deformation of the spray and drift loss is less than cooling towers. The drift loss which occurs settles in an extremely short distance from the unit as contrasted to the concentrated high altitude plume of a cooling tower.

The KOOL-FLOW heat dissipation unit is a unique and innovative water-cooling system designed to meet industrial needs for far more efficient and effective cooling.

MECHANICAL DESCRIPTION

Self-contained, corrosion-resistant stainless steel unit.
Polyurethane foam-filled float.
Floating or fixed design.
5 degree approach to Wet Bulb.
Special heavy-duty, chemical service motor sealed to prevent moisture entry.
High efficiency axial-flow pump with no external bearings.
Patented annular slot nozzle.
High disc solidity axial-flow impeller.
More cooling per applied horsepower.

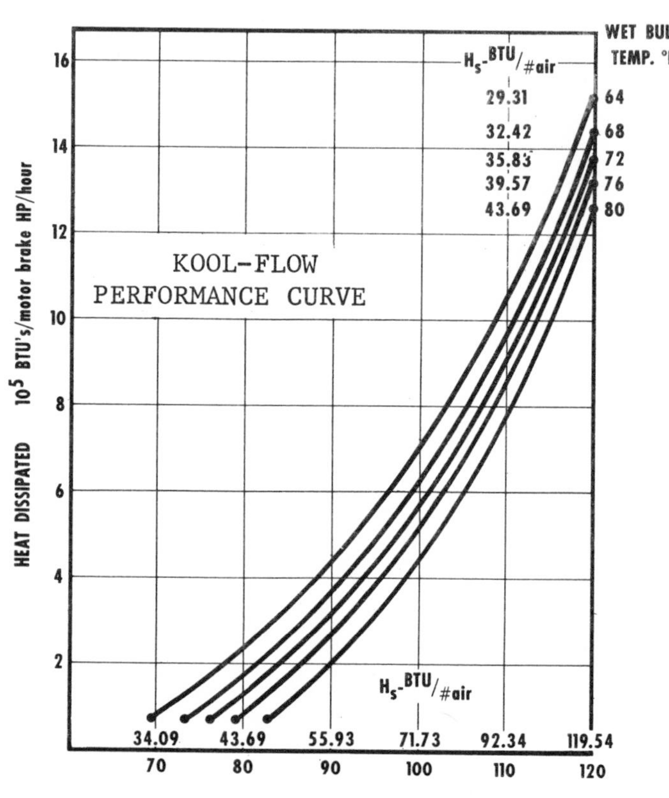

Schutte and Koerting Company

MERCURY REMOVAL PROCESS

The Ventron process of mercury removal or recovery is ideally suited for mercury-using plants where the effluent discharged must be reduced to a low or non-traceable Hg level. Such processes include mercury-cell chlor-alkali production, mercury-catalyzed sulfonation technique for anthra-quinone and other mercury compounding or consuming installations.

Reclaiming of mercury is a distinct possibility because the chemicals required for the system cost about 1/3 the market value of mercury. This unique process can also be used to remove or reclaim other heavy metals such as cadmium, gold, lead, palladium, platinum and silver. Ventron Corp., which operates the largest mercury chemical manufacturing plant in the free world, successfully uses this process to remove in excess of 99.5% of the mercury in their plant effluent.

Treatment of either contaminated liquid or gas streams can be accomplished and, since the process is non-corrosive, expensive corrosion-resistant equipment is unnecessary. Original purchase and maintenance costs are kept to a minimum because of this and the fact that all of the process equipment is standard and readily available.

OPERATION

Waste Water (refer to schematic shown below). First step is to adjust the pH of the mercury-containing effluent to a pre-determined level. Sodium borohydride is then metered into the effluent along with a flocculant to reduce the mercuric compounds to the metallic state. Organic mercury must be pre-treated and converted into the inorganic form.

The precipitated mercury metal is removed by clarification and filtration. Resultant mercury sludge is suitable for distillation without further processing. Additional treatment to lower the residual mercury level even further can be accomplished by treatment with carbonaceous material and then passing the effluent thru a resin bed.

TYPICAL WASTE WATER SYSTEM

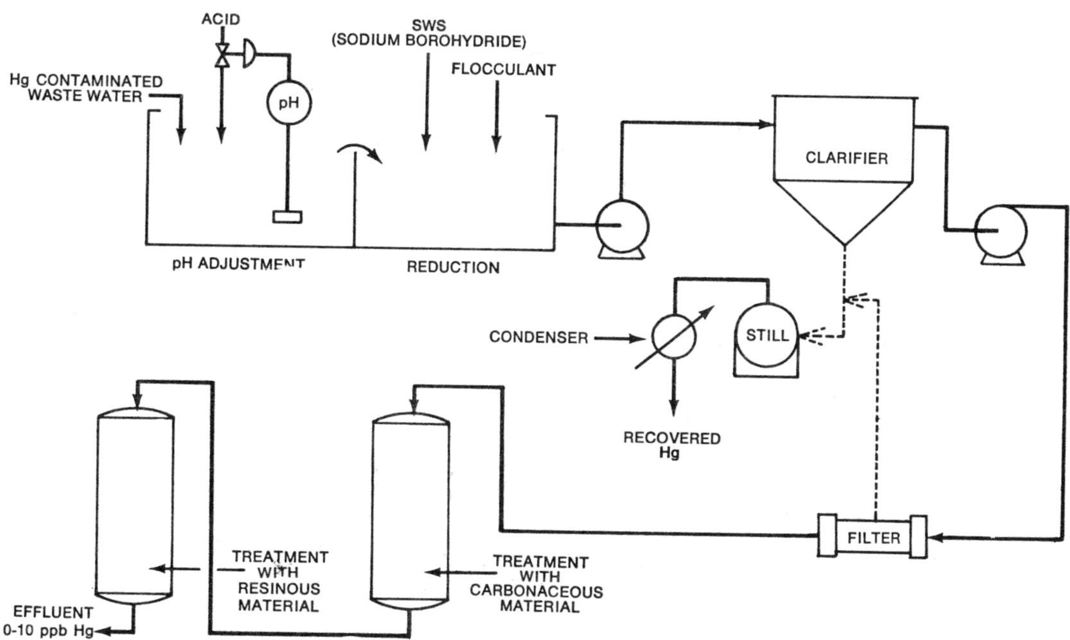

Gas Streams (refer to schematic below) The process just described must be preceded by the following to condense mercury from contaminated gases. Gas is introduced to a Koertrol high energy venturi condenser/scrubber which transfers the mercury in the gas to the water which already contains sodium borohydride. A gas-liquid separator discharges purified gas and the mercury-carrying effluent. The process which follows to remove the mercury is identical to that of waste water treatment.

TYPICAL GAS STREAM SYSTEM

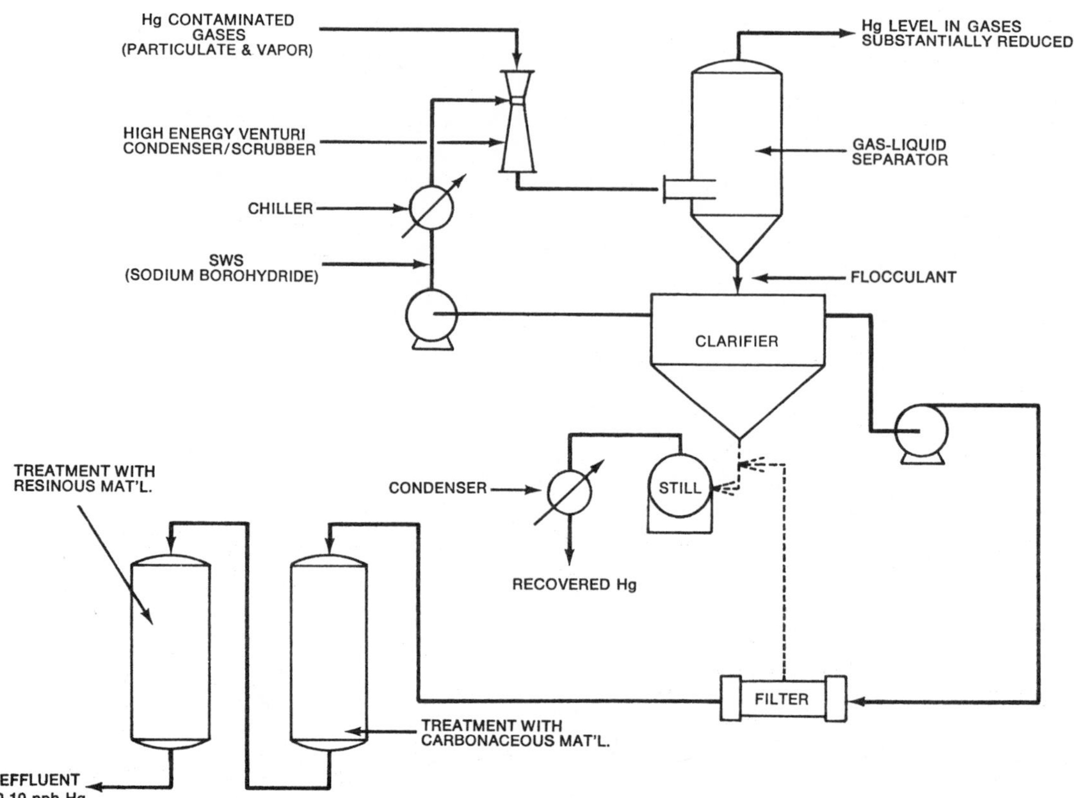

Koertrol can employ the Ventron mercury removal process to assist you in meeting local and federal pollution control codes. Reduction of mercury levels to ranges of 0-10 ppb are entirely possible. Removal criteria are necessarily variable, and local requirements, plant conditions and economic factors will control the final efficiency percentage. The Ventron process is decidedly faster and more efficient than any other technique. Continuing research by Koertrol and Ventron Corp. assures a timely solution to each mercury removal application.

AMETEK's many years of experience in the manufacture of pollution control and chemical process equipment provides well-versed engineering back-up and completely dependable product performance.

BATCH WASTE WATER TREATMENT

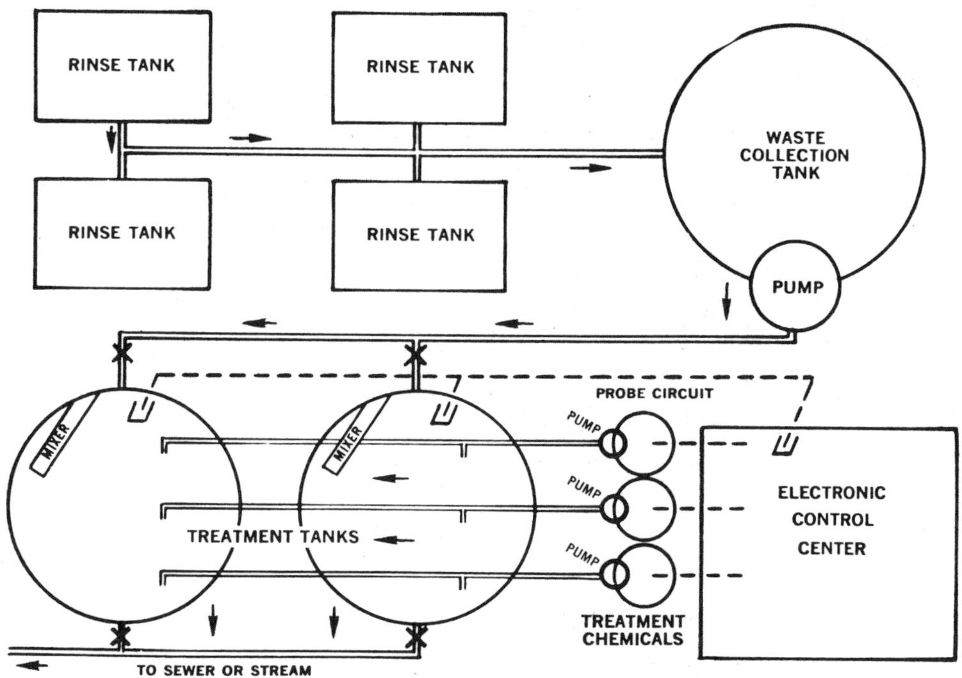

(X—plastic valves)

These semi-automatic, low cost, packaged Batch Waste Water Treatment Systems effectively treat acid, alkaline, cyanide, and chrome waste water before disposal to comply with all legal requirements. They are so low in initial cost and maintenance that no plant, regardless of size, need be without one. Each unit consists basically of an electronically equipped control panel and a combination of chemical storage tanks and treatment tank(s), all in a compact, space-saving unit. Precision engineered controls eliminate the overuse of chemicals and assure treated effluents: Rugged, heavy duty but compact PVC lined steel treatment tanks are completely corrosion resistant. No special floors or pits are required. Simply set down pipe inlet, outlet, chemical supplies and power and the unit is ready for immediate operation. Sethco furnishes all the engineering know how. You are guided all the way from inception of recommendation through State and local approval to final start up by Sethco engineers. Your unit is placed on stream quickly and easily, and your waste water problem is a thing of the past.

SETHCO BATCH WASTE WATER TREATMENT PACKAGE CONSISTS OF:

Precision Engineered Instrumentation

Control panel contains a combination of pH, ORP* (Redox) Meter which automatically follows the pH and oxidation or reduction reactions in the treatment tank. Also furnished are switches for mixer, for automatic drain valve control, for level alarm, and for pumps. Control panel may be mounted in any convenient location within 20 feet of treatment tanks.

**Oxidation-Reduction Potential*

1500 Gallon Treatment Tank

The conical bottom of the treatment tank provides a sludge collection area of 250 gallons or 32 cubic feet. The treatment tank is fitted with a mixer, a high level alarm, an automatically interlocked drain valve for draining liquid to the sewer and a manually locked sludge removal valve. Where the flow rate of the installation dictates, two such tanks may be hooked up to one control panel together with appropriate switching. This is economical and space saving permitting the same control panel to handle both tanks.

Where required, tanks up to 2500 gallon capacity may be furnished.

For cyanide destruction, vented tank covers are provided.

Accessory Equipment Included

a. Pumps for the chemical additives required to accomplish neutralization, chrome reduction and neutralization, cyanide oxidation and neutralization.
b. High quality mixer equipped with totally enclosed chemical plant motor.
c. High level solid state alarm and switching are furnished for each treatment tank.
d. All necessary interconnecting cables for the probes.

PROCEDURE

The liquid to be treated is collected in the treatment tank. When the level is within one foot of the tank top, the high level alarm is activated. The flow is transferred to a duplicate treatment tank or existing storage tank. The mixer is started and the instruments turned on. The treatment procedure is accomplished quickly and easily.

WASTE WATER TREATMENT SYSTEMS

Federal and state laws require the treatment of all acid, alkaline, cyanide, and chromate wastes. This makes it necessary for all metal finishing plants to install effluent treatment systems.

Sethco has designed fully automatic 'packaged' systems which are compact, inexpensive and designed to produce effluents meeting all legal requirements. Each unit consists essentially of an electronically equipped control panel and a combination of chemical storage tanks and a reaction chamber.

Standard Units of 2,500, 6,250 and 12,500 GPH are available for acid/alkaline neutralization, and chrome reduction while standard units of 1,250 and 3,100 GPH are available for cyanide oxidation. By a suitable combination of these units, a complete system can be supplied to deal with mixed effluents.

These units are interchangeable and do not become obsolete. Changing effluent requirements or plant growth can be met simply and inexpensively by rearranging or adding standard units.

Optional accessory equipment, such as flow recorder systems, low level reagent tank alarm systems, reagent tank auto-fill systems and rinse water control equipment make the complete Sethco Waste Water Treatment System fully automatic. Interchangeable instrumentation and valve gear between sizes of units reduce inventories.

Precision engineered automatic controls with proportional characteristics eliminate overuse of chemicals, assure neutralized effluents and greatly reduce the overall size of the system. The resulting compact water treatment system requires so little space that it can be installed at the end of a plating line. The tight instrumentation handles the extreme variations in flow frequently encountered in metal finishing plants.

Sethco Packaged Waste Water Treatment Systems are simple, easy and inexpensive to set up requiring no major installation work. The instruments are supplied in a pre-wired panel, complete with cable ready for further interconnecting to motorized valve , probes, etc. Plant personnel can usually accomplish all necessary work. The tanks* are vinyl coated inside and out and are completely corrosion resistant. They can, if necessary be installed in the ground in place of a sump or alongside tanks.

For intermittent as well as continuous operation, retention and sedimentation tanks with necessary pumps and controls can be supplied.

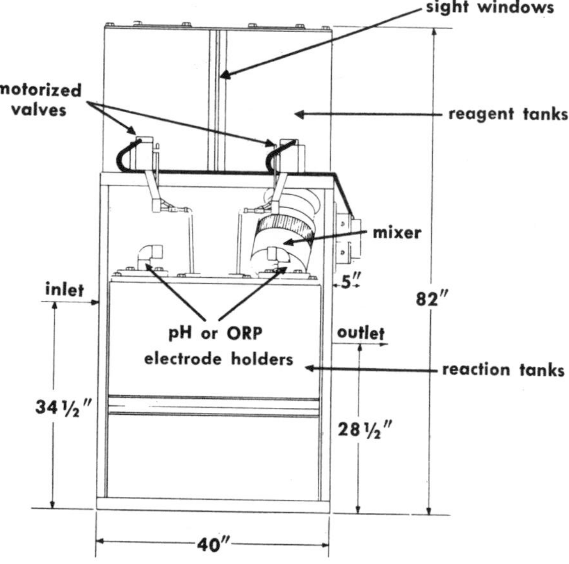

DESCRIPTION

The standard unit consists of an electronically equipped control panel, a rugged vinyl coated reaction chamber* with high speed stirrer, and two storage tanks* for reagents. The reagent tanks are mounted above the main reaction chamber permitting reagent levels to be observed through the sight windows. The high speed stirrer in the reaction chamber assures complete mixing of treatment agents and prevents build up of precipitates.

Flow is rated to 12,500 GPH. It varies depending on the concentration of the effluent. The greater the concentration, the greater the reaction time that is required, the slower the flow.

Effluent is pumped or gravity fed into the reaction chamber. Redox and pH probes in this chamber efficiently control chemical additives. Motorized valves are precision engineered for regulating the flow of treatment chemicals. The valves are controlled by means of a three term controller. This means the desired control point is always achieved regardless of flow or contamination variations.

MIXED EFFLUENT TREATMENT SYSTEM FLOW CHART
Schematic Layout of Typical Treatment
Unit for Cyanide, Chrome and Acid/Alkaline
Waste Waters

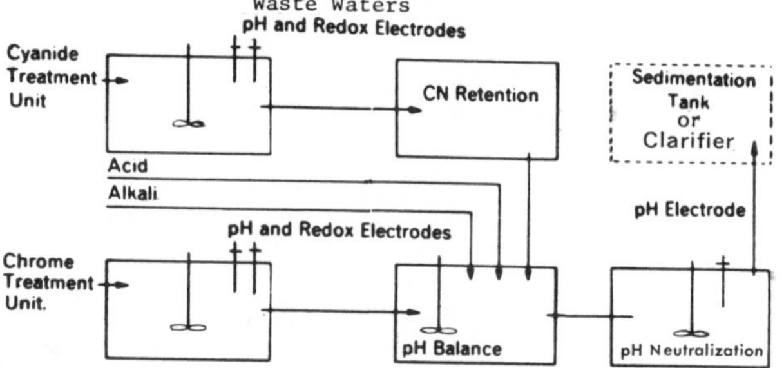

ACID OR ALKALINE WASTES (MODELS 250A, 625A and 1250A)

Flows to 12,500 GPH of acid or alkaline wastes can be neutralized. The pH probe, proportional controller and one motorized valve govern the reagent addition. After neutralization, if solids are below legal limits, discharge can go down the drain. A sedimentation tank, clarifier, centrifuge or filter is used to collect solids if they are above legal limits.

CHROME WASTES (MODELS 250D, 625D and 1250D)

Suitable for flows to 12,500 GPH. The efficient reduction of chromium from hexavalent chrome to its trivalent form requires preliminary adjustment to pH 2.5. This is achieved and maintained by use of the pH probe, proportional controller and related motorized valve governing the acid addition. The Redox probe, proportional controller and related motorized valve in turn govern the addition of Sodium Metabisulphite reagent for the actual reduction process. After breakdown, the effluent is neutralized in a 250A, 625A or 1250A system. If chromium hydroxide solids are below legal limits, discharge can go down the drain. A sedimentation tank, clarifier, centrifuge or filter is used to collect solids if they are above legal limits. (See flow diagram.)

CYANIDE WASTES (MODELS 250C and 625C)

Flows to 3100 GPH can be achieved depending on cyanide contamination and the absence of complex iron or nickel salts. The pH is adjusted to approximately 10.5-11, by means of the pH probe, proportional controller and related motorized valve controlling the addition of Sodium Hydroxide solution. At the same time the cyanide is oxidized to Cyanogen Chloride by Sodium Hypochlorite, the addition of which is governed by the Redox probe, proportional controller and related motorized valve. Overall control is maintained by the Redox probe which responds to excess chlorine. The first reaction by the hypochlorite forms Cyanogen Chloride which is subsequently broken down to Cyanate. This secondary reaction, the hydrolisis of toxic and volatile Cyanogen Chloride, its reaction time depending on pH, temperature and excess chlorine requires a closed retention tank. Its size should be adequate for the 1 hr. retention of Cyanogen Chloride. After retention the effluent is then neutralized in the standard pH neutralization unit Model 250A or 625A. After neutralization, if solids are below legal limits, discharge can go down the drain. A sedimentation tank, clarifier, centrifuge or filter is used to collect solids if they are above legal limits (See flow diagram.)

VF VERTICAL PLATE FILTERS

A LOW COST FILTER.
VERTICAL TANK–VERTICAL PLATE DESIGN REQUIRING A MINIMUM OF FLOOR AREA.
SPARKLER'S ORIGINAL RIGID FILTER PLATE DESIGN ASSURES MAXIMUM CLARITY AND LONG FILTER MEDIA LIFE

The Sparkler VF Filter can be adapted to the varying requirements of the process industry. It is suitable for wet or dry cake discharge, for manual, semi-automatic or fully automatic operation. The basic equipment may be furnished with any of these accessories: fixed overhead sluice for cleaning the filter plates without opening the filter vessel; oscillating overhead sluice; vibration; hydraulically operated quick opening covers and bottom drop-out doors; steam coils or jackets; pump and motor, slurry feed equipment and tank, all mounted on a common portable or fixed base; automated filter.

Sparkler's exclusive rigid filter plate design assures not only low initial cost of the equipment but continuing low maintenance cost. Higher flow rates per square foot of filtration area are available; filter plates have greater internal free flow area; rigid construction; perforated sheets supporting the filter media to reduce surface friction. Rigid non-warping plates provide for quicker and easier cake removal during the cleaning operation.

THE SPARKLER MCRO FILTER

- Engineered for heavy duty service
- Rugged structural steel frame
- Rigid support for the tank and cover
- Stationary cover — unnecessary to break inlet and outlet piping to open filter
- Retractable tank with 4 point suspension
- Perfect meshing of tank and cover — positive seal
- No cake disturbance in opening — plates remain stationary
- Internal self-sluicing
- Dry cake removal either by retracting tank or internal conveyor screw
- While in operation ½ of frame space free for traffic
- Simple to completely automate
- Capacity 10 to 3000 sq. ft. filter area

The combination of overhead frame design, the retractable tank, the single header sluice, and the fixed plate assembly makes the Sparkler MCRO the most versatile filter on the market. Filter areas from 10 to 3000 sq. ft. are no problem — the plates are equally accessible, regardless of size. The sluice is as efficient in a 3000 sq. ft. filter as it is in a 10 sq. ft. unit. Consequently, the MCRO is adaptable to heavy duty filtering in many industries such as sugar refineries, breweries, chemical and food processing plants, petroleum industry, mining industry and water treatment.

CONCENTRATED SLUICE VOLUME RESULTS IN DISINTEGRATED CAKE REMOVAL

- Directs a forceful sluice of liquid to the surface of each plate from every angle.
- Concentrates a large volume at a high velocity over the entire face of every plate.
- Completely sluices the stickiest and most fibrous cakes.
- Dislodges the cakes both by impingement and by flushing.
- Removes and immediately disposes of the cakes one at a time.
- Penetrates through the pores of the filter media (wire or fabric cloth) and prevents "blinding."
- Washes out entire interior of filter tank completely.
- Uses minimum amount sluice liquid (1/3 to 1/2 gal. per sq. ft. filter area).
- Cleans rapidly (2 min. for 600 sq. ft. filter).
- Simple to automate.
- Sluice rate of travel and velocity can be adjusted by fingertip control.
- The Sparkler MCRO is cleaned with equal speed and efficiency by removing the filter cake in a dry or semi-dry state. This is accomplished by any of three ways:

- By manual shocking or scraping.
- By an electric or pneumatic vibrator.
- By compressed air.

As dry cake removal is peculiar to each process, we suggest you consult your Sparkler engineer for detailed information.

*Pat. in process

HOW IT WORKS

This sluice, positioned above the plates, consists of a sliding pipe with a single cross arm sluicing header. This cross header extends over the full width of the filter plates. It has multiple holes, so spaced and of such sizes and at such angles that the high pressure jet streams cover every square inch of the filter plate faces from varying angles as the header is passed over the top of the plates.

During the filtering cycle this header is retracted to the rear of the horizontal filter tank. At the end of the filter cycle, the tank is drained in the closed position, approximately 100# pressure is applied to the sluice mechanism and the pipe pushed all the way forward in the tank and then fully retracted again (one complete cycle) at the rate of 5′ per minute.

SPARKLER FILTER MODEL HRC

The Model HRC combines the advantage of horizontal plate cake stability with automatic cake discharge. It is available in capacities up to 300 sq. ft. of filtering surface.

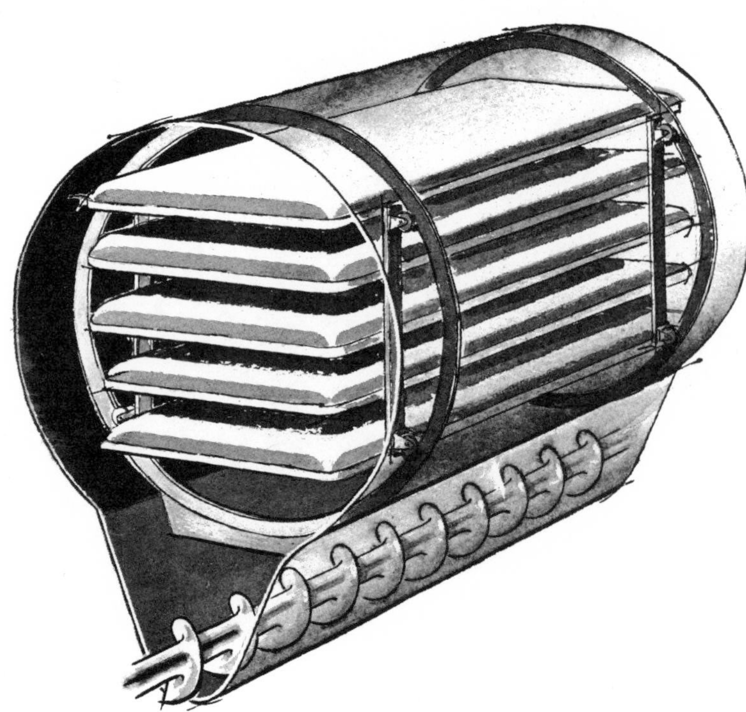

Plates in horizontal position during precoat, filtration, cake washing and drying cycles.

Advantages:

Precoat is uniform and time of cycle is shortened due to horizontal position of plates.

Horizontal position of the plates assures cake stability even with intermittent flow.

Consistent high quality filtration is achieved from start to finish.

After filter cycle, cake may be washed and/or dried for product recovery with no danger of slippage.

During the cake discharge cycle it is unnecessary to open the filter.

Cake is vibrated from plates in vertical position.

Filter may be equipped with cake breakers on trough walls to assure no cake "hang-up".

Cake scroll removes spent cake from filter.

A choice of several media is available: wire mesh, synthetic or natural fiber cloth. Cake can be built up to any thickness, limited only by pressure resistance. This makes longer filtering cycles practical with many products.

During the precoat, filtration and wash cycles the plates are in the horizontal position. To discharge the cake, the plate carriage is rotated inside the tank to a vertical position. The cake scroll located in the trough of the vessel easily removes the solids. Viscous cake can be easily discharged with the addition of cake breakers, an optional feature.

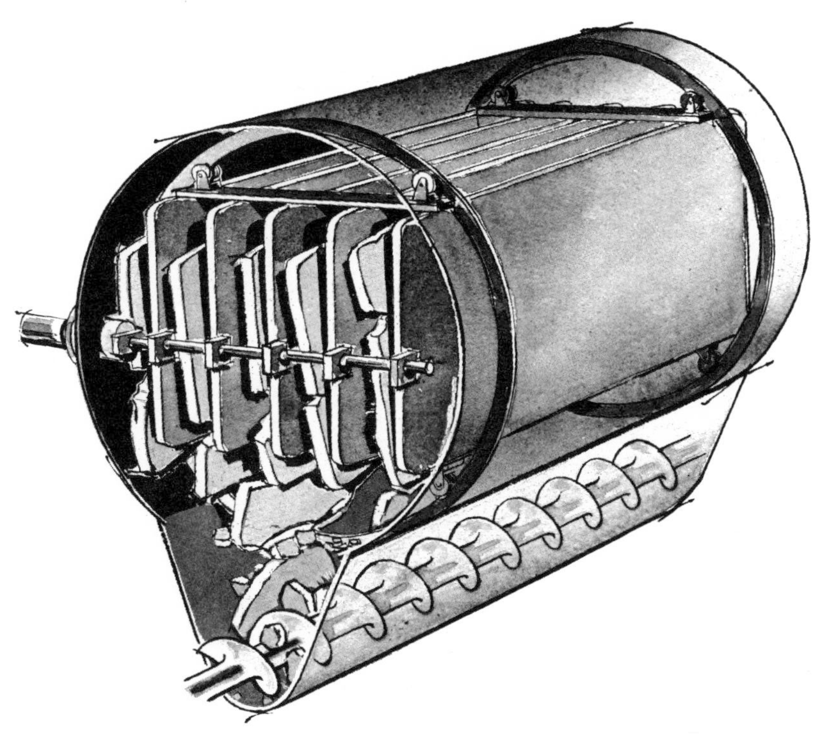

Plates in vertical position during the cleaning cycle.

ENGINEERING DATA

MODEL NO.	FILTRATION AREA SQ. FT.*	CU. FT. CAKE ½″ CLEAR	NUMBER OF PLATES	TANK VOL. U. S. GALS.	EST. NET WT.	APPROX. FLOOR SPACE (in.)	APPROX. HEIGHT O. A. (in.)
HRC-50	62.6	10.4	7	450	4800	105″ x 60″	87″
HRC-100	108.0	18.0	12	575	5000	113″ x 69″	87″
HRC-150	154.8	25.8	12	900	7800	127″ x 74″	107″
HRC-200	206.9	34.5	12	1100	9800	146″ x 74″	107″
HRC-250	258.7	43.0	15	1485	13700	130″ x 90″	120″
HRC-300	310.5	51.7	15	1720	15000	168″ x 90″	120″

*This area includes frame bars

ADVANTAGES:

Unexcelled quality to the end of each cycle.

Higher flow rates per square foot of area.

Totally enclosed filter plates — no leaking or evaporation.

Convenient operation:
a. Portable.
b. Complete with pump, motor and interconnecting piping if desired.

Compact — occupies minimum floor space.

Low cost operation:
a. Low cost paper media and filter aid.
b. Easy and quick to clean — cleaning time reduced 50%.
c. Minimum of unfiltered residue at end of run. (See Scavenger Plate.)

Engineered to use modern, diatomaceous earth filter aids to the best possible advantage.

Patented Scavenger Plate. This feature facilitates batch filtration. At the end of each run, the main outlet valve is closed and Scavenger Plate valve is opened. Residue is then filtered through the Scavenger Plate by means of air or gas pressure through the air vent.

STANDARD HORIZONTAL PLATE CARTRIDGE

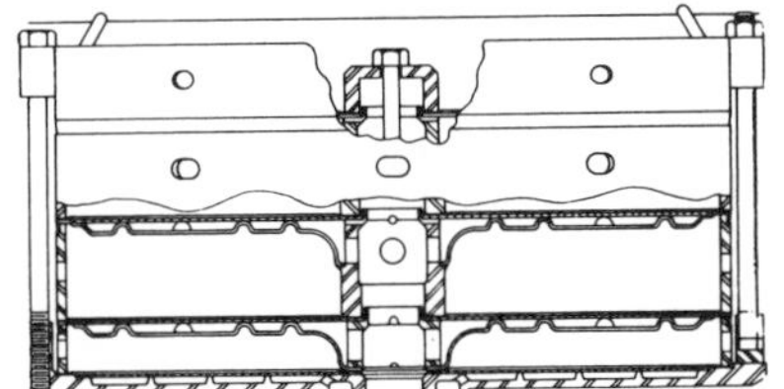

Sparkler horizontal plate filters are particularly suited for fine filtration and polishing. They are equally efficient for continuous or intermittent operation.
Complete recovery of product is obtained by "washing" or by "blow-down" of the cake. Any kind of filter media--fabric, metallic wire cloth, or filter paper is available. All types of filter aids, activated carbons or Fuller's earths are also adaptable with maximum efficiency.
Due to horizontal position of the plates, only a very thin precoat of filter aid is necessary. The cake is evenly distributed, assuring economical operation and uniform clarity.

VR AND V DUAL DISC CARTRIDGES

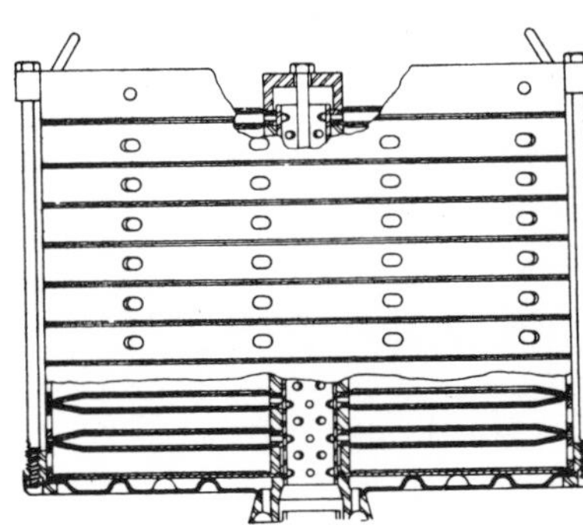

Sparkler VR trap filters are designed for processes which require the removal of all traces of solids. Large area at an unusually small investment, coupled with low operating cost. Only the filter paper or cloth is discarded, thereby offering greater economy than "one-time-use" cartridges. Versatile media selection for varying porosities are available.
The V type filter cartridge is similar to the VR type. The outer ring of the VR cartridge is left out and cloth media is sewn on the V type plate. This is advantageous for multiple use of cloth media.

WASH-OFF CARTRIDGE

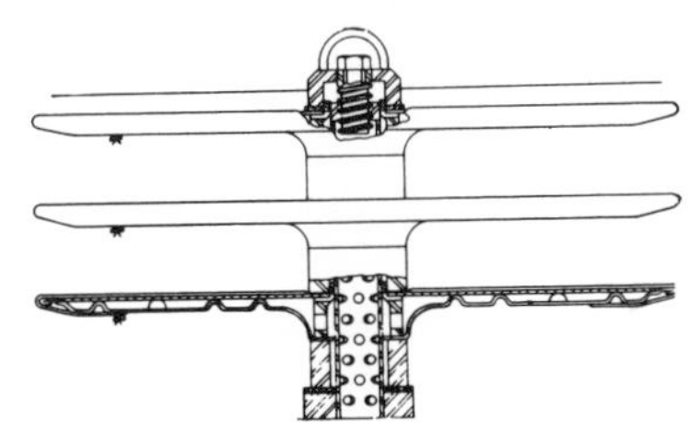

Sparkler wash-off filters offer cake stability, plus low first cost, less labor, fast cleaning, and savings in materials for real filtration economy with no sacrifice in quality. Cake is removed by simply "hosing off" without disassembling the cartridge.
Even thickness, uniform precoating, a total absence of any danger of the cake dropping or cracking is assured. The filter can be completely shut down and filtering resumed without any cake damage.
A semi-permanent synthetic cloth or wire screen filtering media covers the plates which filter through the top side only. The media is available in a variety of materials and weaves suitable to the product filtered.
Can be furnished for use in any 18" or 33" horizontal plate filter tank. A pivot rack for holding cartridge in position during cleaning is available.
This filter is proving very popular in every field where fabric or metallic cloth can be used in place of filter paper.

Surface Separator Systems, Inc.

FLOATING OIL SKIMMER TO RETRIEVE OIL FILMS FROM WATER

Operation: Special rotating cylinders uniquely utilize the adhesive, cohesive, surface tension and water repellent characteristics of oil to separate, collect and pump oil films in one continuous and combined operation. SSS skimmers are relatively insensitive to surface motion and reject floating objects of any significant size. Other than a storage tank, no supplementary facilities are needed.

The action is similar to that when a finger is immersed and then withdrawn from an oil film on a water surface. The coating on the finger is primarily oil with a few clinging droplets of water. If the oil is wiped from the finger and the process repeated a number of times the entire film can be removed. The apparatus used in the Earle System is essentially a continuous "finger" of large size which can rapidly recover oil films with little water inclusion.

SSS skimmers depend upon adherence of oils to the surface of a special revolving cylinder and will recover all types of adherent/coherent supernatant oils and oil/water emulsions. Exclusive of the water content of emulsions, free water inclusions in the recovered oil will be within the range of 2 to 5% when cylinder rotating speeds are properly adjusted. Small solids adhering to or included in the oil film will also be retrieved. SSS skimmers will not pick up large "globs" of heavy greases or highly oxidized or other non-adherent oils or solids. They will not separate out any subsurface oil suspensions or emulsions and no guarantee can be made as to the oil inclusions (ppm) in the final waterway effluent.

The Industrial Skimmer: SSS Floating Industrial Oil Skimmers are both buoyant and stable, thus allowing them to be mounted for automatic adjustment to changing water levels. The recovery rate is governed by the type, physical characteristics and thickness of an oil film but is usually upward of one gallon per minute per lineal foot of recovery cylinder. Where recovery is in excess of normal oil inflow, the unit may be periodically activated and the reserve capacity used for peak loads or emergency situations. Speed adjustment permits retrieval of persistent oils in varying viscosities from light diesel oil to heavy fuel oil.

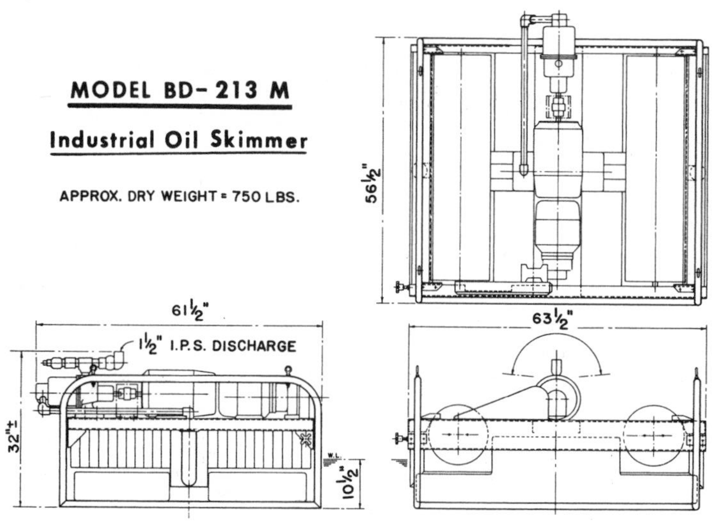

Surface Separator Systems, Inc.

WARNE ANTI-POLLUTION BOOMS

The use of booms to limit pollution and spillage when petroleum and other nonmiscible floating liquids are transferred from ship to shore is now accepted practice throughout the world. Surface Separator Systems, Inc. is the distributor for the William Warne & Co., of Barking, Essex, England, who make Warne Oil Spillage Booms which have been in service at major oil terminals for many years. Originally developed as air inflated devices, the range of Warne Booms has been increased to meet the requirements of a wide range of conditions. Three types of anti-pollution booms are now available:

Type T: This is a heavy-duty type of boom intended for use where it is to be kept permanently afloat and where rough conditions may be expected. It consists of a buoyancy chamber which is filled with a large number of plastic tubes, each sealed at frequent intervals so that there are many hundreds of airtight compartments in each boom section thus ensuring reasonable immunity from damage. From this buoyancy chamber hangs a skirt weighted at its lower end by a ballast chain with anchorage points at frequent intervals. This type is available with either 8 or 16 in. buoyancy chamber.

An alternative form of this type of boom has a buoyancy chamber filled with a series of 10 ft. long non-intercellular foam cylinders with a flexible portion of the boom between each cylinder. This enables it to be stored in concertina form should this be more convenient. It also renders transport to the eventual site simpler.

Type E: This is made as an emergency boom which would normally be kept ashore in protected conditions and only used during an actual spillage emergency. This is of much lighter construction and is air inflated. After use it should be withdrawn from the water, cleaned up and replaced in store again. If circumstances demand that this type of boom be left afloat for a protracted period, occasional topping-up with air may be necessary. Care should be taken in using a boom of this type as it is less robust than the Type T and can more easily be damaged. This type is made with an 8 in. buoyancy chamber.

For use in situations where an air inflated boom is unsuitable, this type can also be supplied with the buoyancy chamber filled with cylinders of non-intercellular foam as mentioned above for Type T. This avoids any necessity for occasional topping-up and is immune from malicious damage.

Type S: This is supplied for use in special circumstances where it is desired to sink the boom to the bottom to enable shipping to pass over. This is particularly suitable for use in calm waters and on such sites as cut-out berths on canals. It is air inflated and control of its position is made by inflating and deflating the buoyancy chamber. The buoyancy chamber of the Type S boom is normally 8 in. Type S cannot be used if the bottom is likely to cause damage or entanglement.

Construction: The outer casing of the buoyancy chamber and the skirt of all booms are constructed of Chloroprene synthetic rubber, specially compounded to provide the maximum resistance against sea and atmospheric conditions, suitably reinforced by synthetic fabrics. Connection between successive lengths of boom is effected in the case of Type T and E by joint plates made of a non-corrodible material.

Surface Separator Systems, Inc.

GENERAL ARRANGEMENT - MODEL 318

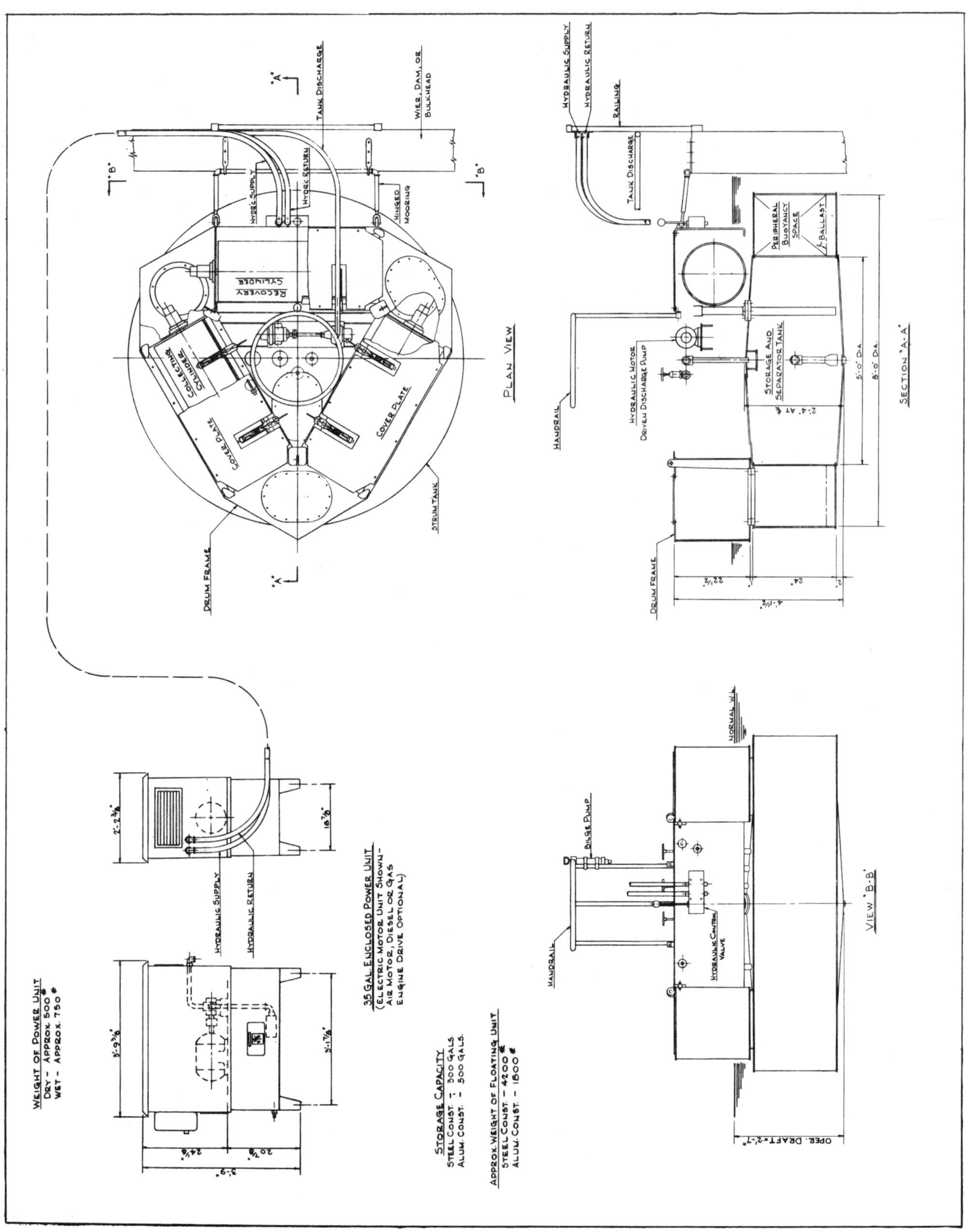

Surface Separator Systems, Inc.

NON-PROPELLED OIL SKIMMER — MODEL 318

This basic unit, or modifications thereof, is intended to be used in the recovery of persistent oil slicks in gravity separators, lagoons, and drainage basins where there is a continuous inflow of oily waste. The installation as shown on the reference plan consists of a self-buoyant unit moored in a fixed position within an oil retention area. Natural currents of air and water will deliver the oil film to the recovery cylinders, and the oil level in the basin will be maintained at a reasonable and acceptable level.

The oil film recovery mechanism is mounted on top of a 300 U.S. gallon oil storage tank. The buoyancy of both sections is such as to support the combined system at the proper operating draft. A hinged mooring connection is utilized to allow for variations in water level, surface motion, and equipment adjustment. The recovery unit and storage tank can be supplied in carbon steel, stainless steel, aluminum, or reinforced plastic.

The Earle system oil recovery device is a three-cylinder, hydraulic powered unit which takes full advantage of the cohesion, viscosity, and surface tension of oils (and their repulsion of water) and will concentrate, recover, separate, and store the oil in one continuous and combined operation. The oil recovery rate, based on an average film thickness of 0.10" (2.50 mm.), a temperature of 70°F., and retrieval of an average grade of Bunker C oil, is about 600 gallons of 95% pure oil per hour. Film thickness, temperature, oil type and viscosity, oil film contact, and cylinder speed all have some effect upon the oil recovery rate. Wave heights up to 9 inches, wind and rain, however, will have practically no effect on the rate of retrieval. The recovered oil is delivered to the storage tank by gravity. To avoid significant changes in draft, this tank operates on a hydraulic displacement system, i.e., it will always be full of water, oil, or a combination thereof when in operating condition. An oil discharge system is installed, including pump, valves, piping, and discharge hose, to deliver the recovered oil to disposal facilities within any reasonable distance from the skimmer.

For fixed installations, such as that shown on the reference plan, the hydraulic power pack will generally be shore based in a location close to the skimmer. The power pack is thus readily available for operation and service. The prime mover for the power pack may be an electric motor, air motor, diesel engine, or gasoline engine as selected for the particular installation.

The design of the skimmer is predicated upon simplicity, economy of operation, reliability of performance, ruggedness for hard usage, safety in hazardous locations, ease of maintenance, and a minimum of repair. The system can be arranged for continuous operation or for intermittent service utilizing automatic, semi-automatic, or manual controls. For use in gravity separators or other installations where separate oil storage facilities are provided, the recovery unit may be made self-buoyant and independent of the storage tank.

For most harbor and marine terminal usages the Model BP-318 self-propelled skimmer will be found the most advantageous. However, in locations where very infrequent oil spills occur and where auxiliary marine equipment is available, the Model 318 may serve for emergency use. In such cases the power pack is mounted directly on the unit to make it completely self-contained. The use of a diesel engine as the prime mover is recommended for this service.

Surface Separator Systems, Inc.

M/V "PORT SERVICE" OIL RECOVERY BARGE

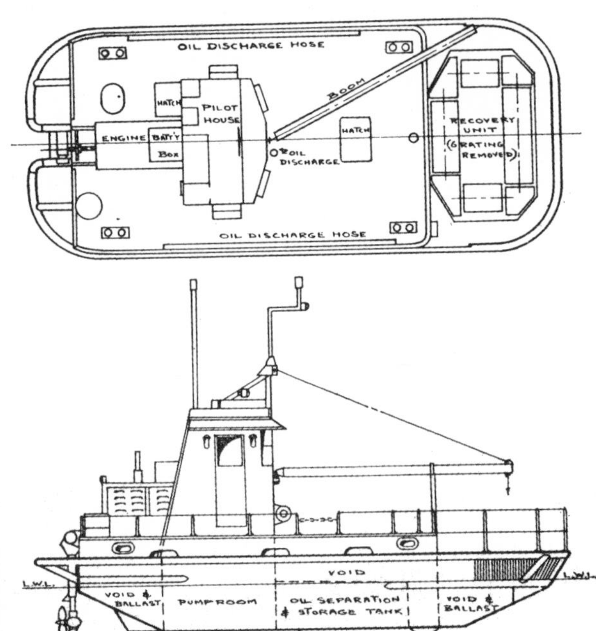

Principal particulars

Length overall	**38 ft 6 in. (11·7 m)**
Length b.p.	**34 ft 0 in. (10·4 m)**
Beam m'l'd	**16 ft 0 in. (4·9 m)**
Depth m'l'd	**5 ft 3 in. (1·6 m)**
Operating draught	**3 ft 3 in. (1·0 m)**
Operating displacement	**41·2 tons**
Oil storage capacity	**2,440 gallons**

The vessel was designed by M. Mack Earle, built by Wiley Manufacturing Company, and is jointly owned and operated by the Maryland Port Authority and the Baltimore City Fire Department. The oil recovery unit was built by the Washington Aluminum Company under license by Surface Separator Systems, Inc. The M/V "Port Service" was placed in operation in the spring of 1963.

In addition to its primary usage in oil recovery, the craft may be utilized for several other harbor services: (1) The boom has a capacity of 2,500 lb. and can be used for removal of logs or other large floating debris, or for buoy handling. (2) The fire pump delivers 100 gpm at 60 psig through two hoses, and this capacity or the pressure may be considerably increased if so desired. The shallow draft and good maneuverability of the vessel thus make it a valuable auxiliary fireboat. (3) The protective screens and the oil recovery unit are readily removable. The sump may then be fitted with a cargo net or wire mesh basket and the barge used for recovery of flotsam. Sweep boards may be easily rigged to increase the capacity for trash collection. (4) Additional auxiliary equipment may also be installed to suit other special requirements.

A self-propelled barge designed specifically for the recovery of oil from the surface of river and harbor waters, by the Earle surface separator system, has been put into service by the Port of Baltimore authorities.

The hull of the vessel is a compartmented pontoon of all-welded, transversely framed mild steel construction with, at the forward end, the oil recovery plant installed in a free-flooding sump divided from the outside water only be a series of grills on three sides. The superstructure comprises a small pilot house mounted over the pump room trunking.

Propulsive power is provided by a 52 hp. Harbormaster outboard unit; the diesel prime mover is also arranged to drive the hydraulic pumps powering the oil recovery system, the water and oil pumps, and the boom winch.

The oil recovery unit is a 4-roller, hydraulically-powered device which makes use of the adhesion, cohesion, viscosity and surface tension of oils, together with their repulsion of water; in one continuous and combined operation it concentrates, separates and recovers oil from the surface of the water in which the barge is floating and passes it to a storage compartment in the hull. When retrieving an average grade of Bunker C oil, with an average film thickness of 0.1" (2.5 mm.) at 70°F. (21°C.), the recovery rate is in the region of 830 gallons of 95% pure oil/hour. Increased oil thickness, lower temperatures, higher oil viscosity and assured oil film contact with the rollers improve the rate of recovery, and vice versa.

Recovered oil is fed by gravity to a 2,440 gallon storage tank. This compartment, in order to avoid significant changes in the draft of the barge and to ensure that the sump is always flooded, operates on the hydrostatic displacement system and is always full of water, or oil or a combination of the two.

Stability and flotability of the oxygenator is demonstrated in test tank at end of access bridge.

Union Carbide Corporation — Linde Division

UNOX SYSTEM WASTEWATER TREATMENT

The UNOX System is Union Carbide's Direct Oxygenation System utilizing oxygen gas to improve the secondary activated sludge wastewater treatment process. The product of a comprehensive program conducted over many years, the UNOX System offers the long sought technology which renders direct oxygenation both practical and economical when compared with conventional air aeration systems. The UNOX System has been tested and proven in numerous pilot scale studies and an FWQA sponsored full-scale operation of a 3 MGD municipal wastewater treatment plant. These tests have demonstrated that direct oxygenation makes practical a high-rate, high-solids, activated sludge process providing reliable, high quality treatment of both municipal and industrial wastes while reducing total treatment costs up to 40%. In addition, the UNOX System offers many other advantages such as reduced land requirements, improved process control, and significantly reduced odor.

The UNOX System is based on the combination of recent advances in gas-liquid contacting and fluid control systems with well-established air separation and wastewater treatment technology. This combination results in an economical supply of enriched oxygen gas which is efficiently mixed and dissolved in the "mixed-liquor" of the activated sludge process. In order to concentrate first on the wastewater treatment technology aspects of the UNOX System, this part of the description will be limited to the equipment which performs the dissolution and mixing steps. The sources of oxygen supply will be discussed later.

Bio-oxidation Process Description

While the equipment design to accomplish the dissolution and mixing steps can vary to meet specific process requirements, an understanding of system operation can be gained through the discussion of a "typical" design. The figure shows a schematic, cross-sectional view of such a design. As indicated, the aeration tank is divided into sections or stages by means of baffles and is completely covered (usually with precast slab concrete) to provide a gas-tight enclosure. The tank dimensions and configuration are not restricted by the UNOX System design since the depth, for example, can vary between 10 and 30 feet. The liquid and gas phases flow concurrently through the system. The feed wastewater, recycled sludge and oxygen gas are introduced into the first stage. The number of stages and the exact method of gas-liquid contacting, however, may vary with the specific application.

In such a system as shown in the figure, the oxygen gas is fed into the first stage at a pressure of only about 1 to 4 inches of water column above ambient. Small recirculating gas blowers in each stage pump the oxygen gas through a hollow shaft to a rotating sparger device at a rate sufficient to maintain the required mixed-liquor D.O. level. The indicated pumping action of the impeller located on the same shaft as the sparger promotes adequate liquid mixing and yields relatively long residence times for the dispersed oxygen gas bubbles.

Gas is recirculated within a stage at a rate usually higher than the rate of gas flow from one stage to another. The successive aeration stages or chambers are connected to each other in a manner which will allow gas to flow freely from stage to stage with only a very slight pressure drop, but yet sufficient to prevent gas backmixing or interstage mixing of the aeration gas. This is accomplished by appropriate sizing of the interstage gas passages. The liquid flow (mixed-liquor) through successive stages is concurrent with gas flow. Each successive stage is essentially identical to the preceding one except that, as a higher proportion of the oxygen demand is met in the initial stages, the required volume of gas to be recirculated in subsequent stages will be less to maintain the desired dissolved oxygen level in the mixed-liquor. Effluent mixed-liquor from the system is settled in the conventional manner and the settled activated sludge is returned to the first stage for blending with the feed raw or settled sewage.

The entire multistage activated sludge unit is fitted with a gas tight cover to contain the oxygen aeration gas. A restricted exhaust gas line from the final stage vents the waste gas to the atmosphere. Oxygen gas is automatically fed to the system on a pressure demand basis with the entire unit operating in effect as a respirometer. A small positive pressure is maintained by the feed gas flow controller. As the organic load and respiration (oxygen demand) of the biomass increases, the pressure tends to decrease and feed oxygen flow into the system increases to re-establish the pressure set point of the controller. Feed oxygen to the multistage system can be controlled on this pressure demand basis by a simple regulator, or differential pressure controller, automatic valve combination.

SCHEMATIC DIAGRAM OF UNOX SYSTEM WITH ROTATING SPARGER (3 STAGES SHOWN)

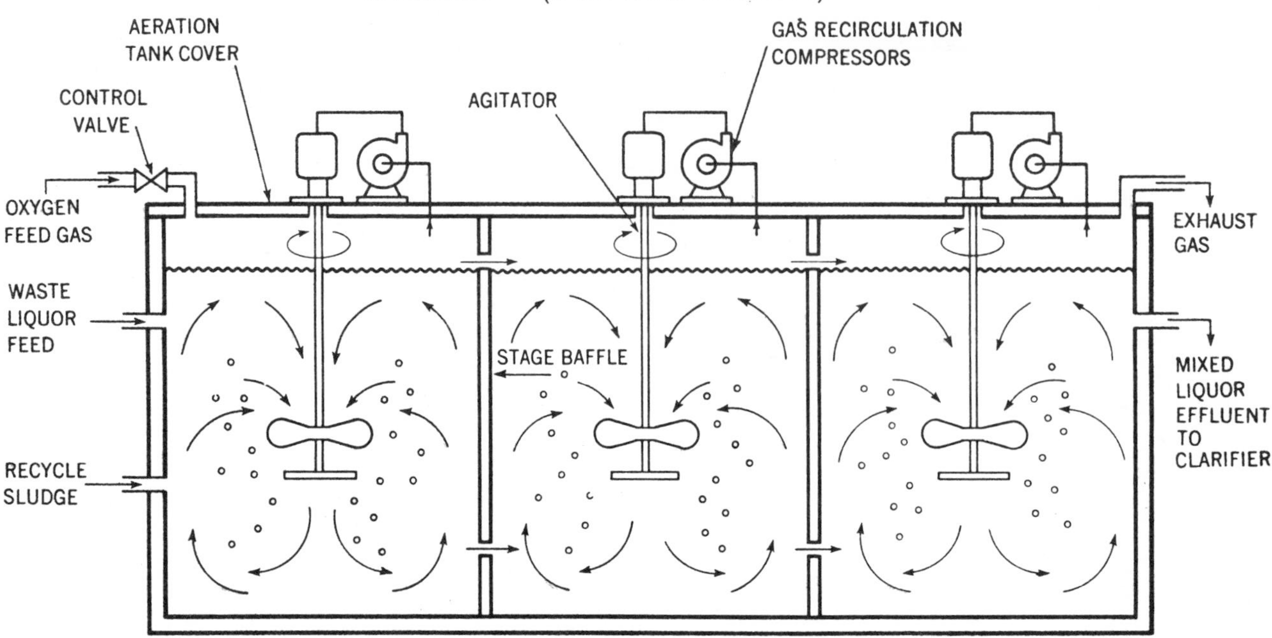

The exact method of gas-liquid contacting employed in the UNOX System can be varied without substantially altering the overall process efficiency. Depending upon specific process conditions, surface aerators can also be used to contact the oxygen gas with the mixed-liquor. This design eliminates the need for gas recirculating compressors and the associated piping. The system operates exactly like that shown in the above figure in all other respects. The required level of bulk fluid mixing required to maintain the sludge in suspension and insure a uniform liquid composition is provided by an efficient pumping, slow speed, low shear agitation impeller much like that used in the design shown in the above figure. These units are designed to provide the necessary liquid mixing in the most efficient form possible.

INDUSTRIAL DESALINATION SYSTEMS

GENERAL

The BW Series conversion system utilizes reverse osmosis to dramatically reduce the disolved solids in brackish or problem water. The system is self-contained and requires only electrical power for operation. Desalting modules are utilized which provide "building block" flexibility within a wide range of product capabilities. The equipment is designed to incorporate the latest developments with minimum effort and expense.

FEATURES

1. Tubular desalting members.
2. Desalting unit made of corrosion resistant material.
3. Safe, automatic, unattended operation.
4. Compact industrially oriented design.
5. Optional conversion accessories available.
6. Flexibility of design plus options allow us to customize the basic BW unit to your specifications.

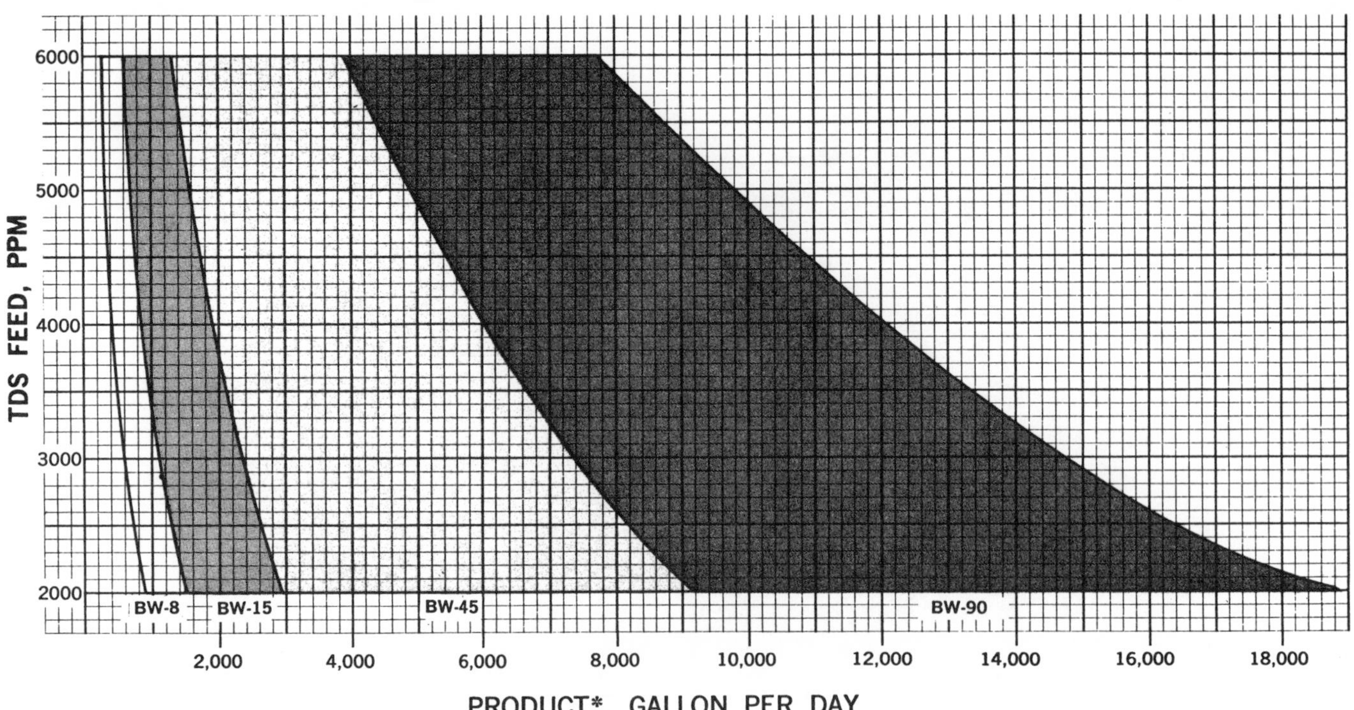

*Less than 500 PPM TDS in product

The operating cost of purifying water by reverse osmosis is extremely low. The water is merely pressurized. It does not have to be boiled, frozen, or treated chemically or ionically as in other water purifying processes.

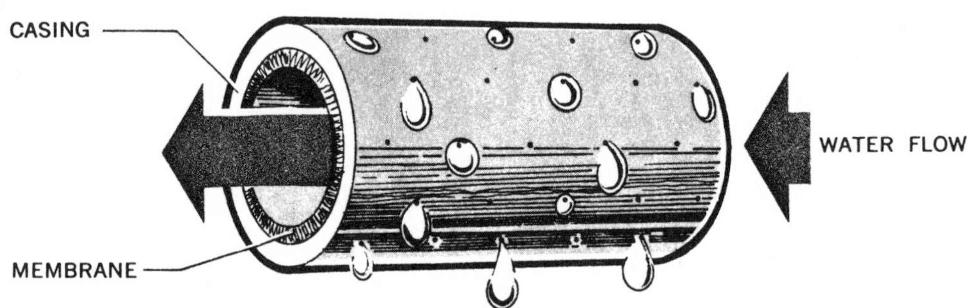

UWC's basic structural unit is a rigid tube enclosing a smaller synthetic osmotic membrane tube made of cellulose acetate. Impure water flows through the tube under pressure; pure water permeates the membrane and flows through small holes in the walls of the tube. The remaining impure water, or effluent, within the membrane tube is carried away.

Vaponics

Vaponics Carbon Filters are designed to remove and/or reduce impurities such as organic coloring, odors and chlorine from waters by passing through a bed of activated carbon supported by layers of washed silica sand and gravel.

Application

Feedwater Pretreatment to:

Water Stills

Demineralizers

Reverse/Osmosis Module

Electrodialysis

Plant Process

Water Recycling & Re-use

Water Waste Treatment

Pollution Control

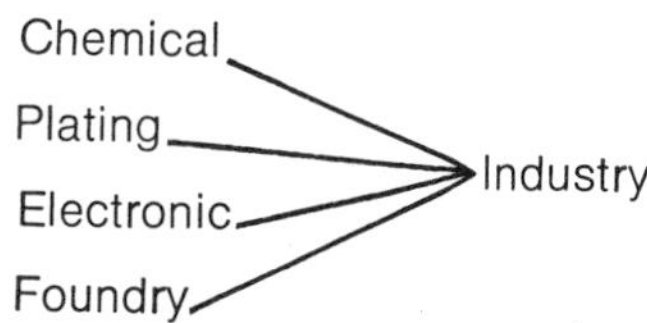

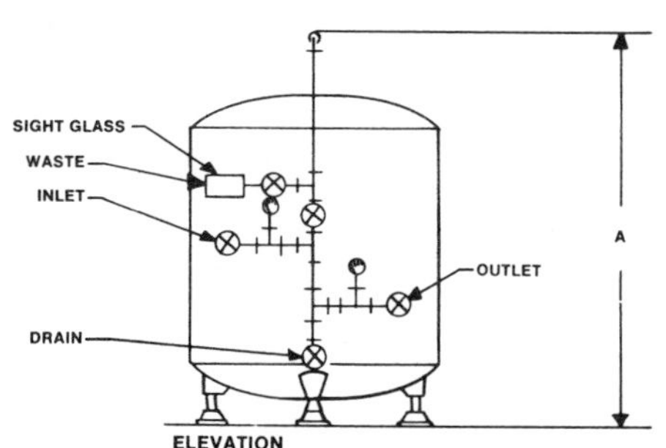

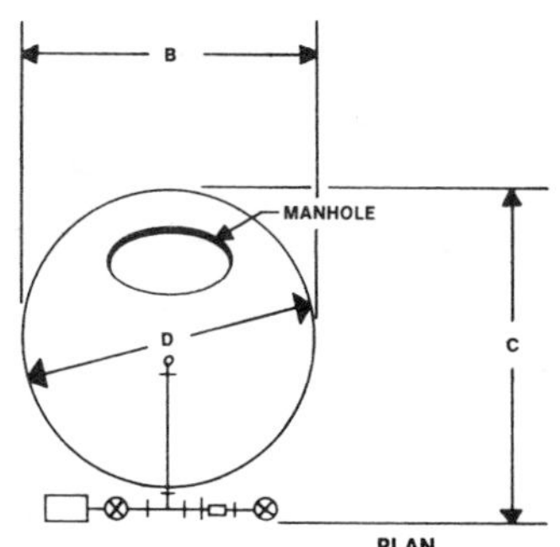

Specifications for Vaponics Carbon Filters

Filter Body: Fabricated of cold rolled steel with dished heads lined with epoxy. Designed to ASME Standards for pressure vessels at 100 psig and tested to 150 psig.

Piping and Valves: All interconnecting piping and valves of red brass or hard copper tubing are furnished. Included as standard equipment are inlet and outlet pressure gauges, backwash sight glass, air vent valve and drain valve.

Filter Bed Media: Consists of a bed of activated carbon supported by five layers of graded washed gravel and sand in the material and depth noted below:

Layer	Material	Depth
Bottom	Coarse Gravel	5″
2nd	Medium Gravel	2½″
3rd	Fine Gravel	2½″
4th	Extra Fine Gravel	3″
5th	Silica Sand	3″
Top	Activated Carbon	20″

CLARIFLOW - RECTANGULAR

The Rectangular CLARIFLOW is functionally similar in design and operation to the Type HC unit. It is often applied where common wall, concrete basin design is desired. In many instances this design can be used to improve an existing concrete basin clarifier, increasing the efficiency by 300%. UNIMIX flocculator units are employed in the rapid flocculation region. The flow leaves this region through multiple diffusion ports, designed to direct mild, rolling energy into the adjacent secondary slow flocculation zone.

Rectangular CLARIFLOW, Jameson Canyon W. T. P., Napa, California.

CLARIFLOW UP-FLOW SOLIDS CONTACT CLARIFIERS

The CLARIFLOW series of solids contact clarifiers includes a variety of designs tailored for circular or rectangular basins. All utilize instant coagulant blending for chemical efficiency and integral flocculation and clarification, combined with good fluid mechanics to avoid density currents and short circuiting. The flocculation phase is enhanced and speeded through the slurry recirculation capability. In all instances, bottom scraper assemblies move settled sludge to a hopper for removal and direct recirculation, where employed. Clarified liquid upflows to a multiple trough (orifice or weir) system at the surface.

CLARIFLOW - TYPES HC & LC

These CLARIFLOW design types, proven over the years, are basically circular in construction, for installation in circular basins, or in square basins, where corner sweeps are employed. A centrally located reaction zone contains an axial flow mixing–recirculation pump, and utilizes multiple tangential diffusers to impart a mild turbulence throughout the flocculation zone. This energy is sufficient to ensure suspension of solids to the extent that "water falling," or density currents are avoided–even when heavy slurry return is practiced.

AEROGEST AEROBIC DIGESTION SYSTEM

The WALKER PROCESS AEROGEST system utilizes either SWIRLMIX aeration or ROLLAER aeration.

Aerobic digestion or auto-oxidation, produces the same sludge volume reduction as does anaerobic on comparable sludge. It is entirely free of odors, requires no outside source of heat, produces the best drying sludge, and has a clear supernatant that does not back-load the plant.

AEROGEST systems use only about twice as much power as an anaerobic digester; and no more digestion capacity. Open-top tank construction for the AEROGEST system is considerably less expensive than for anaerobic.

ROLLAER aeration (SPARJERS operating at 12-ft. submergence in a draft tube) permits the use of deep hopper bottom tanks (old anaerobic digester), because the draft tube which provides a strong overall roll, scours the entire bottom.

AEROGEST system can be used effectively on an existing activated sludge plant to digest the waste activated sludge only; thereby alleviating the operating problems and greatly increasing the capacity of the existing anaerobic digester and greatly reducing the back-load otherwise resulting from septic supernatant.

AEROGEST aerobic digester, Arlingwood Subdivision, Jacksonville, Florida.

CIRCULAR COLLECTORS

WALKER PROCESS circular sludge collector mechanisms are available to meet every municipal and industrial requirement. Included are center pier supported units; bridge supported units; Tractor-drive (peripheral drive) design; SIGHT WELL (combination suction nozzle and plow type) design. Drive selection includes rugged, sealed, oil-bath turntable spur gear and worm gear designs in a full range of sizes. Positive overload protection is provided. Designs permit inclusion of corner sweep attachments and surface skimming assemblies, where required.

The long-life turntable gears are cut and assembled and tested in our factory and are preferably made of 50 thousand pound alloy iron, an excellent turntable gear material. Cast steel gears can also be furnished.

Seven circular collectors, two suction collectors, and digester mixers, Village Creek S. T. P., Fort Worth, Texas.

RECTANGULAR COLLECTORS

Designed and manufactured for maximum operating life and low maintenance with emphasis on the development of wear-resistant materials.

Drive and idler sprockets are offered in Telluriron, with a greater Brinell hardness than ordinary chilled alloy cast iron. Sprockets are also available in alloy steel, flame hardened. Telluriron sprockets and Promal chain assure many years of trouble-free performance.

Rectangular collectors, Cedar Rapids, Iowa, W. T. P.

CLARIFLOW THICKENERS

Tractor drive thickener handling hot strip mill wastewater and fine scale removal, Calif.

This unit applies the Clariflow principle of fluid mechanics to effect the maximum clarification in the smallest possible unit. Clariflow thickener produces a low solids carrying effluent and a high concentration of solids in settled slurry. Includes energy inlet and flocculation center well, low rate overflow weir system, and thickener mechanism. In many cases the unit permits re-use of clear effluent and slurry in further plant processes.

Scraper arms are provided with manual or hydraulic power lifting mechanisms to withdraw from stalling overloads. Walker Process has many years of experience in solving all forms of steel mill water-borne dust problems.

SIGHT WELL CLARIFIER

One of three SIGHT WELL collectors at Battle Creek, Michigan, S. T. P.

The WALKER PROCESS SIGHT WELL clarifier's major purpose is the quick anl uniform withdrawal of return activated sludge from final clarification tanks before the sludge becomes anearobically degraded and denitrified through prolonged blanket detention.

The multiple suction effect ensures that the withdrawal will not break through the blanket (drawing relatively clean return liquor and leaving the seed sludge behind) and that if any one suction nozzle should become clogged it can be readily detected and easily rodded.

The SIGHT WELL clarifier includes an above water turntable drive unit, SIGHT WELL and sludge collector with V-plows positioned continguously along the truss arms to lead thickened bottom sludge into vertical pipes taking suction from the floor at the apex of the V-plows—covering the entire tank floor for effective sludge removal. The suction pipes carry the sludge to the SIGHT WELL attached to the drive cage.

Sludge flows from the SIGHT WELL through a vertically placed pipe conduit located inside the center support pier, which normally also serves as the tank influent.

GRIT SEPARATION UNITS

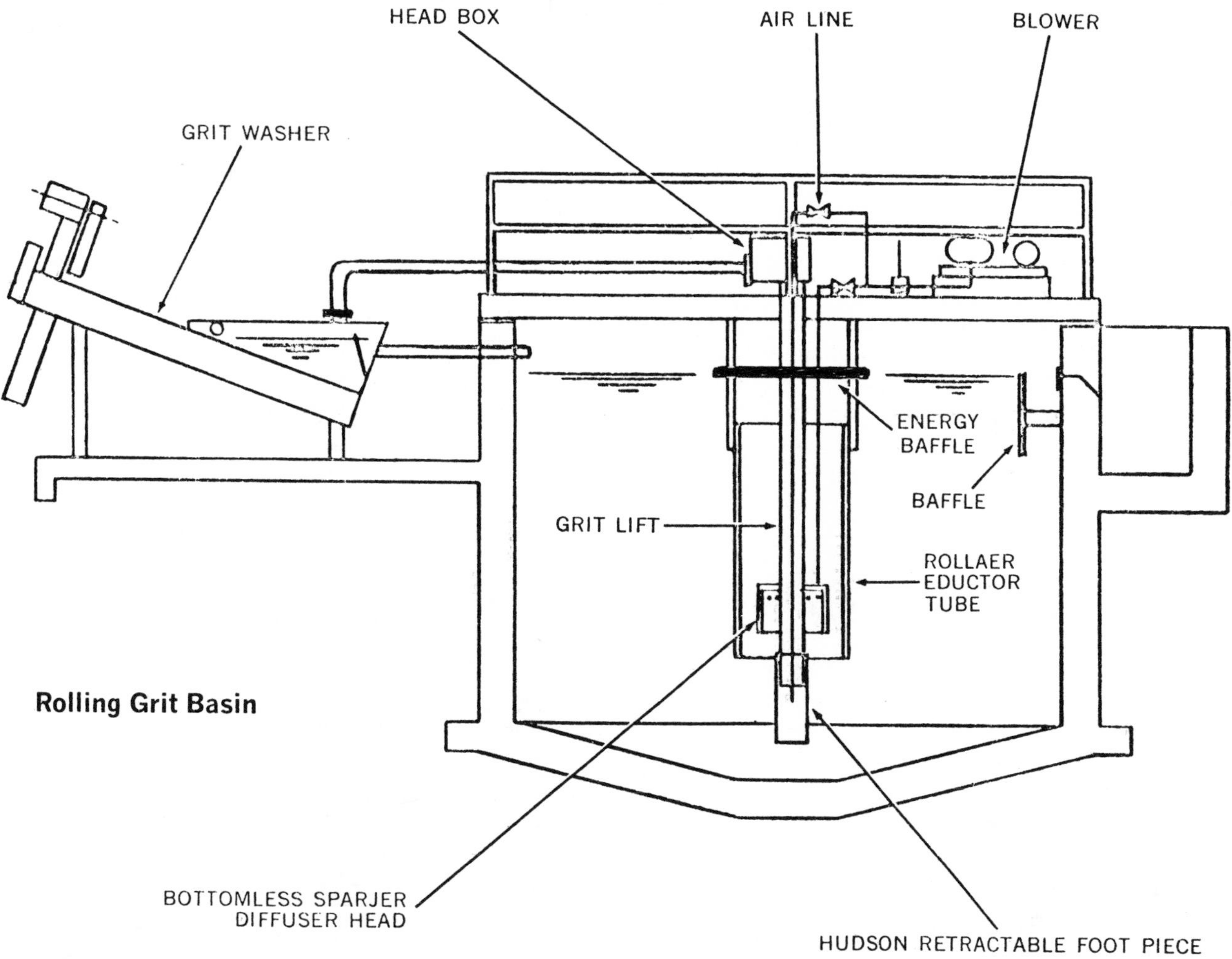

Rolling Grit Basin

Walker Process manufactures a line of rugged grit separation equipment to fit every grit removal situation, whether the setting is shallow or deep and whether the grit is fine or coarse. Included are heavy duty V-bucket type collector-elevators, grit hydro-separators, aerated grit removal basins, grit airlift pumps and the Rolling Grit basin.

The Rolling Grit basin offers an excellent means of pre-aerating plant influent for better overall treatment efficiency. In the Rolling Grit basin, grit is classified on the bottom by the rolling aeration action, is picked up by a grit lift airlift pump and discharged into an inclined screw washer-classifier, and discharged into a box or truck in a dewatered, clean condition.

AerGrit aerated grit basins, featuring MONOSPARJ and DUOSPARJ diffusers and Ease-Out headers are available in a range of sizes to suit practically all basin sizes. Grit hydro-separators include a rugged circular collector mechanism with an influent control and a flight type drag out unit which delivers a clean, washed grit for discharge into a box or truck.

MONOSPARJ - DUOSPARJ - SWIRLMIX

WALKER PROCESS has pioneered in the development and testing of new and improved air diffusion units and systems. Today's MONOSPARJ and DUOSPARJ diffusers and the SWIRLMIX (Complete Mix) aeration system have gained wide acceptance and are very successful on pharmaceutical, phenol and oil refinery wastes and similar situations where high intensity mixing aeration and oxygen transfer efficiency are critical. The aeration equipment is furnished with a wide selection of accessories, as well as in compact plant designs to utilize space requirements.

Aeration aids in treatment of wastes from manufacture of pharmaceuticals at a large Eastern laboratory.

DUOSPARJ AND MONOSPARJ AIR DIFFUSERS

WALKER PROCESS DUOSPARJ and MONOSPARJ diffusers mounted on the bottom of the air header, have been developed to provide air diffusers that do not clog; require little or no maintenance; do not need air filtering systems, yet produce good oxygenation efficiency.

WALKER PROCESS SPARJERS, applied in the proper system and geometry, will produce as much oxygenation (O_2 exchange efficiency) as any fine bubble, porous diffuser. These bottom mounted units with plunge pipes not only are more efficient but are capable of operating at low turn-down in air rates without clogging and never have a back pressure loss at increased air rates exceeding 10 inches of water column.

Designed for today's advanced activated sludge practices, DUOSPARJ and MONOSPARJ units now feature plunge pipes, open bottom tubes projecting 6 inches below each SPARJER orifice cluster, which allows rapid purging of liquor-solids from headers as well as produces increased airlift and O_2 exchange. Spilled air from these pipes rises rapidly to join the main air column, creating a turbulent effect, where the two bubble groups can combine. This results in fine-scale bubbles, better circulation through over-all mixing and greater oxygenation.

These two WALKER PROCESS diffusers supersede the older top mounted SPARJER and now make it possible to select a design which exactly suits your requirements. MONOSPARJ for low air release rates per foot of header and DUOSPARJ for higher rates.

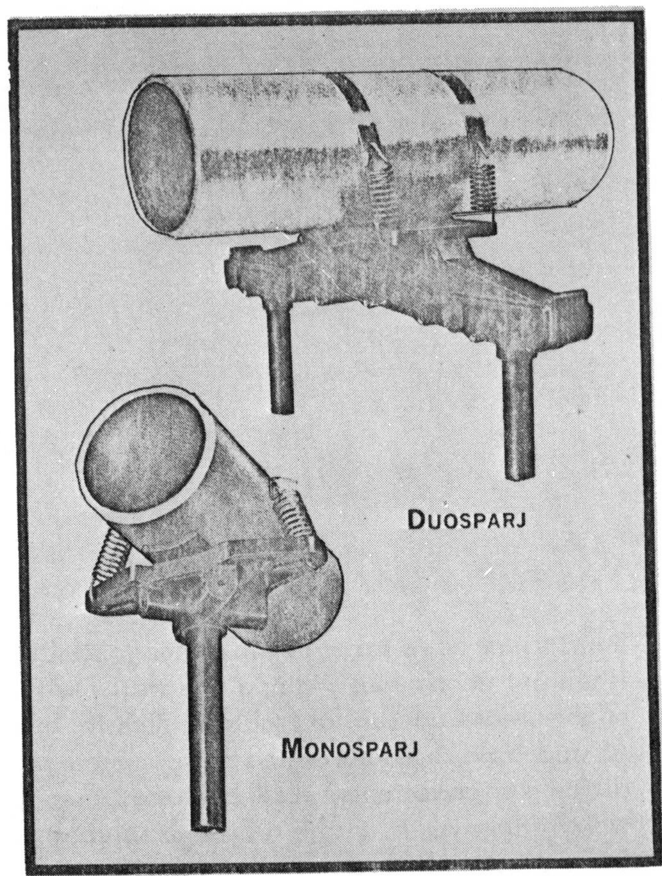

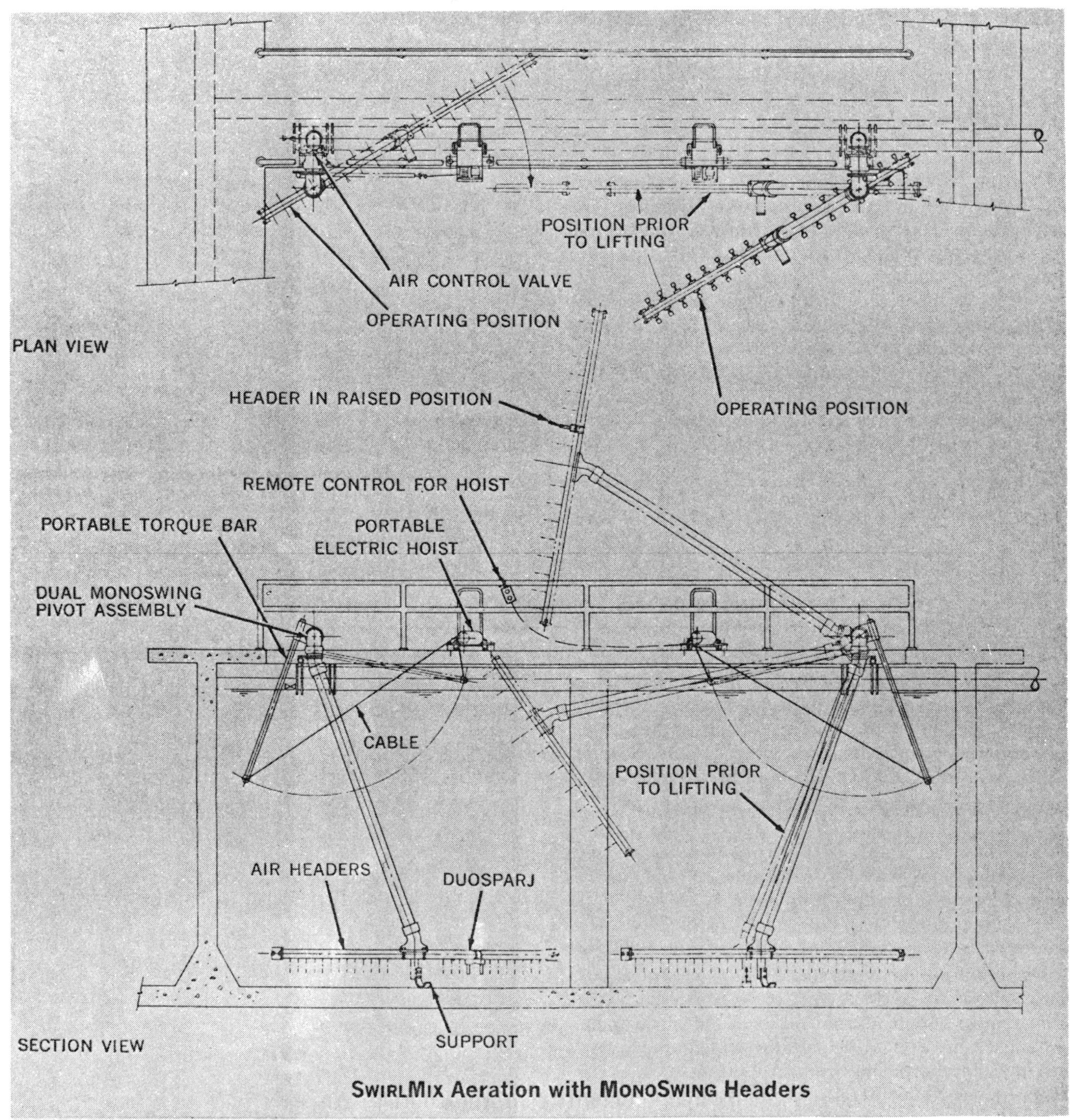

SwirlMix Aeration with MonoSwing Headers

Sanitarians have for some time recognized that the spiral roll regime, which utilizes diffusers lined out contiguously along one wall, left a lot to be desired; and that other configurations might create a more favorable regime for mixing, contact and oxygenation. After many years of study and research on the performance and operating characteristics of many variations of diffuser arrangements, Walker Process engineers have developed the SwirlMix arrangement which improves even the rather complete mixing obtained with the Cross-Roll configuration. SwirlMix aeration ensures the quickest possible actual homogeneous whole tank mixture of food supply (sewage) and microorganisms (activated seed sludge) for positive results.

SwirlMix aeration is a proprietary arrangement of headers so as to concentrate the air release along shorter header lengths, angled in such a pattern that the result is tantamount to a series of mild cyclonic disturbances. This technique, which produces an imbalance and an overall swirling turbulence, is the result of applied fluidics or fluid vector amplification.

SwirlMix diffuser configuration provides a more efficient job of overall mixing and contact, breaks up the clumping tendency of activated sludge-flocs — exposing more contact surface, and uses less air to do the job. This is because of the marked greater efficiency of oxygenation in the swirling resulting from this unique air header arrangement.

MONO-SWING HEADERS

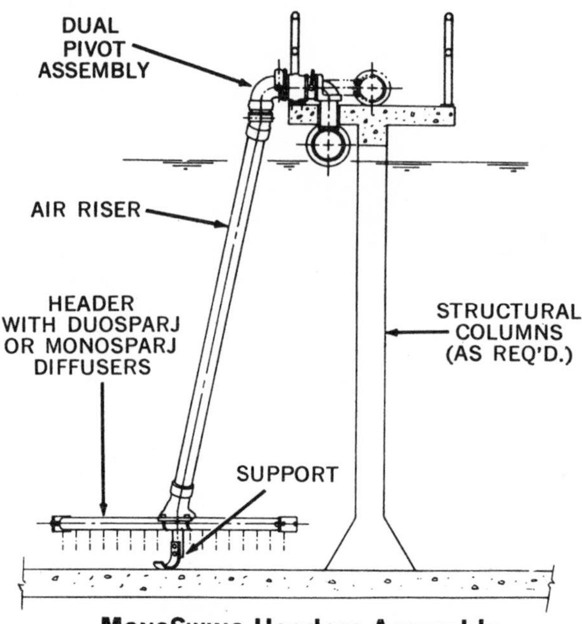

MonoSwing Headers Assembly

The most versatile removable header unit of them all has been developed by Walker Process. This MonoSwing unit (Patent applied for) features a single above water joint pivoted in the axial plane so that all air flow impediments are removed. The absence of a core axle also makes the unit extremely rugged and service-proof.

The unit is also provided for modern wide tank tie-bridge mounting to be used with any one of the SwirlMix variations. An above water simple pivot joint allows the header attitude to be angled to the most effective position (for the tank geometry); and to be raised parallel to the service bridge by a cleverly simple hoist giving perfect access to inspect the diffusers or header proper.

EASE OUT headers, Kiehl, Wisconsin, S. T. P.

EASE-OUT HEADERS

EASE-OUT top removable headers (Patent Nos. 3,242,072 & 3,339,901) provide economical and simple aeration equipment for raising and inspecting the diffusers without draining the aeration tank.

This form of removable header is recommended to the design engineer and the plant operator in lieu of the inaccessible fixed header.

Power savings are also accomplished with EASE-OUTS, since there are no pivotable joints and therefore, less air head loss. There are no moving parts to maintain or replace—and yet the entire header unit can be hoisted by one man.

The top anchor elbow has a special self wrenched Easy-Connector flange with sliding ways to facilitate removal and replacement of the individual headers.

EASE-OUT headers are equipped with BLOW-OFF legs which aid in the clog-free performance of the DUOSPARJ and MONOSPARJ diffusers by rapidly purging liquor-solids from the header upon start-up. The BLOW-OFF legs also reduce the sudden "closed valve" start-up of rotary positive blowers—an important point that saves blower trouble.

JACK-KNIFE HEADERS

WALKER PROCESS JACK-KNIFE headers, also with BLOW-OFF legs (see above) are designed to permit easy inspection of the diffusers. The center and top JACK-KNIFE pivot joints will not bind or wobble because rotating faces are fitted with bronze wearing rings with grease lubrication of the pivot joints.

Walker Process Equipment

DEEP CROSS ROLLAER (SwirlMix or Cross-Roll Systems)

Walker Process Deep Cross Rollaer is especially designed for aeration, aerobic digestion and similar applications when the tanks involved are deeper than the customary 15 feet. Deep Cross Rollaer can be installed in circular or rectangular tanks as deep as 36 feet without any loss in effectiveness, although tank depths 24 to 32 feet are considered more ordinary.

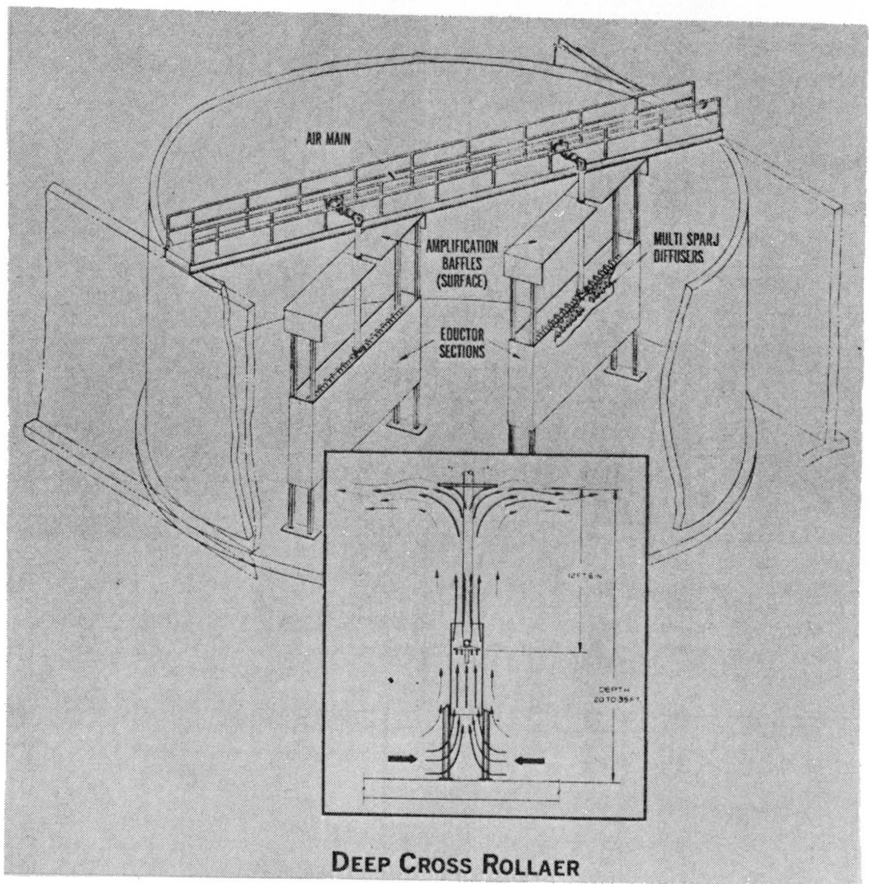

DEEP CROSS ROLLAER

The Sparjer diffuser system is installed just 12 feet below the surface, thereby limiting the air pressure to ordinary 6 1/2 to 7 psi blowers. The eductor tube sections, developed by Walker Process, extend from a short distance above the Sparjer system to within a few feet of the tank bottom. The air diffusion grid with exclusive QuadSparj diffusers operates within the eductor area to induce a strong air lift pumping effect. A patented (No. 3,028,817) energy baffle increases the circulation efficiency to such an extent as to cause the entrained air to remain in the circulation pattern moving downwardly with the current in an unsteady non-laminar flow creating optimum conditions for oxygen transfer.

Deep Cross Rollaer, installed in the SwirlMix configuration, will further increase the mixing intensity and swirling of entire tank contents. There is no short circuiting or bottom deposits.

INTENSAER MECHANICAL SURFACE AERATORS

Walker Process Intensaer surface aerators provide high oxygen transfer efficiency along with good stirring and solids suspension. The latter, like any other surface aerator, providing good inundation is practiced. They are backed by extensive research and development studies conducted in the company's large test tank; plus on-the-job tests at actual, full-scale installations. Walker Process Intensaer aerators feature variable speed drives, tapered blade stainless steel impellers and bridge level adjustment of blade inundation. They are custom built to suit each specific application, and above-water steady bearings are used to resist lateral shock loads and protect the gear reducer against strain.

INTENSAER mechanical surface aerators, Mansfield, Massachusetts, S. T. P.

REELAER SURFACE AERATION

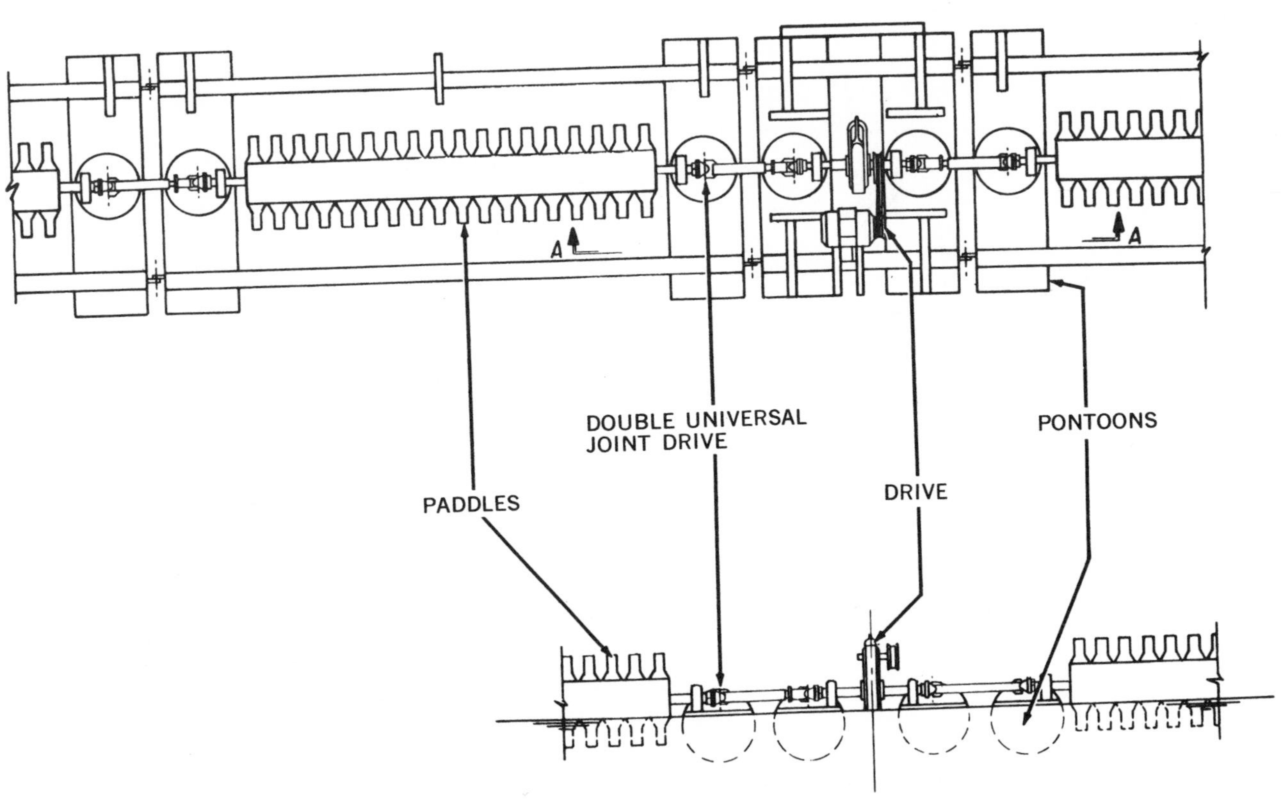

ReelAer, Cage Aerator for Surface Aeration

Walker Process Equipment

SWIRLMIX (Complete Mix and Premix) WASTE WATER TREATMENT PLANT

SWIRLMIX, complete mix plant, Proctor Creek S. T. P., Atlanta, Georgia.

Walker Process is proud of its leadership with over 600 much-imitated SPARJAIR plants now in operation.

Now the SwirlMix plant, another outstanding Walker Process development, brings new and better methods to the design engineer. Ease of maintenance, efficiency of operation and economy have been incorporated as major design factors.

The SwirlMix compact system is a well demonstrated scientific method which: (1) simplifies operation because there are fewer "things to do"—no delicate detention, or biological and return sludge balances to achieve and hold; (2) cuts construction costs with fewer tanks, auxiliaries and connecting lines, commonwall, compact construction is utilized; (3) eliminates odors as there is no primary settling where odors can develop; (4) eliminates the foaming problem, and (5) can stand shocks or overloads of organic material much better than traditional activated sludge or contact stabilization processes.

To improve the complete tank mixing, Walker Process applies SwirlMix diffusion, to ensure the quickest possible actual homogenous mixture of food supply (sewage) and microorganisms (activated seed sludge) for positive results.

SwirlMix diffuser arrangement effects a faster and more efficient job of over-all mixing and contact, and uses less air to do the job. This is because of the marked efficiency of oxygenation and synthesis resulting from this unique, patented air header arrangement. The SwirlMix header arrangement takes advantage of an imbalance to achieve vector amplification resulting in advantageously turbulent stirring in all directions.

SwirlMix compact plants are offered in a range of sizes and capacities for population loads up to twenty thousand persons. They are ideal for complete wastewater treatment for large establishments beyond the reach of city sewers, but also are particularly adapted to use by consulting engineers for economical sewage treatment in municipal and industrial plants, all the way through 20,000 population equivalent.

These plants can also be provided with feeders for heavy metals (iron or aluminum) additions to sequester and reduce phosphates; and/or tertiary dual media (air backwash scour) filters, or microstrainers to remove pin-point and other particulates from the effluent.

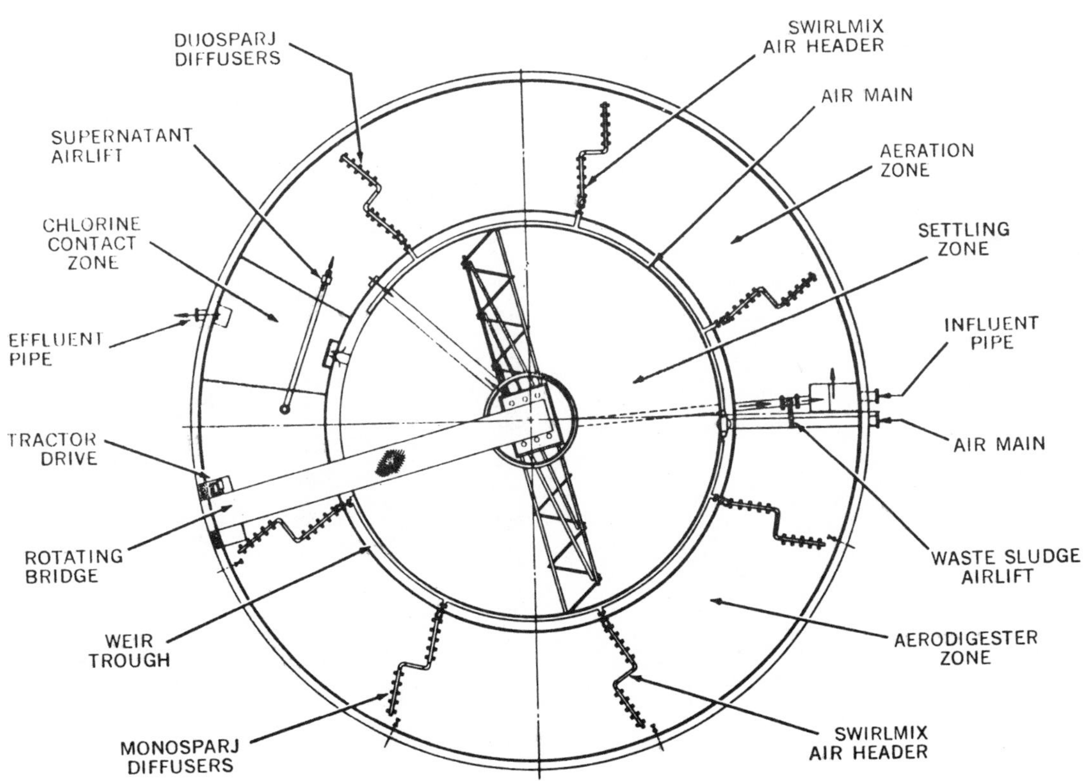

Plan View of SwirlMix, Complete Mix Plant

ROTOCLEAR AUTOMATIC ROTARY SCREEN

Walker Process RotoClear equipment includes a complete line of automatic, rotary screening units to provide the most economical method of solids separation and reclamation waste water treatment. Models are available with bronze, stainless and synthetic filter media, of various porosity, suitable for single treatment and pre-filtration in potable water treatment systems, secondary effluent polishing in tertiary sewage treatment systems, industrial and commercial waste water reclamation and gross solids separation involved in primary industrial and commercial treatment.

With the Walker Process RotoClear unit it is easily possible to obtain up to 95% removal of suspended solids and BOD through treatment of secondary effluents. The self-cleaning Walker Process RotoClear is an efficient means of removing suspended solids from open-channel or gravity flow systems. Suspended solids are trapped by the RotoClear drum surface and flushed into waste hoppers. The effluent from the RotoClear unit can then be re-used or discharged to a receiving stream.

The Walker Process RotoClear unit is completely automatic in operation and economical to install, operate and maintain.

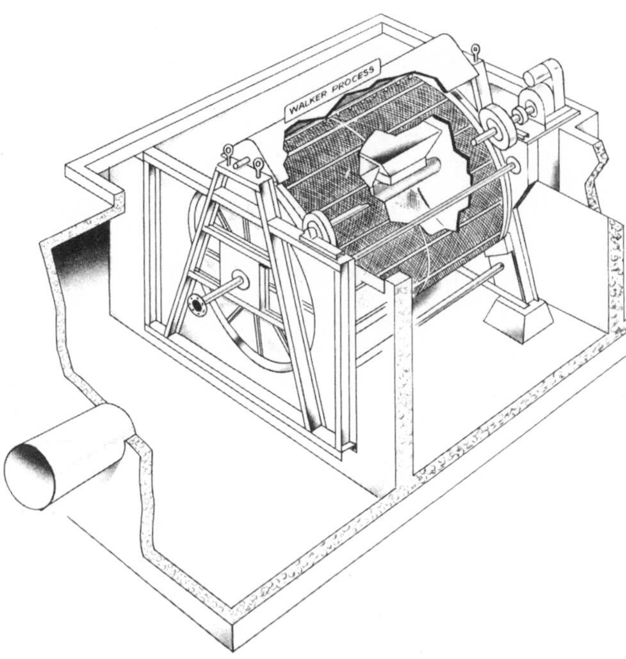

ROTOCLEAR rotary screen unit is designed to use both synthetic media, at considerable cost savings, and regular fine mesh metal media.

"BLOW-OUT" TOWERS

In water treatment, the stripping of hydrogen sulfide, carbon dioxide, and methane is effectively accomplished with Walker Process Blow-Out (forced draft) Towers. H_2S is particularly difficult to remove to a very low chlorine demand value, but the unique Walker Process design (special H_2S-proof metal tray with special plastic raschig rings) does, in practice, achieve just this. Carbon dioxide and methane are likewise stripped out and blown clear to effect absolute minimum residuals.

The forced draft fan and motor handles only clean, dry air, thus protecting it from corrosion. When hydrogen sulfide is to be stripped, aluminum is used for all parts and is extra heavily anodized.

Blow-Out Towers, Orlando, Florida, W. T. P.

AUTO-BELT VACUUM FILTERS

Vacuum filtration systems, designed and manufactured by WALKER PROCESS, include all the necessary chemical conditioning and other controls for sludge dewatering.

The WALKER PROCESS AUTO-BELT vacuum filter incorporates the latest technical advances in the construction of rotary drum type filters. An exclusive, patented, filter cloth tracking mechanism (Patent No. 3,445,003) maintains positive and uniform cloth tension and alignment. This cloth tracking system will not pull, stretch or tear the cloth laterally, making it a most important feature. The discharge roller arrangement allows continuous sludge cake take-off and cloth return. Double spray headers wash the cloth both front and back before returning to the drum.

With the AUTO-BELT vacuum filter, a wide variety of sludges, such as primary, digested, activated primary and many industrial wastes can be processed with equal efficiency.

AUTOBELT vacuum filters, Aurora Sanitary Dist. S. T. P., Illinois.

DIRECTUBE GASLIFTER

The WALKER PROCESS DIRECTUBE GASLIFTER (Patent Nos. 3,225,887: 3,242,071: 3,194,756: 3,187,897 and 3,055,502) has been carefully designed and developed to provide the most efficient and trouble-free digester heating and mixing method available. All of the features have been field tested and proven in many hundreds of plants, to ensure trouble-free installation.

The GASLIFTER intermixes and stirs the digesting contents to diffuse enzymes and sweep away poisons and accelerate biochemical reactions. Organic greases and scum blankets, and capacity-robbing build-up of fine bottom silts are controlled by this positive over-all circulation. At the same time, the DIRECTUBE feature of the GASLIFTER provides the advantage of a convenient, inexpensive and trouble-free method of heating the digester contents—an exclusive feature eliminating the cost and nuisance of an outside heat exchanger with its circulating pumps and intricate piping. The DIRECTUBE GASLIFTER puts heat directly into the circulating sludge and the radial flow from the eductor tube carries the heat to the outer areas of the digester, exactly where it is required, so that the sludge is evenly heated and well distributed.

The GASLIFTER is generally operated intermittently, so that a strong over-all roll of the contents takes place for one or more periods during the day—preferably after new feed (if feed is intermittent).

The intermittent strong over-all roll re-distributes enzymes and substrate—which over a long period tends to stratify due to the ebullition effect of generated gas. The GASLIFTER breaks up this undesirable stratification and distributes heat and thereby maintains ideal anaerobic digestion conditions.

The GASLIFTER increases digester capacity and greatly improves digester operation because it completely circulates and homogenizes the digester contents while operating.

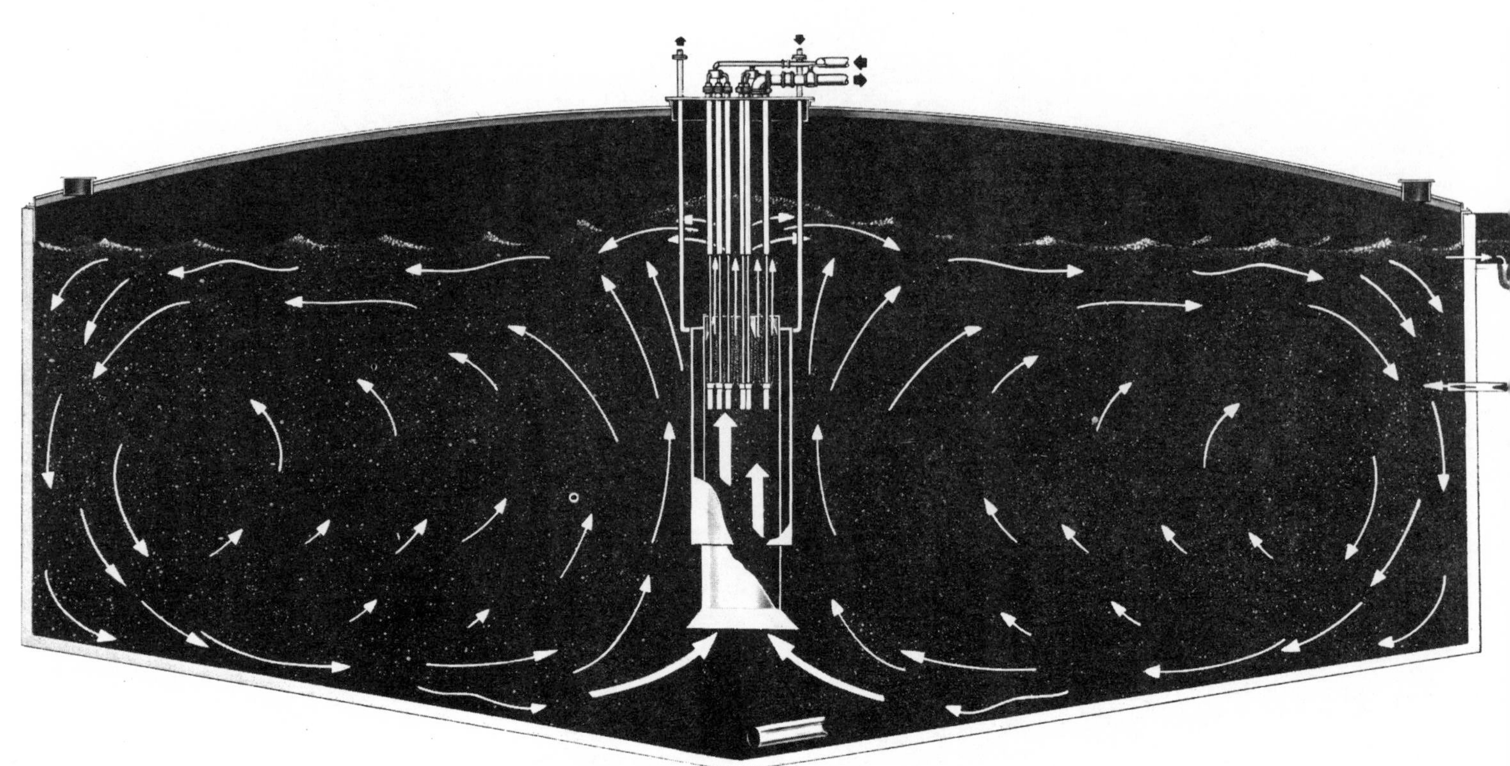

DIRECTUBE GASLIFTER in fixed cover digester.

Walker package sewage treatment plant is supplemented with a Walker tertiary filter at an apartment complex in Atlanta, Georgia.

Present day requirements for wastewater treatment and removal of organic pollutants go beyond the range of standard two-stage processes and require tertiary or third stage treatment. For these applications, as the result of extensive field test work, Walker Process has developed a gravity feed type tertiary filter embodying Camp Filter Underdrain components. This Walker Process Tertiary Filter uses the modern concept of dual media incorporating an air/water backwash system and does not use multiple gravel layers to support the filter sand. Ordinary tertiary filters are extremely difficult to operate since they must handle slimes and algea which are difficult to backwash off the media. To rid the bed of these clinging solids, the backwash system must provide sufficient agitation or "scrubbing" action of each individual filter particle, without carrying the filter media over the backwash trough. In the Walker Process Tertiary Filter, this is accomplished by an air scour cycle followed by a water wash that then flushes the dislodged solids up and over the backwash troughs. This air/water system uses considerably less backwash water than a straight hydraulic system.

COG-BRIDGE RECIPROCATING CARRIAGE COLLECTORS

COG-BRIDGE collectors, R. D. Bollman W. T. P., Contra Costa County, California.

WALKER PROCESS COG-BRIDGE reciprocating sludge collectors are especially designed for sludge and scum removal in rectangular primary tanks. They are applied in municipal and industrial water and waste water pollution control plants; particularly where operators wish to avoid the inevitable chain wear and expense and arduous labor of chain turning and replacement.

Contrasted to the inordinate wear of all of the underwater bearings, pivots and other moving parts attendant to chain collectors, the COG-BRIDGE collector has all the rotating parts above water and enjoys the demonstrated longevity of the sister, circular pier-supported mechanisms.

Even in run-of-the-mill application COG-BRIDGE collectors provide several advantages not obtainable with conventional chain type units:

- Effective, continuous removal of sludge in shortest time.
- Less tank construction cost through elimination of interior walls (carriage can bridge several tanks).
- Factory assembly of "top side" machinery reduces installation costs.
- No underwater bearings–subsurface parts raise above water level for easy service.
- Inspection and lubrication readily accomplished from the bridge.

COG-BRIDGE reciprocating sludge collectors are built for tank widths from 40 feet (12.2M) to well over 100 feet (30.5M). Sludge scraper assemblies suspended from the bridge, move the sludge to the collection hopper, and, on the return run, raise off the tank bottom. Surface skimming devices, if used, then move into position. The scraper mechanism is pivoted and counterbalanced to automatically adjust to the floor slope and exert just the right amount of pressure on the tank floor.

COG-BRIDGE collectors, Manatee Country, Florida, W. T. P.

WALKER PROCESS COG-BRIDGE reciprocating sludge collectors are designed with heavy duty components throughout to give years of trouble-free service with only routine maintenance.

HELITHICKENER SLUDGE THICKENING UNIT

The HELITHICKENER (Patent No. 3,285,415) development is applied: (1) to rectangular basins in place of the usual flight or helicoid cross collector: (2) to circular basins where the usual center sludge hopper is eliminated and, instead, the HELITHICKENER is installed in a radial trough into which the rotating sludge collector scraper arms move the settled sludge.

The HELITHICKENER, with its buoyant large diameter core, and interrupted helical flights with intermediate stirring paddles, produces a greater thickening effect than does any conventional cross collector. It alternately kneads and moves the sludge so as to squeeze out much of the mechanically bound water. The stirring paddles not only arrange the sludge in this compression zone for greater thickness, but also release any gas bubbles (putrefaction or denitrification) to avoid gas bulking.

The HELITHICKENER turns much faster than an ordinary screw conveyor, operating from 10 to 20 rpm by means of the vertical shaft-and-miter gear box drive. No chain sprocket drive could effectively run at near these speeds.

This unit was developed employing flexible underwater jaw (universal) coupling and labyrinth sealed bearings so as to avoid any alignment strain or wear on line bearings or the submersible miter gear box drive.

The WALKER PROCESS Source Thickening System eliminates the need for secondary re-thickeners and open sludge observation wells. Sludge pumping can effectively be automatic, by using a Density Meter in combination with the source thickening hoppers, assuring precise control of sludge density, resulting in more efficient anaerobic digestion or filtering.

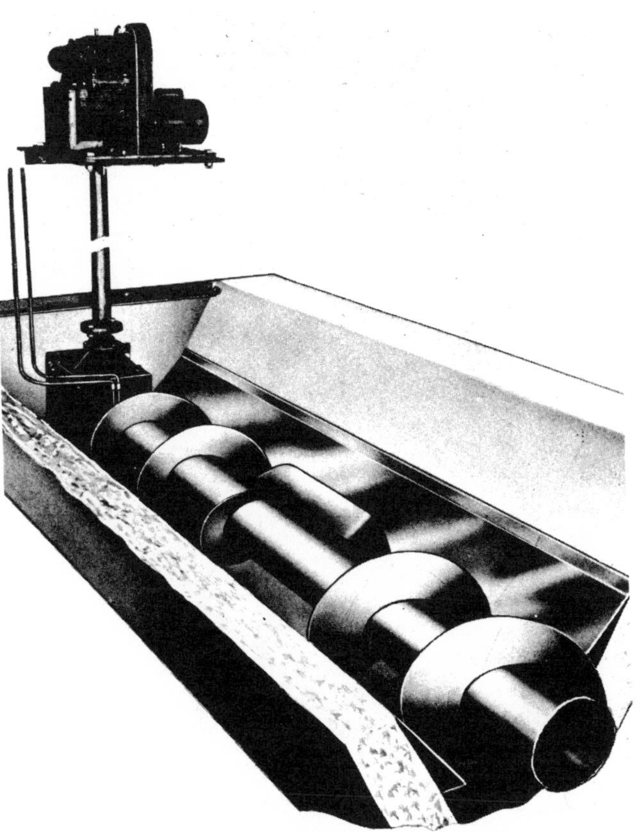

HELITHICKENER in Rectangular Tank

HELITHICKENER in Circular Tank

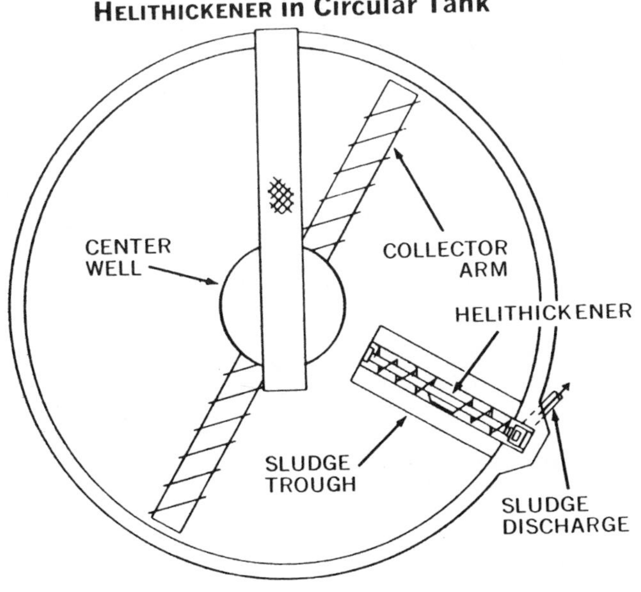

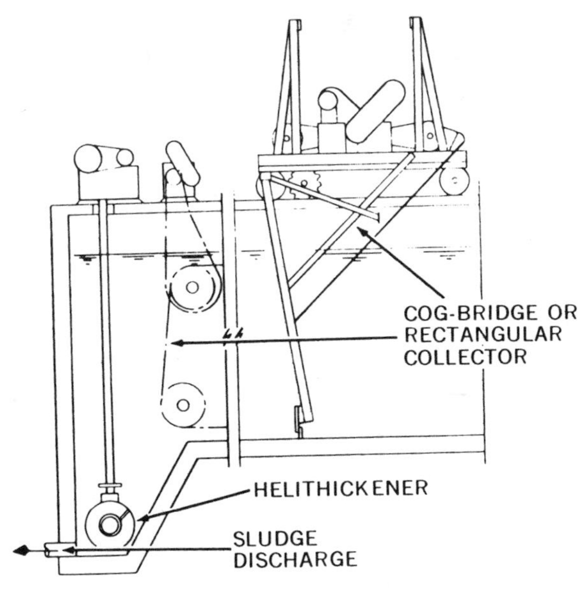

Walker Process Equipment

POLYTHICKENER

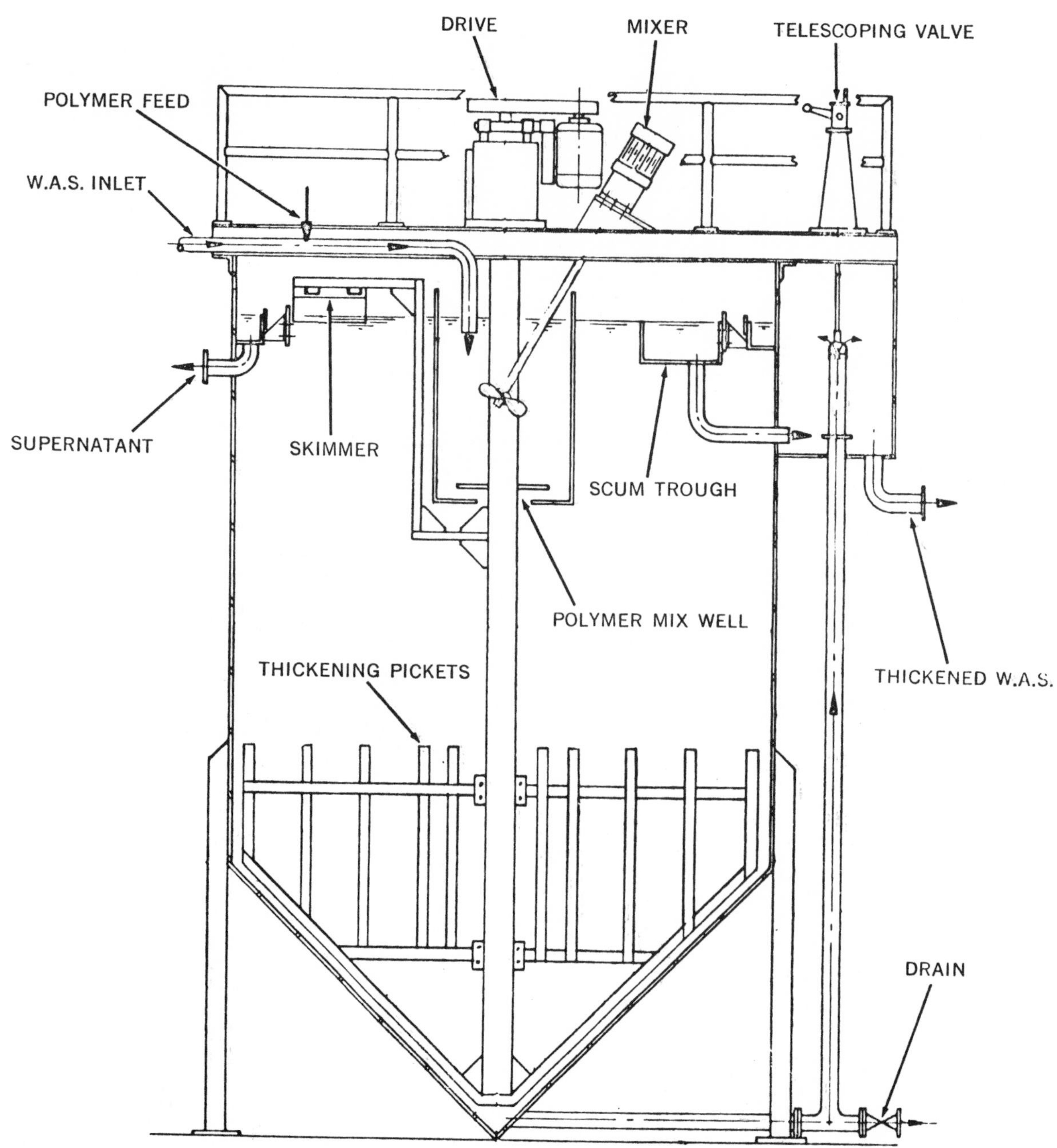

Walker Process PolyThickeners for rethickening waste activated sludge combine chemical (polymer) and mechanical thickening plus storage capacity in one unit. In addition to provisions for chemical preparation and feeding of polymers and a kneading, rotating picket sub-assembly, Walker Process PolyThickeners also include an effective surface skimmer to remove floated (denitrified) sludge mat. The effectiveness of the PolyThickeners, which will thicken waste activated sludge to the same consistency as will air flotation, relies on the overall stroking, or kneading action of vertical paddle flocculation units turning up to (variable) 1/2 fps. This mechanically rearranges the coagulated particles in the compression zone to yield much greater thickness than can be accomplished by simple gravity thickening.

CLARIFLOW

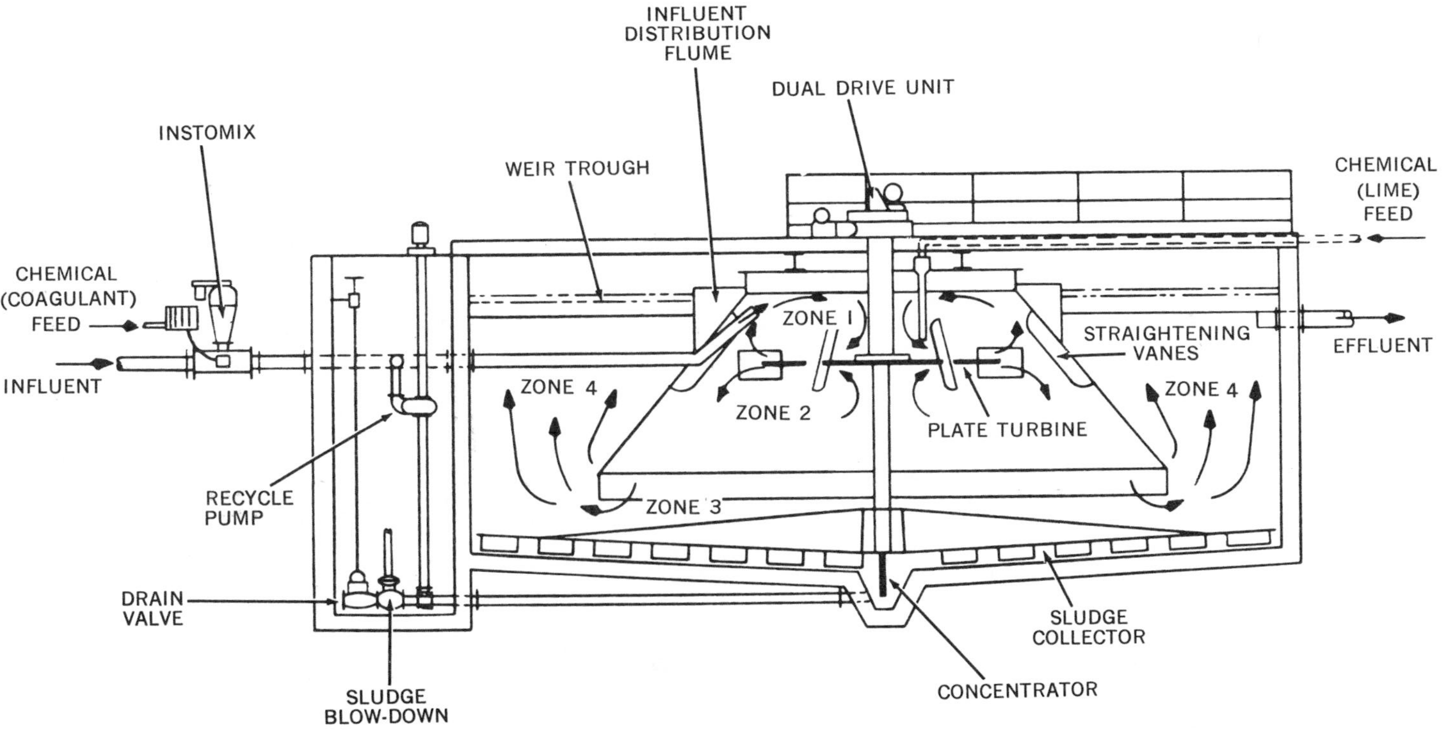

CLARIFLOW Mark I

The MARK I CLARIFLOW design, more recently developed, includes rapid and slow flocculation stages within a common circular or square, vertical rise tank, equipped with positive, bottom scraper slurry removal and control. A slow speed plate turbine provides dual circulation.

Any degree of slurry admixture can be accomplished for solids contact operation by simply varying the external pumping arrangement.

Ideal coagulation is achieved with the least possible chemical feed by including the INSTOMIX installation in the raw water line.

Operational functions are as follows:

1. Coagulant feed is ideally blended in the INSTOMIX.
2. The feed stream containing the coagulated microfloc flows into the Rapid Flocculation region (1) which is formed above the large diameter plate turbine circulator.
3. Displacement flows to the lower, Slow Flocculation region (2) where final floc aggregation takes place.
4. Uniform radial flow under the skirt is ensured by a slight residual mass rotation, controlled by the peripheral straightening vanes, which displaces into the hindered Blanket Zone. (3).
5. Vertical flow of clear water rises uniformly (4) to the multiple trough (orifice or weir) system at the surface. This trough system serves the total area to achieve the greatest clarification efficiency possible.

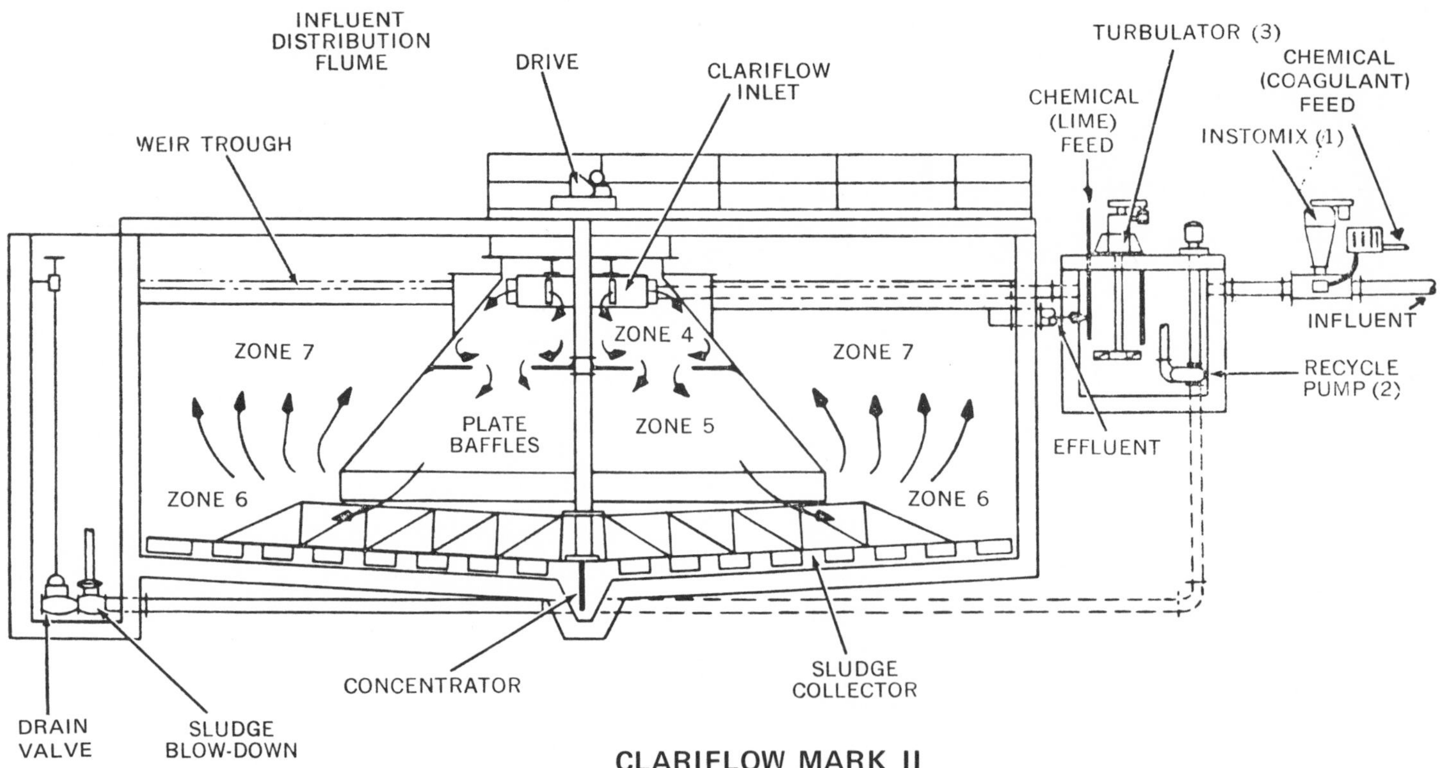

CLARIFLOW MARK II

The Mark II Clariflow system embodies the experience of thirty years of practice in water treatment combined with the latest coagulation technology. All of the steps of instantaneous chemical diffusion, formative flocculation, fresh precipitate admixture, floc aggregation and clarification are carried out as separate, definitive, completely controllable functions as follows:

1. The INSTOMIX 1 is used to blend coagulant in less than 10 milliseconds following pH correction (if used) for ideal coagulation.
2. The return slurry for particle contact is controllably pumped 2 for the best possible admixture and blended with the coagulated microfloc in the TURBULATOR 3 basin.
3. In the TURBULATOR basin, sufficient G-value (well distributed fine scale turbulence) is applied to effect the floc growth during its formative phases. This basin has a low enough retention to affect only the microfloc. It will have a short die-out time to avoid lingering because of short circuiting.
4. Flow progresses to the flocculation zones 4 and 5 in the center compartment of the clarifier where the CLARIFLOW INLET (with its multiple tangential gates) dissipates sufficient energy, carefully directed to cause general roiling in this region. The directed fluid movement ensures that floc particles do not detrain or "waterfall" and excessive density currents are avoided.
5. Outflow, slightly rotating around the skirt, flows uniformly into the hindered blanket zone 6 along the bottom of the tank and thickens to be scraped to the slurry well.
6. The rise of clarified water zone 7 is exceedingly stable and uniform and overflows the multiple trough (weir or orifice) system at the surface. Because of the uniformity of rise, lack of density currents and turbulence, and total area coverage by effluent weirs, this basin design will outperform ordinary plug flow sedimentation tanks by a factor of 3 to 1.

UNIMIX-CLARIFIER FLOCCULATION-CLARIFICATION UNIT

WALKER PROCESS UNIMIX-Clarifiers provide an economical and efficient concentric flocculation and clarification function. They are specially designed for the chemical flocculation and clarification of municipal and industrial water supplies, the handling of unscreened primary sewage, wastes containing gritty matter, and industrial waste waters carrying heavy settleables.

The large diameter, nearly bottomless steel flocculation chamber, designed into the center of the settling tank, permits heavy material to settle and be removed by the collector mechanism. At the same time, UNIMIX slow mix units within the flocculation chamber, promote the aggregation of microfloc particles because of the uniformly distributed fine scale turbulence. The UNIMIX flocculation units have taperlock V-belt drives to permit easy field-changing of speed to meet varying conditions. The flocculation well is expertly baffled so that flocculation energy is confined to the well and does not impair clarification.

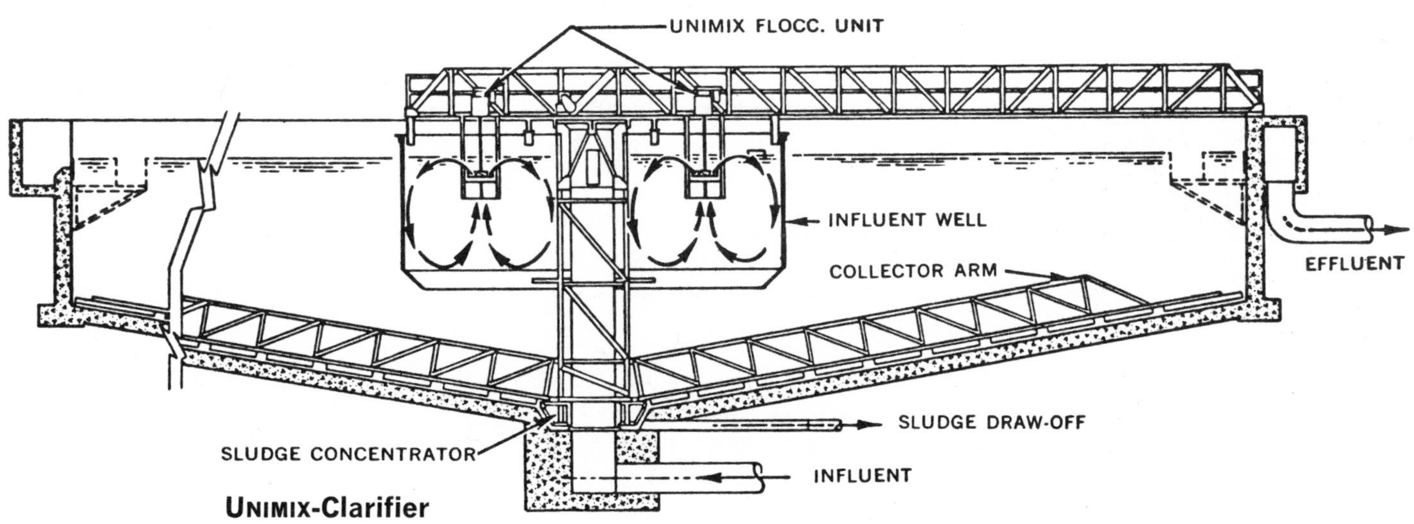

UNIMIX-Clarifier

UNIMIX HIGH ENERGY FLOCCULATORS

WALKER PROCESS UNIMIX flocculation units include designs for either horizontal long shaft, multiple impeller installation or vertical single impeller applications.

The UNIMIX unit is the result of extensive research and tests conducted in the company's own facilities, including a 150,000-gallon test tank at the factory. Full scale tank test studies that include water horsepower readings, tank turnover (agitation) and actual floc development have been intensely researched.

Both horizontal and vertical UNIMIX designs can be applied to a flocculation problem to operate at the highest chemical efficiency possible.

The UNIMIX creates a regime of well distributed small scale turbulence for the most ideal flocculation condition possible; even under conditions of higher (50 to 200 $sec.^{-1}$) G-values.

Development features of the UNIMIX include an exclusive four-blade impeller with large anti-back recycle hub and flow straightening vanes for efficient axial flow pumping action.

Vertical UNIMIX flocculation unit with large hub impeller and straightening vanes.

Horizontal UNIMIX Flocculation

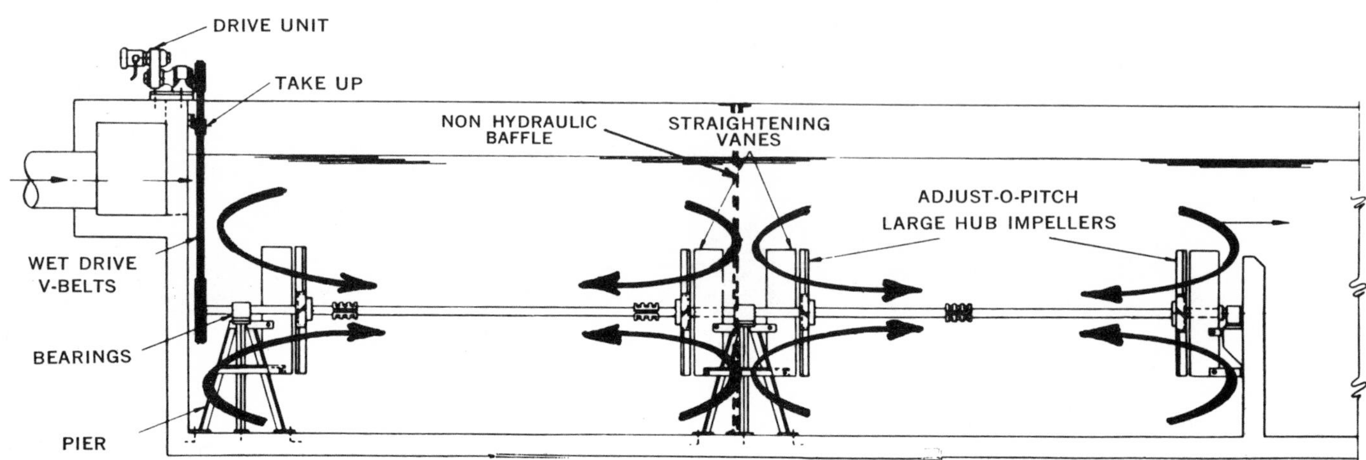

Horizontal UNIMIX flocculation unit combines large hub impellers and straightening vanes, in multiples, to suit almost any tank length and width.

RAPID SAND FILTERS

WALKER PROCESS offers complete, pre-fabricated or job-site erected rapid sand filters. A full range of sizes and capacities in three basic types are available including standard design filters; self-stored backwash water units, and nested, multiple filters with the self-stored backwash water feature. All may be completely automated.

All sand filters feature an air/water backwash, dual media system through the use of CAMP FILTER UNDERDRAINS. This eliminates the need for surface wash systems and graded gravel support layers.

CAMP FILTER UNDERDRAINS are an exclusive WALKER PROCESS product based on years of experience and research and development by T. R. Camp. The system consists of the filter bottom either vitrified clay blocks, poured-in-place concrete or steel plate, fitted with thermoplastic resin nozzles having 0.25 mm aperature slots to retain the filter sand.

The elimination of gravel layers makes air/wash practical, and dual media thereby more feasible. Dual media permits longer and higher rate filter runs without the many disadvantages of operating the filter under high headloss and impending break-through conditions. The coarser layer of anthracite media will absorb and trap particle carry-over from the clarifier so that this discrete material need not be handled by the filter sand. Thus, the filter sand will perform its primary function for a longer epriod before requiring backwashing, and yield as good a filtered water as tho the applied turbidity had been perfect.

The airlift effect of the air/water backwashing technique will thoroughly scour the media and remove the filtered residual in less time and with considerably less backwash water required than with the old surface wash straight hydraulic system.

Vitrified clay block CAMP filter underdrains.

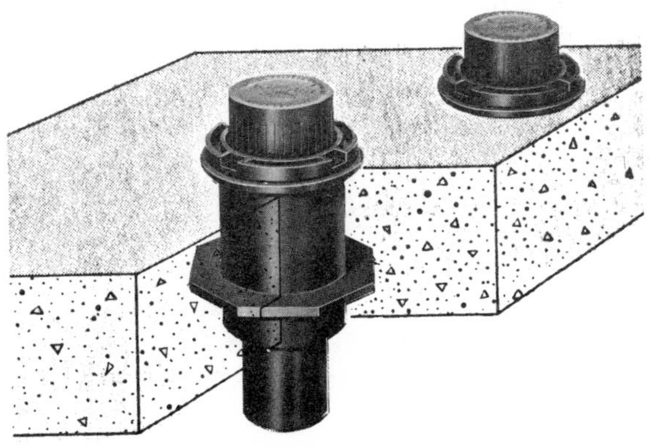

CAMP air/water backwash nozzle for poured-in-place filter bottom installation.

The Warren Rupp Company

SANDPIPER AIR OPERATED DIAPHRAGM PUMP

Warren Rupp offers the Sandpiper air operated diaphragm pumps in 1 1/2", 2" and 3" sizes. The 3" size, for instance, will pump sewage, abrasive slurries, viscous liquids, bilge, etc.

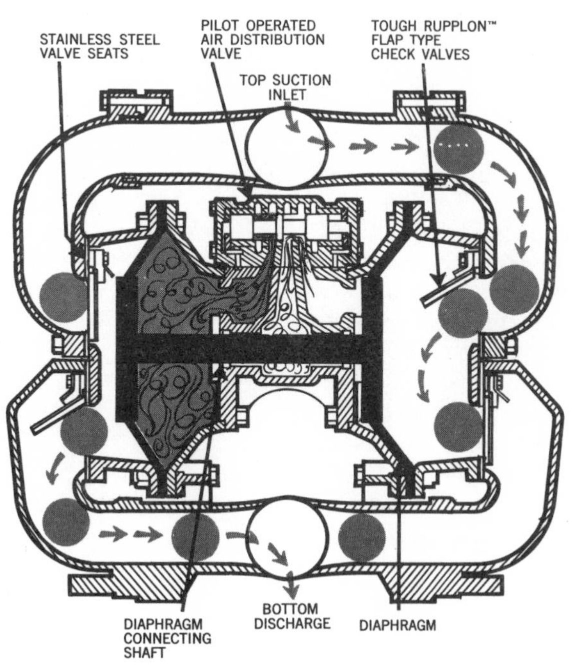

OPERATION:
Inside the two diaphragm chambers, two flexible diaphragms are clamped in sandwich fashion at their outer edges and are connected at their movable central portion by a horizontal connecting shaft. As a result, both diaphragms move simultaneously in a parallel path. The outer part of each diaphragm chamber is fitted with both an inlet and outlet check valve. The two inlet suction valves are connected through a manifold to a common suction inlet on top of the pump, while the two discharge check valves share a similar connection to the discharge outlet at the bottom of the pump.

The air distribution valve controls the alternate pressurizing of one side of the diaphragm chamber and the simultaneous exhausting from the other chamber. As air enters one chamber it forces the diaphragm axially outward on a discharge stroke and forces the other shaft-connected diaphragm inward on a suction stroke. When the diaphragms reach the end of a stroke, the air distribution valve shifts automatically to reverse their movement.

Since the air pressure acts on one entire side of a diaphragm to move against the pressure of the pumped liquid on the other side, the diaphragm is essentially balanced, mechanical problems are effectively eliminated, and it easily operates at heads in excess of 200 feet.

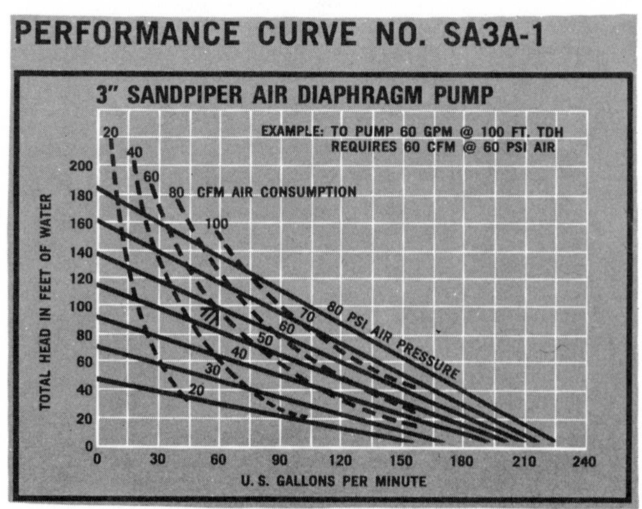

MODEL "CLP" OZONATORS

Welsbach Model CLP ozonators are totally unitized, skid-mounted ozone generators, complete with all auxiliaries, producing from four to eighty pounds of ozone per day from air feed.

The unique design of these ozonators enables them to be put into service quickly, with a minimum of field installation. Once the packaged ozonator is set in place, it requires only installation of dielectrics and connection of piping and wiring to be put into operation. The ozonators are used in many varied applications. Typical installations include:

Exhaust Odor Control	Waste Treatment
Commercial restaurants	Cyanide effluents
Municipal sewage plants	Phenolic effluents
Rendering plants	Tertiary municipal sewage

Operation:

A compressor forces filtered air through one or more coolers and a drier, and then through the stainless steel ozonator, where one percent by weight of the air is converted to ozone. The ozonated air is then conducted to the point of application. Water is used as a coolant to remove excess heat from the ozonator. Ozone production and concentration are controlled by a variable voltage transformer and a flow control valve.

INTRODUCING THE ALL-NEW PATENTED "ZURN-ATTISHOLZ PROCESS"

The all-new Zurn-Attisholz activated sludge treatment process is a two-stage biological system – complete with provisions for phosphate removal within the second stage – designed to remove over 90 percent of the organic impurities from municipal and industrial flows. Each stage utilizes aeration, settling, and a return sludge system and is programmed for optimum utilization of the proper micro-organisms best suited to accomplish a high degree of treatment. The patented process is applicable as a customed-designed treatment plant capable of serving municipalities of any size . . . package plants serving smaller communities generally in the 500 to 2500 population range . . . or designed to remove high strength organic decomposable pollutants in industrial applications.

95% REMOVAL OF ORGANIC POLLUTANTS FROM INDUSTRIAL FLOWS

The "Zurn-Attisholz Process" can be designed to remove high strength organic decomposable pollutants from industrial flows, such as food wastes, animal wastes, and similar substances associated with paper mills and petro-chemical plant discharges. By separating the system into a two-stage process, a bacteria type sludge requiring low dissolved oxygen is developed in the first stage. This is a hardy, rapidly settling sludge which allows for suspended solids concentration of 10,000 to 16,000 P.P.M. The second stage sludge is predominately protozoan which further treats and polishes the effluent from the first stage system. If phosphate removal is desired, chemicals can be added to the second stage to meet this requirement. The "Zurn-Attisholz Process" has inherent advantages over conventional plants, such as lower air usage, lower operating costs, smaller tank volume requirements, shorter detention times, lower capital investment, greater process stability and treatment, and lower cost phosphate removal.

MOST EFFICIENT, ECONOMICAL SYSTEM FOR WASTE WATER TREATMENT

Most conventional waste water treatment plants today consist of one mechanical and one biological stage, but now the "Zurn-Attisholz Process" has created a new method. The mechanical stage is omitted and replaced by a second biological stage, in order to supply – in each stage – the most favorable living conditions for the organisms which eliminate the impurities from the waste water. A further pollution of our waters originates from solved phosphates which, in spite of the biological purification, flow into the rivers and lakes and cause excessive growth of algae. Periodically vegetation and algae perish and decay, thus causing a secondary pollution. For this reason, Zurn has studied the problem of chemical purification and is today in the position to effect a simultaneous flocking-out in the two biological stages.

AVAILABLE AS COMPLETE PACKAGE TREATMENT SYSTEM

The "Zurn-Attisholz Process" is particularly attractive as a complete package system because of simplicity of operation and design, low operating costs and high degree of treatment and reliability. The unique two-stage design enables these package plants to be easily expandable, adaptable for tertiary treatment, and can be designed to eliminate the possibility of objectionable odors to the surrounding environment. The patented "Zurn-Attisholz Process" can be completely factory-assembled for immediate delivery, or shipped in modular sections for erection at your plant site. Easy to install, operate, and maintain, the "Zurn-Attisholz Process" will remove in excess of 90 percent of the suspended solids and B.O.D. assuring a high degree of effluent treatment for hotels, motels, trailer parks, subdivisions, schools, shopping centers, hospitals, industrial complexes or other private, municipal, and industrial treatment systems.

"ZURN-ATTISHOLZ PROCESS" DESIGNED AS A PACKAGE SEWAGE TREATMENT PLANT

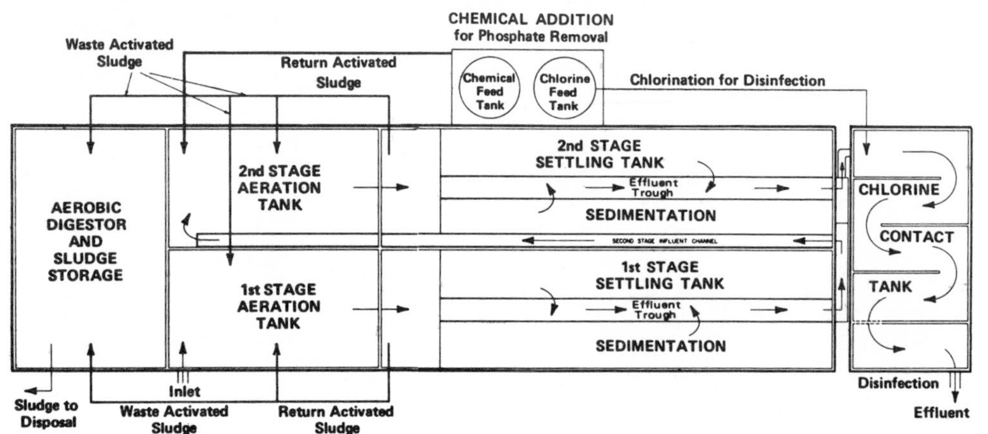

"Zurn-Attisholz Process" includes first and second stage aeration and settling tanks, aerobic digestion and disinfection. Depending upon transportation limitations, the "Zurn-Attisholz System" can be factory-assembled and shipped directly to the plant site for immediate operation.

MUNICIPAL SEWAGE TREATMENT COMPLETE WITH PHOSPHATE REMOVAL CAPABILITIES

The "Zurn-Attisholz Process" is uniquely adaptable to the growing sewage treatment needs of our cities and metropolitan areas by providing a versatile system design. Modular construction enables plant facilities to be enlarged as required, or custom-designed to meet any specific land configuration and can be easily adapted to phosphate removal within the second stage without the addition of costly tankage. With the addition of both phosphate removal and filtration equipment, suspended solids and B.O.D. removal can be increased to 95% or better and the effluent phosphate level reduced to under 1 P.P.M. The treated effluent that is produced is stable and the overall system reliability is maximized by utilizing this unique two-stage process.

80% REMOVAL OF SOLIDS DURING STORM FLOW PERIODS

By changing the flow pattern from a series flow to a parallel flow the "Zurn-Attisholz Process" has the capabilities of preventing serious damage to the receiving waters by providing partial treatment (80% removal through one stage) to the total flow during storm periods. This concept can be easily designed into the system through a simple split flow bypass from the inlet of the first stage to the inlet of the second stage.

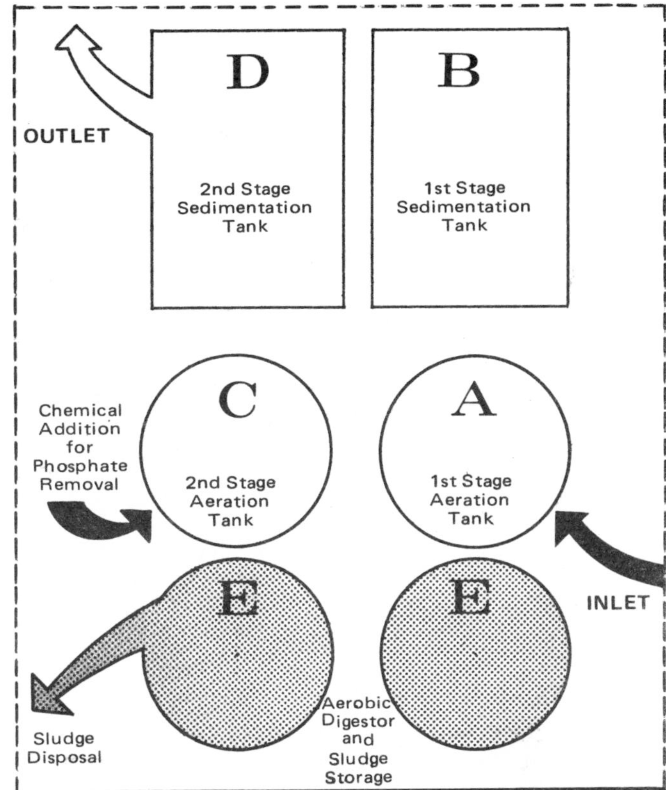

PROCESS FLOW CHART

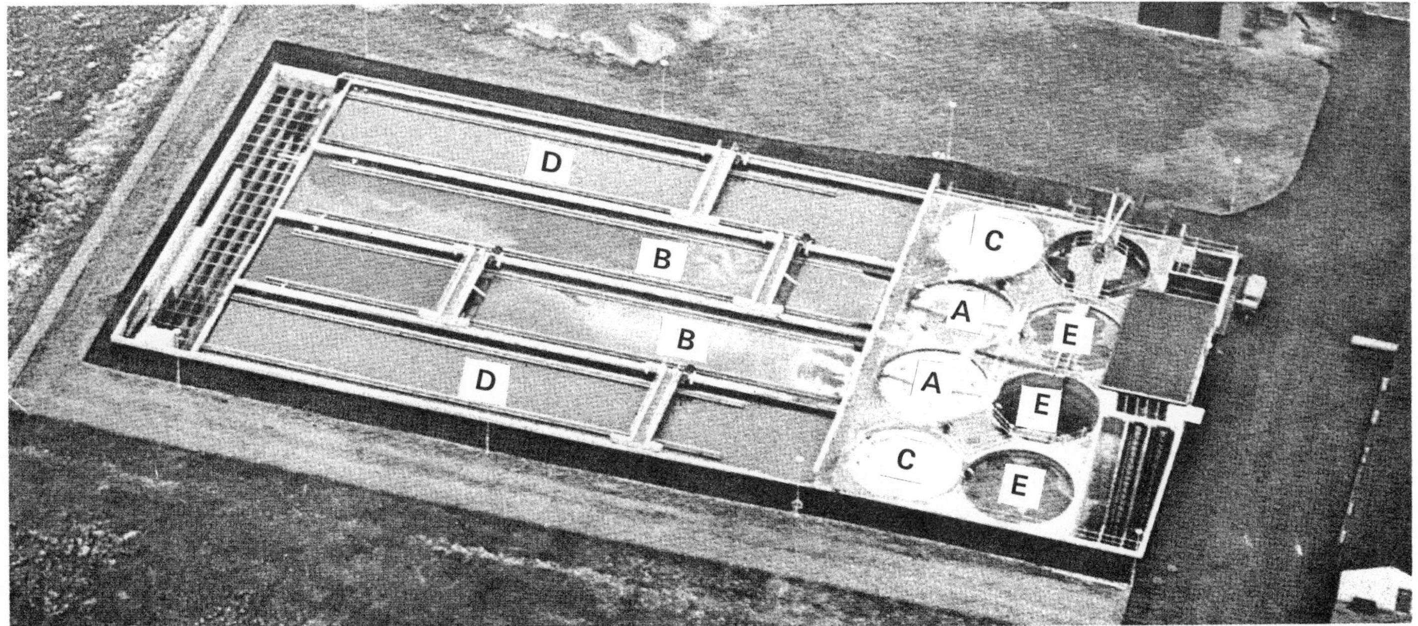

Zurn-Attisholz sewage treatment plant, serving a community of 25,000 residents, was designed as a modular, expandable facility complete with phosphate removal to assure up to 95% suspended solids and B.O.D. removal, while keeping the phosphorous level below 1 P.P.M. in the effluent.

The influent enters (A) 1st stage aeration tank with a detention time of 33 minutes; passes on to (B) 1st stage settling tank with a detention time of 211 minutes; back to (C) 2nd stage aeration tank detention time of 33 minutes; passes on to (D) 2nd stage settling tank with a detention time of 211 minutes with the effluent to the outlet. Sludge is pumping from (A) to (E) aerobic digestor, for concentration, stabilization, and storage prior to dewatering. The detention times are based on a 13 hour dry weather flow rate.

VERTI-MATIC UP-FLOW FILTRATION SYSTEM

The Zurn "Verti-Matic" is an up-flow filtration system specially engineered to remove particulate matter from water, wastewater, and other fluid streams. It represents a major breakthrough in fluid filtration, by employing the up-flow principle of reversing the fluid flow. The Zurn "Verti-Matic" can increase solids holding capacity, duration of the filtration cycle, and create a higher net filter yield. It is capable of flow rates from 2 to 15 GPM per square foot, and dirt holding capacities of between 5 and 10 pounds per square foot of surface area.

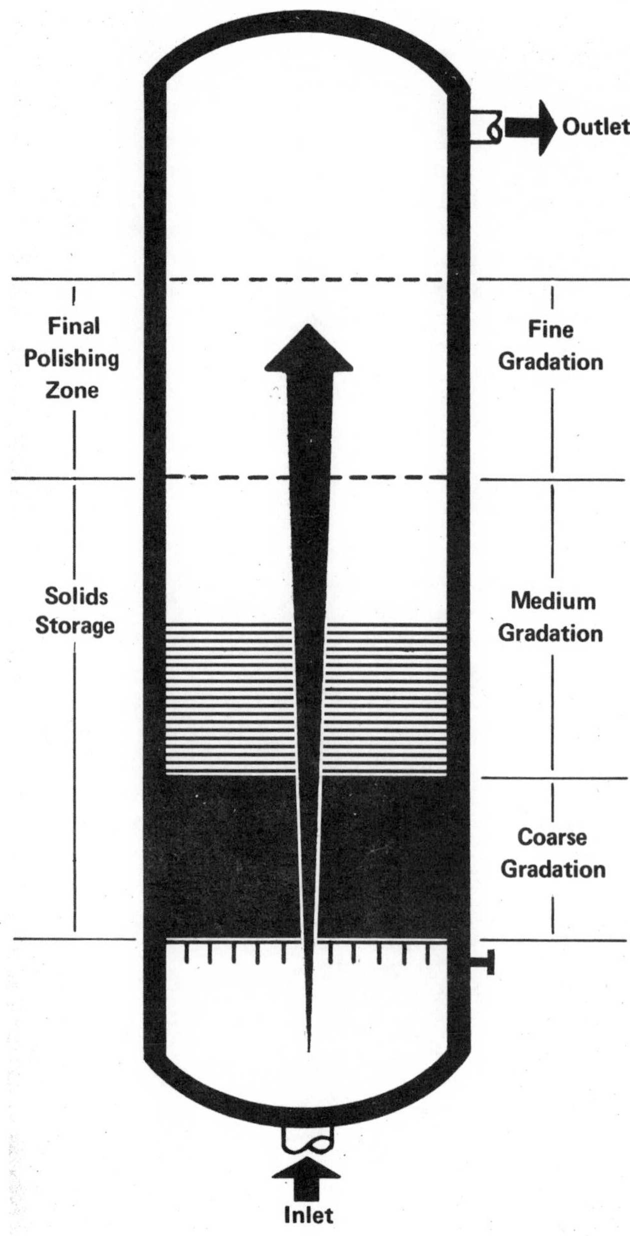

NATURAL GRADATION OF THE FILTER MEDIA

"Verti-Matic" During Filtering Cycle

During the filtration cycle, the raw water containing solids enters the most porous bottom layers of the "Verti-Matic" first, where the greatest penetration of the filter takes place. By traveling upward, the effluent utilizes the entire mass of the filter media, and allows the finest layers at the top to further polish the filtered water. This efficient use of the total filter media provides longer filtration runs between wash cycles and a greater solids holding capacity. With the use of wash water and air the entire filter is cleaned, and then a natural, hydraulic gradation of the filter media takes place, leaving the most porous media on the bottom.

"Verti-Matic" Washing Cycle

To facilitate cleaning the filter media, the Zurn "Verti-Matic" employs an air-assisted wash cycle. The implementation of air increases the effectiveness of the scouring action during the wash cycle and decreases wash water volume. The air bubbles impart an agitation throughout the filter media as they rise up to the top of the filter. The accumulated solids break up and easily allow the wash water to fluidize the individual media grains. The net result is a highly efficient wash cycle, allowing maximum removal of the filtered solids in a minimum period of time. In municipal water treatment applications, it is recommended that chlorine be added during the final stages of the washing cycle.

SLUDGE HEAT TREAT PROCESS

For Municipal Sewage and Industrial Waste Sludges

The new Zurn advanced sludge heat treat process, the most reliable and economical process for sterilizing and conditioning sewage sludges prior to dewatering, utilizes heat exchangers, heat economizers, reactor vessels, a closed circuit high temperature water system, and a vacuum filter to separate the solids from the liquids while producing a dried filter cake (40-50% dry solids) which is easily disposed of through hauling or incineration.

For Sterilizing and Conditioning Sludges Prior to Dewatering

The Zurn sludge heat treat system eliminates chemical additions for conditioning; produces less volume for hauling; more efficient incineration without supplemental fuel for continuous firing; becomes economically feasible to incinerate sludges at plant site; generates higher BTU output to provide heat for closed circuit hot water system; and can be adapted for waste energy recovery to produce productive steam power for various in-plant processes.

ZURN SLUDGE HEAT TREAT PROCESS

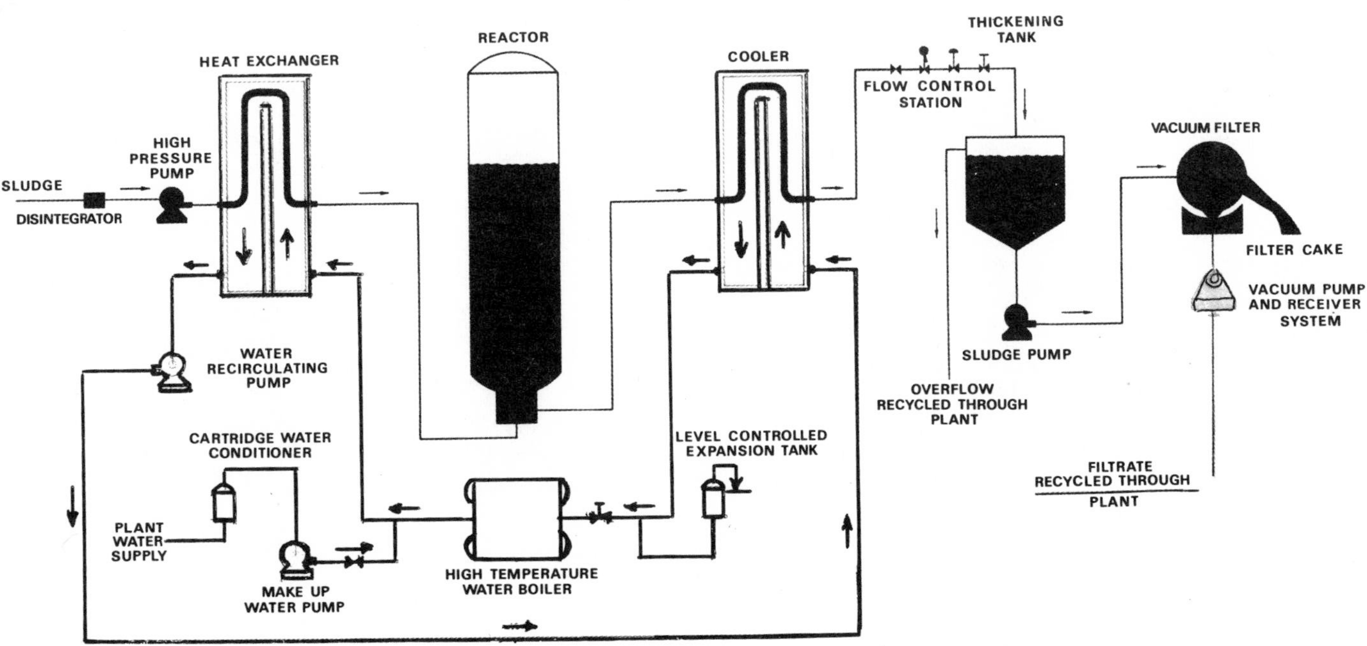

MOST EFFICIENT, ECONOMICAL SLUDGE DEWATERING SYSTEM

TWO-STAGE CARBON ADSORPTION PROCESS

The Zurn two-stage granular carbon bed adsorption process is a counter-current two-bed system which utilizes a simplified piping and valving arrangement. The two-beds are arranged so that the gravity open top contactors are operated as a series roughing up-flow primary contactor and a secondary down flow polishing contactor. When the C.O.D. in the effluent of the secondary bed reaches the predetermined level, the primary bed is emptied of spent carbon and new regenerated carbon is replaced. The valving is changed to the two-bed system, changing the previous secondary down flow bed into the primary up-flow roughing bed while the new carbon in the other bed is utilized as the down flow polishing bed.

The Zurn two-stage method of operation utilizes the carbon to its fullest extent of adsorbing capacity with higher C.O.D.-on-carbon loading. By using the carbon as a polishing contactor, and bringing the adsorbed quantity of C.O.D. up to a predetermined breakthrough rate, and then switching directions and making it the roughing contactor, you can achieve maximum C.O.D. per pound of carbon loading. The higher the adsorption rate the lower the operating costs. The Zurn two-stage system cuts the costs associated with the percentage of carbon lost in handling and transportation of the spent and regenerated carbon, as well as the regenerating losses during the regenerating cycle.

ZURN TWO-STAGE GRANULAR CARBON BED ADSORPTION SYSTEM

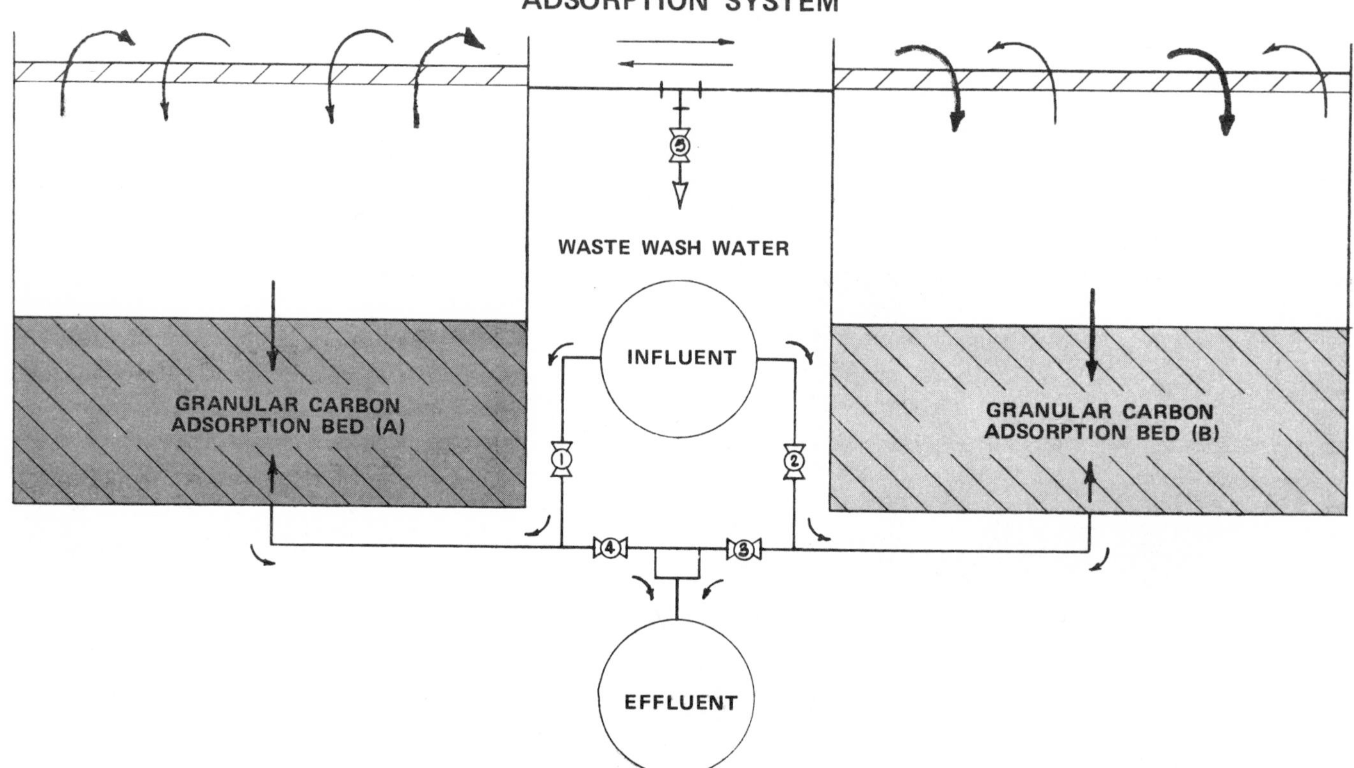

FOR EFFICIENT, LOW COST C.O.D., BOD REMOVAL

PHYSICAL-CHEMICAL WASTEWATER TREATMENT PROCESS

The Zurn Physical-Chemical Wastewater Treatment Process requires less space; provides for higher degree of efficiency; it is less sensitive to unusual loadings or daily flow-variations; provides greater design and operational flexibility; can operate as a chemical plant; has instant on-off capabilities; and has the ability to process stored wastewater. Can be applied to recreational areas, marinas, race tracks, industrial complexes, hotels, motels, service stations, schools, and mobile home parks where variable or intermittent flow conditions exist.

The Zurn Physical-Chemical Wastewater Treatment Process incorporates a combination of time-tested principles and practices of wastewater treatment; activated carbon contact for dissolved organics removal; alum coagulation, flocculation, and settling out of solids; disinfection and oxidation of the effluent flow by chlorine contact; and filtration through sand and gravel. The treated effluent is then ready for discharge into receiving streams. The carbon sludge is thickened and is ready for regeneration and reuse or hauling to a final disposal site.

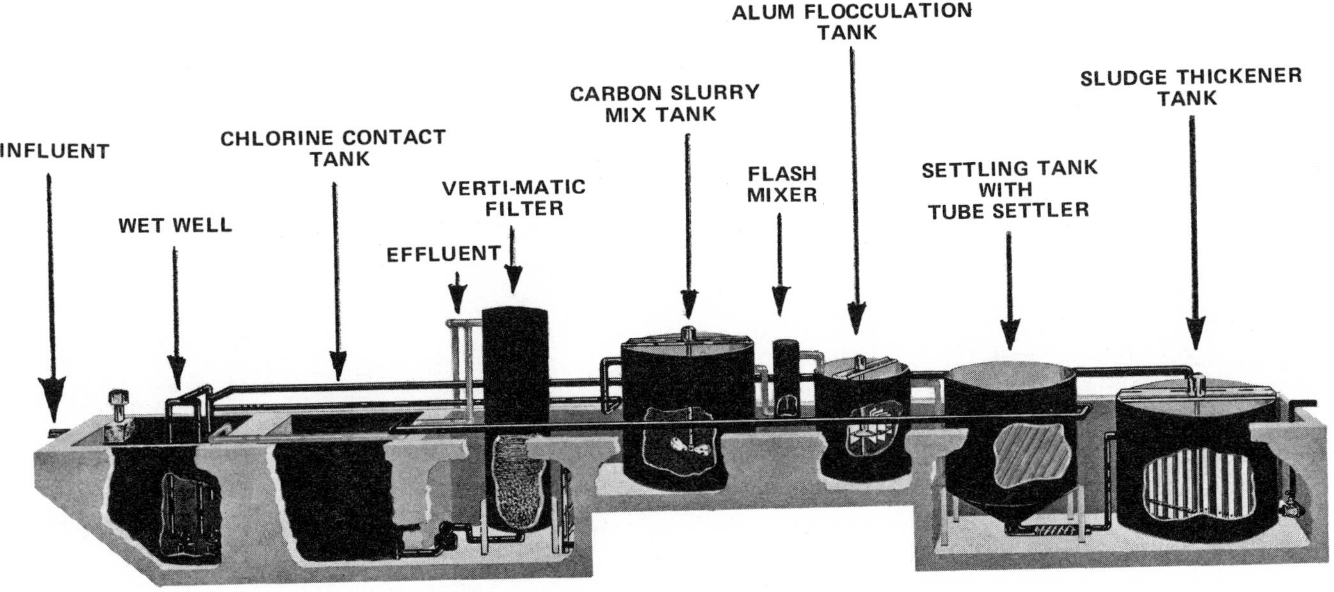

ZURN PHYSICAL-CHEMICAL WASTEWATER TREATMENT PROCESS

"MICRO-MATIC" STRAINING SYSTEM

To meet modern day standards of water quality, Zurn Industries, Inc. has designed the "Micro-Matic," a continuous self-cleaning straining system for the removal of minute contaminants or large entrained solids from open-channel or gravity-flow systems. The heart of the "Micro-Matic" is a horizontal, rotating drum type straining element whose periphery is enveloped with an extremely fine woven fabric. Debris-laden liquid enters the interior of the partially submerged drum through an open end and is strained outward through the straining media. The debris remains on the interior surface of the drum and is transported upward over a stationary waste collector located inside the drum above the water level.

A row of water jets positioned directly above and outside the drum uses a small percentage of product water to flush the debris from the interior surface of the drum into the waste collector and to a drain line. As a result, a thoroughly cleaned screening surface is continuously entering the debris-laden fluid.

The rotating drum, straining fabric, drum suspension, support structure, backwash system, and drive are ruggedly designed for continuous service in the most demanding municipal and industrial environments. Materials have been carefully selected and sections generously sized to provide many years of service.

The Zurn "Micro-Matic" has a superior straining mesh system which utilizes rugged straining panels featuring a unique method of continuously joining the straining fabric to a rigid support material. Panel edges are continuously sealed and clamped to the drum structure for maximum structural stability and ease of individual screen panel installation.

The drum suspension system is designed to overcome the problems inherent in other designs. The three bearing points — two at the open end and one at the closed end — provide stable support for the drum. This permits deep submergence and excellent access to all components.

The infinitely variable drive system avoids heavy inertia loads at starting and assures smooth continuous rotation. It is compatible with automatic controls which regulate drum speed, while providing maximum flow capacity and filtration efficiency.

"MICRO-MATIC" EFFICIENTLY REMOVES ALGAE TASTE AND ODOR

When fitted with 21 to 36 micron straining fabric, the Zurn "Micro-Matic," used in pre-filtration applications, will generally remove 75-85% of the algae and other naturally occurring plankton, which cause taste and odor and has demonstrated an ability to remove 50% of the suspended solids when straining raw water prior to slow or rapid sand filtration. This greatly extends filter runs between backwashing, permits higher flow rates, reduces chemical requirements, helps quality conditions, and generally improves the overall treatment quality and economy.

"MICRO-MATIC" PLUS DISINFECTION

Some water sources are of a quality which permits direct usage without treatment. During periods of storage these sources are often subject to contamination by living organisms which can result in taste and odor problems. The Zurn "Micro-Matic," as a complete filtration device, serves to completely remove these undesirable materials larger than approximately 20 micron and, due to the rapid formation of a "Schmutzdecke," much of the material smaller in size. Under conditions where raw water is relatively free of color and colloidal particles and contains mainly distinct algae and plant organisms, the Zurn "Micro-Matic," fitted with 21 micron fabric plus disinfection generally is the only treatment required before use in water supply systems.

Zurn Industries, Inc.

"MICRO-MATIC" FOR CONTINUOUS NON-CLOGGING OPERATION IN SEWAGE TREATMENT

90% REMOVAL OF SETTLEABLE SOLIDS

In lieu of conventional primary sedimentation, the Zurn "Micro-Matic" can be used as a roughing strainer to remove the gross solids prior to biological waste treatment or as a means of reducing pollutants from receiving streams during storm flow peaks. In this application the "Micro-Matic" can remove approximately 90% settleable and 50% suspended solids and biochemical oxygen demand.

Continuing product development in new dewatering concepts are being studied to make certain that the solids removed from the waste collector will be readily disposable.

NEW STRAINING TECHNIQUES FOR TERTIARY TREATMENT

With new advances in micronic fabric techniques, the Zurn "Micro-Matic" has proven to be an effective means of polishing secondary effluents from conventional waste treatment plants. The "Micro-Matic," when fitted with approximately 20 micron fabric, has reduced suspended solids by 60 to 90%, turbidity by 30 to 60%, and biochemical oxygen demand by 40 to 80% when treatment plants are properly operated.

It should be noted that the size and type of solids, particularly colloidal particles, which occasionally occur, can affect removal efficiency. The small space requirements, low initial and operating costs, automatic and self-cleaning operation make the Zurn "Micro-Matic" an ideal selection for tertiary treatment. Another ideal application for the "Micro-Matic" is for cleaning effluent water prior to use in non-potable water systems and froth spray requirements. Automatic controls can be furnished to vary fabric wasting and drum speed dependent on loading conditions.

EFFICIENT WASTE PROCESSING

In order to eliminate waste treatment plant operational difficulties due to solid discharges from fruit processing plants, paper mills, textile mills and animal processing plants, the Zurn "Micro-Matic" has been effectively used to reduce solids to those levels acceptable to the treatment facilities. When fitted with the proper mesh fabric, the Zurn "Micro-Matic" can recover valuable by-product materials for reuse or recover water for recycling purposes.

CAPACITY

Size	Primary (MGD)	Tertiary (MGD)
4 x 2	0.2 to 0.8	0.1 to 0.3
4 x 4	0.4 to 1.6	0.2 to 0.7
6 x 4	0.7 to 2.8	0.4 to 1.1
6 x 6	1.0 to 4.2	0.6 to 1.6
6 x 8	1.4 to 5.6	1.8 to 2.2
10 x 10	3.0 to 12.0	1.5 to 4.5
10 x 16	4.8 to 19.2	2.4 to 7.2

USE OF THE "MICRO-MATIC" IN SEWAGE TREATMENT

INDEX